Impact of Neuroplasticity on Organizational Effectiveness and Labor Productivity

Bryan Christiansen
https://orcid.org/0000-0001-7282-8497
Southern New Hampshire University, USA

Anatoly Zhuplev
https://orcid.org/0009-0008-5124-5063
Loyola Marymount University, USA

IGI Global
Scientific Publishing
Publishing Tomorrow's Research Today

Vice President of Editorial | Melissa Wagner
Director of Acquisitions | Mikaela Felty
Director of Book Development | Jocelynn Hessler
Production Manager | Mike Brehm
Cover Design | Jose Rosado

Published in the United States of America by
IGI Global Scientific Publishing
701 East Chocolate Avenue
Hershey, PA, 17033, USA
Tel: 717-533-8845 | Fax: 717-533-7115
Website: https://www.igi-global.com E-mail: cust@igi-global.com

Library of Congress Cataloging-in-Publication Data

Names: Christiansen, Bryan, 1960- editor | Zhuplev, Anatoly, 1948-
Title: Impact of neuroplasticity on organizational effectiveness and labor productivity / [edited by] Bryan Christiansen, Anatoly Zhuplev.
Description: Hershey, PA : IGI Global Scientific Publishing, [2026] | Includes bibliographical references and index. | Summary: "This book aims to examine how neuroplasticity influences organizational effectiveness and labor productivity in an era of global hyper-competition and transformation today"-- Provided by publisher.
Identifiers: LCCN 2025015521 (print) | LCCN 2025015522 (ebook) | ISBN 9798337322926 hardcover | ISBN 9798337322933 paperback | ISBN 9798337322940 ebook
Subjects: LCSH: Psychology, Industrial http://id.loc.gov/authorities/subjects/sh85108476 | Neuroplasticity http://id.loc.gov/authorities/subjects/sh85091160 | Organizational effectiveness http://id.loc.gov/authorities/subjects/sh85095526 | Labor productivity http://id.loc.gov/authorities/subjects/sh85073719
Classification: LCC HF5548.8 .I473 2026 (print) | LCC HF5548.8 (ebook) | DDC 158.7--dc23/eng/20250529
LC record available at https://lccn.loc.gov/2025015521
LC ebook record available at https://lccn.loc.gov/2025015522
British Cataloguing in Publication Data
A Cataloguing in Publication record for this book is available from the British Library.

Table of Contents

Detailed Table of Contents

This chapter undertakes a comprehensive exploration of the specific organisational characteristics that are foundational to the cultivation and sustained presence of happiness and well-being within the workplace. Moving beyond the well-trodden path of individual-level determinants of employee satisfaction, this contribution meticulously dissects the systemic elements, deeply ingrained cultural norms, and thoughtfully designed structural frameworks that collectively contribute to the emergence of a positive and flourishing work environment. By examining existing scholarly research and synthesizing diverse theoretical perspectives, this chapter identifies the core organisational attributes that not only nurture employee happiness and a profound sense of well-being but also exert a demonstrably positive influence on critical organisational outcomes.

In an era defined by volatility, uncertainty, complexity, and ambiguity (VUCA), organisational change is no longer episodic—it is a continuous necessity. This chapter explores how organisations can navigate constant change in a volatile, uncertain, complex, and ambiguous (VUCA) world. It examines key drivers of change technological, demographic, environmental, and socio-political and integrates classical and contemporary theories, including Lewin's model, Kotter's framework, and sensemaking. Emphasis is placed on strategic foresight, organisational agility, innovation culture and role of neuroplasticity in change implementation. Real-world case studies (e.g., Microsoft, Patagonia, GM) demonstrate practical applications from organisational and neuroplasticity perspective. The chapter highlights the importance of adaptive resilience, aligning formal and informal systems, empowering middle managers, and fostering psychological safety to enable sustainable transformation and continuous learning in rapidly evolving environments.

This chapter examines one of the most critical yet underexplored aspects of neuroplasticity in organizational contexts: the rigorous measurement of its impact on quantifiable metrics. Building on this publication's core idea that neuroplasticity offers deep insights into individual and team capabilities, this contribution to the body of knowledge addresses the challenge of transforming ideas into tangible evidence. It reviews existing research, points out gaps in current studies, and proposes new frameworks and methods to evaluate how fostering neuroplasticity can enhance effectiveness, productivity, innovation, retention, and business excellence in a competitive global environment. By examining different levels—individual, team, and organizational—and identifying relevant indicators, this chapter aims to provide researchers, practitioners, and students with a clear plan for future research and practical application, transforming the "so what?" question into meaningful insights that support the enablement of neuroplastic, healthy organizations and thriving workforces globally.

This chapter examines the strategic application of neuroplasticity in contemporary enterprises to improve cognitive agility and cultivate future-ready talent. As disruption transforms industries, conventional paradigms of learning and performance are inadequate. This book delineates how restructuring the brain facilitates the transformation of the workplace through an integrated framework of neuroscience, organizational behavior, and human resource innovation. Topics encompass the creation of neuro-friendly workplaces, the utilization of immersive learning technology, the advancement of adaptive leadership, and the assessment of neural growth via performance indicators. By synchronizing learning ecosystems with cerebral processes, companies can transition from static skill training to dynamic, ongoing transformation. This book offers a visionary and pragmatic framework for creating neuroadaptive workplaces that excel in complexity and promote sustainable innovation

The chapter primary goal is to investigate the connections among productivity of labor force, well-being of workforce, and building resilient corporates. It tackles the issue of how corporate resilience might augment workforce happiness and engagement in a dynamic workplace. In order to investigate the links among the constructs, the investigation uses a quantitative investigation approach, collecting data from workforces in different corporates through surveys. Prominent outcomes show that incubating resilient corporates has a beneficial influence on productivity of labor force, which in turn greatly augments well-being of workforce. The outcomes imply that corporates that place a high priority on resilience not only augment operational influence but also manifested a more salubrious workplace for staff members. Future investigation will look at the precise mechanisms underlying these connections as well as how individual differences may mitigate the benefits of resilience

This chapter looks at how ethical leadership affects the well-being of employees at Ethiopian hospitals, focusing especially on how someone's willingness to help others at work (called organizational citizenship behavior) acts as a middleman among the leader's actions as well as the overall happiness of the staff. Using samples from 390 healthcare professionals, the study checked how good the data was for analysis through adequacy, as well as the outcomes showed that the data was good enough to look at the relationships among variables. CFA is done to check if the study's model is correct as well as to make sure the concepts like ethical leadership, OCB, as well as job life are accurately measuring what they aim to look at SEM is used through the AMOS to look at how the different constructs are connected through each other. The outcomes show that ethical headship really helps publics do more for their organization, as well as this has a big positive effect on how happy as well as comfortable workers feel at work. This means that when headships act in an honest as well as fair

This chapter looks into how transformational leadership affects having a growth mindset as well as how it subsequently changes OCB in the Ethiopian healthcare sector. The data are tested by the study through EFA through the KMO measure included in a sample of 385 healthcare professionals. First, the presence of the main concepts is tested using confirmatory factor analysis (CFA). Then, SEM analysis is carried out on AMOS software to see how transformational leadership, growth mindset, organizational citizenship behavior, as well as organizational culture are related to one another. The outcomes suggest that strong leadership leads to employees adopting a growth mindset, which helps them put in more OCB. Moreover, organizational culture is considered a main factor that shapes how employees think as well as behave in their workplace. The authors demonstrate that a flexible workplace culture supports transformational leadership by helping to grow the mindset as well as activity of its members. The findings are added to the existing investigation on leadership as well as

Neurodegenerative Diseases (ND) cause the progressive deterioration of the central nervous sys-tem, leading to impairments in memory, motor skills, cognition, and autonomic functions. When experienced by individuals within the working population, these symptoms can become a burden for both workers and employers. Barriers such as limited funding, expertise, and a fragmented healthcare system can contribute to the underrepresentation of neurodegenerative diseases in Trinidad and Tobago. This study will focus on the implications of ND in the labour market within workplaces in Trinidad and Tobago and explore how workers can be supported through Socially Assistive Robotics (SARs) and Soft Robotics to better manage their symptoms while remaining active at work.

Preface

Edited by Bryan Christiansen and Anatoly Zhuplev

In the modern world of work, organizations are increasingly seeking innovative, science-based solutions to improve employee well-being, optimize performance, and maintain a competitive edge. Among the most promising developments in this pursuit is the integration of neuroscience into organizational theory and practice, particularly through the study and application of neuroplasticity.

Neuroplasticity, the brain's ability to adapt structurally and functionally in response to experiences and environments, is no longer confined to clinical or educational contexts. Its relevance to the workplace from leadership development and stress management to skill acquisition and adaptability has profound implications for how we understand and enhance organizational effectiveness and labor productivity.

This edited volume, *Impact of Neuroplasticity on Organizational Effectiveness and Labor Productivity*, brings together researchers and practitioners from across the globe to explore how neuroplastic principles can be practically applied to build resilient, high-performing, and human-centered organizations. The chapters within span neuroscience, psychology, human resource development, organizational behavior, and leadership studies, providing interdisciplinary insight into how the brain's adaptive potential can be strategically harnessed in diverse professional environments.

From Ethiopia to India, Portugal to Trinidad and Tobago, the contributions in this volume explore culturally and economically diverse contexts, emphasizing the global relevance and adaptability of neuroplastic frameworks. Through empirical studies, theoretical analyses, and practical recommendations, the book offers a roadmap for educators, policymakers, business leaders, and organizational scholars seeking to foster continuous learning, cognitive agility, and sustainable workforce well-being.

CHAPTER OVERVIEW

Chapter 1: The Impact of Neuroplasticity on Skill Acquisition and Cognitive Growth

Rathi Priya Bollu Yashwant, Eng Hwa Wong

A deep dive into how neuroplasticity supports lifelong learning, skill development, and cognitive adaptation—highlighting applications in education, workplace training, and neurorehabilitation, including among neurodiverse individuals.

Chapter 2: The Neural Path to Productivity: How Brain Science Transforms Workplace Efficiency

Muralidhar B, Lavakumar V, Usha Prabhu, N Sathyanarayana, Rahul Varanasi

Linking neuroscience to workplace performance, this chapter analyzes how neuroplasticity, emotional intelligence, and attention systems can be leveraged to enhance decision-making, innovation, and well-being in the modern organization.

Chapter 3: Reprogramming the Mind: Neuroplasticity, Stress Resilience, and Emotional Intelligence in Professional Settings

Nidhi Suresh, Srishti Muralidharan

This chapter presents a neuroscience-informed framework for building stress resilience and emotional intelligence in the workplace, offering practical strategies to develop adaptive teams and psychologically safe organizational cultures.

Chapter 4: Neuroplasticity, Stress, and Resilience in the Modern Workplace

Jossy George, Peter Muttungal, Benny Godwin J. Davidson, Nishi Priya

Exploring the physiological effects of chronic stress on brain function, the chapter presents evidence-based practices such as mindfulness and exercise to promote neuroplastic recovery and resilience in high-pressure work environments.

Chapter 5: Organizational Characteristics of "Happy Organizations"

Evangelina Joseph, Vinita Seshadri

Focusing on structural and cultural attributes rather than individual psychology, this chapter outlines the systems and values that underpin long-term employee happiness and positive organizational outcomes.

Chapter 6: Navigating the Rapid: Organizational Change and Adaptation in the 21st Century

Evangelina Joseph

This chapter examines the neuroscience of change, emphasizing neuroplasticity's role in fostering adaptive leadership and agile cultures amid VUCA (Volatility, Uncertainty, Complexity, Ambiguity) environments.

Chapter 7: Measuring the Immeasurable? Quantifying Neuroplasticity's Impact on Organizational Performance and Workforce Productivity

Joanne Vázquez de Príncipe

A pioneering chapter that explores the challenges and methodologies for measuring the impact of neuroplastic interventions on organizational metrics such as productivity, engagement, innovation, and retention.

Chapter 8: Rewiring the Workplace: Harnessing Neuroplasticity for Cognitive Agility and Skill Mastery in the Age of Disruption

Rismawati Rismawati, Anuradha Ranjan, Kirti Sharma

This chapter provides a strategic framework for using neuroplasticity to future-proof talent development in disrupted industries, incorporating immersive learning technologies and adaptive leadership approaches.

Chapter 9: Building Resilient Organizations: Effect on Workforce Well-being by Mediation of Labor Productivity in Gedeo Zone, Ethiopia

Shashi Kant, Dawit Jabo, Zinabu Wolde, Zerihun Alemu, Rakesh Gautam

Drawing on data from Ethiopian enterprises, this quantitative study demonstrates how organizational resilience directly supports labor productivity and indirectly boosts employee well-being.

Chapter 10: Ethical Leadership Effect on Workplace Well-being through Organizational Citizenship Behavior in Ethiopia's Health Care Sector

Wako Jio, Shashi Kant, Hirut Assegid

This chapter investigates how ethical leadership promotes organizational citizenship behavior (OCB), which in turn enhances workplace well-being, using empirical evidence from Ethiopian healthcare workers.

Chapter 11: Transformational Leadership and Growth Mindset: The Mediating Role of Organizational Culture in Ethiopia's Health Care Sector

Shashi Kant, Wako Jio, Hirut Assegid

Focusing on leadership's psychological impact, this study shows how transformational leaders foster growth mindsets and OCB among healthcare staff—mediated by a supportive organizational culture.

Chapter 12: Supporting Workers with Neurodegenerative Diseases in Trinidad and Tobago

Roshnie Doon

This innovative chapter explores the use of Socially Assistive and Soft Robotics to support workers with neurodegenerative diseases, making a case for inclusive, neuro-adaptive workplace design in developing countries.

Chapter 13: Transitioning into Industry 5.0: Integrating Neuroplasticity into Human Capital Strategies in Trinidad and Tobago

Roshnie Doon

Linking human-machine collaboration with brain-based learning, this chapter outlines how principles of neuroplasticity can be used to develop cognitive agility and strategic adaptability in an Industry 5.0 context.

By spanning disciplines, regions, and industries, this book affirms that neuroplasticity is a transformative principle for rethinking human potential at work. We invite readers to engage with these chapters not merely as isolated studies, but as interconnected explorations into how reshaping minds can reshape organizations—toward more resilient, intelligent, and human-centered futures.

Chapter 1
The Impact of Neuroplasticity on Skill Acquisition and Cognitive Growth

Rathi Priya Bollu Yashwant
https://orcid.org/0009-0003-2526-9903
Taylor's University, Malaysia

Eng Hwa Wong
https://orcid.org/0000-0001-9322-772X
Taylor's University, Malaysia

ABSTRACT

This chapter explores the foundational role of neuroplasticity in skill acquisition and cognitive growth across the human lifespan. Neuroplasticity—the brain's ability to reorganize its structure and function in response to internal and external stimuli—underpins the development of new skills, adaptation to environmental demands, and recovery from injury. Drawing on recent advances in neuroscience, the chapter examines key mechanisms of plasticity, including synaptic strengthening, dendritic remodeling, and network-level reorganization, and links these to practical outcomes in education, professional development, and rehabilitation.The chapter also addresses neuroplasticity in neurodiverse populations and clinical conditions, highlighting both the potential and limitations of plastic change. By integrating empirical findings with applied frameworks, this work provides a multidimensional understanding of how neuroplasticity can be harnessed to foster learning, improve cognitive performance, and promote lifelong adaptability.

DOI: 10.4018/979-8-3373-2292-6.ch001

1. INTRODUCTION

The human brain, with its ~100 billion neurons, is a highly adaptive organ capable of learning, recovery, and transformation (Verma, 2023). Central to these abilities is neuroplasticity, the brain's capacity to reorganize structurally and functionally in response to stimuli throughout life (Asuku et al., 2025; Srijit & Sadhana, 2023). Contrary to earlier beliefs, plasticity continues into adulthood, enabling learning, behavioral flexibility, and recovery from injury or disease (Sagi & Rosenzweig, 2020).

Neuroplasticity operates across biological levels ranging from synaptic changes to large-scale cortical reorganization allowing dynamic adaptation to experience and environment (Marzola et al., 2023). Influenced by genetics, lifestyle, and context, it plays a key role in cognitive enhancement, education, and rehabilitation (Noda, 2024). Across the lifespan, neuroplasticity takes different forms. In early development, rapid synaptogenesis and critical periods shape foundational circuits. In adulthood, plasticity becomes experience-driven, supporting lifelong learning and behavioral refinement (Babcock et al., 2021; Wallace et al., 2023). These mechanisms also aid recovery and are increasingly linked to neurodegenerative and psychiatric conditions (Dinius et al., 2023; Dey et al., 2024; Babiy et al., 2025).

Beyond clinical and educational contexts, neuroplasticity is gaining relevance in organizational development. It underpins lifelong learning, adaptability, and cognitive growth, key assets in evolving workplaces shaped by technology and change (Mateos-Aparicio & Rodríguez-Moreno, 2019; Vinogradov, 2023; Pickersgill et al., 2022). Cognitive training, mindfulness, and continuous learning enhance neuroplastic responses, improving performance, resilience, and career progression (Lenze et al., 2022).

Understanding neuroplasticity offers a foundation for workforce development strategies such as upskilling and talent cultivation. Bridging neuroscience, psychology, and education, it provides a framework for optimizing human potential (Herzberg et al., 2025). Aligning organizational goals with the brain's adaptive capacity fosters innovation, agility, and long-term growth (Alvarado-García et al., 2025; An et al., 2025). This chapter explores the neurobiological basis of neuroplasticity in skill acquisition and cognitive development, integrating insights from neuroscience, psychology, and education. It aims to provide a translational framework for future research and practical applications across diverse domains.

1.1. Research Methodology

Literature was sourced from databases including ScienceDirect, Google Scholar, PubMed, Scopus, Web of Science, and the Cochrane Library, with additional searches via Google. Keywords included neuroplasticity, cognitive growth, skill acquisition, cognitive training, mindfulness, and meditation. Studies from the past 20 years were prioritized, and relevant abstracts and full texts were screened and synthesized.

1.2. Objectives

This chapter aims to:

- Trace the historical development and contemporary understanding of neuroplasticity.
- Examine core mechanisms and types of neuroplasticity involved in learning, memory, and cognitive growth.
- Evaluate the role of neuroplasticity in skill acquisition and cognitive development across contexts.
- Review therapeutic strategies such as cognitive training, physical activity, and mindfulness that enhance neuroplasticity.
- Assess neuroplasticity-based techniques for improving learning in academic and organizational environments.

2. THEORETICAL FOUNDATIONS OF NEUROPLASTICITY

2.1. Historical Perspective of Neuroplasticity

The concept of neuroplasticity has evolved from philosophical speculation to a central tenet of modern neuroscience. In the 19th century, William James introduced the idea of brain "plasticity," describing its capacity to be molded by experience (James, 1890). Around the same time, Santiago Ramón y Cajal revolutionized neuroscience with his Neuron Doctrine and the notion of "cerebral gymnastics" to enhance mental capacity (Ramón y Cajal, 1899/2004; Markram et al., 2011). Eugenio Tanzi and Ernesto Lugaro advanced early theories of synaptic modification and chemical transmission (Berlucchi & Buchtel, 2009). The term "synapse" was coined by Charles Sherrington and Michael Foster in 1897, setting the stage for future discoveries.

In the mid-20th century, Jerzy Konorski formally introduced "neural plasticity" and linked learning to synaptic reorganization (Konorski, 1968). Donald Hebb's influential theory "cells that fire together, wire together" provided a mechanistic explanation for associative learning and memory (Markram et al., 2011). These foundational ideas led to the discovery of long-term potentiation (LTP) and long-term depression (LTD), key mechanisms underlying memory formation.

By the late 20th and early 21st centuries, researchers such as Michael Merzenich, Norman Doidge, and Fred Gage demonstrated that neuroplasticity persists into adulthood, challenging earlier assumptions of a fixed brain (Famitafreshi & Karimian, 2019; Babcock et al., 2021). Contemporary work by Takao Hensch, Mark F. Bear, and Maiken Nedergaard continues to expand our understanding of critical periods, synaptic dynamics, and glial contributions to plasticity. These milestones are summarized in Table 1.

Table 1. **Timeline of Neuroplasticity with Key Scientists and Their Contributions**

Time period	Scientist(s)	Key contribution(s)
19th century	William James (1890)	Introduced concept of brain plasticity and explained how experiences shape neural structure.
19th century	Eugenio Tanzi (1893)	Proposed synaptic strengthening and chemical transmission
19th century	Santiago Ramón y Cajal (1894)	Neuron Doctrine; "cerebral gymnastics" to enhance mental capacity
19th century	Charles Sherrington & Michael Foster (1897)	Coined the term "synapse"
Early 20th century	Ernesto Lugaro (1906–1909)	Linked neural activity to structural change and expanded chemical transmission theory
Mid-20th century	Jerzy Konorski (1948)	Introduced "neural plasticity" and linked learning to synaptic and morphological change
Mid-20th century	Donald Hebb (1949)	Hebbian theory of synaptic change and laid foundation for LTP and LTD
21st century	Michael Merzenich (2000)	Demonstrated adult neuroplasticity; developed brain training programs
21st century	Norman Doidge (2007, 2015)	Popularized neuroplasticity through clinical case studies
21st century	Takao Hensch (2000–present)	Identified critical period mechanisms; reactivation of plasticity in adulthood
21st century	Mark F. Bear (2000–present)	Investigated LTP and synaptic plasticity in learning and memory
21st century	Maiken Nedergaard (2010)	Discovered glymphatic system's role in brain function and plasticity
21st century	Fred Gage (2000–present)	Demonstrated adult neurogenesis in the hippocampus

These findings highlight the multifaceted nature of neuroplasticity and set the stage for a closer examination of its key dimensions across biological, experiential, and technological domains.

2.2. Contemporary Dimensions of Neuroplasticity

Recent research affirms that neuroplasticity is a lifelong process, shaped by environmental conditions, lived experiences, and learning. This dynamic adaptability underpins cognitive function, mental health, and recovery from neurological injury (Milbocker et al., 2023; Fossati et al., 2020; Khaspekov & Frumkina, 2023). Key mechanisms, cellular contributors, and technological advances supporting neuroplasticity are summarized in Table 2.

Table 2. **Key Dimensions of Contemporary Neuroplasticity**

Dimension	Description	Key References
Lifelong Plasticity	Brain adapts and reorganizes throughout life, including adult neurogenesis	Babcock et al., 2021; Vaz et al., 2022
Experience-Driven Changes	Learning and training induce structural and functional synaptic changes	Monday et al., 2018
Environmental Influences	Enrichment enhances plasticity; deprivation impairs it	Milbocker et al., 2023
Glial Cell Involvement	Astrocytes and microglia support synaptic remodeling and neurotrophic signaling	Kaczmarek, 2020; Wang et al., 2016
Glymphatic System	Sleep-dependent waste clearance supports memory and plasticity	Gao et al., 2023; Beschorner & Nedergaard, 2024
Technological Advancements	Neuroimaging and neuromodulation reveal and influence plastic changes	Noda, 2024
Quantifying Cognitive Trajectories	Imaging tools track cognitive reserve and brain network changes	Frizzell et al., 2022; Chen & Goodwill, 2022
Personalized Interventions	Tailored therapies improve outcomes in education and clinical care	Li & Han, 2022; Lenze et al., 2022
Bidirectional Nature	Plasticity can be adaptive or maladaptive	Appelbaum et al., 2023; Pickersgill et al., 2022
Cross-Disciplinary Integration	Applications span neuroscience, psychology, AI, and education	Pickersgill et al., 2022

Collectively, these insights have transformed our understanding of the adult brain's capacity to adapt, reorganize, and regenerate in response to experience, learning, and environmental conditions. To further elucidate this adaptability, the following section explores the core biological mechanisms that drive neuroplastic change at molecular, cellular, and systems levels.

3. MECHANISMS UNDERPINNING NEUROPLASTICITY

Neuroplasticity refers to the brain's capacity to reorganize itself by forming new neural connections and modifying existing ones. This adaptability is shaped by genetic, epigenetic, and environmental influences and underlies essential functions such as learning, memory, sensory processing, and recovery from neurological injury (Herzberg et al., 2025; Asuku et al., 2025). As a lifelong process, neuroplasticity enables individuals to acquire new skills, respond to challenges, and optimize cognitive performance across the lifespan (Wallace et al., 2023). To elucidate how these changes occur, the following section explores core mechanisms at the cellular and molecular levels, including synaptic modifications, gene regulation, neurotransmitter activity, and neurotrophic signaling. Figure 1 shows the major mechanisms of neuroplasticity.

Figure 1. Mechanisms of Neuroplasticity

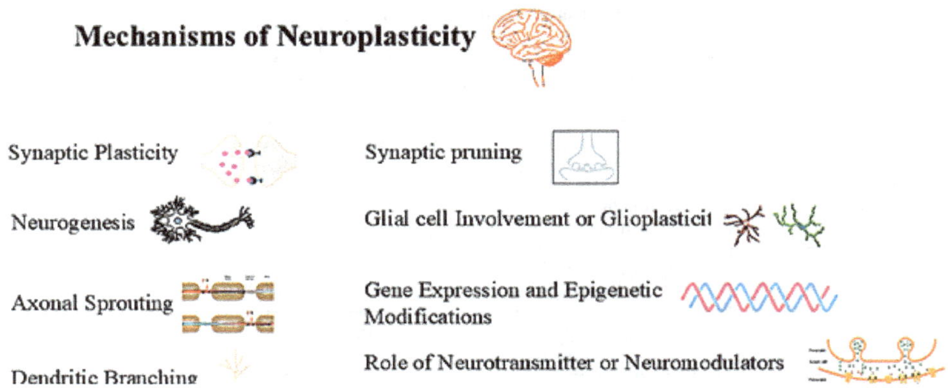

3.1. Synaptic Plasticity

Synaptic plasticity refers to the brain's capacity to modify the strength and structure of synaptic connections in response to experience, activity, and learning (Appelbaum et al., 2023). It serves as a core mechanism for skill acquisition, memory formation, and behavioral adaptation. By enhancing the efficiency of neurotransmission, synaptic plasticity enables the nervous system to fine-tune its responses to stimuli, positioning it as a cornerstone of cognitive flexibility and learning (Sagi & Rosenzweig, 2020). This adaptability is supported by multiple, interrelated processes that balance neural flexibility with circuit stability. Synaptic plasticity is broadly categorized into two complementary forms:

- Functional Plasticity: Refers to changes in synaptic strength and efficiency. Key mechanisms include Hebbian plasticity (e.g., LTP and LTD) and homeostatic plasticity, which maintain overall network stability.
- Structural Plasticity: Involves physical changes in neuronal architecture. Mechanisms include neurogenesis, synaptic remodeling, axonal sprouting, dendritic reorganization, and glioplasticity.

3.1.1. Functional Plasticity

Functional plasticity refers to the brain's capacity to reorganize neural activity in response to experience, learning, or injury without requiring structural alterations (Frizzell et al., 2022). By modulating synaptic strength and efficiency, it enables rapid adaptation to cognitive demands and environmental shifts (Marzola et al., 2023). This form of plasticity supports skill acquisition, memory consolidation, and attentional regulation by refining neural pathways for more efficient processing. During early development, functional plasticity enables the refinement of sensory and motor circuits in response to environmental input, contributing to critical period learning and long-term behavioral outcomes (Hensch, 2005). Core mechanisms such as LTP and LTD modulate synaptic strength by enhancing the efficacy of frequently activated neural pathways while attenuating those that are less engaged (Sagi & Rosenzweig, 2020). These dynamic changes facilitate perception, learning, and executive control (Nudo, 2013).

In clinical contexts, functional plasticity plays a critical role in recovery and compensation. Following stroke, perilesional and contralateral regions may be recruited to restore motor and cognitive functions (Reidler et al., 2018). In neurodegenerative conditions like Alzheimer's disease, adaptive synaptic responses and alternative network engagement help maintain cognitive performance despite progressive structural decline (Popescu et al., 2024; Hidayat, 2023). In psychiatric rehabilitation, targeted cognitive training and behavioral therapies have been shown to induce functional reorganization in prefrontal and limbic circuits, improving emotional regulation and executive functioning in conditions such as depression and schizophrenia (Subramaniam et al., 2014).

Subtypes of Functional Plasticity: Functional plasticity encompasses distinct yet complementary mechanisms that enable the brain to adapt to experience while maintaining network stability.

(i) ***Hebbian Plasticity*** is an activity-dependent process in which the co-activation of presynaptic and postsynaptic neurons strengthens their synaptic connection (Mateos-Aparicio & Rodríguez-Moreno, 2019). It drives two core processes:

- **LTP:** High-frequency stimulation activates NMDA receptor-mediated calcium influx, triggering intracellular signaling cascades that enhance synaptic strength and dendritic spine density. These changes support memory encoding and motor skill consolidation.
- **LTD:** Low-frequency stimulation reduces synaptic efficacy, facilitating the pruning of underused connections and promoting cognitive efficiency (Pinar et al., 2017; Stacho et al., 2022).

Together, LTP and LTD refine behaviourally relevant circuits, enabling adaptation to novel environments and task demands (Chindemi et al., 2022).

(ii) ***Homeostatic Plasticity*** maintains overall network stability by regulating neuronal activity and preventing excessive excitation or silencing. Unlike Hebbian mechanisms, which are synapse-specific, homeostatic processes operate across broader neural populations (Martin-Aragon & Bermejo-Bescós, 2019). Key mechanisms include:

- **Synaptic Scaling:** Uniform adjustment of synaptic strengths to preserve consistent firing rates across neurons.
- **Intrinsic Plasticity:** Modulation of ion channel expression to stabilize neuronal excitability.

These mechanisms balance excitatory-inhibitory dynamics, ensuring sustained learning without functional overload.

(iii) **Complementarity and Cognitive Relevance:** Hebbian and homeostatic plasticity fulfil distinct roles, one driving experience-specific refinement, the other preserving functional equilibrium. Their interplay supports lifelong learning, skill acquisition, and behavioral flexibility, offering a foundation for translational strategies in education, therapy, and cognitive enhancement.

3.1.2. Structural Plasticity

Structural plasticity refers to the brain's capacity to undergo physical changes in response to experience, learning, injury, and developmental transitions (Hidayat, 2023). This process encompasses dendritic spine remodeling, axonal sprouting, glioplasticity, synaptic pruning, and the formation of new synapses (synaptogenesis) and neurons (neurogenesis) (Jajarmi et al., 2024). Such modifications contribute to developmental maturation, cognitive flexibility, and neural repair, enhancing memory and adaptive capacity in response to internal and external stimuli (Hermans et al., 2024). Crucially, structural plasticity underpins therapeutic interventions and neurorehabilitation, particularly in stroke recovery (Li & Han, 2022) and neurodegenerative disorders like Alzheimer's disease, where compensatory changes promote functional resilience (Popescu et al., 2024).

(i) Neurogenesis:

Developmental neuroscience shows that during prenatal stages, neurogenesis occurs rapidly, while postnatal development involves programmed cell death (apoptosis) to refine neural circuits. Neural stem and progenitor cells in the CNS retain lifelong potential for differentiation, maturing into functional neurons in response to stimuli like skill acquisition or injury (Lucassen et al., 2020). In adulthood, neurogenesis is largely confined to the subventricular zone and hippocampal dentate gyrus, where it supports memory, cognitive flexibility, and recovery. This process is enhanced by factors such as physical activity, cognitive stimulation, and social interaction, but can be suppressed by stress, aging, and neurodegenerative diseases (Ahamed et al., 2024; Mishra et al., 2021).

Disruptions in adult neurogenesis are linked to conditions like depression, anxiety, autism, and Alzheimer's and Parkinson's diseases (Fries et al., 2023; Popescu et al., 2024). While evidence supports ongoing neurogenesis in adults, especially in the hippocampus, its extent and functional significance remain debated. Advances in transcriptomics and machine learning have identified proliferating progenitors even in older adults (Simard et al., 2024), though inconsistencies in detection methods highlight the need for standardized protocols and longitudinal studies. Understanding and enhancing neurogenesis holds promise for therapeutic innovation, offering pathways for cognitive restoration and improved mental health.

(ii) Axonal Sprouting:

Axonal sprouting is a key neuroadaptive mechanism wherein surviving neurons extend new branches to restore and reorganize neural circuits following injury or degeneration (Jajarmi et al., 2024). Key molecular regulators include neurotrophic factors such as brain-derived neurotrophic factor (BDNF) and neurotrophin-3, which facilitate post-injury connectivity, particularly in spinal cord lesions. Proteins like Growth-associated protein-43 are essential for axonal guidance and synaptogenesis, contributing to functional recovery (Mascaro et al., 2013). Clinically, axonal sprouting supports cortical reorganization in patients with CNS damage and may signal compensatory responses in neurodegenerative conditions such as Alzheimer's disease (Zhang et al., 2021). Its therapeutic relevance is underscored in disorders like amyotrophic lateral sclerosis, where enhancing sprouting may preserve motor functions (Marshall & Farah, 2021). However, unregulated or aberrant sprouting, known as maladaptive plasticity, can disrupt network integrity, emphasizing the importance of targeted strategies to ensure constructive remodeling (Asuku et al., 2025).

(iii) Dendritic plasticity:

Dendritic plasticity refers to the structural and functional adaptability of dendrites, which are the neuronal extensions that facilitate synaptic communication in response to activity and experience (Nishiyama, 2019). Central to this process is the remodeling of dendritic spines, which originate as filopodia and mature into diverse morphologies, such as mushroom-shaped or bifurcated forms, depending on synaptic input. LTP promotes spine growth and stabilization, whereas LTD drives spine shrinkage or elimination, thereby modulating synaptic strength (Mateos-Aparicio & Rodríguez-Moreno, 2020). Intracellular signaling pathways, particularly those involving Ras GTPases, regulate spine architecture and efficacy (Nishiyama, 2019). While essential for learning and behavioral flexibility, dendritic plasticity requires tight homeostatic control, as its disruption is linked to neurodevelopmental disorders (D'Aquin et al., 2022). In summary, dendritic plasticity is a key mechanism of experience-dependent neural adaptation, with implications for both cognitive enhancement and the pathophysiology of brain disorders.

(iv) Synaptic Remodeling: Formation and Elimination of Synapses

Synaptic remodeling, encompassing both synaptogenesis and synaptic pruning, is a fundamental aspect of neuroplasticity that shapes the neural architecture underlying learning, memory consolidation, and adaptive behavior (Martinez-Morga & Martinez, 2017). During early development, the brain rapidly generates an abundance of synaptic connections, which are subsequently refined through experience-dependent pruning to enhance circuit efficiency. At birth, each of the approximately 100 billion neurons are connected via around 2,500 synapses, increasing to nearly 15,000 per neuron by the age of two to three. This surge represents a peak in connectivity and plasticity, followed by a reduction to about 7,000–8,000 synapses per neuron in adulthood as networks are optimized. Although the rate of synaptogenesis declines with age, it remains responsive to cognitive demands, enriched environments, and learning experiences (Faust et al., 2021; Marzola et al., 2023). Synaptic remodeling is orchestrated by molecular mechanisms including neurotransmitter release, receptor clustering (AMPA and NMDA), and stabilization of postsynaptic structures through cell adhesion molecules that maintain synaptic specificity (Gall et al., 2023).

Alongside dendritic and axonal plasticity, these processes contribute to long-term refinement of neural circuits. Such remodeling supports sensory integration, behavioral flexibility, and skill acquisition, whether in learning a second language, developing problem-solving strategies, or rehabilitating motor functions (Dan, 2019). Therapeutic strategies increasingly target synaptic remodeling to restore neural function in neurodegenerative conditions. Approaches such as neurotrophic modulation, behavioral interventions, and cognitive rehabilitation harness both synaptogenesis and neurogenesis to counteract network disruption and promote functional recovery (Sasmita et al., 2018). In conclusion, synaptic remodeling serves as a cornerstone of lifelong neuroplasticity, offering a promising avenue for interventions aimed at enhancing cognitive resilience and restoring neural integrity in the face of aging and disease.

(v) Glioplasticity:

Glioplasticity encompasses the dynamic and adaptive roles of glial cells, particularly astrocytes and microglia, in modulating synaptic function and restructuring neural circuits (Liu et al., 2023). Astrocytes are instrumental in synaptogenesis and synaptic pruning, maintaining neurotransmitter homeostasis, and regulating NMDA receptor activity, thereby influencing key mechanisms of synaptic plasticity such as LTP and LTD (Khaspekov & Frumkina, 2023). Microglia contribute to synaptic refinement through activity-dependent surveillance, secretion of neurotrophic factors, and direct interactions with neuronal networks (Wang et al., 2016). Both astrocytes and microglia are actively involved in remodeling the extracellular matrix, facilitating structural reorganization in response to environmental stimuli and pathological conditions (Dzyubenko & Hermann, 2023).

Age-related alterations in neurotransmitter plasticity, often governed by epigenetic mechanisms, highlight the potential for targeted, non-invasive interventions in neuropsychiatric disorders characterized by synaptic dysregulation and cognitive decline (Pritchard et al., 2019). Investigating epigenetic regulation at the level of specific neuronal subtypes, rather than whole tissue samples, offers enhanced precision in elucidating the maintenance of neurotransmitter identity and plasticity across the lifespan. These insights underscore the therapeutic promise of glial modulation in preserving neuroplastic integrity and promoting cognitive resilience.

3.2. Genetic and Epigenetic Modulators of Neuroplasticity

Neuroplasticity is shaped by both genetic architecture and epigenetic regulation. Key genes such as BDNF, catechol-O-methyltransferase, and apolipoprotein E modulate neurotrophic support, neurotransmitter metabolism, and synaptic integrity, directly influencing learning, memory, and recovery (Bhaduri, 2024). Variations in these genes, such as reduced BDNF or altered catechol-O-methyltransferase activity, are associated with cognitive deficits and working memory challenges. Epigenetic mechanisms including DNA methylation, histone modification, and microRNA activity adjust gene expression in response to environmental cues. Enrichment experiences or early-life adversity can imprint long-term epigenetic signatures, affecting brain adaptability and resilience (Benatti et al., 2024). Conversely, toxin exposure may disrupt these pathways and impair neurodevelopment. Together, these genomic and epigenomic processes account for individual variability in cognitive capacity and neural flexibility across the lifespan.

3.3. Role of Neurotransmitters in Plasticity

Neurotransmitters are chemical messengers that facilitate synaptic communication, regulating neural excitability, connectivity, and cognitive-emotional processes (Bhat et al., 2021). Key neurotransmitters such as glutamate, GABA, dopamine, serotonin, endorphins, and enkephalins play central roles in neuroplasticity by modulating synaptic strength and responsiveness (Guo et al., 2020; Kaczmarek, 2020). Glutamate and GABA balance excitatory and inhibitory signaling, while dopamine and serotonin influence motivation and affective regulation.

A distinct aspect of plasticity is neurotransmitter switching, where neurons adapt their chemical identity in response to environmental stimuli such as photoperiod changes or learning demands (Pritchard et al., 2019). This process involves synaptic remodeling and is regulated by transcriptional programs, calcium signaling, and epigenetic mechanisms like DNA methylation and histone modification (Demarque & Spitzer, 2020). Age-related decline in these regulatory pathways affects neurotransmitter expression and plasticity (Pritchard et al., 2019). Altered neurotransmitter levels are implicated in neurodegenerative diseases (e.g., Alzheimer's, Parkinson's, Huntington's, amyotrophic lateral sclerosis, epilepsy) and psychiatric conditions (e.g., depression, schizophrenia, autism, addiction, sleep disorders), disrupting synaptic signaling and adaptive brain function (Satarker et al., 2022). Understanding neurotransmitter plasticity offers translational potential for non-invasive interventions aimed at restoring neural communication and cognitive integrity.

3.4. Role of Neurotrophic Factors

Neurotrophic factors, particularly BDNF and nerve growth factor, are vital modulators of neuroplasticity, supporting neuronal survival, differentiation, neurogenesis, and synaptic remodeling (Ball et al., 2022; De Vincenti et al., 2019). They play essential roles in learning, memory, and emotional regulation, and are increasingly recognized for their therapeutic relevance in neurological and psychiatric conditions (Amidfar et al., 2021). Reduced BDNF levels have been associated with impaired hippocampal function and synaptic deficits in depression, while its restoration shows promise for cognitive and emotional recovery. Similarly, nerve growth factor enhances neuronal adaptability and resilience, reinforcing the importance of neurotrophic signaling in maintaining brain health and promoting recovery following stress or injury (Mamczur et al., 2024). Neurotrophic signaling is a core driver of neuroplasticity, offering therapeutic promise for enhancing resilience and recovery in neurological and psychiatric conditions.

4. FUNDAMENTAL TYPES OF NEUROPLASTICITY

Classifying plasticity by outcome and mechanism provides a comprehensive framework for understanding how the brain navigates development, disease, and recovery, and informs targeted interventions in both clinical and educational contexts.

4.1. Developmental Plasticity

Developmental neuroplasticity describes the brain's heightened capacity to adapt and reorganize during early life and adolescence in response to experiences and environmental influences (Faust et al., 2021; Marzola et al., 2023). In early childhood, this plasticity facilitates rapid synapse formation, laying the groundwork for essential abilities such as language acquisition, sensory processing, and emotional regulation. As development progresses, a process known as synaptic pruning refines neural networks by removing underused connections and reinforcing frequently activated ones, thereby improving cognitive efficiency. Adolescence represents a second, more targeted phase of plasticity, in which the brain continues to mature, particularly in the prefrontal cortex, while pruning occurs in regions associated with sensory and emotional processing. Although the brain's plasticity gradually diminishes with age, it remains active throughout adulthood in a more regulated form, influenced by factors such as learning, therapeutic interventions, and enriched environments. This lifelong adaptability is shaped by a combination of age, experience, and contextual factors (Monday et al., 2018; Wallace et al., 2023).

4.2. Adaptive Neuroplasticity

Adaptive plasticity refers to the brain's continuous ability to reorganize and modify its structure and function in response to behavioral demands, experiences, and environmental changes. It encompasses processes such as synaptic remodeling, dendritic branching, and cortical reorganization (Faust et al., 2021; Marzola et al., 2023), which are essential for learning, memory formation, skill development, and recovery from neurological injuries. These mechanisms enable the brain to fine-tune motor and sensory functions and highlight its capacity to adjust dynamically throughout life, both in healthy individuals and in therapeutic or rehabilitative settings (Monday et al., 2018; Wallace et al., 2023).

4.2.1. Subtypes of Adaptive Plasticity

Adaptive plasticity encompasses several distinct but overlapping mechanisms through which the brain reorganizes in response to experience and activity:

(i) **Use-Dependent Plasticity:** Neural circuits strengthen or weaken based on repeated use or disuse.

Examples:

- Motor skill refinement through repetitive practice (e.g., typing, playing instruments)
- Language maintenance via continued use of a second language
- Constraint-induced movement therapy in stroke rehabilitation

(ii) **Activity-Dependent Plasticity:** Driven by neuronal firing and synaptic activity, primarily through LTP and LTD.

Examples:

- Memory formation via hippocampal LTP
- Sensory development shaped by early visual exposure
- Auditory learning through repeated sound exposure

(iii) **Experience-Dependent Plasticity:** Reflects changes shaped by individual learning and environmental exposure, integrating use- and activity-dependent mechanisms.

Examples:

- Skill acquisition in music or sports
- Environmental enrichment enhancing hippocampal synaptic density

- Psychotherapy and cognitive training promoting reorganization in emotional and cognitive networks

(iv) **Experience-Expectant Plasticity:** Occurs during critical developmental windows, requiring specific inputs to shape neural circuits.

Examples:

Visual system development through early patterned light exposure
Language acquisition during sensitive periods
Social bonding shaped by early caregiver interaction

4.2. Reactive Plasticity

Reactive plasticity, also known as injury-induced plasticity, refers to the brain's innate ability to reorganize neural circuits to compensate for lost functions or reduce deficits following CNS damage, such as stroke, TBI, or neurodegenerative diseases (Puderbaugh & Emmad, 2023). After injury, the brain activates various neurobiological mechanisms to support recovery (Nudo, 2013), including:

- **Equipotentiality**: In children, functions like language or motor control can shift to undamaged regions (Finger, 2010).
- **Vicariation**: In adults, unrelated brain areas may assume lost functions (Carrera & Tononi, 2014).
- **Diaschisis**: Remote brain regions may experience temporary functional suppression due to disrupted connectivity or neurochemical changes (Reidler et al., 2018).

At the cellular level, recovery involves:

- **Reactive synaptogenesis:** Surviving neurons form new synaptic connections.
- **Axonal sprouting:** Intact neurons extend axons to reconnect with lost targets.
- **Neurogenesis and gliogenesis:** In regions like the hippocampus, injury may stimulate neurogenesis and gliogenesis, though their integration remains under study.

These responses are regulated by molecular signals such as BDNF, inflammatory cytokines, and extracellular matrix remodeling enzymes, which coordinate repair and plasticity.

4.3. Maladaptive Neuroplasticity

Maladaptive plasticity refers to neural reorganization that disrupts normal functioning or reinforces pathological patterns rather than promoting adaptation or recovery (Pickersgill et al., 2022). Though rooted in the brain's inherent flexibility, these changes can have detrimental outcomes across various domains. To contextualize the functional relevance of maladaptive neuroplasticity across domains, the following table outlines core mechanisms, clinical and behavioral manifestations, and associated real-world implications. This structured overview synthesizes key examples across neurological, psychiatric, occupational, and educational settings.

Table 3. Manifestations of Maladaptive Plasticity Across Contexts

Domains	Examples	Mechanisms	Impacts	References
Clinical	Phantom limb pain	Aberrant cortical remapping in somatosensory cortex	Persistent pain and distorted body representation	Pickersgill et al., 2022
Clinical	Use-dependent dystonia	Repetitive motor activity inducing dysfunctional motor circuitry	Involuntary movements and loss of fine motor control	Pickersgill et al., 2022
Clinical	Chronic pain syndromes	Hyperexcitability in nociceptive pathways; altered spinal/cortical activity	Heightened pain sensitivity without ongoing injury	Fuchs et al., 2015
Psychiatric	Addiction	Reinforcement of compulsive reward-seeking circuits	Habitual substance use; cognitive rigidity	Grilli & Zaed, 2023
Psychiatric	Obsessive-compulsive disorder	Pathological strengthening of obsessive neural loops	Intrusive thoughts and compulsive behaviors	Appelbaum et al., 2023
Occupational	Burnout	Reinforced stress circuitry from sustained cognitive overload	Emotional exhaustion; reduced executive function	Yam et al., 2023; Pickersgill et al., 2022
Educational	Rote-driven learning fatigue	Rigid instructional repetition dampens engagement circuits	Reduced cognitive flexibility; surface-level processing	Cantor et al., 2019

While neuroplasticity is inherently adaptive, adverse or repetitive conditions can reinforce dysfunctional patterns. Understanding the factors that steer plasticity toward maladaptive outcomes is crucial for developing interventions that promote recovery and resilience. This awareness is vital not only for clinical strategies but also for designing cognitively sustainable practices in educational and workplace settings.

5. NEUROPLASTICITY AND SKILL ACQUISITION

Neuroplasticity is a foundational principle in understanding how the brain learns, adapts, and develops new competencies. Contemporary neuroscience confirms that plastic changes are not limited to early development or clinical recovery, they extend across the lifespan, supporting the acquisition of new skills, cognitive flexibility, and professional growth (Park & Fine, 2020; Asuku et al., 2025). Skill learning involves both structural modifications such as synaptogenesis, dendritic branching, and white matter remodeling, and functional reorganization of neural circuits that encode, consolidate, and refine performance (Gallo et al., 2025).

(i) Phases of Plasticity in Skill Learning

Skill acquisition progresses through dynamic stages. Early learning activates widespread cortical networks, while advanced stages involve localized refinement and increased neural efficiency. Neuroimaging reveals linear increases in white matter for faster interregional communication and quadratic changes in gray matter for optimized local processing (Emmenegger, 2025). These adaptations occur across age groups, including spinal cord plasticity, which fine-tunes motor patterns via reflex modulation (Randelman et al., 2021).

(ii) Prior Experience and Metaplasticity

Individual experience shapes plastic potential through Metaplasticity, the brain's ability to adjust its responsiveness based on prior activity. Trained individuals, such as musicians, show enhanced neuroplastic responses, enabling faster skill acquisition (Altenmüller & Furuya, 2016). This supports personalized training approaches tailored to neuroplastic profiles (Parija & Singh, 2023).

(iii) Mechanisms Driving Skill Acquisition

Skill learning relies on synaptic mechanisms like long-term potentiation (LTP), synaptogenesis, and dendritic remodeling (Monday et al., 2018; Asuku et al., 2025). Repeated, focused practice strengthens relevant circuits, while intrinsic (e.g., proprioception) and extrinsic (e.g., coaching) feedback refines performance. Techniques such as deliberate practice, interleaved learning, and structured rest enhance neural adaptation and reduce fatigue (Klutz, 2023; Statsenko et al., 2025).

(iv) Neuroimaging Evidence

Neuroimaging confirms training-induced plasticity across domains. fMRI shows increased activation in sensorimotor and associative regions during skilled task performance (Frizzell et al., 2022). Motor training boosts gray matter in the sensorimotor cortex, while speech therapy improves white matter tract integrity (Whelan et al., 2021). Diffusion Tensor Imaging reveals strengthened corticospinal pathways, and cognitive training enhances network efficiency and executive function (Karbach & Schubert, 2013).

(v) Critical and Sensitive Periods

Plasticity peaks during critical periods in early development, when experience induces lasting neural changes (Ismail et al., 2017). Early exposure to music or language can lead to profound adaptations. Sensitive periods, extending into adolescence and adulthood, support continued learning with reduced plasticity, informing educational and therapeutic strategies (Park & Fine, 2020; Herzberg et al., 2025).

(vi) Adaptive vs. Maladaptive Plasticity

While neuroplasticity enables learning and recovery, it can also reinforce maladaptive patterns. Repetitive, unvaried practice may entrench dysfunctional circuits, as seen in use-dependent dystonia or phantom limb pain. Understanding the context and direction of plastic changes is crucial for promoting adaptive outcomes (Pickersgill et al., 2022; Grilli & Zaed, 2023).

(vii) Cognitive Skill Development

Neuroplasticity supports higher-order cognitive skills like reasoning, decision-making, and language. Bilingual individuals show adaptations in the hippocampus and prefrontal cortex, enhancing memory, attention, and flexibility (Gallo et al., 2025). These mechanisms contribute to academic success and professional competence.

(viii) Expertise and Brain Adaptation

Expertise reflects long-term neural adaptation. Experts in music, mathematics, or sports exhibit increased cortical thickness, network efficiency, and specialized activation patterns, hallmarks of sustained, domain-specific practice (Reddy, 2025).

(ix) Lifelong Learning and Workforce Adaptability

In dynamic work environments, neuroplasticity underpins adult learning and career transitions. Cognitive training, motivation, and exposure to novel challenges enhance adaptability and resilience (Wallace et al., 2023). These principles inform personalized learning platforms and workplace training programs (Chen & Goodwill, 2022).

Neuroplasticity drives skill acquisition across the lifespan, enabling learning, adaptation, and expertise. Understanding its mechanisms informs interventions that enhance resilience, optimize performance, and support recovery.

6. NEUROPLASTICITY AND COGNITIVE GROWTH

Neuroplasticity supports the development and refinement of cognitive functions such as attention, memory, and executive control across the lifespan. Its influence varies with age and individual profiles, offering insights into adaptive mechanisms and targeted interventions, particularly for neurodiverse populations. This section begins by examining age-related dynamics of plasticity, followed by its influence on core cognitive domains, and concludes with sociocultural factors that shape individual trajectories of cognitive growth.

6.1. Age-Related Dynamics of Plasticity

Neuroplasticity persists across the lifespan, though its expression and efficiency vary with age. Understanding these dynamics is essential for optimizing learning, cognitive development, and rehabilitation (Hiskens, 2024). As summarized in Table 3, early childhood represents a peak in plasticity, marked by rapid synaptogenesis, cortical reorganization, and heightened sensitivity to environmental input—making it an ideal window for foundational learning and skill acquisition (Fandakova & Hartley, 2020).

Although plasticity gradually declines from its childhood peak, adults retain significant adaptive capacity, especially when learning is goal-directed and reinforced through repetition and feedback. In older adulthood, neuroplasticity is reduced due to diminished neurogenesis, lower synaptic density, and age-related neural degradation (Famitafreshi & Karimian, 2019). However, lifestyle factors such as physical activity, enriched environments, and cognitive engagement can mitigate decline and preserve function (Dinius et al., 2023).

Recent findings affirm that while the rate of change slows with age, the brain's capacity for meaningful adaptation remains intact. This highlights the importance of sustained cognitive stimulation and active living to harness neuroplastic potential throughout life (Mora, 2013; Marzola et al., 2023).

Table 4. **Age-Related Cognitive Growth and Neuroplasticity Across the Lifespan**

Age Group	Neuroplasticity Characteristics	Cognitive Development Milestones	Representative Examples
Infancy (0–2 years)	Rapid synaptogenesis; peak plasticity, critical periods for sensory and language input.	Acquisition of basic sensory processing and early language comprehension.	Recognition of caregivers, initiation of vocalization, and development of motor coordination.
Early Childhood (3–6 years)	Onset of synaptic pruning and high learning receptivity.	Vocabulary explosion, symbolic reasoning, and foundational literacy, and numeracy skills.	Learning grammar rules, engaging in pretend play, and early reading and counting.
Middle Childhood (7–12 years)	Continued synaptic refinement and strengthening of neural networks for focused tasks.	Logical reasoning, increased attention span, and consolidation of academic knowledge	Grasping mathematical operations, reading fluency, and developing memory strategies
Adolescence (13–18 years)	Heightened plasticity in the prefrontal cortex, hormonal influences, and socio-emotional neural remodeling.	Abstract thinking, identity formation, and executive function enhancement.	Goal-setting, improved impulse control, and hypothetical reasoning.
Young Adulthood (19–30 years)	Peak efficiency in cognitive control and learning capacity, and adaptive plasticity remains robust	Complex problem-solving, skill acquisition, and adaptability to new environments.	Pursuing higher education, job training, and expanding interpersonal relationships
Middle Adulthood (31–60 years)	Gradual decline in synaptic plasticity, preservation of crystallized intelligence, and experience-based adaptation	Expertise refinement, strategic thinking, and emotional regulation.	Mastery in career roles, mentoring, and integrating life experiences.
Older Adulthood (60+ years)	Decrease in synaptic density and processing speed, and reliance on compensatory neural mechanisms.	Retention of accumulated knowledge and potential decline in fluid intelligence.	Lifelong learning, memory aid utilization, and participation in cognitive stimulation activities

6.2. Cognitive Functions Influenced by Plasticity

Neuroplasticity underlies core cognitive functions such as attention, memory, reasoning, and executive control. Through dynamic structural and functional changes, the brain adapts to stimuli, recovers from injury, and optimizes performance across domains (Flegal et al., 2019; Appelbaum et al., 2023). Structural plasticity including dendritic spine remodeling and neurogenesis, and functional plasticity such as LTP and LTD support the encoding, retrieval, and refinement of information (Khaspekov & Frumkina, 2023; Klutz, 2023).

Cognitive training programs have demonstrated measurable neuroplastic effects, including:

- **Increased gray matter density in the hippocampus**, enhancing memory retention and processing efficiency.
- **Strengthened connectivity in the prefrontal cortex**, supporting attention regulation and executive control.
- **Plastic changes in the frontoparietal network**, induced by complex tasks like reasoning and problem-solving, promoting flexible thinking and synaptic efficiency.

These findings highlight the brain's capacity to adaptively rewire in response to cognitive demands. Targeted interventions can harness this potential to enhance mental agility, resilience, and overall cognitive performance.

6.3. Neuroplasticity and Sociocultural Diversity

Neuroplasticity does not manifest uniformly across individuals; rather, it is shaped by sociocultural contexts and cognitive diversity. Factors such as educational systems, linguistic exposure, and stress environments significantly influence neural development and adaptability (Sarto-Jackson et al., 2017). Cross-cultural research demonstrates that multilingualism, immersive learning, and culturally enriched environments enhance executive function and cognitive flexibility through multisensory stimulation and distributed neural engagement (Gkintoni et al., 2025). Population-based studies involving multilingual children, neurodivergent individuals, and marginalized learners reveal distinct patterns of plasticity, shaped by both biological predispositions and lived experiences. These findings highlight the importance of inclusive neuroscientific frameworks that account for sociocultural variability and support tailored, neuroplasticity-informed interventions aimed at promoting lifelong cognitive and emotional well-being (Lin et al., 2024).

7. NEUROPLASTICITY IN NEURODIVERSE POPULATIONS

Neuroplasticity underpins cognitive and emotional adaptation across neurodiverse populations, including individuals with attention-deficit/hyperactivity disorder, autism spectrum disorder, learning disabilities, mood and anxiety disorders, and those recovering from brain injuries (Vinogradov, 2023). Although plasticity may follow atypical developmental trajectories, its capacity for reorganization remains central to both challenge and recovery. Table 4 below summarizes key neural features and evidence-based interventions across common neurodiverse conditions.

Table 5. **Neuroplasticity in Neurodiverse Populations**

Clinical Condition	Neuroplastic Changes Observed	Effective Interventions	References
Attention-Deficit/ Hyperactivity Disorder	Atypical maturation of prefrontal cortex, basal ganglia, and cerebellum; reactive and adaptive plasticity	Neurofeedback, cognitive training, Cognitive Behavioral Therapy (CBT), mindfulness-based therapies	Tamon et al., 2024
Autism Spectrum Disorder	Excessive synaptic connectivity in sensory and social circuits; heightened early-life plasticity	Structured behavioral interventions, sensory integration therapy	Tamon et al., 2024
Learning Disabilities (e.g., Dyslexia, Dyscalculia)	Atypical connectivity in language and numerical processing networks	Targeted educational strategies, compensatory training, multisensory learning	Holmes et al., 2021
Mood Disorders (e.g., Depression)	Reduced hippocampal volume; disrupted prefrontal-limbic connectivity	CBT, physical exercise, enriched environments, mindfulness	Guo et al., 2020; Grilli & Zaed, 2023
Anxiety Disorders	Amygdalar hyperactivity; impaired top-down regulation	CBT, exposure therapy, relaxation training, mindfulness	Grilli & Zaed, 2023
Brain Injury (e.g., Stroke, TBI, Tumor-related damage)	Functional reorganization; reparative and reactive plasticity	Timely and specific rehabilitation, enriched environments, task-specific training	Duffau, 2025; Kennedy, 2021; Li & Han, 2022; Dey et al., 2024

These insights underscore the importance of individualized, context-sensitive approaches that harness neuroplastic potential to foster resilience, reduce comorbid symptoms, and promote lifelong cognitive and emotional well-being across the lifespan (Lin et al., 2024).

8. ENHANCING NEUROPLASTICITY FOR SKILL ACQUISITION AND COGNITIVE GROWTH

Recent research across cognitive neuroscience, psychology, and medicine has identified a wide range of modifiable factors that promote neuroplasticity throughout life. Lifestyle and therapeutic interventions such as physical activity, cognitive training, mindfulness, and social engagement have been shown to enhance plasticity across cognitive, emotional, and functional domains. Tables 5 and 6 summarize evidence-based strategies and their underlying neurobiological mechanisms, illustrating how these interventions activate distinct neural pathways to support brain adaptability. Collectively, these approaches offer promising avenues for lifelong brain health, improved learning and performance, and recovery in both neurotypical and neurodiverse populations (Lin et al., 2024).

Table 6. Neuroplasticity-Enhancing Interventions: Types, Examples, and References

Type of Intervention	Examples	References
Physical exercise	Aerobic exercise, resistance training, walking, or cycling	Pickersgill et al., 2022
Cognitive stimulation and training	Working memory training and problem-solving tasks	Au et al., 2020; Dinius et al., 2023
Social interaction	Group activities, peer support, and community engagement	Hagstrom et al., 2021
Environmental enrichment	Novel learning tasks, dynamic workspaces, and sensory stimulation	Mishra et al., 2021
Caloric restriction and Healthy diet	Mediterranean diet, omega-3 rich foods, and antioxidant intake	Pickersgill et al., 2022
Sleep hygiene	Optimized sleep routines, reduced screen exposure, and circadian alignment	Gao et al., 2023; Beschorner & Nedergaard, 2024
Music therapy	Instrumental training, rhythmic entrainment, and music-assisted relaxation	Herholz & Zatorre, 2012
Mindfulness-based practices	Mindfulness-based stress reduction (MBSR) or focused attention training	Lenze et al., 2022; Alvarado-García et al., 2025
Meditation	Breath awareness, loving-kindness meditation, and body scan meditation	An et al., 2025
Non-invasive brain stimulation	TMS and tDCS	Carrarini et al., 2025

Table 7. **Evidence-Based Interventions to Enhance Neuroplasticity: Neurobiological Mechanisms and Cognitive Outcomes**

Intervention	Neurobiological Mechanisms	Cognitive Outcomes
Physical Exercise	↑ BDNF, neurogenesis, and cerebral perfusion	Improves memory, attention, and executive function.
Cognitive Training	↑ Synaptogenesis and ↑ network efficiency	Enhanced working memory, reasoning, and problem-solving.
Social Interaction	↑ Prefrontal-limbic connectivity and ↓ stress hormones	Better emotional regulation, resilience, and memory.
Environmental Enrichment	↑ Dendritic complexity and ↑ synaptic plasticity	Enhances learning capacity and cognitive flexibility.
Caloric Restriction	↑ Mitochondrial efficiency and ↓ oxidative stress	Slows down cognitive aging and improves synaptic integrity.

continued on following page

Table 7. Continued

Intervention	Neurobiological Mechanisms	Cognitive Outcomes
Healthy Diet	↓ Neuroinflammation and ↑ neurotrophic signaling	Improves attention, memory, and processing speed.
Sleep Hygiene	↑ Glymphatic clearance and ↑ memory consolidation	Improves focus, decision-making, and emotional stability.
Music Therapy	↑ Cross-modal integration and auditory-motor coordination	Enhances verbal fluency, working memory, and mood.
Mindfulness	↑ Cortical thickness and Default Mode network regulation	Improves attentional control and reduces stress.
Meditation	↑ Connectivity in attention and emotion-regulation networks	Increases metacognitive awareness, stress resilience, and working memory capacity.
TMS / tDCS	↑ Cortical excitability and LTP	Accelerates recovery, and improves learning and mood.

9. Neuroplasticity-Based Techniques to Enhance Learning in Organizational Environments

Organizations seeking to enhance employee learning and development can adopt a range of evidence-based strategies that foster adaptive neural change. Grounded in the principles of neuroplasticity, these approaches improve cognitive function, support emotional regulation, and facilitate skill acquisition within dynamic work environments (Forget & Le Pertel, 2024). By aligning training initiatives with current neuroscience insights, modern workplaces can optimize learning outcomes and overall performance. The following figure 2 and table 7 summarize key neuroplasticity-based techniques applicable in organizational settings.

Figure 2. Neuroplasticity-Based Techniques for Enhancing Learning in Organizational Environments

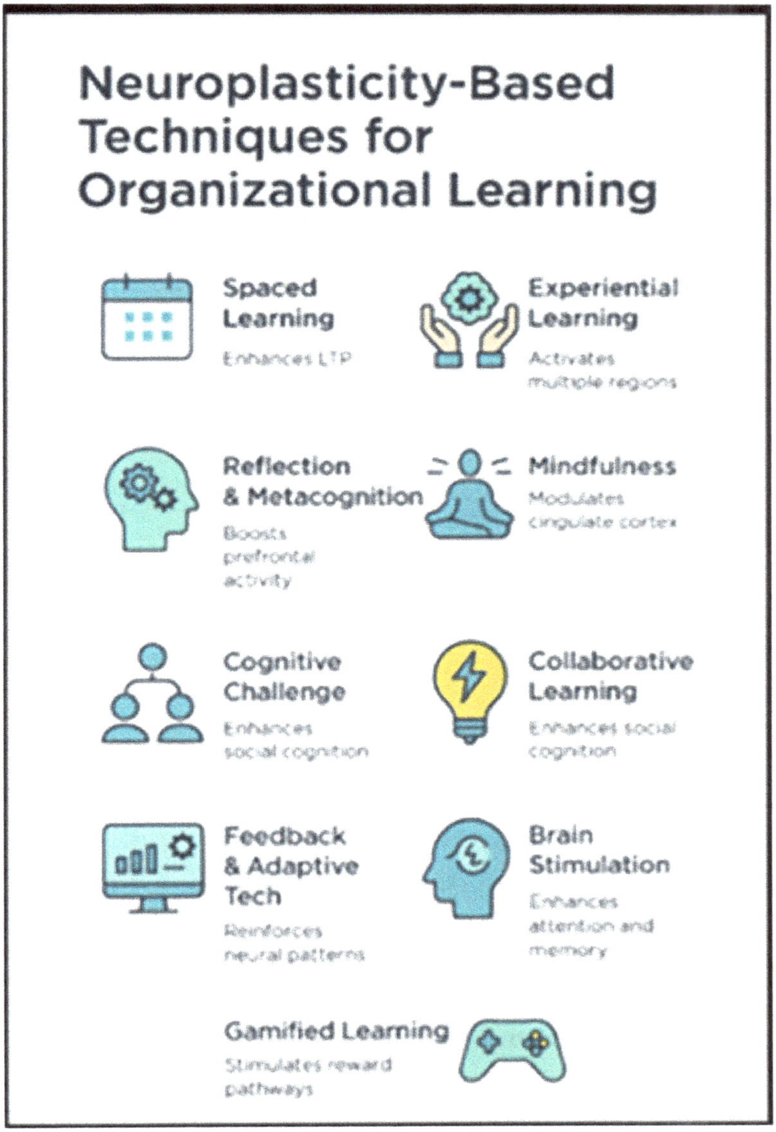

Table 8. **Neuroplasticity-Based Techniques for Organizational Learning**

Technique	Neural Mechanism	Application Example	Key References
Spaced & Repetitive Learning	Strengthens synaptic connections via LTP; improves long-term retention	Weekly refreshers, spaced e-learning modules	Narkhede et al., 2024
Multimodal & Experiential Learning	Activates multiple brain regions; enhances encoding and transfer	Job rotations, simulations, hands-on workshops	Karim et al., 2021
Reflection & Metacognition	Boosts prefrontal activity; supports executive function and self-regulation	Learning journals, goal-setting sessions	Forget & Le Pertel, 2024
Mindfulness & Stress Regulation	Reshapes anterior cingulate and amygdala circuits; improves focus and resilience	Guided meditation, stress management training	An et al., 2025; Alvarado-García et al., 2025
Cognitive Challenge & Novelty	Stimulates cortical restructuring; promotes flexible thinking	Hackathons, problem-solving sprints	Singha, 2024
Social & Collaborative Learning	Activates mirror neuron systems; enhances social cognition and communication	Peer mentoring, team-based projects	Chen & Goodwill, 2022
Feedback & Neuroadaptive Technologies	Reinforces correct neural patterns; supports personalized learning	AI-driven platforms, performance dashboards	Ooi, 2025
Brain Stimulation Techniques	Enhances attention and memory via tDCS/TMS when paired with cognitive tasks	tDCS with memory training (experimental use)	Carrarini et al., 2025
Gamified Learning	Stimulates reward pathways; reinforces adaptive neural patterns	Interactive apps, point-based modules	Parija & Singh, 2023

10. STRATEGIES FOR MEASURING NEUROPLASTICITY IN EDUCATIONAL AND ORGANIZATIONAL SETTINGS

Assessing neuroplasticity in applied contexts requires multidimensional strategies that capture both structural and functional brain changes. These approaches are increasingly relevant in educational and organizational settings where cognitive adaptability is essential. Key strategies include:

- **Cognitive and Behavioral Assessments**:

Tools such as working memory tasks, attention-switching paradigms, and executive function batteries can track adaptive changes over time.

- **Neuroimaging and Electrophysiological Techniques**:

EEG, fMRI, and near-infrared spectroscopy are increasingly used to monitor neural activation patterns during learning or task performance (Maier, 2024).

- **Longitudinal Performance Metrics**:

Tracking skill acquisition, error correction, and adaptability across training phases provides indirect but meaningful indicators of plasticity.

- **Technology-Enhanced Learning Analytics**:

Adaptive learning platforms and cognitive training apps offer real-time data on engagement, retention, and neural responsiveness (Dalexander, 2025).

- **Multisensory and Immersive Learning Environments**:

Virtual Reality/Augmented Reality simulations and embodied learning modules stimulate distributed neural networks, allowing for experiential evaluation of plasticity (Gkintoni et al., 2025).

Together, these strategies support the development of personalized interventions and evidence-based practices that enhance learning, performance, and cognitive resilience.

11. STRATEGIC IMPLICATIONS FOR ORGANIZATIONAL AND EDUCATIONAL SYSTEMS

Neuroplasticity research provides actionable insights for enhancing performance, resilience, and inclusivity in professional and educational settings. Emphasizing adaptability and contextual sensitivity is key to fostering sustained cognitive health and innovation.

11.1 Organizational Applications

Neuroplastic workforce development involves personalized learning, psychological safety, and growth mindset cultivation through mentorship, cross-functional training, and challenge-based learning (Forget & Le Pertel, 2024). Plasticity-friendly environments featuring novelty, feedback, and reflection boost creativity and problem-solving. Investments in ideation labs and scheduled "thinking time" support divergent thinking (Ooi, 2025). Neuroplasticity-based interventions like adaptive learning platforms, emotional regulation strategies, and mindset training enhance cognitive flexibility, efficiency, and well-being. Models such as Microsoft's growth mindset framework show improved productivity and reduced burnout (Badshah, 2020).

11.2. Educational and Civic Integration

Lifelong plasticity supports inclusive policies for adult learning, reskilling, and mental health. Sociocultural influences on plasticity highlight the need for differentiated instruction and culturally responsive pedagogy (Sarto-Jackson et al., 2017). Curriculum reforms emphasizing engagement, emotional regulation, and adaptive strategies foster academic resilience and civic agency. As neuroenhancement tools enter learning and workspaces, ethical governance is essential to protect autonomy, mental privacy, and equitable access (Jwa & Poldrack, 2022).

12. TOWARD A TRANSLATIONAL FRAMEWORK FOR NEUROPLASTICITY RESEARCH AND APPLICATION

12.1. Individual Variability and Personalized Applications

Contemporary research underscores the substantial influence of individual variability on neuroplastic responses, shaped by genetic predispositions, epigenetic regulation, and modifiable lifestyle factors such as physical activity, nutrition, and cognitive engagement (Noda, 2024; Marzola et al., 2023). These interindividual differences necessitate the development of personalized strategies in educational and therapeutic contexts to optimize intervention efficacy and long-term outcomes. Building on the mechanistic foundations outlined in earlier sections, this synthesis aims to bridge foundational insights with translational relevance.

12.2. Neuroenhancement and Ethical Considerations

While neuroenhancement tools such as nootropics, neurofeedback, transcranial stimulation, and brain-computer interfaces offer promising avenues for cognitive improvement, their real-world application presents significant challenges. In high-pressure environments, these technologies may contribute to cognitive overload, stress, and burnout (Reddy, 2025). Their integration into educational and organizational settings also raises ethical concerns, including autonomy, mental privacy, informed consent, and equitable access. Risks such as coercive optimization, neuro-surveillance, and unequal distribution of enhancement tools may deepen existing social and cognitive disparities (Jwa & Poldrack, 2022; López-Silva et al., 2024). In Latin America, neuroethical discourse is advancing, with growing attention to neurorights legislation. For instance, Herrera-Ferrá et al. (2022) highlight culturally specific concerns in Mexico, while López-Silva (2025) discusses Chile's emerging frameworks to protect cognitive liberty and ensure fair access to neurotechnologies. To ensure responsible use, neuroplasticity-based interventions must be grounded in empirical evidence, ethical safeguards, and inclusive design principles.

12.3. Future Research and Translational Integration

Advancing neuroplasticity research requires longitudinal studies that explore how plastic changes support learning and adaptation across diverse contexts. Future work should prioritize methodological rigor, ecological validity, and cross-sector relevance. Interdisciplinary collaboration especially among neuroscience, education, organizational psychology, and bioethics is essential to translate theory into scalable, context-sensitive interventions (Lin et al., 2024).

12.4. Emerging Domains and Policy Integration

Fields like epigenetics offer new insights into how environmental factors influence gene expression linked to neuroplasticity. Simultaneously, the rise of AI and brain-computer interfaces in cognitive enhancement introduces both opportunities and ethical challenges, particularly around autonomy, data governance, and mental privacy (Drigas & Sideraki, 2024). Regional efforts, such as Chile's proposed neurorights legislation, reflect the growing role of developing countries in shaping neuroethical policy (López-Silva, 2025). These initiatives emphasize the need for culturally grounded responses to emerging cognitive technologies. To promote cognitive resilience and lifelong learning, neuroplasticity-informed frameworks must guide public policy across education, workforce development, and mental health anchored in ethical principles and inclusive design.

12.5. Integrative Outlook

This chapter proposes a multidimensional framework that integrates structural, functional, and developmental aspects of neuroplasticity with emerging technological and ethical considerations. By examining the interplay between biological mechanisms, individual variability, and societal needs, it offers a foundation for designing evidence-based, ethically responsible, and context-sensitive interventions. Researchers, educators, clinicians, and policymakers are encouraged to adopt this framework to foster interdisciplinary collaboration and translational impact. To ensure global relevance, future research should incorporate culturally diverse perspectives, particularly from regions like Latin America, where socio-environmental factors and resource constraints shape both neuroplastic outcomes and intervention feasibility (Galvez-Contreras et al., 2022).

13. CONCLUSION

Neuroplasticity is a core principle in understanding human cognitive adaptability, offering a dynamic framework for optimizing learning, behavior, and development across the lifespan. Contemporary neuroscience demonstrates the brain's capacity to reorganize in response to experience and intervention, informing innovative strategies in education, healthcare, and workforce development. Aligning environments and practices with neuroplastic principles fosters resilience, adaptability, and lifelong growth. This chapter presents a multidimensional framework integrating biological mechanisms, emerging technologies, and ethical considerations. By emphasizing individual variability, translational relevance, and societal impact, it supports inclusive, evidence-based approaches across disciplines. As neuroplasticity continues to shape our understanding of human potential, interdisciplinary collaboration and ethical stewardship will be essential for guiding its responsible application.

REFERENCES

Ahamed, A. A., Olanrewaju, O. O., Wilberforce, E. J., Gbayisomore, T. J., Anene-Ogbe, M. C., Farounbi, G., & Ayodele, Y. J. (2024). Adult Neurogenesis: A Review of Current Perspectives and Implications for Neuroscience Research. *Journal of Neuroscience and Neurological Disorders*, *8*(2), 106–114.

Altenmüller, E., & Furuya, S. (2016). Brain plasticity and the concept of metaplasticity in skilled musicians. *Progress in Motor Control: Theories and Translations*, 197-208.

Alvarado-García, P. A. A., Soto-Vásquez, M. R., Infantes Gomez, F. M., Guzman Rodriguez, N. M., & Castro-Paniagua, W. G. (2025). Effect of a mindfulness program on stress, anxiety, depression, sleep quality, social support, and life satisfaction: A quasi-experimental study in college students. *Frontiers in Psychology*, *16*, 1508934. PMID: 40012941

Amidfar, M., Réus, G. Z., de Moura, A. B., Quevedo, J., & Kim, Y. K. (2021). The role of neurotrophic factors in pathophysiology of major depressive disorder. *Major Depressive Disorder: Rethinking and Understanding Recent Discoveries*, 257-272.

An, D. C. V. T., Lan, P. T. Q., & Hai, N. T. T. (2025). Neuroscience of mindfulness and meditation: Examining the impact on brain activity, neural plasticity, and mental health. *American Journal of Psychiatric Rehabilitation*, *28*(4), 292–300.

Appelbaum, L. G., Shenasa, M. A., Stolz, L., & Daskalakis, Z. (2023). Synaptic plasticity and mental health: Methods, challenges and opportunities. *Neuropsychopharmacology: Official Publication of the American College of Neuropsychopharmacology*, *48*(1), 113–120. PMID: 35810199

Asuku, A. O., Arinde, E. S., Saliu, S. B., Oyerinde, T. O., Olajide, T. S., Afolabi, H. O., & Kehinde, A. (2025). Neuroplasticity: Harnessing the Brain Ability to Heal itself. In *Advancing Medical Research Through Neuroscience* (pp. 95-118). IGI Global Scientific Publishing.

Babcock, K. R., Page, J. S., Fallon, J. R., & Webb, A. E. (2021). Adult hippocampal neurogenesis in aging and Alzheimer's disease. *Stem Cell Reports*, *16*(4), 681–693. PMID: 33636114

Babiy, Z., Frey, B. N., Bieling, P. J., McCabe, R. E., & Green, S. M. (2025). Mindfulness-Based Interventions and Neuroplasticity: A Review of Network Connectivity in Healthy and Clinical Samples. *Mindfulness*, •••, 1–14.

Badshah, A. (2020). *Purpose Mindset: How Microsoft Inspires Employees and Alumni to Change the World*. HarperCollins Leadership.

Benatti, B. M., Adiletta, A., Sgadó, P., Malgaroli, A., Ferro, M., & Lamanna, J. (2024). Epigenetic Modifications and Neuroplasticity in the Pathogenesis of Depression: A Focus on Early Life Stress. *Behavioral Sciences (Basel, Switzerland)*, *14*(10), 882. PMID: 39457754

Beschorner, N., & Nedergaard, M. (2024). Glymphatic system dysfunction in neurodegenerative diseases. *Current Opinion in Neurology*, *37*(2), 182–188. PMID: 38345416

Bhaduri, A. (2024). Chimeric brain organoids capture human genetic diversity. *Nature*. PMID: 38926554

Cantor, P., Osher, D., Berg, J., Steyer, L., & Rose, T. (2019). Malleability, plasticity, and individuality: How children learn and develop in context. *Applied Developmental Science*, *23*(4), 307–337. DOI: 10.1080/10888691.2017.1398649

Carrarini, C., Pappalettera, C., Le Pera, D., & Rossini, P. M. (2025). Non-invasive brain stimulation in cognitive sciences and Alzheimer's disease. *Frontiers in Human Neuroscience*, *18*, 1500502. PMID: 39877800

Chen, S. A., & Goodwill, A. M. (2022). Neuroplasticity and adult learning. In *Third international handbook of lifelong learning* (pp. 1–19). Springer International Publishing.

Chindemi, G., Abdellah, M., Amsalem, O., Benavides-Piccione, R., Delattre, V., Doron, M., & Muller, E. B. (2022). A calcium-based plasticity model for predicting long-term potentiation and depression in the neocortex. *Nature Communications*, *13*(1), 3038. PMID: 35650191

D'Aquin, S., Szonyi, A., Mahn, M., Krabbe, S., Gründemann, J., & Lüthi, A. (2022). Compartmentalized dendritic plasticity during associative learning. *Science*, *376*(6590). PMID: 35420958

Dan, B. (2019). Neuroscience underlying rehabilitation: What is neuroplasticity? *Developmental Medicine and Child Neurology*, *61*(11), 1240. PMID: 31576561

Demarque, M., & Spitzer, N. C. (2020). *Neurotransmitter phenotype plasticity: from calcium signaling to functional consequences*. Academic Press., DOI: 10.1016/B978-0-12-823672-7.00017-X

Dey, S., Arya, A., Raut, A., Katta, S., & Sharma, P. (2024). Exploring the Role of Neuroplasticity in Stroke Rehabilitation: Mechanisms, Interventions and Clinical Implications. *International Journal For Multidisciplinary Research, 6*(2).

Dinius, C. J., Pocknell, C. E., Caffrey, M. P., & Roche, R. A. (2023). Cognitive interventions for memory and psychological well-being in aging and dementias. *Frontiers in Psychology, 14*, 1070012. PMID: 36818134

Drigas, A., & Sideraki, A. (2024)... *Brain Neuroplasticity Leveraging Virtual Reality and Brain–Computer Interface Technologies., 24*. Advance online publication. DOI: 10.3390/s24175725 PMID: 39275636

Duffau, H. (2025). Neuroplasticity in diffuse low-grade gliomas: Backward modelling of brain-tumor interactions prior to diagnosis is needed to better predict recovery after treatment. *Current Neurology and Neuroscience Reports, 25*(1), 1–12. PMID: 39786618

Dzyubenko, E., & Hermann, D. M. (2023, May). Role of glia and extracellular matrix in controlling neuroplasticity in the central nervous system. *Seminars in Immunopathology, 45*(3), 377–387. DOI: 10.1007/s00281-023-00982-y PMID: 37052711

Emmenegger, T. (2025). Development and Application of Quantitative MRI Processing Tools to Track Structural and Functional Changes along the Neuraxis Following Spinal Cord Injury (Doctoral dissertation, University of Zurich).

Famitafreshi, H., & Karimian, M. (2019). Can adult neurogenesis Be considered as neuroplasticity. *Archives of Neuroscience, 7*, e79812.

Fandakova, Y., & Hartley, C. A. (2020). Mechanisms of learning and plasticity in childhood and adolescence. *Developmental Cognitive Neuroscience, 42*, 100764. PMID: 32072937

Faust, T. E., Gunner, G., & Schafer, D. P. (2021). Mechanisms governing activity-dependent synaptic pruning in the developing mammalian CNS. *Nature Reviews. Neuroscience, 22*(11), 657–673. PMID: 34545240

Finger, S. (2010). Recovery of function: Redundancy and vicariation theories. In *Handbook of Clinical Neurology* (Vol. 95, pp. 833–841). Elsevier., DOI: 10.1016/S0072-9752(08)02151-2

Flegal, K. E., Ragland, J. D., & Ranganath, C. (2019). Adaptive task difficulty influences neural plasticity and transfer of training. *NeuroImage, 188*, 111–121. PMID: 30521951

Forget, M., & Le Pertel, N. (2024). Enhancing Neuroplasticity and Promoting Brain Health at Work: The Role of Learning and Memory in Workplace Performance. In *Learning and Memory-From Molecules and Cells to Mind and Behavior*. IntechOpen.

Fossati, G., Matteoli, M., & Menna, E. (2020). Astrocytic factors controlling synaptogenesis: A team play. *Cells*, *9*(10), 2173. PMID: 32993090

Fries, G. R., Saldana, V. A., Finnstein, J., & Rein, T. (2023). Molecular pathways of major depressive disorder converge on the synapse. *Molecular Psychiatry*, *28*, 284–297. PMID: 36203007

Frizzell, T. O., Phull, E., Khan, M., Song, X., Grajauskas, L. A., Gawryluk, J., & D'Arcy, R. C. N. (2022). Imaging functional neuroplasticity in human white matter tracts. *Brain Structure & Function*, *227*, 381–392. PMID: 34812936

Fuchs, X., Bekrater-Bodmann, R., & Flor, H. (2015). Phantom Pain: The Role of Maladaptive Plasticity and Emotional and Cognitive Variables. In Pickering, G., & Gibson, S. (Eds.), *Pain, Emotion and Cognition*. Springer., DOI: 10.1007/978-3-319-12033-1_12

Gall, C. M., Le, A. A., & Lynch, G. (2023). Sex differences in synaptic plasticity underlying learning. *Journal of Neuroscience Research*, *101*, 764–782. PMID: 33847004

Gallo, F., Voits, T., Rothman, J., Abutalebi, J., Shtyrov, Y., & Myachykov, A. (2025). Experience-dependent neuroplasticity in the hippocampus of bilingual young adults. *eNeuro*. PMID: 40379481

Galvez-Contreras, A. Y., Vargas-de la Cruz, I., Beltran-Navarro, B., Gonzalez-Castaneda, R. E., & Gonzalez-Perez, O. (2022). Therapeutic approaches for ADHD by developmental stage and clinical presentation. *International Journal of Environmental Research and Public Health*, *19*(19), 12880. DOI: 10.3390/ijerph191912880 PMID: 36232180

Gao, Y., Liu, K., & Zhu, J. (2023). Glymphatic system: An emerging therapeutic approach for neurological disorders. *Frontiers in Molecular Neuroscience*, *16*, 1138769. PMID: 37485040

Gkintoni, E., Vassilopoulos, S. P., & Nikolaou, G. (2025). Brain-inspired multisensory learning: A systematic review. *Biomimetics*, *10*(6), 397. DOI: 10.3390/biomimetics10060397 PMID: 40558367

Grilli, M., & Zaed, I. (2023). Neuroplasticity in cognitive and psychological mechanisms of anxiety. *Frontiers in Molecular Neuroscience*, *16*, 1241279. PMID: 37465365

Guo, X., Rao, Y., Mao, R., Cui, L., & Fang, Y. (2020). Common cellular and molecular mechanisms and interactions between microglial activation and aberrant neuroplasticity in depression. *Neuropharmacology, 181*, 108336. DOI: 10.1016/j.neuropharm.2020.108336 PMID: 32980387

Hagstrom, C., Heppner, A. G., & Kunde, K. N. (2021). The Effects of Socialization on the Progression of Dementia.

Hensch, T. K. (2005). Critical period plasticity in local cortical circuits. *Nature Reviews. Neuroscience, 6*(11), 877–888. DOI: 10.1038/nrn1787 PMID: 16261181

Herholz, S. C., & Zatorre, R. J. (2012). Musical training as a framework for brain plasticity: Behavior, function, and structure. *Neuron, 76*(3), 486–502. PMID: 23141061

Hermans, E. J., Hendler, T., & Kalisch, R. (2024). Building resilience: The stress response as a driving force for neuroplasticity and adaptation. *Biological Psychiatry*. PMID: 39448004

Herrera-Ferrá, K., Nicolini, H., & Giordano, J. (2022). Professional attitudes toward the use of neuromodulatory technologies in Mexico: Insight for neuroethical considerations of cultural diversity. *CNS Spectrums, 27*(3), 255–257. PMID: 33298232

Herzberg, M. P., Nielsen, A. N., Luby, J., & Sylvester, C. M. (2025). Measuring neuroplasticity in human development: The potential to inform the type and timing of mental health interventions. *Neuropsychopharmacology: Official Publication of the American College of Neuropsychopharmacology, 50*(1), 124–136. PMID: 39103496

Hidayat, R. (2023). Study of Brain Neuroplasticity Analysis in Response to the Learning Process. https://doi.org/DOI: 10.59345/sjn.v1i2.91

Hiskens, M. I. (2024). Glua1 Ubiquitination in Synaptic Plasticity and Cognitive Functions: Implications for Neurodegeneration. *The Journal of Neuroscience: The Official Journal of the Society for Neuroscience*, •••, 44. PMID: 38233219

Holmes, J., Guy, J., Kievit, R. A., Bryant, A., Mareva, S., & Gathercole, S. E., & CALM Team. (2021). Cognitive dimensions of learning in children with problems in attention, learning, and memory. *Journal of Educational Psychology, 113*(7), 1454. PMID: 35855686

Ismail, F. Y., Fatemi, A., & Johnston, M. V. (2017). Cerebral plasticity: Windows of opportunity in the developing brain. *European Journal of Paediatric Neurology, 21*(1), 23–48. PMID: 27567276

Jajarmi, M. R., Oladi, M., Araeynejad, F., Darrehkordi, H., & Nourani, M. R. (2024). Axonal Sprouting in Neural Repair and Plasticity: Mechanisms and Implications. *Life Sciences Student Journal*, *2*(1), 1–8.

Jwa, A. S., & Poldrack, R. A. (2022). Addressing privacy risk in neuroscience data: From data protection to harm prevention. *Journal of Law and the Biosciences*, *9*(2), lsac025. Advance online publication. DOI: 10.1093/jlb/lsac025 PMID: 36072418

Kaczmarek, B. L. J. (2020). Current views on neuroplasticity: what is new and what is old? 18(1), 1–14. https://doi.org/DOI: 10.5604/01.3001.0013.8808

Karbach, J., & Schubert, T. (2013). Training-induced cognitive and neural plasticity. *Frontiers in Human Neuroscience*, *7*, 48. DOI: 10.3389/FNHUM.2013.00048 PMID: 23437015

Karim, A. K. M., Proulx, M. J., de Sousa, A. A., & Likova, L. T. (2021). Neuroplasticity and crossmodal connectivity in the normal, healthy brain. *Psychology & Neuroscience*, *14*(3), 298. PMID: 36937077

Kennedy, N. C. (2021). The role of neuroplasticity in stroke nursing. *British Journal of Neuroscience Nursing*, *17*(Sup2), S20-S25.

Khaspekov, L. G., & Frumkina, L. E. (2023). Molecular mechanisms of astrocyte involvement in synaptogenesis and brain synaptic plasticity. *Biochemistry (Moscow)*, *88*(4), 502–514. PMID: 37080936

Klutz, D. (2023). Neurofeedback for Cognitive Enhancement Intervention and Brain Plasticity. *Journal of Biomedical and Sustainable Healthcare Applications*, 3(1), 045-055.

Lenze, E. J., Voegtle, M., Miller, J. P., Ances, B. M., Balota, D. A., Barch, D., & Wetherell, J. L. (2022). Effects of mindfulness training and exercise on cognitive function in older adults: A randomized clinical trial. *Journal of the American Medical Association*, *328*(22), 2218–2229. PMID: 36511926

Li, L., & Han, J. (2022). Neuroplasticity and stroke recovery: From mechanisms to therapeutic interventions. *Frontiers in Cellular Neuroscience*, *16*, 872123.

Lin, D. J., Backus, D., Chakraborty, S., Liew, S. L., Valero-Cuevas, F. J., Patten, C., & Cotton, R. J. (2024). Transforming modeling in neurorehabilitation: Clinical insights for personalized rehabilitation. *Journal of Neuroengineering and Rehabilitation*, *21*(1), 18. PMID: 38311729

Liu, Y., Shen, X., Zhang, Y., Zheng, X., Cepeda, C., Wang, Y., Duan, S., & Tong, X. (2023). Interactions of glial cells with neuronal synapses, from astrocytes to microglia and oligodendrocyte lineage cells. *Glia*, *71*(6), 1383–1401. DOI: 10.1002/glia.24343 PMID: 36799296

López-Silva, P. (2025). Neurorights and cognitive liberty in Latin America: Ethical and legal challenges. *Neuroethics*, *18*(1), 23–36.

López-Silva, P., Wajnerman-Paz, A., & Molnar-Gabor, F. (2024). Neurotechnological applications and the protection of mental privacy: An assessment of risks. *Neuroethics*, *17*(2), 31. DOI: 10.1007/s12152-024-09565-2

Lucassen, P. J., Fitzsimons, C. P., Salta, E., & Maletic-Savatic, M. (2020). Adult neurogenesis, human after all (again): Classic, optimized, and future approaches. *Behavioural Brain Research*, *381*, 112458. DOI: 10.1016/j.bbr.2019.112458 PMID: 31899214

Mamczur, M., Szczepański, M., Feret, D. M., Kuliga, M., Zapasek, D., Słowik, J., Bajak, M., Sowa, D., Inglot, J., & Inglot, J. (2024). Neuroplasticity in Depressive Disorders: The Role of BDNF in Linking Pharmacotherapy and Physical Activity. *Quality in Sport*, *36*, 56457.

Markram, H., Gerstner, W., & Sjöström, P. J. (2011). A history of spike-timing-dependent plasticity. *Frontiers in Synaptic Neuroscience*, *3*, 4. DOI: 10.3389/fnsyn.2011.00004 PMID: 22007168

Marshall, K. L., & Farah, M. H. (2021). Axonal regeneration and sprouting as a potential therapeutic target for nervous system disorders. *Neural Regeneration Research*, *16*(10), 1901–1910. PMID: 33642358

Martin-Aragon, S., & Bermejo-Bescós, P. (2019). Homeostatic Plasticity and Therapeutic Approaches in. *Biogenic Amines in Neurotransmission and Human Disease*, 23.

Martinez-Morga, M., & Martinez, S. (2017). Neuroplasticity: Synaptogenesis during normal development and its implication in intellectual disability. *Revista de Neurología*, *64*(s01), S45–S50. PMID: 28256686

Marzola, P., Melzer, T., Pavesi, E., Gil-Mohapel, J., & Brocardo, P. S. (2023). Exploring the role of neuroplasticity in development, aging, and neurodegeneration. *Brain Sciences*, *13*(12), 1610. PMID: 38137058

Mascaro, A. L. A., Cesare, P., Sacconi, L., Sacconi, L., Grasselli, G., Mandolesi, G., Maco, B., Knott, G., Huang, L., De Paola, V., Strata, P., & Pavone, F. S. (2013). In vivo single branch axotomy induces GAP-43–dependent sprouting and synaptic remodeling in cerebellar cortex. *Proceedings of the National Academy of Sciences of the United States of America, 110*(26), 10824–10829. DOI: 10.1073/PNAS.1219256110 PMID: 23754371

Mateos-Aparicio, P., & Rodríguez-Moreno, A. (2019). The impact of studying brain plasticity. *Frontiers in Cellular Neuroscience, 13*, 66. PMID: 30873009

Milbocker, K. A., Smith, I. F., & Klintsova, A. Y. (2023). Maintaining a Dynamic Brain: A Review of Empirical Findings Describing the Roles of Exercise, Learning, and Environmental Enrichment in Neuroplasticity from 2017-2023. *Brain Plasticity*. Advance online publication. DOI: 10.3233/bpl-230151 PMID: 38993580

Mishra, A., Patni, P., Hegde, S., Aleya, L., & Tewari, D. (2021). Neuroplasticity and environment: A pharmacotherapeutic approach toward preclinical and clinical understanding. *Current Opinion in Environmental Science & Health, 19*, 100210.

Monday, H. R., Younts, T. J., & Castillo, P. E. (2018). Long-Term Plasticity of Neurotransmitter Release: Emerging Mechanisms and Contributions to Brain Function and Disease. *Annual Review of Neuroscience, 41*, 299–322. PMID: 29709205

Narkhede, K., Patil, R., Kasat, A., Gawas, A., & More, P. (2024, March). Unveiling the Art of Effective Learning through Spaced Repetition and Evidence-Based Techniques. In *2024 IEEE International Conference on Contemporary Computing and Communications (InC4)* (Vol. 1, pp. 1-6). IEEE.

Nishiyama, J. (2019). Plasticity of dendritic spines: Molecular function and dysfunction in neurodevelopmental disorders. *Psychiatry and Clinical Neurosciences, 73*(9), 541–550. PMID: 31215705

Noda, Y. (2024). Exploring Optimal Neuroplasticity: Neuromodulation for Enhanced Human Well-Being. *Journal of Integrative Neuroscience, 23*(8), 155. PMID: 39207072

Nudo, R. J. (2013). Recovery after brain injury: Mechanisms and principles. *Frontiers in Human Neuroscience, 7*, 887. PMID: 24399951

Ooi, R. W. G. (2025). *From Neurons to Organisations: Awakening Regenerative Leadership through Neuroplasticity*. AI, and Integrative Consciousness.

Parija, R., & Singh, N. T. (2023, August). Neuroplasticity Changes to Configure Human Talent. In *2023 International Conference on Circuit Power and Computing Technologies (ICCPCT)* (pp. 942-947). IEEE.

Park, W. J., & Fine, I. (2020). New insights into cortical development and plasticity: From molecules to behavior. *Current Opinion in Physiology*, *16*, 50–60. PMID: 32923755

Pickersgill, J. W., Turco, C. V., Ramdeo, K., Rehsi, R. S., Foglia, S. D., & Nelson, A. J. (2022). The combined influences of exercise, diet and sleep on neuroplasticity. *Frontiers in Psychology*, *13*, 831819. PMID: 35558719

Pinar, C., Fontaine, C. J., Triviño-Paredes, J., Lottenberg, C. P., Gil-Mohapel, J., & Christie, B. R. (2017). Revisiting the flip side: Long-term depression of synaptic efficacy in the hippocampus. *Neuroscience and Biobehavioral Reviews*, *80*, 394–413. PMID: 28624435

Popescu, B. O., Batzu, L., Ruiz, P. J. G., Tulbă, D., Moro, E., & Santens, P. (2024). Neuroplasticity in Parkinson's disease. *Journal of Neural Transmission (Vienna, Austria)*, *131*(11), 1329–1339. PMID: 39102007

Pritchard, R., Chen, H., Romoli, B., Spitzer, N. C., & Dulcis, D. (2019). Photoperiod-induced neurotransmitter plasticity declines with aging: an epigenetic regulation? *bioRxiv*, 563213.

Randelman, M., Zholudeva, L. V., Vinit, S., & Lane, M. A. (2021). Respiratory training and plasticity after cervical spinal cord injury. *Frontiers in Cellular Neuroscience*, *15*, 700821. PMID: 34621156

Reddy, K. J. (2025). Neuroplasticity and Technology. In *Innovations in Neurocognitive Rehabilitation: Harnessing Technology for Effective Therapy* (pp. 137–169). Springer Nature Switzerland.

Reidler, P., Thierfelder, K. M., Fabritius, M. P., Sommer, W. H., Meinel, F. G., Dorn, F., & Kunz, W. G. (2018). Thalamic diaschisis in acute ischemic stroke. *Stroke*. PMID: 29523650

Sagi, Y., & Rosenzweig, N. (2020). Plasticity in the adult brain: From synapses to circuits. *Nature Reviews. Neuroscience*, *21*(9), 535–549.

Sarto-Jackson, I., Larson, D. O., & Callebaut, W. (2017). Culture, neurobiology, and human behavior: New perspectives in anthropology. *Biology & Philosophy*, *32*(5), 729–748.

Sasmita, A. O., Kuruvilla, J., & Ling, A. P. K. (2018). Harnessing neuroplasticity: Modern approaches and clinical future. *The International Journal of Neuroscience*, *128*(11), 1061–1077. PMID: 29667473

Satarker, S., Bojja, S. L., Gurram, P. C., Mudgal, J., Arora, D., & Nampoothiri, M. (2022). Astrocytic Glutamatergic Transmission and Its Implications in Neurodegenerative Disorders. [Google Scholar] [CrossRef] [PubMed]. *Cells*, *11*, 1139. PMID: 35406702

Simard, S., Rahimian, R., Davoli, M. A., Théberge, S., Matosin, N., Turecki, G., & Mechawar, N. (2024). Spatial transcriptomic analysis of adult hippocampal neurogenesis in the human brain. *Journal of Psychiatry & Neuroscience: JPN*, *49*(5), E319–E333. PMID: 39414359

Singha, R. (2024). Stress, Resilience, and brain performance. In *Building Organizational Resilience With Neuroleadership* (pp. 14–29). IGI Global.

Srijit, D., & Sadhana, R. (2023). Neuroplasticity of the Brain: An Uncharted Sea with Great potential?. Medicine@ Health (Universiti Kebangsaan Malaysia), 18(1), 4-8.

Stacho, M., & Manahan-Vaughan, D. (2022). The intriguing contribution of hippocampal long-term depression to spatial learning and long-term memory. *Frontiers in Behavioral Neuroscience*, *16*, 806356. PMID: 35548697

Statsenko, Y., Kuznetsov, N. V., & Ljubisaljevich, M. (2025). Hallmarks of brain plasticity. *Biomedicines*, *13*(2), 460. DOI: 10.3390/biomedicines13020460 PMID: 40002873

Subramaniam, K., Luks, T. L., Garrett, C., Chung, C., Fisher, M., Nagarajan, S., & Vinogradov, S. (2014). Intensive cognitive training in schizophrenia enhances working memory and associated prefrontal cortical efficiency. *NeuroImage*, *99*, 281–292. DOI: 10.1016/j.neuroimage.2014.05.057 PMID: 24867353

Tamon, H., Fujino, J., Itahashi, T., Frahm, L., Parlatini, V., Aoki, Y. Y., & Cortese, S. (2024). Shared and specific neural correlates of attention deficit hyperactivity disorder and autism spectrum disorder: A meta-analysis of 243 task-based functional MRI studies. *The American Journal of Psychiatry*, *181*(6), 541–552. DOI: 10.1176/appi.ajp.2023.23020171 PMID: 38685858

Vaz, A., Ribeiro, I., & Pinto, L. (2022). Frontiers in neurogenesis. *Cells*, *11*(22), 3567. PMID: 36428996

Verma, R. (2023). Neuroplasticity: Evolving Concept in Neurology. *Journal of Clinical Research and Applied Medicine*, *2*(3), 52–53.

Vinogradov, S. (2023). Harnessing neuroplasticity. *Psychiatry Research*, *330*, 115607. PMID: 38000206

Wallace, R., Olson, D. E., & Hooker, J. M. (2023). Neuroplasticity: The Continuum of Change. *ACS Chemical Neuroscience*. Advance online publication. DOI: 10.1021/acschemneuro.3c00526 PMID: 37676498

Wang, F., Yuan, T., Pereira, A.Jr, Verkhratsky, A., & Huang, J. H. (2016). Glial cells and synaptic plasticity. *Neural Plasticity*, *2016*, 5042902. PMID: 27293902

Whelan, B.-M., Theodoros, D., McMahon, K. L., Copland, D. A., Aldridge, D., & Campbell, J. (2021). Substrates of speech treatment-induced neuroplasticity in adults and children with motor speech disorders: A systematic scoping review of neuroimaging evidence. *International Journal of Speech-Language Pathology*, •••, 1–14. PMID: 34030526

Yam, K. C., Tang, P. M., Jackson, J. C., Su, R., & Gray, K. (2023). The rise of robots increases job insecurity and maladaptive workplace behaviors: Multimethod evidence. *The Journal of Applied Psychology*, *108*(5), 850. PMID: 36222634

Zhang, L., Qi, Z., Li, J., Li, M., Du, X., Wang, S., & Deng, Y. (2021). Roles and mechanisms of axon-guidance molecules in Alzheimer's disease. *Molecular Neurobiology*, *58*, 3290–3307. PMID: 33675023

Chapter 2

The Neural Path to Productivity:
How Brain Science Transforms Workplace Efficiency

L. B. Muralidhar
https://orcid.org/0000-0003-3453-613X
Jain University, India

V. Lavakumar
https://orcid.org/0000-0002-3539-9280
GITAM School of Business, GITAM University, India

Usha Prabhu
https://orcid.org/0000-0001-6219-5358
Jain University, India

N Sathyanarayana
https://orcid.org/0000-0002-4185-7751
Jain University, India

Rahul Varanasi
https://orcid.org/0000-0003-2407-7654
Jain University, India

ABSTRACT

This chapter explores the transformative impact of neuroscience on workplace productivity. By examining how brain function influences decision-making, memory, focus, and stress management, it offers valuable insights into optimising employee performance. The role of neuroplasticity in continuous learning, the significance of

DOI: 10.4018/979-8-3373-2292-6.ch002

emotional intelligence in leadership, and the mechanisms of attention and cognitive control are discussed. Additionally, the chapter delves into motivation, reward systems, and the impact of stress on productivity. It emphasises strategies for fostering creativity, innovation, and overall well-being through brain-based practices. The integration of neurotechnology and AI is also considered, forecasting a future where workplaces utilise neuroscience to create adaptive and productive environments.

1. INTRODUCTION: BRIDGING NEUROSCIENCE AND WORKPLACE PRODUCTIVITY

In the tangled, increasingly cognitively saturated dynamics of hybrid workforces, productivity can no longer be viewed through the sterile lens of linear behaviours or mechanical measures; the nuance of neurocognitive flexibility, social-emotional intelligence, and decision-making amidst volatility yields it. The rise of neuroscience within organisational theory represents a structural inflexion between behavioural surface readings and biologically grounded analyses of motivation, stress regulation and moral agency (Lieberman, 2013).

Dominant theories, be it Taylorism or Herzberg's models, were successful but empirically acknowledged observable incentives while ignoring the complex neural fabric of affective regulation, attentional coherence, and moral cognition. Neurobiological findings, particularly those grounded in the delicate balance between a hyperactive amygdala and an underactive prefrontal cortex, indicate how environmental cues and extrinsic overjustification heavily influence stress, motivational drive, and cognitive flexibility (Deci et al., 1999; Rock & Schwartz, 2006).

Thus, Leadership needs to transition from command and control toward creating psychologically congruent workplaces where affective safety, empathy and regulatory consistency are nurtured (Boyatzis et al., 2015). However, although promising, translational models such as SCARF (Rock, 2008) may also reduce the nuanced emotional quality to uncontextualized neuro-correlates, ignoring cultural diversity or environmental stimulus fidelity (Lindebaum, 2013; Poldrack, 2006). This is even more evident when autonomy-priming, based in Western epistemologies, is transferred into collectivist contexts, triggering an unanticipated backlash (Chiao et al., 2009).

What is not sorted by the corporate initiatives of Google or Unilever is the asymmetry of infrastructural access, ethical standardisation and cognitive data governance. Neuroadaptive technologies add a layer of complexity—engaging questions of surveillance, consent and interpretative malpractice to those of cognitive performance-as biometric extraction. In this climate, while integration of neuroscience should not be directed to optimisation only, it should drive a re-humanising of organisational life, where a neurocognitive variety is revered and not designed out of a job (Gallup, 2017).

The structure of this chapter intends to provide a balance between theoretical intuition and pragmatic usage. It summarises the key problem, objectives, and a theoretical framework connecting neuroscience with workplace behaviour. Sections that follow consider emotional intelligence, motivation, and environmental design, followed by practical applications and ethical considerations. A visual meta-model is introduced to unify concept relationships, and finally, the chapter concludes with a futuristic view. The aim is to provide an orientation for the reader, academic, practitioner, or layperson, so that they might navigate from theory to application throughout the chapter.

2. STATEMENT OF THE PROBLEM

The changing requirements of contemporary workplaces call for insight into how neuroscience can enhance workers' productivity through practices that differ from traditional management. Traditional organisational behaviour theories emphasise external incentives and observable behaviour over neurobiological processes that control choice, attention, memory, and emotional states. The increasing complexity and hybridisation of workplaces present an opportunity to leverage organisational practices by applying neuroscience, thereby increasing the chance to enhance cognitive flexibility, creativity, and productivity levels. There are, however, no practical, theory-driven, and culturally aware applications of brain science by organisations, and therefore, brain science is not leveraged effectively to drive innovation, well-being, and sustainable performance. This research aims to fill the gap by investigating how neuroscience can effectively maximise workplace efficiency, decision-making, and stress management, while considering the cultural and ethical implications across multinational settings. The challenge lies in developing a neurobiologically based model that organisations can embrace to holistically tackle productivity, employee well-being, and cognitive health.

3. OBJECTIVES

- To investigate the role of neuroscience in enhancing workplace productivity
- To explore the implications of neuroplasticity and emotional intelligence in leadership and employee development
- To examine the cultural and ethical challenges of applying neuroscience in multinational organisations

4. THEORETICAL AND CONCEPTUAL FRAMEWORK

A theoretical model informed by work from cognitive neuroscience, organisational behaviour, and stress theory is being used. This model can uncover how neurobiological processes affect leadership efficacy and employee functionality. It reconceives productivity not as a fixed attribute or machine-like output but as the result of a series of fluidic neural processes of attention, memory, and emotional regulation. Neuroplasticity bursts the view of fixed intelligence, showing that experiential design is the road to developing adaptive, high-performing teams (Doidge, 2007; Dweck, 2006).

Emotional intelligence, rooted in neuroanatomical systems linked to empathy and regulation, becomes less an abstract competency than a biologically grounded leadership imperative (Boyatzis, 2018). Similarly, chronic stress interferes with prefrontal control and emotional integration, disrupting performance and decision-making, unless mitigated by systemic interventions (McEwen, 2007; Arnsten, 2009). Therefore, organisational design should shift from behavioural discriminators toward neuroresponse — stress regulation, emotional fitness, and environmental scaffolding — to shore up sustainable cognition and resilient leadership.

Cross-cultural neuroscience elucidates how culturally constructed cognitive and affective constructs shape culture-specific neural activation patterns. For instance, research reveals that medial prefrontal regions are engaged in individualist culture participants when performing self-referential tasks. However, collectivist culture participants show greater involvement of the temporoparietal junction related to social perspective taking (Zhu et al., 2007). This has significant implications for leadership development and motivational interventions in multinational corporate settings where a 'one size fits all' Western-oriented solution can be confronted by cultural incongruence and thus resistance. Research by Chiao et al. (2009) and Kitayama & Park (2014) reveal that cultural values modulate the neural correlates of reward, stress and empathy — making the need for culture-specific neuro-leadership frameworks possible.

Mindfulness, neurofeedback, and environmental design can complement these efforts by enhancing cognitive control, emotional self-regulation, and creativity, ultimately improving workplace productivity.

5. ORIGIN AND EVOLUTION OF THE KEY CONCEPTS

The history of neuroscience in the organisational context represents a reconceptualisation of behaviourist orthodoxy toward a more integrated neurocognitive paradigm and, hence, an expanded view of performance, leadership, and well-being.

However, early management theories focused on external and visible motivators and did not consider the intricate neural processes of attention, emotion, and choice. This epistemological shift was catalysed in the 1990s with the arrival of neuroimaging, which laid

The foundation for Neuroplasticity, once overlooked, provides evidence to design learning ecosystems that value flexibility over fixed competencies (Doidge, 2007). After years of controversy, emotional intelligence, for example, becomes incontrovertibly empirical in neural findings demonstrating that empathy is connected directly to regulation by the executive (Goleman, 1995). Once depicted as a means of motivation, stress is now understood to inhibit prefrontal action and result in chronic burnout (Arnsten, 2009). Moreover, as neurotechnology and AI are embedded into workplace design, they present new affordances while engendering urgent ethical questions around consent, surveillance, and the commodification of cognitive life.

With few resources, SMEs are trialling low-cost neuro interventions – from mobile-based cognitive load assessment using EEG (artificial boredom is Killing Productivity), to mindfulness-based stress reduction (MBSR) courses for small teams. Case studies of SMEs in Scandinavia and Southeast Asia (Andersson & Lagerkvist, 2022; Kumar et al., 2020) show that incorporating neuroscience-informed HR practices – such as scheduling shifts with circadian rhythms or holding team retrospectives based on emotions – can lead to quantifiable gains in resilience and speed of decision making without significant cost. Such efforts prove the scalability of neuro-informed practices beyond big business.

6. THE NEUROSCIENCE OF WORK: UNDERSTANDING THE BRAIN'S ROLE IN PROFESSIONAL ENVIRONMENTS

As 21st-century bodies and corporate bodies navigate the neurobiological costs of unyielding output, a neuroscientific interface exposes chronic stress, attentional scattiness, and moral violation not as abstractions of management but as actual disruptions of prefrontal-amygdala-hippocampal circuits necessary for planning, affect regulation, and moral judgment (Arnsten, 2009; McEwen, 2007).

Instead of rational optimisation engines, brains under stress and pressure come with the price of reduced coherence, decision fatigue, and the frustration of impulsive reasoning, which undermines both performance and moral clarity. The valorisation of multitasking and urgency at the level of personal and organisational consciousness, frequently misrecognised as efficiency, propels neurocognitive depletion, especially for neurodivergent people who are invisibly excluded by environments structurally prejudiced against attentional diversity. Neuroplasticity, so often cast as a personal attribute, has a growing structural dependence—financial resources, emotional safety, cognitive pace—that define the extent of one's learning, resilience, and creativity (Fields, 2005).

Restoring cognition becomes an ethical design requirement, not an optional benefit. Interventionally, "Quiet Times" by Intel, which aimed at establishing uninterrupted periods of focused work, resulted in reduced physiological stress markers and greater measured productivity gains (Rosen, 2016). Similarly, Boyatzis and Jack (2018) Longitudinal Neural imaging studies showed that managers trained in mindfulness and EI learned to maintain more stable prefrontal-limbic synchrony, while improving team cohesion. These understandings call for a change in our perspective – from 'reactive performance indicators' to pro-active neuro-aligned design – grounded in theories of education that are based on brain science (Immordino-Yang & Damasio, 2007) that meet attentional rhythms, are emotionally resonant, and contextually equitable.

Scaffolding mechanisms, such as spaced repetition and metacognitive prompts (Rohrer, 2012), should be integrated into working ecologies rather than superficially applied. Physical exercise and contemplative activities enhance plasticity (Kempermann et al., 2010; Tang et al., 2015), but are ineffective when overextended ecologies exist. The promise of tDCS or measures of enhancement under neurocapitalism (Hamilton et al., 2011) is deeply concerning when it comes to consent, erasure of norms, and reducing cognition to an exploitable commodity. The OECD's lifelong learning frameworks, infused with neuroplasticity, provide a thoughtful blueprint; however, adoption is patchy, influenced by infrastructural inequalities. Organisational neuroscience must move beyond performance engineering and embrace ecologically valid, ethically sound, and culturally sensitive designs, where cognitive flourishing is not dependent on privilege but is built into the system.

7. EMOTIONAL INTELLIGENCE AND BRAIN FUNCTION: A NEUROBIOLOGICAL PERSPECTIVE

Emotional intelligence (EI), once dismissed as managers' bling, has now become neuroscientifically credible as a capacity that is not unitary or fixed, but one that emerges from dynamic neural systems implicated in intra-personal sensitivity, behavioural coordination and social attunement (Pessoa, 2008; Craig, 2009). The interaction between prefrontal regulatory regions and subcortical affective hubs (which includes the amygdala, the hippocampus and the anterior cingulate) makes EI biologically contingent, not biologically invariable or uniformly enacted.

There is also additional evidence from mirror neuron activity, grounding empathy on neural mimicry, which enables affective reciprocity (Iacoboni, 2009). However, the ongoing split between ability-based and behavioural models of EI continues to hide its ecological and plastic nature mediated, as neuroscience reveals, by both neurodevelopmental trajectory and sociocultural conditioning. EI is not a cosmetic veneer under high-pressure, but a regulatory imperative for robustness, as revealed by neuroimaging associations with dampened amygdala reactivity and increased medial prefrontal activation under stress (Decety & Jackson, 2006)

However, its organisation is frequently perverted into responsibilisation—best witnessed by, for example, resilience programmes that obscure structural toxicity—in which neuroplastic becoming is commodified, and is there anything we can do without commercialising it, increasing output rather than promoting wellbeing (Lindebaum, 2009; Fineman, 2000). Corporate offerings such as Google's "Search Inside Yourself" or EI-based diplomacy training embody this tension: neurovalidated but complicit in performative ethics. However, they are using an EI model based on the assumption that it is universal and can be developed in anyone, regardless of neurodiversity and cultural differences in how we express emotional states (Hofstede, 2001), which may seek to defy Westernised EI categories.

Similarly, attention misconstrued as a finite personal resource requires conceptual reshaping as a neuroenergetic process susceptible to partitioning by overtasking, especially in digital, metric-driven contexts (Posner & Petersen, 1990; Dux et al., 2009). The models of attentional stamina now emphasise emotional and environmental modulation, with interventions such as mindfulness or a biophilic environment being able to restock cognitive resources (Tang et al., 2015; Browning et al., 2014).

However, these mismatches are inequitably allocated and offered to professional elites, while frontline workers face fragmented workflows and no restoration infrastructure. Even more horrifying is the advance of biometric surveillance, which transforms cognition into extractable data, damaging inequality neurologically through algorithmic micro-regulation. As a countermodel, Nordic work cultures are notorious for prioritising decompression over presence, whilst meta-analyses show that cognitive and attentional training can only be practical if the ecological contingencies are met (Chiesa et al., 2011; Bernstein & Turban, 2018).

Contesting epistemologies, resource depletion, and motivation-driven affordance underpin the ontological issue of whether EI and attention are individual traits or systemic effects. A growing consensus shifts them to be relationally situated, neurally malleable capacities sensitive to structural care, not optimisation. To institutionalise EI and attention ethically is not to gamify affect or make focus measurable, but to re-describe the institution as a locus of cognitive dignity—where emotional presence is shielded and depth of thought is held culturally durable.

These results confirm that emotional intelligence can be considered a behavioural competency and a neurobiological leadership resource. Moving from the neuroanatomy of emotion to the general dynamics of the workplace, we examine how motivation, stress, and structure interact with the brain on the road to higher achievements.

8. MOTIVATION AND REWARD SYSTEMS: A NEUROSCIENTIFIC APPROACH

The neuroscientific reconceptualisation of motivation, stress, and creativity undermines the behavioural architecture that long supported management theory, exposing a territory where affective, cognitive and somatic systems congeal in forms irreducibly impertinent to linear causality or reductive design. Motivation is no longer a Pavlovian outflow in response to reinforcement stimuli. However, it is embodied in the struts and levers of the mesolimbic dopamine circuitry (in the ventral tegmental area and nucleus accumbens), closely modulated by serotonergic, noradrenergic and opioid systems, and their shaping of emotion tonality, impulsivity and arousal (Berridge & Robinson, 2016; Schultz, 2016).

In contrast, intrinsic motivation triggers medial prefrontal pathways of autonomy and mastery, whereas extrinsic rewards activate the lateral cortical systems that are susceptible to overjustification effects, thereby watering down agency and blurring moral vision (Deci & Ryan, 2000; Murayama et al., 2010). The cognitive costs of coercive incentives are not just demotivation; they are structural fatigue, neural deflation, and ethical shortsightedness disguised as dedication (McEwen, 2007). Stress aggregated temporally and transgenerationally restructures the motivational architecture in its image – raising the amygdala, squishing hippocampal function, rubbing reflective control down the drain – such that urgency becomes normative and critical thought pathologised (Arnsten, 2009; Zannas & West, 2014).

Conversely, creativity relies on the states this model stifles: liminality, associative drift, and oscillatory synchrony with DMN, ECN, and salience networks (Beaty et al., 2016). It does not flourish under time pressure or gamified competition but in ambiguity-rich settings where trust comes before performance and doubling back is temperate, not punished (Jauk et al., 2012). However, current systems — algorithmic nudging, performative innovation dictates, and biometric surveillance — trigger a limbic defensiveness and calcify the creative neural flows.

There are autonomy-based models at work in individualist societies, but they do not thrive where harmony norms hold sway (Hofstede, 2001); the same motivational infrastructure that precludes and excludes precarious workers, desiring and neurodivergent minds, and generational cohorts opted for psychic coherence, not metrics. With time, reflective control decamps the dorsal striatum, leaving behind a habit without control (Schwabe & Wolf, 2013). Neuroscience reveals not just cognitive slackery but institutional injustice: motivation needs to be conserved, not extracted, which requires spatial largesse, narrative fit and epistemic concern. Only in such nurtured ecosystems can motivation and innovation flourish not as managed recitals, but as ethically, neurologically coherent manifestations of human autonomy.

9. THE ROLE OF SLEEP, NUTRITION, AND PHYSICAL ACTIVITY IN BRAIN EFFICIENCY

This myth of the brain as a solitary machine of cognitive command falls apart when we consider its biological dependence—on sleep, food, movement – disruptions of which compromise the intellect's acuity, ethical compass, and ability to regulate itself. Incorrectly viewed as optional, sleep enables glymphatic clearance and hippocampal-prefrontal integration, preserving foresight and moral judgment (Walker, 2017; Xie et al., 2013). Nutrition asymmetry and sedentism, reinforced by institutional patterns, deregulate gut-brain communication, inhibit neurogenesis, and shorten executive function (Gómez-Pinilla, 2008; Ratey, 2008).

However, workplaces misconstrue chronic exhaustion as evidence of dedication, embedding burnout as an operational baseline and wellness as a tokenistic — a decoy for organisational inertia. Even if elite companies introduce sleep pods or circadian policies, the precariat remains excluded and petrified in their neurobiological breakdown behind performance conditioning. Theoretical frameworks differ, as resource theories portray health as refilled, neuroplastic models focus on structural remodelling, and critical structuralism indicates that systematic denial of neuro-sustainability leads to subjugated identities.

Technology further problematises this terrain, where wearables and BCIs promise optimisation but frequently precipitate cognitive surveillance, inactivating the empowering/epistemic extraction divide (Zander & Kothe, 2011). Neural information — as rich with intention and affect as it is — is now threatened with becoming a commodity at the mercy of employers, muddying the distinction between enhancement and intrusion and stretching further inequalities in power. The growing neurorights debate—promoted using Chilean legal revision—argues for the freedom of thought, although its reception worldwide is sporadic (Ienca & Andorno, 2017).

Cultural dissonance adds nuance: Western fetishisation of individual optimisation against collectivist caution about neuro-intrusiveness. Yet the access gaps are glaring—executives get biofeedback, and low-agency workers get neuro-extractive performance systems rooted in opaque calculus. Enhancement devices such as neurofeedback or tDCS are only temporarily effective unless they are based on relational and ecological integrity (Enriquez-Geppert et al., 2017).

Worse, they can induce learned helplessness, disconnect intuition from action and encourage compliance over authenticity. Alternatively, in other words, what is lost is not just cognitive spontaneity but the dialogic, ethical, co-regulated processes that give labour any meaning. Absent anticipatory ethics and the realisation of participatory governance, the rush to neuro-enhancement risks refashioning organisations as throughput devices. There is no equivalent between performance on the one side and "activation" on the other; flourishing thrives on coherence, dignity, and symbol thickness values maintained not through voltage, but with care, inclusion and transdisciplinary stewardship.

10. FUTURE DIRECTIONS: BRAIN SCIENCE, AI, AND THE EVOLVING WORKFORCE

As neuroscience and AI converge, it may not merely be enhancing productivity-mode, but an epistemological recalibration of the nature of human work as it consolidates neural feedback into adaptive systems that act as tools of personalised cognitive load, attention modulation, and fatigue management, such as those enabled by the use of mobile EEG and fNIRS. These machine-learning-trained systems promise a performance edge but risk offloading cognitive agency, introducing significant tension with autonomy, consent, and neuroprofiling. The rise of 'neuro rights' – demands for cognitive liberty and psychological continuity – Constituted another public fear of corporate exploitation, for control of personal experiences as a workplace asset, whose vulnerabilities to commodification and ultimate value were asymmetrical.

As AI-led cognitive calibration brings dividends in personalised learning and attentional synchronisation, it risks marginalising metacognitive self-regulation, generating fragile dependencies not adjusted to novelty, ambiguity, or surprise. The black-box nature of algorithmic inference exacerbates this tension, wherein workers are dissociated from interpretive control over data-driven decisions that impact their autonomy. Theoreticians of cognitive enhancement celebrate this transformation as one that will free us for creativity and resilience. At the same time, critical theorists oppose the calculus of mind and argue that reducing cognition to variables diminishes human unpredictability, an anchor of dignity and adaptability in a chaotic world.

Studies have shown that since COVID-19, hybrid, remote work impacts the distribution of cognitive load and emotional regulation. In a neuroergonomic work conducted by Gao et al. (2023), higher theta power was observed related to cognitive fatigue and enhanced autonomy-based neural coherence in the prefrontal cortex for the remote worker, implying mixed contributions. Moreover, neurofeedback interventions for distributed teams have shown potential to cope with digital fatigue and to increase focus (Kellermann et al., 2022). Such insights may yield cognitively sustainable schemes: Hybrid policies sensitive to neurobiological rhythms, not arbitrary work intervals.

While corporate organisations may have most of the story on neuro-enhancement, there are also prospects for application in developing countries and the non-corporate sectors. In Kenya, Odhiambo and Kiptoo (2021) piloted neuroplasticity-focused low-cost interventions—such as reflective journaling and peer-based emotional debriefing—in public education, which increased teacher retention and student achievement. Singapore's Ministry of Manpower, too, has put neuroethics into practice in its occupational safety policies, incorporating brain-based guidelines in construction and transport sectors, among others. These cases suggest that neuro-informed workplace transformation is not the exclusive domain of advanced economies but is applicable across cultures and economies when designed inclusively.

Emerging trends in neuro-organisational research reveal a complex landscape of opportunities and challenges that are reshaping the modern workplace. One key issue is **cognitive overload**, where excessive task-switching and digital saturation reduce working memory capacity and impair decision-making (Sweller, 2011; Qin et al., 2009). The rapid development of **neuroadaptive systems** and brain-computer interfaces introduces pressing **neuroethical and privacy concerns**, as these technologies risk infringing on cognitive liberty without proper governance (Ienca & Andorno, 2017).

New directions in neuro-organisational research point to a nebulous environment of opportunities and challenges that researchers suggest are reshaping the modern workplace. Cognitive overload is characterised by too much task-shifting and digital saturation, which drain working memory and compromise reasoned decision-making (Sweller, 2011; Qin and colleagues, 2009). A looming neuroethical and privacy concern arises from the fast-paced proliferation of neuroadaptive systems and BCIs, threatening cognitive freedom without safeguards (Ienca & Andorno, 2017).

At the same time, the construct of psychological safety has been neuroscientifically validated with trust and emotional security activating the networks necessary for learning and collaboration (Edmondson, 1999; Boyatzis et al., 2015). Related to this is the substantial progress in AI-neurointegration, merging EEG and fNIRS with machine learning to monitor cognitive states in real time (enhancing learning outcomes, but also with ethical considerations about surveillance and manipulation) (Cinel et al., 2019; Gao et al., 2023). Finally, focusing on neurodiversity and equity requires us to consider designing inclusive workplaces for diverse cognitive profiles and life experiences, particularly in an increasingly hybrid and globalised work world. Such trends require organisations to progress technologically, ethically, inclusively, and evidence-based.

Given these considerations, integrating AI and neuroscience into organisational practices should be done with adherence to ethical values, inclusivity, and human-centred design. Organisations need to adopt clear policy frameworks for collecting and using neural data, make neuroenhancement technologies available equitably, and create conditions that respect psychological safety and personal autonomy. Through the integration of technological innovation and ethical stewardship, it becomes possible to realise the benefits of neuroadaptive systems while preventing damage to the dignity and rights of all workers.

To counteract ethical hazards and technocratic excesses, nascent models like Chile's neurorights code and Singapore's SkillsFuture project show that governance and justice can be reconciled with cognitive novelty (Ienca & Andorno, 2017; OECD, 2016). Rather than repudiating neurotechnology altogether, organisations can embrace transparency around data protocols, engage in participatory design, and develop inclusive roll-out strategies to design neuroadaptive systems that empower rather than control. These case studies state that scalable, value-driven models that place innovation above all else and respect human dignity are possible.

Whereas the neuroscience of the Default Mode and Executive Control Networks offers valuable insight about ideation and self-regulation (Raichle, 2015), managers might do so by organising uninterrupted work cycles and even permitting reflective time that fosters neural synchrony and cognitive resilience.

11. IMPLICATIONS

Based upon these factors, AI and neuroscience must be integrated into organisational practices with respect for ethical values, inclusivity, and human-centred design. Organisations should implement clear policy guidelines for collecting and utilising neural information, provide neuroenhancement technologies to employees on an equitable platform, and ensure conditions honouring psychological safety and individual autonomy. Integrating technological innovation and ethical stewardship makes it feasible to capture the advantages of neuroadaptive systems while avoiding harm to the dignity and rights of all employees.

By applying neurobiological principles to training and growth, managers can promote a culture of continuous learning, resilience, and creativity. On a practical level, this can result in more targeted interventions, such as mindfulness programmes, neurofeedback, and stress management, that are directly aligned with brain processes. Applying neuroscience-based practices positively affects employee wellbeing, generating greater engagement, innovation, and long-term organisational success. However, managers must balance ethical considerations to ensure neurotechnology is applied responsibly, without intruding upon employees' privacy and generating a culture of inclusion, not manipulation.

High-resource institutions may adopt neuroadaptive technology and biometric interfaces; however, resource-limited workplaces in education, healthcare, agriculture, and informal sectors can benefit from low-cost, scalable methods to use neuroscientific findings. Studies indicate that simple interventions, such as scheduled breaks, light exposure regulations, and peer support networks, can significantly enhance cognitive functioning and stress recovery (Jacka et al., 2010; Berman et al., 2008). For instance, a rural health centre-based study in India found that supplementing adapted check-ins on emotional intelligence and flexible schedules helped improve decision-making and decreased burnout amongst community health workers without needing fancy neurotechnology (Patel et al., 2021). Such adaptations illustrate that emotional intelligence, stress regulation, and the principles of neuroplasticity can be operationalised through organisational culture and technology.

To transition from theory to practice, organisations should follow a structured implementation roadmap for neuro-informed interventions:

1. **Audit cognitive ergonomics:** Evaluate attention design, decision fatigue, and stress triggers across workflows.
2. **Pilot low-cost interventions:** Introduce brief mindfulness sessions, environmental design changes (e.g., biophilic layouts), and emotion-focused check-ins.
3. **Train leadership on emotional intelligence and neuroplasticity:** Use evidence-based modules that integrate role-play, journaling, and feedback loops.
4. **Integrate ethical safeguards:** Establish policies to protect cognitive data and ensure informed consent in any neurotech usage.
5. **Evaluate longitudinally:** Use neuropsychological and organisational metrics (e.g., turnover, innovation scores, fNIRS data) to track long-term impact.

While the conversations around neuroscience-informed workplaces revolve around high-tech or well-resourced corporate spaces, the logic behind them can be applied across industries, with careful adjustment. Emotional intelligence, for example, can be developed during organised group reflection or peer debriefing, which do not involve digital tools. Neuroplasticity, likewise, is facilitated by developing repeated low-cost habits that encourage learning and reflection, like writing or feedback circles (Dweck, 2006). Resource-poor environments can mediate stress through adaptive schedules, safe contexts for communicating, and spatial alterations, such as exposure to natural light – rallying forces that are cost-effective, but powerful on the brain (Berman et al., 2008; Edmondson, 1999).

Even resource-constrained organisations can effectively incorporate neuroscience-based principles via non-costly, non-complex interventions. For example, "Mindfulness Minutes", where individuals spend just five minutes of quiet breathing or focused attention before meetings, can downregulate stress and increase emotional regulation. Aligning flexible schedules with the various chronotypes of employees promotes better sustained alertness and mitigates cognitive fatigue to improve overall productivity. Explorations and conversations around emotional intelligence—like the peer-led group reflections into workplace struggles—encourage affective processing and empathic discourse, creating a healthier team dynamic.

Exposure to nature, including indoor plants or short outdoor walks, has also been found to restore cognitive function and executive attention (Berman, Jonides,

& Kaplan, 2008). Lastly, sleep hygiene education programmes focusing on the neurological significance of rest for employees can positively impact decision-making and emotional regulation, underpinning workplace well-being (Walker, 2017). These practical approaches show how neuroscience can be used in any institution without the need for technology or funds.

Science-based interventions do not have to be resource-heavy. Five-minute mindfulness sessions can regularise stress; flexible scheduling along the circadian rhythm can augment attention; peer-based EQ dialogues can foster team cohesion; exposure to green views or daylight supports cognitive recovery (Berman et al., 2008); and basic sleep hygiene training encourages emotional regulation and decision-making (Walker, 2017). These strategies can be integrated into schools, hospitals, or small business settings with no specialised equipment, and they illustrate that we can structurally support mental and emotional well-being wherever we are. This staged model contributes to keeping interventions context-sensitive, ethically defensible, and organisationally embedded.

11.1 Translating Neuroethics into Scalable Organisational Practice

These legitimate and important critiques of neuro-reductionism and cognitive surveillance must be countered with theory-to-practice-focused, solution-oriented paths forward. Inclusive stakeholder participation, cross-cultural ethical auditing, and flexible infrastructural design can be embedded with neuroscience in workplace policy (Davidson & McEwen, 2012). For example, instead of opposing neurotechnologies outright, organisations should adopt a tiered consent model, where employees have to voluntarily participate in neuro-monitoring and know what level of data gathering and feedback mechanism is in place.

Within multicultural environments, context-sensitive neuro-integration models based on modulations (like culturally responsive pedagogy) can avoid Western-centric application of neuroleadership. Adjusting emotional intelligence programs for collectivist standards, such as focusing on collective duty rather than personal expression, results in more authentic systems of cognitive processing (Chiao et al., 2009; Hofstede, 2001). Similarly, creating ecological scaffolds, such as rhythm-informed work processes, sensory modulation environments, and biophilic design, may promote neuroplastic responses in various worker profiles without needing persistent cognitive or surveillance activity (Browning et al., 2014).

Public-private collaborations can also help to foster cross-sector neuroethical norms, ensuring that the use of neuroscience in the workplace does not cross into coercion and/or mirror structural injustices. Neuro-inclusion audit practices will need to be scalable, able to audit for the impact on neurodivergent employees of policies and tools, to turn 'feeling crap' into 'specific and measurable injustice' and then abstract, more historically based critique towards frameworks for restorative justice and cognitive dignity (Lindebaum, 2013).

International initiatives, such as new transnational companies, offer effervescent models to extend the reach of neuroscience-inspired approaches outside high-income corporate environments. For example, the OECD's pilots on neuroscience-informed education and in the workplace in countries such as the Netherlands and Finland show that contextualising neuroplasticity and cognitive ergonomics can improve employee well-being and labour adaptability across different socio-economic contexts (OECD, 2016). Likewise, the WHO 'Mental Health at Work' policy framework provides locally relevant 3P checks for mental wellbeing, discussing how to incorporate a psycho-socially safe culture alongside stress-reduction into the fabric of organisations (WHO, 2020).

These cross-cutting, policy-sensitive approaches reinforce the chapter's claim that neurobiological knowledge must be enacted situationally—the ethical infrastructures, inclusive design, and process mechanisms that onboard autonomy and organisational feeling. The chapter's scope broadens from theoretical idealism to rooted, globally relevant policy practice by adding such geographically and institutionally diverse initiatives.

12. META FRAMEWORK

Figure 1. Meta Framework

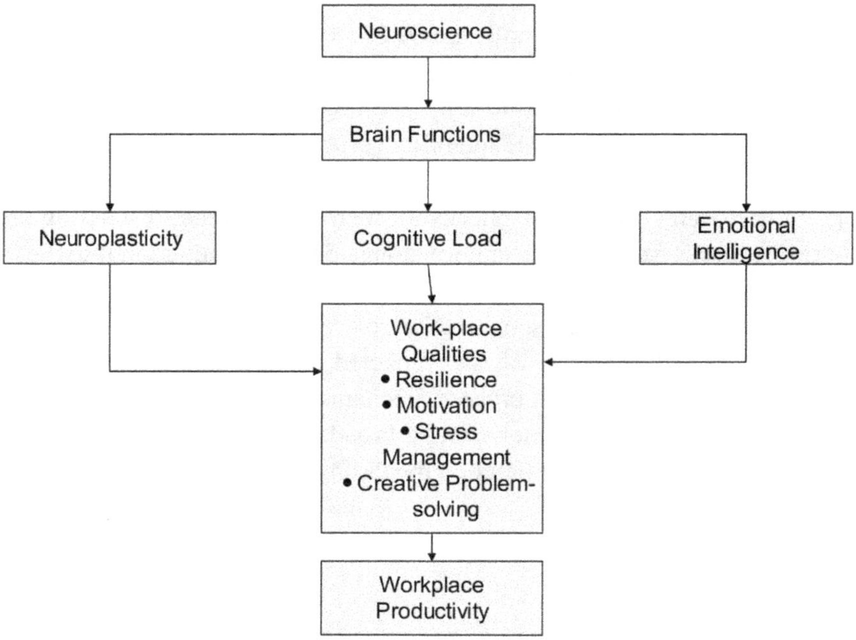

Source: Authors

The model outlined in Figure 1 describes elaborate correlations between neuroscience, brain function, and workplace productivity. Fundamentally, it describes how neuroscience influences brain function, which, in turn, affects cognitive load and emotional intelligence. Neuroplasticity suggests that the brain's ability to adapt and reorganise can enhance cognitive abilities, which immediately impacts workplace attributes such as resilience, motivation, stress management, and creative problem-solving. All these attributes, when combined, are related to workplace productivity.

The interconnection is transparent here: neuroplasticity enables cognitive load management, which affects emotional intelligence, and all these factors enhance critical workplace attributes. As a result, workplace productivity is affected by the combined effort of these aspects. This model differs from earlier models by placing great importance on the brain's adaptable ability (neuroplasticity) and its direct correlation with emotional intelligence and cognitive load. Whereas such models might emphasise emotional intelligence or cognitive behaviour separately, this model brings them together through integrating recent neuroscientific understandings and presents an integrated picture of how mental processes and workplace outputs are coupled.

To translate neuroscience-informed insights into managerial utility, the following stepwise framework is proposed for practical adoption:

Step 1: Diagnose Cognitive Landscape

Begin with organisation-wide assessment tools that measure stress levels, attentional fragmentation, and emotional climate using non-invasive surveys and biometric feedback (Kabat-Zinn, 2013; Edmondson, 1999).

Step 2: Design Neuro-Responsive Interventions

Implement modular programs tailored to cognitive demands—e.g., stress reduction via mindfulness, attention-sustaining architecture, and emotional intelligence training linked to real-world scenarios (Goleman, 1995; Berman et al., 2008).

Step 3: Create Feedback Loops

Set up closed-loop systems where employees receive aggregated cognitive performance data for self-reflection and development. Use tools like self-reporting apps, brief neurofeedback dashboards, or pulse surveys (D'Mello & Graesser, 2015).

Step 4: Scale with Contextual Sensitivity

Adapt initiatives for cross-cultural and neurodivergent inclusivity by engaging diverse voices in design and feedback cycles (Chiao et al., 2009; Hofstede, 2001). Avoid one-size-fits-all approaches.

Step 5: Sustain Through Policy Integration

Embed cognitive health metrics into HR policy and performance reviews—not as productivity markers, but as well-being indicators—to institutionalise supportive environments (Gallup, 2017; McEwen & Morrison, 2013).

This framework ensures that neuroscientific insights evolve beyond conceptual rhetoric into structured, reproducible, and culturally attuned management practice.

12.1 Making Brain Science Actionable: A Guide for Mid-Level Leadership

While senior executives and academics might concentrate on strategic policy or metaphysical implications, mid-level practitioners need practical implications for leading teams, coaching, and communication. For example, knowing that chronic multitasking diminishes decision-making quality is not just conceptual knowledge; it is also an argument for designing meetings with an agenda, minimising task switching, and developing in break intervals between work (Qin et al., 2009; Sweller, 2011).

Likewise, emotional intelligence – a concept more often taught in the abstract – can be made concrete by training team leads to pick up on emotional cues, provide empathetic feedback, and build trust via transparent feedback loops (Goleman, 1995). Weekly mood check-ins, biophilic breakout spaces and low-stimulus zones are not abstract metrics, but immediately actionable ways to positively impact cognitive well-being. To bridge the gulf between scientific rigour and managerial utility, organisations could benefit from a set of "Neuro-Insight Cards": a curated pack of simplified ideas derived from neuroscience, associated with everyday leadership practices. They democratise brain science: making it available without expertise.

13. CONCLUSION: BRAIN SCIENCE AS A CATALYST FOR TRANSFORMING WORKPLACE EFFICIENCY

Incorporating brain science into organisational systems is not a technological superimposition, but a philosophical pivot—requiring the shedding of industrial logics that prefer linearity, speed, and control over cognitive coherence and affective density. Neuroscientific revelations confirm that creativity, cooperation, and moral discernment emerge from fragile neurobiological ecologies—mutually constructed by sleep, emotional coherence, and attentional synchrony—not from a flight from rationality (Davidson & McEwen, 2012). From a design perspective, cognitive ergonomics should rise from the level of a design afterthought to that of an ethic, inflaming critical reappraisals of digital flow, architecture and sensory load (Norman, 1993).

Learning systems, too, must reject transactional metrics and embrace the relational scaffolding of neuroplasticity, embedding challenge and safety as symbiotic preconditions for meaningful adaptation. However, this imperative remains stifled by tokenistic wellness initiatives that individualise systemic burdens while glorifying burnout as valorised productivity. AI and neurotechnologies offer potential to synchronise design with brain rhythms, but without anticipatory ethics, inclusive governance, and robust neurorights (Ienca & Andorno, 2017), they risk entrenching surveillance and neuro-inequity.

A genuinely brain-responsive organisation does not celebrate intellectual uniformity but recognises neurodiversity as a state of systemic intelligence. Restoring is no longer counterfactual, but is a part of excellence. When we repose emotional labour, long dismissed as a light matter of the conversation heart, we re-imagine this work as mentally dense with moral reflection and adaptive work. This shift will depend on integration with broader policy structures, including mental privacy, cognitive equity, and brain health within public and education ecosystems.

Empirical research consistently shows that the neuro-aligned environment enhances innovation (Schneider, Barbera, & Holc, 2016) and retention (Schneider & Barbera, 2011) and that the cognitively injurious space expedites detachment and departure (Frazier et al., 2017; Baicker et al., 2010). However, systemic inertia and injustice still obstruct change. By collective moral will and interdisciplinary foresight, organisations can evolve into ecosystems where brain dignity is constitutive, not conditional, of economic purpose and human value.

REFERENCES

Andersson, M., & Lagerkvist, C. J. (2022). Neuroeconomics and SMEs: Experimental insights into decision fatigue and cognitive control. *Small Business Economics, 58*(3), 911–928.

Arnsten, A. F. (2009). Stress signalling pathways that impair prefrontal cortex structure and function. *Nature Reviews. Neuroscience, 10*(6), 410–422. DOI: 10.1038/nrn2648 PMID: 19455173

Baicker, K., Cutler, D., & Song, Z. (2010). Workplace wellness programs can generate savings. *Health Affairs, 29*(2), 304–311. DOI: 10.1377/hlthaff.2009.0626 PMID: 20075081

Beaty, R. E.. (2016). Creativity and the default network: A functional connectivity analysis of the creative brain at rest. *Neuropsychologia, 91*, 381–389. DOI: 10.1016/j.neuropsychologia.2016.08.015 PMID: 25245940

Berman, M. G., Jonides, J., & Kaplan, S. (2008). The Cognitive Benefits of Interacting With Nature. *Psychological Science, 19*(12), 1207–1212. DOI: 10.1111/j.1467-9280.2008.02225.x PMID: 19121124

Berman, M. G., Jonides, J., & Kaplan, S. (2008). The cognitive benefits of interacting with nature. *Psychological Science*. PMID: 19121124

Bernstein, E. S., & Turban, S. (2018). The impact of the 'open' workspace on human collaboration. *Philosophical Transactions of the Royal Society B, 373*(1753), 20170239. https://doi.org/DOI: 10.1098/rstb.2017.0239

Berridge, K. C., & Robinson, T. E. (2016). Liking, wanting, and the incentive-sensitisation theory of addiction. *The American Psychologist, 71*(8), 670–679. DOI: 10.1037/amp0000059 PMID: 27977239

Boyatzis, R. E. (2018). *The Competent Manager: A Model for Effective Performance*. Wiley.

Boyatzis, R. E., & Jack, A. I. (2018). The neuroscience of coaching: Actionable insights for sustainable leadership. *Journal of Management Development, 37*(3), 209–219.

Boyatzis, R. E., Rochford, K., & Taylor, S. N. (2015). The role of the positive emotional attractor in vision and shared vision: Toward effective leadership, relationships, and engagement. *Frontiers in Psychology, 6*, 670. DOI: 10.3389/fpsyg.2015.00670 PMID: 26052300

Browning, W. D., Ryan, C. O., & Clancy, J. O. (2014). *14 Patterns of Biophilic Design*. Terrapin Bright Green LLC.

Chiao, J. Y.. (2009). Cultural neuroscience: Cultural influences on brain function. *Progress in Brain Research, 178*, 287–304. DOI: 10.1016/S0079-6123(09)17820-9 PMID: 19874977

Chiesa, A., Calati, R., & Serretti, A. (2011). Does mindfulness training improve cognitive abilities? A systematic review of neuropsychological findings. *Clinical Psychology Review, 31*(3), 449–464. DOI: 10.1016/j.cpr.2010.11.003 PMID: 21183265

Cinel, C., Valeriani, D., & Poli, R. (2019). Neurotechnologies for Human Cognitive Augmentation: Current State of the Art and Future Prospects. *Frontiers in Human Neuroscience, 13*, 13. DOI: 10.3389/fnhum.2019.00013 PMID: 30766483

Craig, A. D. (2009). How do you feel now? The anterior insula and human awareness. *Nature Reviews. Neuroscience, 10*(1), 59–70. DOI: 10.1038/nrn2555 PMID: 19096369

D'Mello, S. K., & Graesser, A. C. (2015). Feeling, thinking, and computing with affect-aware learning technologies. In Calvo, R. A. (Eds.), *The Oxford Handbook of Affective Computing*. Oxford University Press.

Davidson, R. J., & McEwen, B. S. (2012). Social Influences on Neuroplasticity: Stress and Interventions to Promote Well-being. *Nature Neuroscience, 15*(5), 689–695. DOI: 10.1038/nn.3093 PMID: 22534579

Decety, J., & Jackson, P. L. (2006). The functional architecture of human empathy. *Behavioral and Cognitive Neuroscience Reviews, 3*(2), 71–100. DOI: 10.1177/1534582304267187 PMID: 15537986

Deci, E. L., Koestner, R., & Ryan, R. M. (1999). A meta-analytic review of experiments examining the effects of extrinsic rewards on intrinsic motivation. *Psychological Bulletin, 125*(6), 627–668. DOI: 10.1037/0033-2909.125.6.627 PMID: 10589297

Deci, E. L., & Ryan, R. M. (2000). The "what" and "why" of goal pursuits: Human needs and the self-determination of behaviour. *Psychological Inquiry, 11*(4), 227–268. DOI: 10.1207/S15327965PLI1104_01

Doidge, N. (2007). *The Brain That Changes Itself: Stories of Personal Triumph from the Frontiers of Brain Science*. Penguin.

Dux, P. E.. (2009). Training improves multitasking performance by increasing information processing speed in the human prefrontal cortex. *Neuron, 63*(1), 127–138. DOI: 10.1016/j.neuron.2009.06.005 PMID: 19607798

Dweck, C. S. (2006). *Mindset: The New Psychology of Success*. Random House.

Edmondson, A. (1999). Psychological Safety and Learning Behaviour in Work Teams. *Administrative Science Quarterly*, *44*(2), 350–383. DOI: 10.2307/2666999

Edmondson, A. (1999). Psychological safety and learning behaviour in work teams. *Administrative Science Quarterly*.

Enriquez-Geppert, S.. (2017). Neurofeedback as a treatment intervention in ADHD: Current evidence and practice. *Current Psychiatry Reports*, *19*(3), 22. DOI: 10.1007/s11920-017-0771-8 PMID: 31139966

Fields, R. D. (2005). Making memories stick. *Scientific American*, *292*(2), 74–81. DOI: 10.1038/scientificamerican0205-74 PMID: 15715394

Fineman, S. (2000). *Emotion in Organisations* (2nd ed.). Sage Publications.

Frazier, M. L.. (2017). Psychological safety: A meta-analytic review and extension. *Personnel Psychology*, *70*(1), 113–165. DOI: 10.1111/peps.12183

Gallup. (2017). *State of the Global Workplace*. Gallup Press.

Gao, R., Chen, C., & Ma, Y. (2023). Remote work and cognitive fatigue: A neuro-ergonomic investigation. *Human Factors*, *65*(2), 287–302.

Goleman, D. (1995). *Emotional Intelligence: Why It Can Matter More Than IQ*. Bantam Books.

Gómez-Pinilla, F. (2008). Brain foods: The effects of nutrients on brain function. *Nature Reviews. Neuroscience*, *9*(7), 568–578. DOI: 10.1038/nrn2421 PMID: 18568016

Hamilton, R., Messing, S., & Chatterjee, A. (2011). Rethinking the thinking cap: Ethics of neural enhancement using noninvasive brain stimulation. *Neurology*, *76*(2), 187–193. DOI: 10.1212/WNL.0b013e318205d50d PMID: 21220723

Hofstede, G. (2001). *Culture's Consequences: Comparing Values, Behaviours, Institutions, and Organisations Across Nations* (2nd ed.). Sage Publications.

Iacoboni, M. (2009). *Mirroring People: The Science of Empathy and How We Connect with Others*. Picador.

Ienca, M., & Andorno, R. (2017). Towards new human rights in the age of neuroscience and neurotechnology. *Life Sciences, Society and Policy*, *13*(5). Advance online publication. DOI: 10.1186/s40504-017-0050-1 PMID: 28444626

Immordino-Yang, M. H., & Damasio, A. (2007). We Feel, Therefore We Learn: The Relevance of Affective and Social Neuroscience to Education. *Mind, Brain and Education: the Official Journal of the International Mind, Brain, and Education Society*, *1*(1), 3–10. DOI: 10.1111/j.1751-228X.2007.00004.x

Jacka, F. N.. (2010). Association of Western and traditional diets with depression and anxiety in women. *The American Journal of Psychiatry*, *167*(3), 305–311. DOI: 10.1176/appi.ajp.2009.09060881 PMID: 20048020

Jauk, E.. (2012). The relationship between intelligence and creativity: New support for the threshold hypothesis using empirical breakpoint detection. *Intelligence*, *40*(5), 379–391. DOI: 10.1016/j.intell.2012.05.002 PMID: 23825884

Kabat-Zinn, J. (2013). *Whole Catastrophe Living: Using the Wisdom of Your Body and Mind to Face Stress, Pain, and Illness*. Bantam.

Kellermann, T.. (2022). Neurofeedback training for hybrid workers: Improving focus and reducing fatigue. *Frontiers in Human Neuroscience*, *16*, 912345.

Kempermann, G.. (2010). Activity-dependent regulation of adult neurogenesis. *Cold Spring Harbor Perspectives in Biology*, *2*(6), a001891. DOI: 10.1101/cshperspect.a001891

Kumar, P.. (2020). Mindfulness interventions in Indian SMEs: A pilot neuro-organisational study. *Asian Journal of Business Ethics*, *9*(1), 33–47.

Lieberman, M. D. (2013). *Social: Why Our Brains Are Wired to Connect*. Crown Publishers.

Lindebaum, D. (2009). Rhetoric or Remedy? A Critique of Developing Emotional Intelligence. *Academy of Management Learning & Education*, *8*(2), 225–237. DOI: 10.5465/amle.2009.41788844

Lindebaum, D. (2013). Ideology in organisational cognitive neuroscience studies and other blind spots: Critical reflections on the neural turn in organisational research. *Academy of Management Review*, *38*(3), 482–503. DOI: 10.5465/amr.2012.0097

McEwen, B. S. (2007). Physiology and neurobiology of stress and adaptation: Central role of the brain. *Physiological Reviews*, *87*(3), 873–904. DOI: 10.1152/physrev.00041.2006 PMID: 17615391

McEwen, B. S., & Morrison, J. H. (2013). The Brain on Stress: Vulnerability and Plasticity of the Prefrontal Cortex over the Life Course. *Neuron*, *79*(1), 16–29. DOI: 10.1016/j.neuron.2013.06.028 PMID: 23849196

Murayama, K., Matsumoto, M., Izuma, K., & Matsumoto, K. (2010). Neural basis of the undermining effect of monetary reward on intrinsic motivation. *Proceedings of the National Academy of Sciences of the United States of America, 107*(49), 20911–20916. DOI: 10.1073/pnas.1013305107 PMID: 21078974

Norman, D. A. (1993). *Things That Make Us Smart: Defending Human Attributes in the Age of the Machine*. Addison-Wesley.

Odhiambo, J. M., & Kiptoo, M. T. (2021). Neurocognitive resilience in East African classrooms: A policy and pedagogical analysis. *African Journal of Educational Studies, 11*(2), 118–137.

OECD. (2016). *Skills for a Digital World: 2016 Ministerial Meeting on the Digital Economy Background Report*. OECD Publishing.

OECD. (2016). *Skills for a Digital World: 2016 Ministerial Meeting on the Digital Economy Background Report*. OECD Publishing., DOI: 10.1787/5jlwz83z3wnw-

Patel, V., Saxena, S., & Maj, M. (2021). Bridging mental health gaps in low-resource settings: Strategies for global equity. *Global Health Journal (Amsterdam, Netherlands)*.

Pessoa, L. (2008). On the relationship between emotion and cognition. *Nature Reviews. Neuroscience, 9*(2), 148–158. DOI: 10.1038/nrn2317 PMID: 18209732

Poldrack, R. A. (2006). Can cognitive processes be inferred from neuroimaging data? *Trends in Cognitive Sciences, 10*(2), 59–63. DOI: 10.1016/j.tics.2005.12.004 PMID: 16406760

Posner, M. I., & Petersen, S. E. (1990). The attention system of the human brain. *Annual Review of Neuroscience, 13*, 25–42. DOI: 10.1146/annurev.ne.13.030190.000325 PMID: 2183676

Qin, S.. (2009). Acute psychological stress reduces working memory-related activity in the dorsolateral prefrontal cortex. *Biological Psychiatry, 66*(1), 25–32. DOI: 10.1016/j.biopsych.2009.03.006 PMID: 19403118

Raichle, M. E. (2015). The Brain's Default Mode Network. *Annual Review of Neuroscience, 38*, 433–447. DOI: 10.1146/annurev-neuro-071013-014030 PMID: 25938726

Ratey, J. J. (2008). *Spark: The Revolutionary New Science of Exercise and the Brain*. Little, Brown Spark.

Ratey, J. J. (2008). *Spark: The Revolutionary New Science of Exercise and the Brain*. Little, Brown Spark.

Rock, D. (2008). SCARF: A brain-based model for collaborating with and influencing others. *NeuroLeadership Journal, 1*(1), 44–52.

Rock, D., & Schwartz, J. (2006). The neuroscience of leadership. *Strategy & Business, 43*, 70–79.

Rohrer, D. (2012). Interleaving helps students distinguish among similar concepts. *Educational Psychology Review, 24*(3), 355–367. DOI: 10.1007/s10648-012-9201-3

Rosen, C. (2016). *Intel's "Quiet Time": A case study in attentional design.* Harvard Business Review Digital Articles.

Schultz, W. (2016). Dopamine reward prediction error coding. *Dialogues in Clinical Neuroscience, 18*(1), 23–32. PMID: 27069377

Schwabe, L., & Wolf, O. T. (2013). Stress and multiple memory systems: From 'thinking' to 'doing'. *Trends in Cognitive Sciences, 17*(2), 60–68. DOI: 10.1016/j.tics.2012.12.001 PMID: 23290054

Sweller, J. (2011). Qin, S. et al. (2009); Raichle, M. E. (2015)

Tang, Y. Y., Hölzel, B. K., & Posner, M. I. (2015). The neuroscience of mindfulness meditation. *Nature Reviews. Neuroscience, 16*(4), 213–225. DOI: 10.1038/nrn3916 PMID: 25783612

Walker, M. P. (2017). *Why We Sleep: Unlocking the Power of Sleep and Dreams.* Scribner.

World Health Organisation. (2020). *Mental Health and Psychosocial Considerations during the COVID-19 Outbreak.* Retrieved from https://www.who.int/docs/default-source/coronaviruse/mental-health-considerations.pdf

World Health Organisation. (2020). *Mental health in the workplace.* Retrieved from https://www.who.int/mental_health/in_the_workplace/en/

Xie, L.. (2013). Sleep drives the clearance of metabolites from the adult brain. *Science, 342*(6156), 373–377. DOI: 10.1126/science.1241224 PMID: 24136970

Zander, T. O., & Kothe, C. (2011). Towards passive brain–computer interfaces: Applying brain–computer interface technology to human–machine systems in general. *Journal of Neural Engineering, 8*(2), 025005. DOI: 10.1088/1741-2560/8/2/025005 PMID: 21436512

Zannas, A. S., & West, A. E. (2014). Epigenetics and the regulation of stress vulnerability and resilience. *Neuroscience, 264*, 157–170. DOI: 10.1016/j.neuroscience.2013.12.003 PMID: 24333971

Zhu, Y., Zhang, L., Fan, J., & Han, S. (2007). Neural basis of cultural influence on self-representation. *NeuroImage*, *34*(3), 1310–1316. PMID: 17134915

KEY TERMS AND DEFINITIONS

Cognitive Load: Cognitive load refers to the mental effort required to process information or complete a task. When the brain is overwhelmed with too much information or tasks at once, cognitive load increases, making it harder to focus, remember, and perform efficiently. Managing cognitive load is crucial for maintaining mental clarity and achieving high performance.

Mindfulness: Mindfulness is being fully present and aware of the current moment, without judgment. It is about observing our thoughts, feelings, and surroundings without getting caught up in them. Mindfulness helps reduce stress, improve focus, and increase emotional resilience, making it a powerful tool for enhancing overall well-being in the workplace.

Neurofeedback: Neurofeedback is a technique where people can learn to control their brain activity by receiving real-time feedback from a device that monitors brainwave patterns. It is like training the brain to be more focused or relaxed, and it has been shown to help improve attention, emotional regulation, and stress management.

Neuroplasticity: Neuroplasticity refers to the brain's remarkable ability to change and adapt. It is like a mental workout that strengthens and forms new connections in response to our experiences, learning, and even injury recovery. The brain is not as rigid as once thought—it can keep growing and evolving, no matter how old we are.

Neuroscience: Neuroscience studies how the brain and nervous system work. It helps us understand everything from how we think and feel to how we behave and learn. Neuroscience provides insights into how biology influences our thoughts, emotions, and actions by exploring the brain's structure and function.

Workplace Productivity: Workplace productivity is not just about how much work gets done, but how effectively and efficiently it is achieved. It is about achieving quality results by balancing focus, creativity, energy, and well-being. Productive workplaces foster environments where employees can work smart, not just hard, by optimising their mental and emotional resources.

Chapter 3
Reprogramming the Mind:
Neuroplasticity, Stress Resilience, and Emotional Intelligence in Professional Settings

Nidhi Suresh
https://orcid.org/0009-0004-7609-7774
PES University, India

Srishti Muralidharan
https://orcid.org/0009-0005-0798-7404
PES University, India

ABSTRACT

This chapter highlights the role of neuroplasticity, the capacity of the brain to reorganize itself in response to experience, that can be applied to maximize organizational effectiveness and individual performance. It considers the neurological underpinnings of emotional intelligence and social neuroscience in forming effective communication and collaboration, applying neuroplasticity to reduce work stress and develop resilience, and how the brain processes feedback of praise, criticism, and errors. Based on evidence from neuroscience, social psychology, and organizational behavior, the chapter proposes a coherent framework for developing interpersonal competence, adaptive teams, and psychologically safe workplaces. The chapter also demonstrates how companies can apply neuroscience-based programs to develop emotionally intelligent cultures that foster cognitive flexibility, stress management, and sustainable performance.

DOI: 10.4018/979-8-3373-2292-6.ch003

INTRODUCTION

Modern companies and organisations of the 21st century are undertaking a seismic shift. Traditional workplace settings, which prioritised efficiency, task delegation, and uniformity, are gradually adopting broader, dynamic paradigms that foster learning agility, emotional awareness, and flexibility. Gone are the days when fixed routines and technical skills guaranteed success. In their place today, adaptability, emotional regulation, and collaborative intelligence have emerged as the hallmarks of professional growth and success.

In the current era, a modern workplace is a complex and dynamic environment, demanding that individuals possess a wide range of skills to thrive. Beyond technical knowledge and expertise, the ability to communicate efficiently, collaborate with colleagues, and adapt swiftly to changes has become paramount. These skills and capabilities are not just innate talents; they are deeply rooted in neuroplasticity, the remarkable capacity of the brain to adapt, learn, and change.

Neuroplasticity, defined as "the brain's ability to change and adapt as a result of experience" (Doidge, 2007), is a transformative mechanism of the brain that challenges the long-held belief that adult brains are immutable and unchangeable. In contrast to this previously held belief, the brain is now viewed as a dynamic system that can be constantly shaped and refined through social and emotional experiences. This biological ability to grow thus becomes the neurological basis of what we now consider lifelong learning and behavioral adaptability in professional settings.

Neuroplasticity is a dynamic, ongoing process that enables the brain to adapt to new experiences, acquire new knowledge, and recover from traumatic injuries, especially to the brain. This unique characteristic of the human brain helps us in continuously modifying individual behavior as a response to experiences and environmental stimuli.

In a workplace setting, neuroplasticity plays a crucial role in shaping an individual's interpersonal and communication skills. Moreover, neuroplasticity also helps individuals respond to feedback and thus learn and grow within the structural setup of their organizations. Understanding the mechanism and working of neuroplasticity provides invaluable insights to improve the organizational setup of a company, thus fostering effective teamwork while also promoting corporate success.

The study of the brain, especially within the fields of social psychology, cognitive psychology, and neuroscience, provides invaluable insights that have real-world applications. Social neuroscience, an integration of the fields of social psychology and neuroscience, explores diverse topics that examine the biological foundations and environmental influences that can impact an individual's interpersonal relationships, both in personal and professional contexts.

As globalization and technological advances challenge conventional work cultures, working professionals face increasing cognitive demands, emotional pressures, and social complexities. To keep up with the ongoing developments in organizational structures, current professionals must remain mentally flexible and agile to navigate the hybrid work environments, manage cross-cultural teams, and respond to continuous digital simulations.

Workplaces today are not only shaped by social and emotional dynamics but increasingly by digital technologies that extend the reach of neuroscience into organizational life. Artificial intelligence (AI), neurofeedback, and virtual reality (VR) are emerging as tools to personalize learning, regulate stress, and enhance resilience at scale. For example, AI-driven coaching systems provide tailored performance insights, while VR simulations allow employees to practice coping strategies in immersive environments (Wang & Zhang, 2023; Cipresso, Riva, & Gaggioli, 2022). Similarly, neurofeedback offers workers real-time data on brain activity, training them to self-regulate under pressure (Thibault, Lifshitz, & Raz, 2020). These innovations signal a shift from one-size-fits-all approaches to deeply individualized methods for cultivating adaptability and well-being in the workplace.

However, these opportunities are tempered by profound ethical challenges. Algorithmic feedback platforms, while reducing some human biases, can overwhelm the prefrontal cortex with constant notifications, triggering stress rather than supporting growth (Buettner, 2021; Gaskell, 2022). As organizations embrace neuroscience-informed tools, they must balance innovation with responsibility, ensuring that practices remain transparent, consent-driven, and aligned with human dignity. Only then can the promise of neuroplasticity and AI be harnessed not just for efficiency, but for building workplaces that are humane, resilient, and forward-looking.(1- review)

As Norman Doidge eloquently puts it, “The brain is not hardwired; it is softwired by experience.” To tackle issues such as workplace stress, cognitive overload, and burnout, neuroplasticity offers a hopeful narrative suggesting that the right practices and interventions can train the brain to thrive in times of complexity.

Keeping these perspectives in mind, this chapter aims to shed light on the fascinating interplay between neuroplasticity and organizational effectiveness. This chapter specifically focuses on three major domains: (1) Emotional Intelligence and Social Neuroscience: The Brain’s Role in Communication and Collaboration, (2) The Power of Feedback: How the Brain Responds to Praise, Criticism, and Learning from Mistakes, and (3) Neuroplasticity, Stress, and Resilience in the Workplace.

This chapter also explores the connection between emotional intelligence, social neuroscience, and the dynamic interplay of communicative and collaborative interactions. By exploring the neural mechanisms underlying these aspects, this chapter aims to provide deep insights into how organizations can cultivate a psychologically sound and productive work culture. This understanding can thus help inform effective strategies and interventions that the corporation can implement to foster organizational growth.

THE NEUROSCIENTIFIC FRAMEWORK OF EMOTIONAL INTELLIGENCE

Emotional intelligence (EI), the ability to recognize, interpret, and regulate one's emotions, is a crucial factor that enables an individual to navigate the organization's social landscape. It encompasses an individual's awareness of their own and others' emotions in a social setting. High EI suggests that an individual may be better equipped to build strong and emotionally stable interpersonal relationships. Emotional intelligence also empowers individuals to resolve conflicts effectively and work collaboratively toward common goals. This ability also helps them foster a positive and supportive work culture, promoting trusting and open communication among colleagues.

Defined by Goleman (1995) as the ability to recognize, understand, and manage one's own emotions as well as the emotions of others, emotional intelligence has been extensively associated with leadership success, effective communication, and conflict resolution. The field of neuroscience sheds more insight into this concept, suggesting that EI is not merely a soft skill but also a biologically rooted capability of the human brain.

The neural basis of emotional intelligence lies in the interplay between the emotional centers of the brain and the executive functions of the prefrontal cortex. (Boyatzis et al., 2013). Social neuroscience, a subfield of psychology that investigates the underlying neural mechanisms that influence social behavior, provides invaluable insights into the brain systems that help improve and develop emotional intelligence. Scientific evidence in this domain of study identifies a few major brain regions, such as the amygdala, prefrontal cortex, and mirror neuron systems, that are crucial in the development of emotional intelligence.

The amygdala plays a central role in emotional reactivity in the case of stressful and emotionally stirring experiences, while the prefrontal cortex is responsible for decision-making and impulse control. The mirror neuron system works in tandem with the other two brain regions to facilitate empathy and social learning. These structures work together to foster El and are constantly shaped and refined through experiences and intentional practice.

People with higher EI are generally more successful in meeting environmental demands and pressures (Bar-On, 2006, p. 16). Scientific evidence highlights the effectiveness of structured training programs in strengthening emotional intelligence among employees. In an organizational context, this suggests that leaders and employees alike can develop and cultivate a higher EI by practicing intervention techniques such as reflection, mindfulness, and empathy.

One such instance is that mindfulness meditation has helped increase the gray matter density in the prefrontal cortex, thus decreasing activation in the amygdala and leading to better emotional regulation and stress management (Hölzel et al., 2011). Neuroscientific research highlights the importance of mindfulness practices in increasing gray matter density, specifically in the anterior cingulate cortex and prefrontal areas, which are associated with self-control and emotional modulation. Compassion training has also been shown to activate regions of the brain that support altruism and empathy (Lutz et al., 2008). This indicates that the organizations can efficiently employ these techniques to improve their overall productivity by building an emotionally intelligent workforce.

This chapter examines how neural systems, when equipped with emotional intelligence, can foster effective communication, trust, and cooperation among colleagues and thus contribute to successful leadership and overall productivity in the organization. For instance, research reports that leaders with high EI have a well-functioning prefrontal cortex, amygdala, and mirror neuron system. This suggests that they can inspire and motivate their teams, thus fostering a positive work environment for all employees. These findings also demonstrate that the right tools and a supportive environment can help individuals unconsciously rewire their neural pathways to favour adaptive and collaborative work environments.

Analyzing the Importance of Feedback and Its Neurological Significance

Feedback is another potent driver of neuroplasticity in a workplace setting. In addition to emotional intelligence, the way feedback is received and processed by people is a critical factor in determining the growth and performance of employees in a workplace. Feedback, even if it comes as praise, criticism, or simply information about performance, offers useful input that the brain uses to adjust behavior, improve skills, and learn from previous errors. The brain's response to feedback involves a complex network of neural structures, which include the reward circuitry, the anterior cingulate cortex, and the prefrontal cortex.

Positive feedback, such as praise or recognition, activates the reward circuitry, which includes the ventral striatum and dopamine pathways. This type of activation leads to the release of dopamine, a neurotransmitter that actively reinforces desired behaviour among individuals, thus evoking a sense of pleasure and motivation to continue that desired action.

Research indicates that employees who receive persistent positive reinforcements are highly motivated, self-assured, and dedicated to their work. Moreover, studies also suggest that businesses that implement structured recognition and reinforcement programs can enhance their employee engagement and productivity. This proves to be a prime example of leveraging reward learning through neurobiological stimulation. In contrast, harsh, ambiguous, or overly critical feedback can activate the brain's threat circuits, which include the amygdala and the insula. This, in turn, can result in increased defensiveness and reduce the motivation to learn and work.

Kulger and DeNisi (1996) emphasized that feedback is most effective when it focuses on the task rather than the employee. This viewpoint aligns with Carol Dweck's (2006) concept of the "growth mindset," which suggests that employees' abilities are not predetermined or fixed; rather, they develop through effort and learning. Furthermore, employees who perceive feedback as a growth opportunity, rather than as a judgment, portray greater improvements in performance and productivity. This demonstrates how a shift in communication can produce significant changes in brain function and also improve professional outcomes.

The workplace itself needs to be reconceived as a neurobiological environment. Every interaction, practice, and cultural norm can either foster or inhibit cognitive expansion. Autonomy, mastery, purpose, and relatedness are more than just motivational constructs: they are linked to neural activity in areas associated with reward processing, emotional regulation, and executive control (Rock, 2008). On the other hand, environments that elicit fear, uncertainty, or social exclusion activate stress responses that suppress higher-order thinking.

THE INTERPLAY BETWEEN STRESS, RESILIENCE, AND NEUROPLASTICITY

Stress, a state of worry or mental tension caused by a difficult situation, has evolved to become an inevitable and unavoidable part of modern professional life. Research indicates that while stress can be managed, chronic stress, when left unmanaged, poses the greatest threat to the overall health of an individual. Moderate to severe stress appears to increase the growth of several sectors of the amygdala, while effects in the hippocampus and prefrontal cortex tend to be opposite (Davidson & McEwen, 2012). Moreover, research conducted by McEwen suggests that prolonged stress exposure can lead to elevated cortisol levels, which in turn impair hippocampal functioning and might disrupt memory consolidation, thus suppressing neurogenesis (McEwen, 2000).

While stress is harmful, it can also be harnessed. This largely depends on the individual's capabilities to tackle and efficiently deal with stressful situations. However, employing coping mechanisms rooted in cognitive behavioural and mindfulness techniques can prove to be highly effective in managing stress. This point gives rise to the concept of "stress resilience," which describes the brain's ability to recover from challenges and sometimes even emerge stronger than before.

Resilience refers to a dynamic process encompassing positive adaptation within the context of significant adversity (Luthar, Cicchetti, & Becker, 2000). Other researchers suggest that flexible neural mechanisms in the prefrontal cortex and anterior cingulate cortex are mainly responsible for the development of resilience. This indicates that modern organisations can implement mindfulness programs to foster resilience and also encourage realistic goal-setting to prevent burnout among employees.

Many progressive organizations prioritize psychologically aligned practices in their training program designs, which can foster leadership development and team-building activities. Activities such as experiential methods, storytelling, and reflection can thus activate deep learning circuits. When combined with neuroscience education, behavioral coaching-based emotional intelligence workshops create lasting behavioral change. In addition, regular mindfulness practices incorporated into the workday — meditation rooms, breathing breaks, mindful leadership training — can reset employees' stress responses and extend their attention span.

The idea of lifelong learning takes on new significance in the light of neuroplasticity. Far from being an optional pursuit, ongoing learning becomes a biological necessity. This is also supported by the fact that acquiring new skills enhances the growth of dendritic spines and synaptic power, especially in the prefrontal cortex and hippocampus. Jaeggi et al. (2008) have demonstrated that training working memory can even result in increases in fluid intelligence—a cognitive ability previously believed to be fixed in adulthood.

In addition, adult learning is more and more multimodal. Online platforms, microlearning modules, virtual simulations, and game environments activate multiple sensory pathways, maximizing information retention and skill transfer. Peer coaching and mentorship programs enable social learning, activating mirror 8 and strengthening prosocial behavior. These innovations not only enhance knowledge acquisition but also foster neurological health by keeping the brain active and plastic.

In short, the professional brain is not a static instrument—it is an adaptive, responsive organ that is shaped by experience and intentional learning. Neuroplasticity provides a compelling model to explain how work environments, learning experiences, and social dynamics shape the brain. Emotional intelligence, once seen as an intangible trait, is now understood as a trainable ability rooted in neural networks. Stress, though possibly detrimental, can be converted to a growth-inducing stimulus with resilience training. Feedback, when empathetically and with intent delivered, is a neural signal for change. And above all, the workplace itself can be made a locus of mental renewal, innovation, and collaborative human advancement.

This chapter invites professionals, educators, and leaders to redefine success not just as performance results, but as neurological successes. Through the incorporation of neuroscience in management, training, and team-building, we can create workspaces where minds are not only productive but also satisfied, adaptable, and thriving.

EMOTIONAL INTELLIGENCE AND SOCIAL NEUROSCIENCE: THE BRAIN'S ROLE IN COMMUNICATION AND COLLABORATION

Today's work environments are intense ecosystems where occupational success depends equally on communication and subsequent accomplishments. While technical proficiency is of immense importance, emotional intelligence is crucial for identifying individuals who can stand out in terms of team performance and leadership.

Increasingly, the importance of emotional intelligence underscores a wider societal change among organizations. With businesses becoming more globalized, multicultural, and diverse, there is also a huge demand for leaders and employees who are emotionally agile, empathetic, and psychologically flexible, particularly in high-pressure situations.

Organisations have their companies and offices established all around the world, spanning multiple nations and time zones. The employees and teams are no longer limited to a regional geographic community but are dispersed across continents. Such a situation further reinforces the importance of collaboration and communication across cultural as well as emotional borders for the sustainability and effectiveness of the organization.

The impact of emotionally intelligent leaders is increasingly being supported by research and empirical facts. Several psychometric tests, psychological models, and theories suggest that leaders who score high in emotional intelligence create a more productive, inclusive, and engaged work environment. This evidence not only shows a rise in employee morale but also substantiates a marked increase in quantifiable outcomes such as greater customer satisfaction, stronger organizational commitment, and enhanced competency.

At its heart, the convergence of social neuroscience and emotional intelligence presents a new but appealing story of hope. It highlights the qualities that we most admire in colleagues, including compassion, curiosity, adaptability, and authenticity, while also suggesting that they are not only learnable but also biologically supported by evidence.

An in-depth understanding of how the brain functions, interprets, responds to, and thus shapes our social experiences can help us design organizations that not only understand the importance of emotion but thrive and prosper by leveraging its advantages. The interconnection between emotional intelligence and neuroplasticity, therefore, bridges the gap between science and humanity, enabling us to cultivate workplace cultures that are not only competent and analytical but also compassionate and cooperative. Backed by social neuroscience research, it is now evident that emotional intelligence and interpersonal skills aren't just "soft skills"; they are deeply embedded in the brain's architecture and can be developed and improved through deliberate effort.

Salovey and Mayer (1990) first introduced the term emotional intelligence, thus defining it as the ability to perceive, use, understand, and manage emotions in oneself and others. Goelman (1995) popularized the concept of EI and brought it to the forefront of public and organisational consciousness. He outlined five essential components of EI, which include: self-awareness, self-regulation, motivation, empathy, and social skills. He emphasized that these abilities were not innate nor fixed at birth. Instead, these components are competencies that are linked to specific neural pathways that are particularly responsible for emotional regulation, executive functioning, and social cognition.

Moreover, Goleman's research also highlighted that Emotional Intelligence is twice as important as IQ in predicting outstanding performance. This is a testament to the need for incorporating and practicing emotional intelligence in our everyday lives to improve both our professional and personal lives. Goleman's research led to the development of various intervention techniques and theoretical frameworks that could guide EI nourishment in workplace settings.

GOLEMAN'S EMOTIONAL COMPETENCE FRAMEWORK

Goleman developed the Emotional Competence Framework in 1998. It has been one of the most widely employed models for EI development, particularly in a work environment. This model breaks down the concept of EI into four broad categories: self-awareness, self-management, social awareness, and relationship management. The categories also include 12 competencies such as emotional self-awareness, accurate self-assessment, self-confidence, emotional self-control, transparency, adaptability, achievement orientation, initiative, empathy, organizational awareness, influence, and conflict management. The model highlights that each of these abilities can be traced back to specific neural pathways and thus can be trained and developed over a period of time. For instance, self-regulation, which includes prefrontal activity, also excites activity in the limbic system, particularly stimulating the amygdala.

One other example of this is emotional self-awareness. Emotional self-awareness, developed through deliberate and continuous practice, allows leaders to gain an awareness of the emotional condition of the employees and thereby adjust and refine their decisions so that they are in accordance with the provided social context.

One of the fundamental roles of self-awareness is precise self-evaluation that allows an individual to know exactly their strength and weaknesses. This real assessment of one's self-awareness is the foundation for establishing progressive leadership, thereby promoting practices that allow for improved communication and respect among the employees.

In the domain of self-management, Goleman highlights competencies such as emotional self-control and adaptability. They are very vital to assist professionals in staying calm and composed under pressure. Social awareness, on the other hand, has empathy and organizational awareness as crucial factors to decipher emotional signals and deal with group life.

Finally, relationship management, a field that requires the leaders to be extremely adept in, comprises influencing, conflict management, and collaboration. These factors not only empower the leaders to manage conflicts and make sound decisions, but also empower them to motivate and inspire their workers to maximize their competence and achieve their established goals.

This model, presented by Goleman, has been implemented in many organizations across the globe to improve their leadership domains. Johnson & Johnson is one such organization. This organization incorporated Goleman's model in its leadership development phase and found extremely positive outcomes. The leaders thus scored high on EI competencies, outperforming their peers in metrics and engagement scores (Boyatzis & Goleman, 2001). With tools like 360-degree feedback and individualized coaching, the company achieved a 20% boost in employee satisfaction and improved retention (Cherniss, 2010).

A meta-analysis by O'Boyle et al. (2011) involving 191 studies concluded that emotional intelligence, as defined by Goleman's model, accounted for 13.95% of variance in job performance across different companies. This study highlights the compatibility of emotional intelligence with intelligence, moreover suggesting that EI's increasing influence on an individual's personality traits exceeds the influence of intelligence.

This meta-analysis also highlights that individuals with high emotional intelligence engage in collaborative behaviour and maintain productive relationships with peers and supervisors. These outcomes thus emphasize and reinforce the integration of EI-based assessments and training programs in a professional environment, especially in job roles that require leadership, teamwork, and interpersonal conflict resolution.

Given the present working conditions, Goleman's Emotional Competence Framework is particularly useful in framing and monitoring leadership developmental programs, customer service training, as well as team-building activities. It is highly applicable to mid-level to senior leaders in a company who work towards encouraging and motivating their teams' progress and success. This framework also plays an important role among HR professionals, helping them redefine and alter current training and intervention techniques and focus on enhancing team dynamics through frontline managers. This framework also finds applications in the fields of education, healthcare administration, businesses, and nonprofit management, which require effective communication and interpersonal skills crucial for the success of the organization.

DUNBAR'S SOCIAL BRAIN HYPOTHESIS

The Social Brain Hypothesis, proposed by Robin Dunbar, provides an evolutionary perspective to better understand the evolutionary implications in the context of complex social interactions. This theory posits that the need to manage increasingly complex social interactions is the primary evolutionary drive for the expansion and development of the human neocortex.

This theory introduces the concept of "Dunbar's Number", which indicates that humans can comfortably maintain 150 stable social and interpersonal relationships. This number is the cognitive limit that is determined by the size of the human brain and its processing capacity. This number also represents the maximum group size with which humans can maintain meaningful contact while still ensuring mutual trust and obligation.

Dunbar's hypothesis is also supported by various comparative studies across primate species. This theory is employed in organization science as well and is used to inform decisions regarding team sizes, communication structures, and social cohesion strategies within companies and businesses. The theory underscores how neural architecture is shaped, formed, and wired to encourage collaboration, bonding, and form cohesive workforces, particularly in companies that require collaboration and communication between employees and customers.

Neuroimaging research by Lewis et al. (2011) showed a positive correlation between the volume of the ventromedial prefrontal cortex—a central part of the neocortex—and the ability of individuals to manage bigger social networks. The research, published in NeuroImage, used structural MRI scans of 40 participants and discovered that individuals with larger social groups showed higher gray matter density in this area, indicating that neuroanatomy plays a significant role in helping complex social cognition. This offers strong empirical evidence in favor of Dunbar's hypothesis that cognitive ability, especially that of the neocortex, is a determinant of the number of significant social relationships an individual can maintain.

In neuroscience, the prefrontal cortex, especially the ventromedial and dorsolateral areas, facilitates complex social cognition, including empathy for others' viewpoints, reading social cues, and regulating interpersonal dynamics. When this area is overworked, workers might develop burnout or emotional exhaustion, especially in jobs involving constant social interaction, like customer service or sales.

To overcome these limitations, modern-day companies are using network analysis software to chart out loose social relationships and maximize collaboration effectiveness. Boston Consulting Group (BCG), for instance, applies such tools to refine team configurations in ways that preserve social bandwidth but achieve high performance. They discovered that small, stable teams did better than large, less cohesive ones, confirming the applicability of the hypothesis.

In addition, digital evolution has brought in new factors. Presently, virtual communication channels raise connection volume but can decrease emotional quality. To overcome any issues that might arise from decreased in-person interactions, companies are now educating their employees in "digital empathy" to maintain meaningful interaction even in remote or hybrid setups.

In a 2017 internal study of corporate teams at Salesforce, managers who cultivated stronger informal networks, echoed Dunbar's principle of social cognitive limits, and demonstrated superior performance in producing innovative output, and portrayed an increased engagement in cross-functional collaboration. The managers not only facilitated smoother interdepartmental communication but also secured higher scores in emotional intelligence assessments conducted by the company's leadership and wellness programs. As one Salesforce team lead reported, "I realized that building authentic relationships across teams wasn't just good for morale—it became a catalyst for innovation."

The Social Brain Hypothesis, given by Dunbar, thus helps us understand how companies can increase cooperation and improve team structure. This is particularly relevant to professionals who are in social-facing roles, for example, sales, public relations, and human resources. This principle can be used by the HR departments to set optimal team sizes that promote cohesion; while learning and development professionals can tailor social learning strategies concerning the cognitive bandwidth of team members. By understanding the neurological constraints on relationship management, more effective interpersonal networks and collaborative environments aligned with human cognitive architecture can be created by very large organizations, thus fostering employee welfare and increased productivity.

Businesses like Microsoft and Google have shown that deliberate development of emotionally intelligent leadership has a direct effect on employee involvement and cooperation. For example, Nadella's cultural transformation at Microsoft highlighted growth mindset and empathy, and that produced quantifiable improvements in innovation and morale (Dweck, 2006; Edmondson, 2019). Google's "Search Inside Yourself" initiative similarly folded mindfulness training and emotional intelligence development into leadership education, highlighting the nexus between neuroscience and organizational effectiveness (Tan, 2012).

Technology also offers another double-edged element in this discussion. On the positive side, computer programs and AI-driven platforms now facilitate customized EI training, immediate feedback, and even virtual reality environments that allow workers to rehearse communication skills under duress (Cipresso, Riva, & Gaggioli, 2022; Wang & Zhang, 2023). Conversely, relentless connectivity and algorithmic surveillance can impose new burdens on the prefrontal systems of the brain, overwhelming cognitive capacity and inducing stress responses instead of learning enhancement (Buettner, 2021; Gaskell, 2022). Therefore, the role of technology needs to be understood as both a stimulant for development and a possible test for neuroplasticity in contemporary workplaces.

However, scholars have raised concerns about the implications of monitoring brain activity or emotional states at work, warning that without clear safeguards such practices may infringe on privacy and autonomy (Ienca & Andorno, 2020). These discussions remind us that while neuroscience provides powerful frameworks for enhancing communication and collaboration, its use must be grounded in fairness, transparency, and respect for human dignity. By incorporating case studies, cultural perspectives, technological opportunities and risks, and ethical considerations, the scope of emotional intelligence and social neuroscience in organizations becomes not only broader but also more balanced and practically meaningful. (2-review)

ROCK'S SCARF MODEL

Integrative Neuroscientist Evian Gordon says that the "minimize danger and maximize reward" principle is an overarching and organizing principle of the brain (Gordon, 2000). This concept aligns with the approach-avoid principle of the brain that has existed for several decades. This principle highlights the response of the brain when an individual encounters a 'good' or 'bad' stimulus. If the stimulus is associated with positive emotions or rewards, then it is likely to elicit an 'approach' response in the brain, where the individual engages with the stimulus. However, if the stimulus is associated with negative emotions or punishments, it is most likely to cause an 'avoid' response in the person, causing the person to disengage and move away from the stimulus.

This approach-avoid response is a survival strategy that can help individuals decode a stimulus quickly and make quick decisions. This concept has been elaborated, modified, and improved in the SCARF model given by David Rock in 2008.

The SCARF model, developed by David Rock, is a neuroscience-based framework that helps explain how people respond to social interactions in the workplace.

It stands for Status, Certainty, Autonomy, Relatedness, and Fairness—five key domains that influence our brain's threat and reward systems, much like physical safety or danger would.

This model is deeply rooted in the belief that the human brain is highly sensitive and instinctive in responding to social threats and rewards. Based on neuroscientific findings and evidence, Rock proposed that the human brain processes social pain and pleasure just like it processes physical pain and rewards. Each of the key domains of the SCARF model taps into the mechanism of the threat-reward systems, which in turn have a direct influence on motivation, engagement, and collaboration in a professional setting.

The five domains of the SCARF model altogether explain how everyday interactions can either activate the brain's threat or reward response and thus elicit either disengagement or cooperation. These domains are not merely theoretical constructs but are grounded in the neural processes that manage emotional and behavioural reactions, specifically in social contexts. These domains can be visualized and interpreted as levers that can be adjusted to reduce perceived threats and thus enhance motivation and engagement in work environments.

Status

Michael Marmot, in his book The Status Syndrome: How Social Standing Affects Our Health and Longevity, strongly implied that status is the most significant contributor and determinant of human longevity and health.

Status is one's relative importance to others. When employees feel valued and recognized, their brain elicits a reward or 'approach' response, thus increasing their intrinsic drive, leading to increased collaboration, engagement, and productivity. Conversely, if an employee perceives demotion or criticism, it can trigger a threat or 'avoid' response that leads to increased disengagement and defensiveness.

For instance, even though getting fewer likes on a social media post than what is usual may seem trivial, it can trigger a threat response in the brain. The human brain interprets that situation as a reduction in social standing comparable to physical pain, as the dorsal anterior cingulate cortex (dACC), which processes social pain, lights up. These circuits do activate even with a small drop in perceived social validation. Activation induces a degree of anxiety and self-doubt until suitable reassurance and comparison resolve the issue.

Functional MRI studies show that the perceived reduction in status activates the same areas in the brain that are activated by physical pain. Studies indicate that the dorsal anterior cingulate cortex (dACC), a region implicated in the experience of physical pain (Eisenberger & Lieberman, 2004), is activated by a threat response. In contrast, the striatum, located in the basal ganglia, is a region associated with reward and is stimulated when there is a perceived increase in status. Therefore, recognizing an employee's contributions in meetings and assigning leadership roles on a rotating basis can go a long way in activating these reward responses and thus build engagement and productivity.

Certainty

Certainty, the second domain of the SCARF model, is the brain's need to predict the future. The brain is a pattern-recognition machine that is constantly trying to predict the near future (Rock, 2008). The brain craves certainty; it loves to accurately know what could happen at every single moment, thus making prediction possible. Without certainty and prediction, the brain must employ the prefrontal cortex to understand and predict moment-to-moment experiences, which requires using excessive energy resources and time.

The lack of accurate information or a sudden occurrence of unanticipated events creates a sense of anxiety as the brain struggles to cope with the ambiguity that arises in the situation. This ambiguity or uncertainty generates an 'error' response in the orbital frontal cortex (OFC) (Rock, 2008). This increases the cognitive load on the prefrontal cortex (PFC), triggering the stress response in the individual, causing the person to reallocate their energy and cognitive resources from their goal to the 'error'.

When an employee doesn't provide all the information regarding a particular project, it can, in turn, trigger the activation of the OFC as it struggles to fill in the gaps with possible predictions. This is equivalent to having a repeated notification on your mobile that reminds you, saying, "There's an issue with your account. Tap to learn more." The brain's prefrontal cortex immediately kicks in, trying to predict the severity of the issue. Your attention remains hijacked until the uncertainty is resolved. This ambiguity triggers the OFC and the amygdala, disrupting the clarity regarding the situation and thus increasing stress until certainty is established.

Therefore, providing clear goals, roadmaps, and regular updates can reduce cognitive load and help employees remain calm and focused. This is why transparent leadership communication is so vital in times of organizational change. Transparency can not only help combat the brain's cumbersome engagement to predict every second of one's life, but can also reduce the reliance on threat-based processing from the amygdala.

Autonomy

The third domain of the SCARF model talks about the importance of autonomy. Autonomy is the perception of exerting control over one's environment; a sensation of having choices (Rock, 2008). This indicates that when people have choices, even insignificant or minute ones, they tend to feel more empowered and less threatened. Research indicates that this ability to make independent decisions without any others' influence reduces stress and enhances an individual's intrinsic motivation and sense of power.

Studies by Donny et al. (2006) suggested that the inescapable and uncontrollable stress can be highly destructive, whereas the same stress interpreted as escapable is significantly less destructive. (Donny et al, 2006). This highlights that the degree of control that a person can exert over a stress factor determines the influence that the stressor exerts over the person's functioning.

The experience of choice and control over actions activates the ventromedial prefrontal cortex (vmPFC), which regulates emotional responses and helps maintain resilience. Lack of autonomy leads to heightened activity in the amygdala, increasing stress.

Consider an example: When your phone's operating system suddenly changes without any warning, icons rearrange, and the settings alter, you feel a loss of immediate control. The amygdala activates itself when it perceives this lack of power, thus triggering stress. On the contrary, the ventromedial prefrontal cortex (vmPFC) is activated when users are able to customize settings or choose dark mode as opposed to light mode. This promotes certain feelings of control, ownership, and calm in the individual, equipping them with an enhanced sense of autonomy.

Google's "20% time" policy is a well-known example of promoting autonomy in the workplace. This policy permits employees to spend a portion of their time working on self-directed and self-initiated projects that are unrelated to their job roles assigned by the company. This sense of control activates the ventromedial prefrontal cortex (vmPC), which plays a significant role in emotional regulation and stress resilience. This policy aligns with the view of autonomy presented by the SCARF model that emphasizes that autonomous engagement in certain activities leads to an increase in intrinsic motivation, which in turn heightens innovation, focus, and well-being.

Relatedness

Relatedness, the fourth domain of the SCARF model, indicates our innate drive to feel connected and safe with others. Relatedness is deeply rooted in the concepts of in-group and out-group dynamics that are embedded in our neural structure. Humans have an innate need to belong, be it at home, among members of their community, or in a professional setting. This need for affiliation and belongingness is supported by the activation of the ventral striatum and oxytocin-mediated bonding circuits when individuals experience social inclusion. However, social exclusion can in turn activate both the dACC and the anterior insula.

For instance, being removed from a group chat without any explanation can make one react with more visceral emotions than usually expected when one leaves a group chat on their own. The brain does activate both dACC and anterior insula as it processes social exclusion similarly to physical pain. Conversely, the ventral striatum, as well as oxytocin release, are initiated by a friendly group invitation or an emoji-filled birthday message, reinforcing social bonds and feelings of trust.

Humans have an innate need to belong, and this is supported by activation in the ventral striatum and oxytocin-mediated bonding circuits when individuals experience social inclusion. Social rejection, on the other hand, activates the dACC and anterior insula. Team-building initiatives, peer mentoring, and inclusive communication styles strengthen relatedness and reduce the neurobiological markers of social threat.

Fairness

Fairness, the fifth and final domain of the SCARF model, is the feeling of justice in social and work relationships and processes. It is a highly powerful regulator of human behavior, controlling both cooperation and trust, and most importantly, engagement in general. Fairness is an extremely effective activator of our brain's intrinsic evaluation system, whether we perceive an environment as psychologically safe or socially risky.

The sense of injustice engages the anterior insular cortex, a brain area linked to moral disgust, pain, and visceral emotional responses (Sanfey et al., 2003). It is not a metaphorical response—unfairness elicits responses that are akin to those that are caused by bodily injury because it engages the dorsal anterior cingulate cortex (dACC) and the amygdala, both of which are involved in detecting threats and processing emotional distress (Tabibnia & Lieberman, 2007).

Furthermore, the hypothalamic-pituitary-adrenal (HPA) axis, which governs the body's response to stress, gets triggered when people face injustices. Repeatedly unfair situations can raise cortisol levels and undermine long-term cognitive and emotional functioning, resulting in lowered motivation and burnout.

Conversely, fair treatment activates the ventral striatum, a part of the brain linked to reward and positive reinforcement. This is the major reason why fair and honest management practices not only reduce stress but also foster loyalty, increase dopamine-mediated motivation, and reinforce cooperative behaviors.

Consider a situation: Suppose the telephone company surprises new customers by granting them a premium plan, saving it for them and not the regular customers. The feeling of injustice activates the insular cortex and the hypothalamic-pituitary-adrenal (HPA) axis, a system that modulates stress hormones such as cortisol.

Conversely, observing clear prices for everyone triggers reward systems such as the striatum, affirming trust within the system and minimizing emotional discomfort.

Apart from personal experience, fairness or unfairness can be conveyed among groups by the mirror neuron system so that humans can feel the emotional state of another vicariously. When a colleague is unfairly treated, witnesses can feel vicarious threat, causing a reduction in trust in leadership and thus reducing the morale in the group. This illustrates that consistent, equitable leadership behavior is not only crucial for individual well-being but also for healthy group processes.

A perception of fairness in the workplace is more than money. It's the allocation of promotions, the manner of providing constructive criticism, and the extent to which policies are applied consistently. Employees are constantly interpreting manager and peer signals in an attempt to determine if they're being treated fairly. When people perceive they're being treated unfairly, such as when they see favoritism or a process that lacks transparency, it can result in disengagement, passive defiance, or even outright conflict.

On the other hand, when organizations commit to open performance measures and formal feedback systems, employees feel very high job satisfaction and less interpersonal conflict. For example, Buffer, a tech company, openly publishes salary information and the mathematical formulas that are used to determine compensation. This radical openness eliminates uncertainty and generates a profound sense of justice among employees.

Cultural variation is also crucial in influencing the way emotional intelligence and social cues are understood at the workplace. For instance, Gunkel et al. (2021) established that workers from collectivist cultures tend to focus on relatedness and harmony, whereas those from individualist cultures value autonomy and straightforward communication. This corresponds with SCARF domains, in which relatedness and status potentially weigh greater in collectivist cultures, while autonomy and certainty can predominate in individualist cultures. By recognizing such cultural aspects, companies can fit feedback models, recognition schemes, and collaboration frameworks to be inclusive and impactful in multisectoral workforces.(2-review)

In total, the SCARF model presents a powerful integration of neuroscience and leadership practice, and how social threats and rewards in the social domain affect workplace behavior and emotional well-being. By rendering the neural foundation of Status, Certainty, Autonomy, Relatedness, and Fairness transparent, SCARF gives leaders and organizations practical strategies to reduce stress, build psychological safety, and enhance emotional intelligence. As part of the larger mind reprogramming theme, the SCARF model informs us that the workplace is not just an environment for cognition but a highly neurobiological and emotional system. From that understanding comes the ability to construct resilient, empathetic, and high-achieving cultures that are in harmony with the social architecture of the brain.

All three theories discussed under this topic—Goleman's Emotional Competence Framework, Dunbar's Social Brain Hypothesis, and Rock's SCARF Model—are collectively a solid foundation to understand how emotional intelligence is not just a set of skills but an intensely embedded neurobiological process. These frameworks show us that our social architecture of the brain is strongly related to our performance, well-being, and partnering within commercial settings.

Goleman's framework encapsulates the reality that self-awareness, self-regulation, motivation, empathy, and social skills are all acquired skills that are rooted in some brain processes, such as those controlled by the prefrontal cortex, amygdala, and mirror neuron systems. This makes emotional intelligence not a trait but an acquired dynamic capacity associated with neuroplasticity and moment-by-moment control of emotions.

Dunbar's Social Brain Hypothesis extends this by demonstrating how the pressures of social living shaped cognitive evolution in the human lineage. It stresses the need for coordination of diverse social relationships—a need in complex working environments—combined with the localization of brain areas like the neocortex as being paramount to the coordination and comprehension of social hierarchies and cooperation.

The SCARF model takes this research one step further by incorporating social neuroscience and dividing it into actionable areas—Status, Certainty, Autonomy, Relatedness, and Fairness. Each area maps to a reward or threat response in the brain, and the model offers practical solutions for enhancing engagement, reducing stress, and creating psychological safety in the workplace. Rock's SCARF model directly indicates that every social interaction can engage neural circuits that build or destroy collaboration and performance.

Collectively, these theories come together to bear witness to a fundamental truth: emotional intelligence and effective social functioning are not "soft skills," but are biologically grounded competencies required to impart leadership, resilience, and organizational effectiveness into individuals. By harnessing these findings, professionals can rewire their brains to become more adaptable, trusting, and productive, transforming workplaces into ecosystems where the brain's need for connection, control, and meaning is respected and optimized.

THE POWER OF FEEDBACK: HOW THE BRAIN RESPONDS TO PRAISE, CRITICISM, AND LEARNING FROM MISTAKES

"We do not learn from experience. We learn from reflecting on experience." — John Dewey.

Feedback is not merely an instructional tool but a biological and emotional experience. Every word of compliment or critique we receive triggers a cascade of neural responses and activity. Whether it's a boss giving a performance evaluation, a peer commenting on a presentation, or an internal voice judging one's work, feedback engages intricate brain networks evaluating social safety, self-esteem, and learning capacity.

Decades of social neuroscience and cognitive psychology research in the past few decades have increasingly established that feedback has a dual nature. It can be an effective motivator that promotes learning and development, or a psychological landmine that undermines confidence and performance. This dual nature is echoed in the research of Kluger and DeNisi (1996), who, in their classic Feedback Intervention Theory, contend that "more than one-third of feedback interventions actually decreased performance." Their meticulous meta-analysis had examined more than 600 effect sizes and established that feedback's success depended not only on its content but also on its context and delivery.

"Criticism, like rain, should be gentle enough to nourish a man's growth without destroying his roots.". This quote by Frank A. Clark beautifully sums up the importance of criticism and feedback, specifically highlighting their role in motivating or derailing a person's performance. Criticism or feedback, when worded to highlight the error of the person in a motivating way, can improve the person's performance. However, receiving critical feedback from senior professionals may prove to be demotivating to employees, thus reducing their morale and performance.

This section of the chapter delves into the neurobiological processes by which feedback impacts learning, emotion, and resilience. It describes how praise engages the reward system of the brain, whereas criticism, particularly when badly worded, can activate areas associated with social pain and threat. Importantly, this section also explains how these responses are not fixed, but instead can be influenced through emotionally intelligent delivery, clear goal orientation, and psychological safety.

The emerging field of social cognitive neuroscience has helped validate these intuitions by showing that feedback physically changes the brain. fMRI studies have indicated that negative feedback activates the amygdala and anterior insula, the same regions involved in detecting threats and experiencing pain (Eisenberger & Lieberman, 2004). Conversely, positive feedback and compliments activate the ventral striatum and prefrontal cortex, releasing dopamine and enhancing memory, motivation, and behavioral flexibility (Lieberman & Eisenberger, 2009).

But what determines whether feedback will be perceived as a reward or a threat? Much of this will depend on framing, delivery, and context of the feedback. Boyatzis and McKee (2005) argue that "People learn best in a state of hope, not fear." Stemming from this context, in a business setting, emotionally intelligent leadership, where feedback is framed in a way that aligns with someone's goals and potential, rather than as a judgment, can induce the neurological conditions for neuroplasticity and growth.

Feedback Intervention Theory (FIT)

Feedback, though appearing simple, is a multifaceted interaction between information, emotion, and neurobiology. At the heart of how human behavior is affected by feedback lies the Feedback Intervention Theory (FIT), constructed by Kluger and DeNisi in 1996. Their ground-breaking meta-analysis combined findings from more than 130 studies and 600 effect sizes to explore what makes feedback work. Their conclusion was both shocking and illuminating: over a third of feedback interventions had the paradoxical effect of making people worse. This finding fundamentally challenges the underlying presumption that all feedback, by its purpose, improves performance.

Kluger and DeNisi suggested that feedback works not because of its content, but because of where it focuses the recipient's attention. Feedback interventions, they argued, operated on three hierarchical levels: task learning, task motivation (or self-regulation), and meta-self (identity or self-worth).

They found that feedback was most likely to improve performance when it refocused an individual's attention on the task, on precise actions, strategies, or outcomes. But when feedback turns the individual's attention inward to his sense of self, particularly when it suggests personal weakness, inadequacy, or incompetence, defensive responses are activated, motivation is diminished, and learning is compromised.

In their own words, "Feedback interventions are less effective, and sometimes harmful, when they shift attention away from the task to the self" (Kluger & DeNisi, 1996, p. 254). This statement describes the fine line between feedback that fosters growth and feedback that threatens psychological safety.

The cognitive science behind this theory is now substantiated by research from neuroscience. When criticism is directed at personal identity—e.g., labelling a person as "lazy" or "incompetent"—it activates the amygdala, the brain's threat response system. This releases a chain of physiological responses, including the release of cortisol, a stress hormone that disables the prefrontal cortex (PFC), which is in charge of logical reasoning, planning, and emotional control. High levels of cortisol under conditions of psychological stress can severely impair executive function, resulting in reactive rather than reflective behavior (Arnsten, 2009).

In contrast, feedback that focuses on behavior, strategy, or results tends to activate regions such as the dorsolateral prefrontal cortex and anterior cingulate cortex (ACC), which are responsible for attention regulation, error detection, and adaptive learning. As Boyatzis et al. (2012) explain, such brain activation forms the neurological basis for goal-directed action and self-regulation, making the learner a better reflector, adjuster, and developer.

However, neuroscience shows that the emotional tone of feedback is just as crucial as its cognitive target. If feedback—no matter whether it's on the task—is judgmental, dismissive, or negative in tone, it might, even when task-focused, still get processed as a social threat, particularly if it threatens a person's sense of status or fairness. This understanding is consistent with David Rock's SCARF model (2008), which names status, certainty, autonomy, relatedness, and fairness as areas that influence the brain's threat-reward response. According to Rock, "Once the brain senses a social threat, cognitive bandwidth is curtailed. The person's capacity for learning from feedback is impaired."

This knowledge has far-reaching implications for leadership, education, and coaching. Feedback is not objective—it passes through filters of emotion and neurology that amplify or dampen its intended effect. Consequently, framing, timing, and tone of feedback delivery become driving variables in determining effects.

A compelling use of FIT principles in the real world can be observed in Adobe's Check-In system. In 2012, Adobe ditched its old yearly performance reviews, which were perceived as being stressful, formal, and unsuccessful. Rather, the company implemented regular, casual feedback discussions among managers and employees. These check-ins centered on goal advancement, future growth, and open discussion, not scoring or judging. Adobe's ex-Chief Human Resources Officer, Donna Morris, said: "We wanted to shift from managing performance to building people." Post-implementation statistics reported a 30% decrease in voluntary turnover and enhanced employee morale (Adobe HR, 2018).

Adobe's method embodied FIT in practice. By keeping the feedback task- and development-focused, and not using evaluative language regarding personality or traits, the company fostered a low-threat, high-trust environment that supported growth. In addition, by making the check-ins ongoing, the feedback remained timely, relevant, and emotionally manageable.

Another example is McKinsey & Company, where feedback is not a once-a-year ritual but an ongoing conversation. Following each client engagement, consultants are asked to offer upward, downward, and peer feedback. Instead of offering general praise or complaint, McKinsey instructs team members to employ behavioral anchors—definite, descriptive words that describe actions in relation to results. This keeps focus on the job, not on the individual's intrinsic personality traits. Feedback may include: "During the last client call, your summarizing clarified a problem that was confusing.". One is to back off a bit so the clients have time to take in your points." These types of remarks strike a balance between acknowledgment and development focus, maintaining learner engagement without eliciting threat.

To apply FIT-guided feedback in any organization, there are a few vital principles. First, observe behavior, not one's assumptions regarding character. For instance, rather than "You're careless," use "This report was deficient in key data points—let's go over how we can prevent that from happening again." Second, offer actionable recommendations that define next steps instead of simply telling the recipient what went wrong. Third, promote self-reflection by posing open-ended questions like "What was the most difficult part of this activity?" which keeps the recipient mentally active and emotionally centered.

Lastly, be aware of the setting—giving constructive feedback in private and with compassion can dramatically lower defensiveness and create room for conversation. FIT also endorses the idea of "feedforward"—a tactic made famous by executive coach Marshall Goldsmith (2007). Rather than dwelling on past error, feedforward concentrates on improving in the future, which complements the reward-based learning systems of the brain.

A manager might say, "In your next presentation, try leading with a story to create more engagement," rather than "Last time, your introduction was boring." This subtle shift in focus from past failure to future opportunity changes the emotional tone and neurological response to feedback.

Finally, FIT teaches us that feedback isn't about performance—it's about managing attention. As a leader, educator, or colleague, our language has the power to shift someone's attention toward learning or away from it. The science is clear: when feedback honors emotional context, attends to behavior, and is given with developmental purpose, it is not merely effective—it is transformative.

Dweck's Mindset Theory

Perhaps the most revolutionary concept in contemporary psychology is that the way we see our intelligence and capability has a profound impact on the way we receive feedback and learn from failure. This is the central premise of Mindset Theory, formulated by Stanford University professor of psychology Carol Dweck. Based on it, the theory establishes two fundamental belief systems: fixed mindset and growth mindset.

"In a fixed mindset, mistakes are threatening; in a growth mindset, they are opportunities." This quote by Carol Dweck accurately sums up his ideology regarding the concept of feedback and its importance in two belief systems: the fixed mindset and the growth mindset.

Individuals with a fixed mindset perceive their talents as unchangeable—that intelligence, talent, or creativity are fixed traits that cannot be altered substantially. For this reason, they see feedback, particularly criticism, as a direct threat to their ego. Errors are seen not as a chance to learn but as a reflection of inadequacy. On the other hand, those with a growth mindset think that their talents can be developed through effort, learning, and persistence. To them, feedback is a guide to growth, and errors are a natural and even essential part of the learning process.

Dweck (2006) clarifies that: "Believing that your qualities are carved in stone—the fixed mindset—creates an urgency to prove yourself over and over. In the growth mindset, failure can be a painful experience. But it doesn't define you. It's a problem to be faced, dealt with, and learned from."

Neuroscience confirms this difference. A review of Moser et al. (2011) employed electroencephalography (EEG) to measure the differences in the way individuals with various mindsets dealt with errors. The researchers discovered that individuals who had a growth mindset had greater error-related positivity (Pe) signals—a brainwave pattern related to conscious awareness of errors and attention towards mistakes. This suggests that individuals with a growth mindset were more inclined to remain interested and learn from their mistakes, instead of evading or disavowing them. By contrast, individuals with a fixed mindset showed less neural activity to errors, indicating withdrawal of the psychological self from the learning process.

In the same manner, functional MRI (fMRI) studies identify that when participants are led to assume intelligence can be changed, there is higher activation in the dorsolateral prefrontal cortex (dlPFC) and anterior cingulate cortex (ACC) when receiving feedback. These areas of the brain participate in cognitive control, conflict monitoring, and performance modification—ingredients for feeding back into future behavior (Mangels et al., 2006).

These findings underscore a critical truth: our beliefs about learning change how our brains respond to feedback. If feedback is perceived as a judgment on immutable traits, the brain goes into defense mode, activating the amygdala and reducing cognitive resources for learning. But when feedback is interpreted as guidance toward growth, the brain becomes more plastic, more open, and more capable of integrating that input into future action.

This neuroscience-based understanding of mindset has taken hold in educational and organizational development. Schools, businesses, and athletic teams have started to move from performance-oriented cultures toward learning-oriented ones. In the workplace, this translates to reframing mistakes and criticism not as accusations but as useful data points on a journey of long-term growth.

A powerful example of the application of mindset theory is provided by Microsoft's organizational turnaround during CEO Satya Nadella's leadership. As Nadella began his tenure in 2014, he presented the idea of transitioning from a "know-it-all" to a "learn-it-all" culture. Fundamentally, this change was an effort to establish a company-wide growth mindset. Workers were challenged to try things out, to take risks, and to view failure as a source of learning, not embarrassment.

As Nadella puts it himself: "Empathy is at the center of everything we do. If you want to have an innovative company, you have to have a culture that allows people to take risks, and that means allowing them to fail."

The organization redesigned its feedback and performance systems to facilitate this cultural change. Traditional annual reviews were supplemented with more regular coaching conversations, and managers were educated to provide feedback that focused on progress, strategy, and learning rather than inherent performance. Within two years, Microsoft documented profound gains in employee engagement, cross-team collaboration, and market innovation, confirming that mindset changes, facilitated by feedback, can change brain and behavior at scale.

Mindset theory has also influenced educational feedback practices. In classrooms and corporate learning settings, trainers are encouraged to avoid person-based praise (e.g., "You're so smart") and instead offer process-based praise (e.g., "Your approach to solving that was creative and persistent"). This subtle but crucial difference promotes resilience and encourages learners to associate effort with outcomes, which Dweck calls "the power of yet". Instead of saying, "I can't do this," a growth mindset student would say, "I can't do this—yet."

The implications for delivering feedback are huge. Emotionally intelligent leaders and teachers can create growth mindsets by framing feedback in a way that helps motivate autonomy, focuses on learning goals rather than performance goals, and recognizes effort without over-pathologizing failure. For instance, a manager could say: "I can see how much thought went into that project. Something we can improve next time is the simplicity of your data visuals—let's brainstorm some new ways to do it next time?"

This type of feedback retains the individual's feeling of competence and autonomy and opens them up to a process of constructive iteration.

It's important to point out, though, that embracing a growth mindset culture is not necessarily about sugarcoating errors or shying away from criticism. It's about using criticism in a way that leads to further engagement, rather than avoidance. As Dweck explains, a growth mindset isn't about being "positive" all the time—it's about being curious, open, and persistent, even when things get tough.

Organisations such as IDEO, the global design company, put this philosophy into practice with their "fail often, fail early" credo. Designers are invited to post work-in-progress, get quick feedback, and look at iterations as a natural part of the design process. This feedback culture fosters psychological safety, a key building block of learning and experimentation (Edmondson, 2019).

In short, Mindset Theory redefines how we view the power of feedback. Instead of as a performance evaluation, we realize that feedback is a neural signal for adjustment. With skillfully worded language, emotionally smart delivery, and an eye for process and progress, feedback can be the pathway from error to expertise. For individuals and organizations alike, learning a growth mindset makes room in the mind and the brain for reflection, resilience, and long-term performance.

Self-Determination Theory (SDT)

"To be motivated is to act with a sense of volition and full endorsement of one's actions."

— Ryan & Deci (2000)

Feedback may always be viewed as an apparatus of correction or assessment, but the motivational value that it carries is the key to making it a driver of development or disengagement. Self-Determination Theory (SDT), founded by psychologists Edward Deci and Richard Ryan in the 1980s, presents a strong framework for understanding how feedback combines with our intrinsic psychological needs. It is specified by SDT that for people to be motivated in a sustainable and autonomous manner, three core psychological needs have to be met: autonomy (the desire to experience control over one's behaviors), competence (the desire to experience efficiency and efficacy), and relatedness (the desire to experience a sense of belonging).

Feedback, therefore, is not only informative—it's either supportive or damaging of these fundamental needs. When feedback encourages autonomy, acknowledges competence, and enhances social contact, it supports intrinsic motivation and increases involvement. When it is controlling, excessively critical, or alienating, it creates amotivational states, disengagement, and even defiance.

In their landmark review, Ryan and Deci (2000) state: "Social environments that support psychological needs facilitate internalization, autonomous regulation, and effective performance." This implies that leaders, teachers, and peers who give feedback in a manner that promotes autonomy, competence, and relatedness can significantly raise the chances that the feedback will be utilized, not because it is obligatory, but because of intrinsic motivation.

Autonomy: The Foundation of Engagement

Autonomy is the feeling of volition and ownership of one's actions. Imposed feedback, delivered in a top-down, directive, or authoritarian style, compromises autonomy and engages the amygdala of the brain, which triggers a state of defensiveness. Feedback, on the other hand, that provides choice, promotes reflection, or solicits partnership enhances the ventromedial prefrontal cortex (vmPFC), a brain area responsible for self-regulation, meaning-making, and intrinsic motivation.

A meta-analysis conducted by Deci, Koestner, and Ryan (1999) concluded that controlling feedback will undermine intrinsic motivation and autonomy-supportive feedback will strengthen it. Their autonomy-supportive feedback would be something like:

"Here's one possible way to do it—what's your idea?" or "Let's look at some possibilities together and figure out what works for you."

These messages acknowledge the recipient's agency and communicate respect, bringing the brain into a reward state, not a threat state.

Competence: Confirming Growth and Capacity

Competence involves feeling competent and capable at what one is doing. Feedback that tells one only what is wrong or deficient without providing constructive guidance depletes this need. It communicates that one is in some way failing without offering a way to succeed. Neuroscientific research has indicated that perceived incompetence is linked with lower ventral striatal activation, the reward center of the brain, and higher amygdala activity, which is an indication of discouragement and low self-efficacy (Peters et al., 2017).

Feedback identifying areas for improvement but confirming strengths, on the other hand, enhances dopamine activity and facilitates prefrontal learning circuits. This might be expressed as:

"Your thoughts during that meeting were innovative—now let's smooth out how they're delivered to suit the client's requirements," or "I can see diligent effort here; with a little more precision of structure, this could truly excel."

This kind of feedback preserves the individual's feeling of competence while giving directions towards mastery, something so important to both motivation and learning.

Relatedness: The Role of Social Connection in Feedback

Relatedness is the desire to be cared about, respected, and bonded with others. Impersonal, cold feedback or feedback given in solitude can induce social pain, which activates the dorsal anterior cingulate cortex (dACC) area—the same area used for the experience of bodily pain (Eisenberger & Lieberman, 2004). Conversely, warm, personalized, and relational feedback activates the release of oxytocin and the social brain network, strengthening trust and cooperation.

This principle is particularly applicable to leadership. A manager who gives constructive feedback in an affirming tone, at the appropriate time, and in a confidential environment increases relatedness and promotes emotional openness. Baumeister and Leary (1995) in one of their studies discovered that individuals are inclined to incorporate feedback when given within the context of a cherished relationship, as contrasted with an objective assessment.

Google's in-house research effort, Project Oxygen, found that the best managers exhibited a few similar behaviors—their primary one being that they provided actionable feedback while promoting autonomy and empathy. Feedback wasn't micromanaging; feedback was coaching.

Managers were taught to pose reflective questions, give feedback specific to the development objectives of the individual, and develop trust-based conversations. Consequently, teams working for strongly rated managers had higher ratings on psychological safety, engagement, and performance measures (Garvin et al., 2013).

Here's a sample phrasing that has been utilized in Google's management training:

"You've shown great progress in handling complex tasks—how do you feel about stretching your leadership in this upcoming project?"

This brief sentence reinforces competence, invites autonomy, and acknowledges the relationship, exactly in line with Self-Determination Theory.

Another illustrative case is Airbnb, where leadership encourages a radical listening culture and constant feedback. Workers are encouraged to give "clear is kind" criticism to one another, always from a perspective of respect and shared intent. As part of their onboarding, recruits are exposed to key communication principles that enable psychological safety and open back loops. Former Chief Ethics Officer of Airbnb, Rob Chesnut, said:

"Feedback works when people feel seen, heard, and valued. It's not about being nice—it's about being honest in a human way."

By connecting feedback to relatedness and purpose, Airbnb makes sure that even tough conversations reinforce the social fabric of the company instead of tearing it apart.

While these strategies highlight the value of human-centered approaches to feedback, many organizations are now turning to digital platforms and AI-driven tools to enhance feedback delivery. This digitalization introduces both powerful opportunities and complex challenges for the brain and the workplace.

Digitalization of feedback brings with it both opportunities and ethical concerns. Technologically facilitated platforms now enable real-time performance feedback, which can minimize bias and enhance accuracy (Gaskell, 2022). Yet, persistent alerting and algorithmic surveillance can activate the threat circuitry of the brain by overloading the prefrontal cortex, elevating stress levels instead of encouraging development. fMRI studies prove that disruptions caused by notifications activate error-related signals in the anterior cingulate cortex (ACC), diverting attention and impairing learning (Hermans, 2025).

In addition, AI-provided feedback is not always accepted as trustworthy or inspiring. Hein (2024) discovered that employees who received AI-generated reviews had lower acceptance and lower motivation than those who were given human feedback, and that social distance mediated this effect. This resonates with the significance of relational trust, which cannot be accomplished by algorithms. A recent study further stated that employee-AI–AI collaboration can lead to higher emotional exhaustion and even counterproductive work behaviors when over-relied on because workers are socially isolated and have decreased psychological safety (PMC, 2025).

The ethical consequences of data gathering, privacy, and fairness in automatic systems cannot be ignored. Urquhart, Laffer, and Miranda (2022) also discovered that employees were concerned about the validity of affective computing systems and how emotional data could be exploited by organizations. Likewise, a journal report in the Journal of Business Ethics highlighted that AI-based performance systems need to ensure transparency and periodic auditing for bias in order to retain fairness and trust (2024). Disconcertingly, a mere 30% of the world's workers today are aware of ethical AI standards in their workplace, which highlights the need for more open communication and governance (IBE, 2025).

A good case in point on how to balance these forces can be observed at Microsoft. In the process of Satya Nadella's cultural shift, the company tested out AI-enabled feedback platforms to give employees instant views of their collaboration and productivity habits. But Microsoft intentionally combined these systems with human mentoring and manager-initiated discussions, with the result that the data were contextualized and not used in punitive fashion. Employee surveys inside the company indicated workers depicted higher trust in feedback procedures if algorithmic conclusions were presented along with tailored advice, as opposed to automated ratings. This mixed strategy permitted the technology to increase efficiency but at the same time accommodate the brain's social and affective requirements, especially autonomy and fairness—two important areas of the SCARF model (Rock, 2008).

These advancements demonstrate that although technology may be used to streamline and scale back feedback, its real value lies in the integration of digital insights with emotionally intelligent, human-centered practices. This equilibrium not only enhances learning but also builds resilience, building a bridge between feedback, stress regulation, and the brain's ability for neuroplastic change.(4- Review)

Feedback is not merely a professional exchange of information—it is a deeply personal experience that can shape how individuals think, feel, and grow. Through the lens of neuroscience, we know today that the brain does not receive feedback passively; it reacts instinctively, decoding it through neural circuits linked to threat and reward.

Praise engages the brain's reward mechanisms, supporting motivation and learning, whereas criticism, especially when inappropriately worded, can activate regions like the amygdala and anterior cingulate cortex, inducing stress, defensiveness, or withdrawal. This neurobiological responsiveness underscores the need for feedback to be undertaken with thoughtfulness and compassion, particularly in high-stakes contexts like workplaces, schools, and coaching relationships.

The integration of key psychological theories—Feedback Intervention Theory, Mindset Theory, and Self-Determination Theory—has illuminated the multiple layers at play when feedback is delivered and received. Kluger and DeNisi's work reminds us that feedback is effective only when it guides attention toward task improvement, not personal identity. Carol Dweck's Mindset Theory places another layer, highlighting the fact that people with a growth mindset are neurologically and emotionally more open to learning through feedback since they view effort and correction as avenues to mastery. Self-Determination Theory further sheds light on this by illustrating how feedback works best when it nourishes a person's sense of autonomy, competence, and relatedness. Collected together, these approaches introduce feedback not as a fixed performance device, but as a flexible, relationship-focused process that shapes learning, behavior, and identity formation.

This theme is particularly beneficial for leaders, teachers, human resource professionals, and coaches—those whose work is dependent on leading others to higher awareness and performance. In these situations, feedback is a kind of communication that can both reinforce fear and self-doubt or create confidence and accountability. Its delivery tone, timing, context, and style decide whether it evokes motivation or resistance.

Within organizations that are aware of these neurological and psychological principles, practices of feedback evolve from punitive or evaluative to collaborative and future-oriented. Firms such as Microsoft, Google, and Adobe have set the standard for this shift by designing systems that value real-time mentoring, psychological safety, and personalized development, allowing workers not only to enhance their performance but also to develop deeper self-knowledge and flexibility.

Ultimately, the strength of feedback is its ability to reprogram the mind, not by requiring transformation, but by enabling it. When feedback is given with emotional intelligence, couched in language of growth and learning, and placed within the context of mutual respect, it reframes the brain's encoding of experience. It builds neural connections for resilience, grows the brain's ability for reflection and problem-solving, and bolsters motivation. By doing so, feedback ceases to become an instrument—it becomes a transformational tool that helps people and organizations alike grow with intent, purpose, and understanding.

NEUROPLASTICITY, STRESS, AND RESILIENCE IN THE WORKPLACE

In today's fast-paced professional environments, stress is not an occasional inconvenience—it is a chronic companion for many. From tight deadlines and performance evaluations to economic uncertainty and organizational change, the workplace is increasingly understood to be a source of high-order psychological stress. But amidst all that turmoil is a remarkable truth: the human brain is not immutable or static. It is plastic, malleable, and resilient. That capacity for adaptation and change is nothing but the property of neuroplasticity, and it offers not just a buffer against stress but a blueprint for building resilience.

Neuroplasticity is the brain's capacity to alter and adapt through the creation of new neural paths based on learning, experience, and the environment. Long thought to be developmentally hardwired, neuroplasticity has been proven through recent decades of research to be an ongoing process. It is the biological process for learning new information, recovering from trauma, and overall adaptability. In the context of workplace stress and resilience, neuroplasticity is the idea that the emotional and cognitive habits stress creates can be reprogrammed—if the proper methods are used.

Knowledge of the effects of stress on the brain is the secret to unlocking its potential for resilience. Acute stress, in moderation, will enhance concentration and performance—a psychological principle referred to as the Yerkes-Dodson law (1908). Chronic stress does the opposite. Chronic stress results in allostatic load, a cumulative wear and tear on the body and brain, particularly the prefrontal cortex (self-regulation and decision-making) and the hippocampus (memory and learning). Allostatic load elevates cortisol levels, which can reduce dendritic connections in the hippocampus and decrease neurogenesis, interfering with the brain's ability to create and store new memories (McEwen, 2007).

Neuroscientific research has indicated that stress also strengthens maladaptive neural patterns. For instance, repeated exposure to negative feedback or high-stress situations has been demonstrated to strengthen circuits within the amygdala, the center of fear within the brain, making a person more reactive and emotionally dysregulated. However, through the utilization of conscious cognitive strategies, mindfulness practice, and emotionally intelligent leadership, these patterns can be rewired.

This is where resilience comes in—not as an innate trait, but as an active, trainable repertoire of reactions. Resilience is the ability to adapt successfully to adversity, trauma, or stress. It is backed by neuroplastic adaptation in brain regions like the ventromedial prefrontal cortex (vmPFC), which assists in modulating emotional reactions, and the anterior cingulate cortex (ACC), which facilitates cognitive flexibility and conflict resolution. As Davidson and McEwen (2012) explain, "Resilience is reflected in the capacity to adapt to challenges, and this adaptive capacity is rooted in neural plasticity."

The workplace is a strong setting for the development or degradation of resilience. Stressors such as role ambiguity, toxic leadership, and poor communication can decrease resilience and disrupt neuroplastic functioning. But if organisations have good practices in place, such as feedback systems that enhance autonomy and development, work-life balance practices, and team-based practices of emotional regulation, they receive adaptive brain responses that enhance resilience.

This section aims to explore where these concepts intersect in real organizational life. We will explore how stress impacts the brain's architecture, how neuroplasticity underpins recovery and learning, and how resilience can be fostered through active development through targeted interventions. Drawing on psychological and neuroscientific evidence, we will also explore case studies and organizational practices leveraging these concepts to create healthier, more agile, and more human workplaces.

Along the way, this chapter doesn't just report, it empowers. It illustrates that regardless of how frozen or reactionary our present emotional patterns might feel, the brain is still a work in progress. With the right tools, right support, and right mindset, people and organizations can rewire for resilience.

Neurovisceral Integration Model (NIM)

The high-pressure environment of the modern workplace requires not only cognitive adaptability and emotional competence but also biological ability to withstand and rebound from stress. Among the models that place the nervous system in explicit conversation with emotional processes is the Neurovisceral Integration Model (NIM) of Julian Thayer and Richard Lane (2000). Central to this model is the hypothesis that emotion regulation, cognition, and behavior are inextricably linked with the brain-body connection, governed by the central nervous system as well as by the autonomic nervous system, and in particular, the parasympathetic branch.

NIM adopts the premise that efficient self-regulation—whether in a boardroom, under a project deadline, or in disagreement with others—demands an organized physiology. The model dissects out a network of brain regions known as the central autonomic network (CAN), which involves the prefrontal cortex (PFC), anterior cingulate cortex (ACC), insula, amygdala, and brainstem nuclei. The network coordinates a balance between arousal and relaxation, vigilance and reflection, action and inhibition. The CAN regulates the heart through the vagus nerve and, thus, heart rate variability (HRV)—a measure that has come to be used as a biological marker of resilience.

Heart rate variability (HRV) is the variation in the time intervals between heartbeats. More variability is not a marker of instability, as one might guess, but of resiliency and flexibility. High HRV is a marker of a nervous system that can adapt well to both stress and recovery, toggling easily between sympathetic activation (fight-or-flight) and parasympathetic restoration (rest-and-digest). People with high resting HRV are more effective at regulating their emotions, stay calm under stress, and bounce back quickly after stressful episodes—all markers of psychological resiliency.

Neurovisceral integration also suggests that the ventromedial prefrontal cortex (vmPFC) is the master regulatory center for this process. When operating at its best, the vmPFC facilitates downregulation of amygdala activity, modulation of emotional reactivity, and flexible decision-making. This is particularly important in the workplace when high-stakes decisions, team conflict, or pressure to perform will hijack the brain's threat response networks. Those with tighter PFC-amygdala coupling are less likely to become entangled in emotional overdrive and more likely to stay in charge.

This model is particularly applicable in chronic job stress situations, where repeated overload is a cause of dysregulation of the CAN. Chronic activation of the hypothalamic-pituitary-adrenal (HPA) axis over time can lead to reduced HRV, increased cortisol, and disrupted function of the PFC and the hippocampus. The outcome is a cascade of maladaptive processes: emotional exhaustion, burnout, reduced cognitive control, and disrupted social regulation. To McEwen (2007), this process—allostatic load—is the resulting burden of uncontrolled stress on the brain and the body.

But that is not the whole story. Neuroplasticity in the brain allows individuals to retrain these systems under intentional practice and outside guidance. Methods that increase HRV and maximize CAN regulation, such as mindfulness meditation, slow diaphragmatic breathing, and cognitive reappraisal, have been shown to increase the vmPFC, emotion regulation, and adaptive responses to stress. For instance, mindfulness training has been shown to enhance gray matter density in the ACC and PFC (Hölzel et al., 2011), areas of central importance to executive function and affect modulation.

Tang et al. (2009) found that even five days of integrative body-mind training enhanced HRV, enhanced activation in the insula, and lowered cortisol levels in the participants. These results support the NIM's argument that self-regulation is an intact-system process, rather than willpower or attitude.

In the workplace, these findings are now creating an increasing number of neuroscience-based interventions. For example, firms like Aetna, SAP, and Google have introduced mindfulness-based stress reduction (MBSR) initiatives, and they have experienced reductions in job stress, as well as improvements in attention, teamwork, and resilience. In-house metrics at Aetna indicated that workers who underwent mindfulness training reported a 28% decrease in stress and a 20% enhancement in sleep quality, as well as quantifiable benefits in productivity (Aetna Mindfulness Report, 2016).

HRV biofeedback training is yet another tangible application of the NIM. This is the methodology of using computer programs or wearable units to educate individuals in self-controlling their heart rates and breathing to maximize emotional and physiological equilibrium. The U.S. military and high-performing sporting units have used biofeedback training to enhance resilience, pressure poise, and high-stakes decision-making accuracy.

One of the most powerful features of the NIM is that it bridges abstract emotional concepts, like resilience, to measurable biological processes. It bridges mind and body, offering leaders, coaches, and healthcare providers a sound foundation to design interventions that improve psychological and physiological health.

From a leadership point of view, knowledge of the NIM gives managers the power to identify stress dysregulation in their teams and respond in advance. Leaders can lead by calm, reflective example, promote psychologically safe team cultures, and facilitate the use of micro-practices such as breathwork or brief reflection breaks on high-stakes projects. These small changes not only decrease collective stress but also recalibrate the nervous systems of the team for resilience.

Briefly, the Neurovisceral Integration Model offers a compelling model for understanding the neurobiological basis of workplace resilience. It teaches us that resilience is not an emotional or mental state in the first place—it is an embodied, acquired ability based on the dynamic interaction between the brain and the body. By the use of specific interventions that increase vagal tone, form prefrontal regulatory circuits, and promote mind-body awareness, companies can build teams not only that survive stress but flourish in the face of adversity.

Barbara Fredrickson's Broaden-and-Build Theory of Positive Emotions

While most organizational psychology attention is placed on stress reduction, another equally strong strategy is in maximizing the brain's ability for positive adaptation. This is the core idea of Barbara Fredrickson's Broaden-and-Build Theory of Positive Emotions (2001), wherein positive emotions are not transient states of pleasure but are biological markers that broaden the person's awareness and build long-term internal strengths—emotional, psychological, and cognitive—that foster resilience over the long term.

Whereas negative emotions, which are designed to constrict attention and prepare the body for direct action (e.g., fight-or-flight), broaden our scope of perception and cognition. Emotions like joy, gratitude, hope, and inspiration elicit a widened field of attention and encourage more integrative, flexible, and creative thinking. Over time, this mental flexibility enables one to develop enduring personal strengths—problem-solving skills, social support, optimism, and emotional control—that can be tapped when under stress or adversity.

Neuroscientific studies confirm this model. Research indicates that positive affect is associated with greater activation in the prefrontal cortex, especially the left dorsolateral PFC, an area linked to executive control and approach behavior (Ashby, Isen, & Turken, 1999). Positive emotions also increase dopaminergic activity in the brain reward system, enhancing memory consolidation, attention, and learning, essential components of neuroplasticity. Repeated encounters with positive emotion strengthen these neural circuits, establishing a feedback loop that increases resilience.

A remarkable illustration is the research of Tugade and Fredrickson (2004), where they discovered that participants who experience and nurture positive emotions regularly heal faster from cardiovascular reactivity due to stress. These subjects had higher vagal tone, a marker of parasympathetic nervous system health and capacity for emotion regulation. This physiological resistance is indicative of a brain-body system conditioned to change in more flexible ways in response to shifting demands—signature features of psychological and neurobiological resilience.

The Broaden-and-Build has important implications for workplace applications. To cultivate positive emotions in high-stress workplaces is not to be superficially cheerful—it is to foster conditions for neural flexibility, social connection, and adaptive learning. Leaders who build cultures of psychological safety, gratitude, and purpose are implicitly cultivating neural networks that favor exploration over defensiveness and collaboration over solitude.

In the workplace, this framework encourages strength-based feedback, celebration of small wins, and gratitude practices. Companies such as Zappos and Google implement rituals, like peer-to-peer appreciation programs or "thank-you walls," that promote daily doses of positivity. The cumulative effect is not only improved morale, but enhanced psychological flexibility and problem-solving capacity, key components of resilience.

Another rich illustration is that of the "Three Good Things" intervention, a brief gratitude practice that Martin Seligman and his colleagues have created. Employees are asked to think back and write down three positive things they experienced at the end of each day at work. This small practice has been demonstrated to boost happiness and diminish burnout in the long run (Seligman et al., 2005). The process of bringing attention to the positive redirects the attentional systems of the brain away from threat and into reward-oriented processing, rewiring default stress-reactive neural pathways into optimism and resilience.

As a leadership principle, it takes more than creating "positive vibes" to leverage the Broaden-and-Build theory. It demands thoughtful feedback construction, such as starting development conversations with an affirmation of strength, or injecting language that holds possibility and growth. It involves acknowledging effort, not achievement, and rewarding learning from mistakes instead of penalizing them. These leadership actions send implicit messages to the brain that the setting is safe, significant, and development-fostering conditions that naturally stimulate cortical growth and emotional strength.

Notably, this theory also guides organizations in responding to post-traumatic workplace trauma, including layoffs, mergers, or crises. In such situations, promoting positivity isn't about disregarding pain—it is about facilitating cognitive reappraisal and psychological recovery. Positive emotions serve as psychological scaffolding, enabling individuals to reflect, reframe, and reconnect with purpose even in the midst of adversity.

From a neuroplasticity point of view, each experience of positive emotion not only reverses the physiological impact of stress but also builds possibilities for new neural connections. Repetitive over time, these experiences begin to shape more resilient emotional set-points and build the brain's ability to learn from failure, recover from adversity, and remain solution-focused in uncertainty.

Even physical space design can embody this theory. Open, bright offices, views of nature, collaborative lounges, and resting areas have all been found to improve mood and cognitive openness, precepts based on Broaden-and-Build. Airbnb, Salesforce, and Patagonia use these principles in that they deliberately build environments that promote trust, positivity, and social support.

To embed this framework within an organization, organizations can implement positive psychology training into leadership development, provide strengths-based communication workshops, develop appreciation rituals, mutual celebration, and reflective storytelling, and get leaders to role-model gratitude, curiosity, and emotional openness.

This way, the workplace is a resilience incubator where emotional flexibility, mental clarity, and social bonding are not accidents but are intentionally fostered through neuropsychological design.

The Broaden-and-Build Theory thus provides a dynamic framework with which to see positive emotions remodeling the brain and behavior. By expanding the scope of awareness and constructing psychological resources, positive affect fortifies the neural structure of resilience. Organationally, that implies culture is not merely an HR issue—it's a neurobiological ecosystem that can reinforce rigidity or facilitate growth. And by cultivating daily routines that create positive emotion, companies can be places where neuroplasticity happens and stress becomes strength.

Seligman's PERMA Model

Resilience in high-stakes, high-pressure environments is frequently framed as a person's ability to recover from adversity. But an emerging body of psychological and organizational research indicates that resilience is less a function of grit or will and more significantly shaped by context. The PERMA model, developed by psychologist Martin Seligman (2011), provides an overarching framework for building environments supporting psychological flourishing and, consequently, the neurobiological ability to be resilient. PERMA is an acronym for Positive Emotion, Engagement, Relationships, Meaning, and Accomplishment—five columns standing in combination as the conditions under which humans can truly thrive.

What makes PERMA different from other models of well-being is that it combines subjective well-being and objective functioning. It recognizes that people flourish not only when they feel good, but when they are fully absorbed, socially connected, meaning-cued, and able to pursue and attain valued goals. These states don't merely maintain morale—they induce neuroplastic change, particularly in areas of the brain responsible for emotional regulation, goal striving, memory, and social cognition.

Positive Emotion, which, as discussed in Fredrickson's Broaden-and-Build Theory, expands attention and facilitates adaptive coping. These positive emotions, such as gratitude and hope, activate the dopaminergic reward pathways and ventral striatum, supporting the reinforcement of approach-oriented behavior. When workplaces emphasize appreciation and optimism through recognition programs, positive feedback loops, or simply supportive leadership styles, they help shift the neural baseline from vigilance to trust. This not only decreases reactivity to stress but facilitates a receptivity to learning and change, a critical trait within changing organizational environments.

The second pillar, engagement, is the state of being fully absorbed and challenged in work that leverages one's talents. Through a neuroscientific lens, engagement is associated with activation in default mode networks (meaning-making) and task-positive networks like the dorsolateral prefrontal cortex, responsible for focus and executive function. Staff who claim greater engagement have higher cognitive resilience, improved working memory, and lower emotional exhaustion. Google's "20% time" practice, where employees can devote a portion of their workweek to side projects that interest them, embodies this rule. By matching work with intrinsic motivation, it increases dopamine, flow states, and eventually, neuroplasticity in connection with the mastery of skills.

Relationships, the third component, are arguably the best resilience predictor. Social support shields against stress effects and promotes emotional recovery. At a brain level, high-quality interpersonal relationships enhance oxytocin release, engage the anterior insula and medial prefrontal cortex, and decrease activity in the amygdala—the threat center. Psychologically safe teams—where individuals can speak freely, take risks, and share dissent safely—facilitate emotional regulation at a group level. Amy Edmondson's study of psychological safety (2019) indicates that highly relational-trust teams perform better than those with superior individual ability but lower social cohesion.

Meaning is the fourth type and is the perception that one's work has value—that what one does helps create something bigger than oneself. Meaning engages the default mode network and activates areas such as the vmPFC, which are responsible for self-referential thinking, moral judgment, and identity formation. Individuals with a high sense of meaning are more resilient in the midst of failure because they are more likely to reinterpret adversity as temporary and meaningful. Companies such as Patagonia, which officially align their commercial endeavors with environmental activism, give employees a story beyond profit. Evidence indicates that staff who perceive the alignment of their work with values demonstrate greater tolerance for stress, increased post-traumatic growth, and more long-term commitment (Steger et al., 2006).

Lastly, Accomplishment strengthens feelings of efficacy and momentum. Accomplishment isn't about outcomes—it's about goal-directed achievement, overcoming obstacles, and getting feedback to affirm progress. Neurobiologically, these moments activate the striatum and strengthen goal-related neural connections, which are key to motivation and cognitive resilience. Organizations that establish well-defined, attainable goals, along with autonomy-supportive feedback, enable employees to develop adaptive habits that create psychological momentum even under stressful contexts.

What makes the PERMA model so effective in the workplace is that it is culturally flexible, scientifically proven, and practically feasible. IBM, for example, incorporated PERMA principles into its global well-being program, teaching managers to apply strengths-based coaching, acknowledge improvement, and establish environments where staff were made to feel valued and challenged. Assessment demonstrated burnout declines, improved engagement scores, and enhanced team resilience in restructuring periods (IBM HR Report, 2018).

A second use is from the US Army Comprehensive Soldier Fitness Program, which employed PERMA-linked training to increase psychological resilience in soldiers. Soldiers trained in positive psychology, relationship skills, and meaning-making had decreased post-deployment symptoms of stress and improved cognitive functioning under stress (Reivich et al., 2011). Such findings are immediately applicable to high-stress professional settings, including healthcare, finance, or technology fields, where both cognitive and emotional resilience are imperative.

At the organizational level, PERMA calls for a move away from reactive, problem-focused interventions towards proactive well-being design. Instead of waiting for burnout to develop, PERMA asks leaders to consider the following: Are we providing our employees with opportunities to experience joy, flow, meaningful connection, purpose, and success? If not, performance interventions might be overlooking the more profound systemic conditions that support authentic, long-term resilience.

To apply PERMA in practice, organizations can perform strengths assessments and match roles accordingly, implement peer recognition systems to boost Positive Emotion, provide coaching and mentoring to increase Engagement and Meaning, offer team-oriented training in emotional intelligence and active listening to improve Relationships, and develop goal-setting and goal-tracking systems for personal and professional development that promote a sense of Accomplishment.

When these dimensions are woven into a fabric on purpose, the outcome is a cultural infrastructure that promotes neuroplasticity, protects against chronic stress, and develops resilience not as an individual characteristic, but as a collective organizational strength.

In short, the PERMA model provides companies with an effective blueprint for building workplaces in which resilience is not a matter of chance. Every component—Positive Emotion, Engagement, Relationships, Meaning, and Accomplishment—engages neural circuits promoting adaptability, dampens threat reactions, and strengthens psychological resilience. As the work environment increasingly evolves into a sophisticated and challenging setting, models such as PERMA allow leaders to shift from crisis management to cultivating resilience, transforming professional spaces into environments where individuals can actually thrive.

Resilience is usually depicted as a personal quality—an inner strength one is either born with or without. But neuroscience paints a different, more encouraging picture: resilience isn't just psychological, but deeply biological. It's a product of neuroplastic processes that fashion the way we react to stress, the way we bounce back from adversity, and the way we emerge stronger from challenge. Here, we have discussed how organizations can tap into such adaptive capacity through purposeful frameworks that bridge biology and behavior, and emotion and environment.

The Neurovisceral Integration Model reminds us that resilience is a bodily experience. The body and brain are constantly in dialogue, with the prefrontal cortex, amygdala, and vagus nerve participating in a regulatory loop that affects our reactivity to stress and recovery. Through techniques such as breath regulation, mindfulness, and self-awareness of emotions, employees can remodel these systems, resulting in more effective decision-making, emotional regulation, and interpersonal communication in high-stakes settings.

The Broaden-and-Build Theory builds on this perspective by demonstrating that positive emotions are not excesses but valuable resources. Happiness, appreciation, and interest open up cognitive and emotional conduits, making people more adaptable, imaginative, and social. These feelings don't just feel good—positive emotions rewire the reward and learning systems of the brain, building a substratum for long-term resilience. At work, this means cultures of appreciation, learning from mistakes, and opportunities embedded in challenges.

Lastly, the PERMA Model provides a system-level design for creating contexts in which well-being is durable. By integrating Positive Emotion, Engagement, Relationships, Meaning, and Accomplishment into the life of organizations, businesses can build the psychological scaffolding that undergirds resilience at scale. Each of the PERMA domains maps onto brain systems engaged in motivation, memory, executive function, and affect regulation, rendering it a psychological and biological template for flourishing in adversity.

Collectively, these models change the paradigm: resilience is no longer a reactive capacity to recover from burnout. It is a proactive, neuroplastic adaptation process developed through purposeful practice, relational support, and environment design. For leaders, educators, and professionals, this awareness reveals potent new avenues for creating workplaces that not only cope with stress but convert it into strength, creativity, and connection.

In an age in which change is a given and stress is unavoidable, the organizations best prepared for the future will not be those with the toughest corners, but the most adaptable minds and the best-nourished nervous systems. Neuroplasticity brings about this change, and the proper workplace practices bring it to life.

CONCLUSION AND FUTURE DIRECTIONS

As work environments continue to change, organizations will increasingly grapple with the dual forces of digital overload and hybrid work arrangements, in addition to the psychological demands of being perpetually connected. Studies indicate that digital stress affects attention and decision-making, and highly flexible work arrangements can paradoxically heighten pressure when boundaries are blurred (Buettner, 2021; Langer, König, & Papathanasiou, 2021

Applications of neuroscience in organizational contexts are already beginning to emerge as responses to these pressures. Virtual reality simulations, for example, are being designed as stress inoculation tools that allow workers to practice coping mechanisms in immersive, controlled environments (Cipresso, Riva, & Gaggioli, 2022).

AI-powered personalized coaching platforms are offering workers customized advice for enhancing self-regulation and adaptability (Wang & Zhang, 2023). These systems hold potential to deliver support at scale but must be balanced with human connection to avoid undermining trust and motivation (Gaskell, 2022). Similarly, neurofeedback technologies are gaining traction as methods for training resilience and concentration by providing individuals with immediate feedback on brain activity, thereby enhancing their ability to self-regulate under stress (Thibault, Lifshitz, & Raz, 2020).

While these innovations offer exciting opportunities, they also bring serious ethical challenges. Constant digital monitoring can overwhelm employees' prefrontal cortex, activating stress circuits rather than supporting growth, and broader concerns about consent, equity, and the risk of manipulation remain critical (Kolbjørnsrud, Amico, & Thomas, 2021). Some scholars have even suggested the institution of new human rights to safeguard mental privacy and cognitive autonomy in response to advances in neurotechnology (Ienca & Andorno, 2020). It is only by facing these issues directly that organizations can ensure neuroscience-driven practices strengthen, rather than erode, trust and psychological safety. By adopting such a forward-thinking attitude, organizations can take the opportunity to leverage the brain's neuroplastic capacity not only to develop stronger and more flexible employees but also to create more human, balanced, and forward-looking workplaces. (7- Review)

In many ways, this sets the stage for the deeper insights explored throughout this chapter. With an ever-more demanding and emotionally complex working life, the ability to adjust, connect, and succeed is no longer optional; there is no choice but to be able to do it. This chapter has explored the interlinked themes of emotional intelligence, feedback, and resilience using the frameworks of neuroscience and psychology. Taken collectively, these themes provide an integrated model for "reprogramming the mind" at work, demonstrating how human potential is not just influenced by training or environment, but by the brain's active ability to grow—its neuroplasticity.

At its essence, the chapter reiterates that human potential is not absolute. The brain is not an unchanging organ but a dynamic system constantly transforming itself with regard to thoughts, emotions, social interactions, and environments. In light of this understanding, characteristics such as resilience, self-awareness, empathy, adaptability, and motivation are not innate attributes but acquireable qualities. When individuals practice intention, like mindful communication, emotionally responsive feedback, or stress-reduction skills, they are reconfiguring their neural circuits for greater functioning.

Social and emotional functioning are found to be rooted profoundly in brain systems like the amygdala, prefrontal cortex, and mirror neurons. These systems influence how people think about threats, rewards, fairness, and belonging in collaborative work interactions. The way we receive feedback, manage emotions, or work together is influenced by these systems, yet is also capable of change. Minor changes in the way individuals provide feedback, recognize effort, or handle conflict can greatly impact motivation, emotional safety, and group relations. Cultures in organizations that actively promote autonomy, emotional safety, and respectful communication allow the brain's natural learning systems to thrive.

Stress, a significant obstacle to well-being and productivity, is also approached from the neurobiological perspective. Chronic stress interferes with cognitive processes such as memory and decision-making, but it is possible to counteract it, according to research. Mindfulness, regulation of breathing, and engagement of positive emotions activate mechanisms that correct the balance in the nervous system. Instead of viewing stress resilience as something immutable, the chapter presents it as a combination of trainable responses that have biological support.

The real-world implications are far-reaching. Leaders can apply this information to construct psychologically safe teams; HR professionals can re-engineer performance cultures to encourage growth rather than fear; coaches and teachers can cultivate individuals' emotional and cognitive flexibility; and employees themselves can learn to more effectively manage their emotional and mental resources. These methods are not theoretical ideals—they are based on actual organizational examples and scientific frameworks that demonstrate how focused interventions can enhance well-being, learning, collaboration, and innovation.

Rather than a list of methods, this chapter is a call to a cultural and intellectual transformation: away from reactive, fear-based performing environments toward psychologically diverse ecosystems that foster human development. Emotional intelligence is not merely understanding oneself; it's about co-regulating with others. Feedback is not only judgmental; it's developmental. And resilience is not a mere individual characteristic; it is a social, learnable ability formed by emotion, cognition, and biology.

Ultimately, to rewire the mind is to understand that each moment of interaction—a compliment, a criticism, a collaboration meeting, a silent moment of introspection—holds the ability to create neural patterns. In organizations that are built on this awareness, individuals are not just getting through the demands of contemporary work—they are growing from them. By integrating leadership, culture, and communication with the way the brain really functions, we design a place where potential is no longer abstract but real.

REFERENCES

Adobe, H. R. (2018). The Check-In: Reinventing performance reviews. Adobe Corporate White Paper.

Aetna. (2016). Mindfulness report: Internal wellness study findings. https://www.aetna.com

Arnsten, A. F. T. (2009). Stress signalling pathways that impair prefrontal cortex structure and function. *Nature Reviews. Neuroscience*, *10*(6), 410–422. DOI: 10.1038/nrn2648 PMID: 19455173

Ashby, F. G., Isen, A. M., & Turken, U. (1999). A neuropsychological theory of positive affect and its influence on cognition. *Psychological Review*, *106*(3), 529–550. DOI: 10.1037/0033-295X.106.3.529 PMID: 10467897

Bar-On, R. (2006). The Bar-On model of emotional-social intelligence (ESI). *Psicothema*, *18*(Suppl), 13–25. PMID: 17295953

Boyatzis, R., & Goleman, D. (2001). *Emotional Competence Inventory*. Hay Group.

Boyatzis, R. E., Goleman, D., & Rhee, K. (2013). Emotional intelligence: Issues and common misunderstandings. *Competency & Emotional Intelligence Journal*, *10*(3), 22–25.

Boyatzis, R. E., Rochford, K., & Cavanagh, K. V. (2012). Emotional intelligence competencies in engineer's effectiveness and engagement. *Career Development International*, *17*(3), 177–193. DOI: 10.1108/13620431211241058

Buettner, R. (2021). Digital stress in the workplace: A neuroIS perspective. *Information Systems Frontiers*, *23*(2), 531–543. DOI: 10.1007/s10796-020-10075-2

Cherniss, C. (2010). Emotional intelligence: Toward clarification of a concept. *Industrial and Organizational Psychology: Perspectives on Science and Practice*, *3*(2), 110–126. DOI: 10.1111/j.1754-9434.2010.01231.x

Cipresso, P., Riva, G., & Gaggioli, A. (2022). Virtual reality as a tool for stress management: Toward stress inoculation training in the workplace. *Frontiers in Psychology*, *13*, 856791. DOI: 10.3389/fpsyg.2022.856791

Davidson, R. J., & McEwen, B. S. (2012). Social influences on neuroplasticity: Stress and interventions to promote well-being. *Nature Neuroscience*, *15*(5), 689–695. DOI: 10.1038/nn.3093 PMID: 22534579

Doidge, N. (2007). *The brain that changes itself: Stories of personal triumph from the frontiers of brain science*. Penguin.

Dweck, C. S. (2006). *Mindset: The new psychology of success*. Random House.

Edmondson, A. C. (2019). *The fearless organization: Creating psychological safety in the workplace for learning, innovation, and growth*. Wiley.

Eisenberger, N. I., & Lieberman, M. D. (2004). Why rejection hurts: A common neural alarm system for physical and social pain. *Trends in Cognitive Sciences, 8*(7), 294–300. DOI: 10.1016/j.tics.2004.05.010 PMID: 15242688

Emotional intelligence (EI) | EBSCO. (n.d.). EBSCO Information Services, Inc. https://www.ebsco.com/research-starters/health-and-medicine/emotional-intelligence-ei

Fredrickson, B. L. (2001). The role of positive emotions in positive psychology: The broaden-and-build theory of positive emotions. *The American Psychologist, 56*(3), 218–226. DOI: 10.1037/0003-066X.56.3.218 PMID: 11315248

Gaskell, J. (2022). The rise of AI in workplace performance feedback. *Harvard Business Review*.

Goldsmith, M. (2007). *What got you here won't get you there*. Hachette Books.

Goleman, D. (1995). *Emotional intelligence: Why it can matter more than IQ*. Bantam Books.

Gordon, E. (2000). *Integrative neuroscience: Bringing together biological, psychological, and clinical models of the human brain*. Harwood Academic Publishers.

Hein, I. (2024). Human versus AI-generated feedback: Social distance mediates impact on motivation. *Frontiers in Organizational Psychology, 3*, 1–13. DOI: 10.3389/forgp.2024.1468907

Hermans, E. J. (2025). Neurocognitive mechanisms of resilience under acute stress. *Biological Psychiatry, 97*(4), 300–312. DOI: 10.1016/j.biopsych.2024.12.015 PMID: 39448004

Hölzel, B. K., Carmody, J., Vangel, M., Congleton, C., Yerramsetti, S. M., Gard, T., & Lazar, S. W. (2011). Mindfulness practice leads to increases in regional brain gray matter density. *Psychiatry Research: Neuroimaging, 191*(1), 36–43. DOI: 10.1016/j.pscychresns.2010.08.006 PMID: 21071182

Human Resources, I. B. M. (2018). *Global wellbeing program evaluation report*. IBM Corporation.

Ienca, M., & Andorno, R. (2020). Towards new human rights in the age of neuroscience and neurotechnology. *Life Sciences, Society and Policy*, *16*(1), 5. DOI: 10.1186/s40504-020-00101-2 PMID: 28444626

Jaeggi, S. M., Buschkuehl, M., Jonides, J., & Perrig, W. J. (2008). Improving fluid intelligence with training on working memory. *Proceedings of the National Academy of Sciences of the United States of America*, *105*(19), 6829–6833. DOI: 10.1073/pnas.0801268105 PMID: 18443283

Journal of Business Ethics. (2024). Ethical implications of AI in employee performance evaluations. Journal of Business Ethics, 189(2), 433–445. https://doi.org/ DOI: 10.1007/s10551-023-05430-1

Kluger, A. N., & DeNisi, A. (1996). The effects of feedback interventions on performance: A historical review, a meta-analysis, and a preliminary feedback intervention theory. *Psychological Bulletin*, *119*(2), 254–284. DOI: 10.1037/0033-2909.119.2.254

Kolbjørnsrud, V., Amico, R., & Thomas, R. J. (2021). Algorithmic management: Opportunities, challenges, and the path forward. *California Management Review*, *63*(3), 5–23. DOI: 10.1177/0008125621992574

Langer, M., König, C. J., & Papathanasiou, M. (2021). Highly flexible working arrangements and their effects on stress and performance: The role of digital overload. *Applied Psychology*, *70*(3), 1011–1042. DOI: 10.1111/apps.12306

Lewis, P. A., Rezaie, R., Brown, R., Roberts, N., & Dunbar, R. I. (2011). Ventromedial prefrontal volume predicts understanding of others and social network size. *NeuroImage*, *57*(4), 1624–1629. DOI: 10.1016/j.neuroimage.2011.05.030 PMID: 21616156

Lieberman, M. D., & Eisenberger, N. I. (2009). Pains and pleasures of social life. *Science*, *323*(5916), 890–891. DOI: 10.1126/science.1170008 PMID: 19213907

Lutz, A., Brefczynski-Lewis, J., Johnstone, T., & Davidson, R. J. (2008). Regulation of the neural circuitry of emotion by compassion meditation: Effects of meditative expertise. *PLoS One*, *3*(3), e1897. DOI: 10.1371/journal.pone.0001897 PMID: 18365029

McEwen, B. S. (2000). Effects of adverse experiences for brain structure and function. *Biological Psychiatry*, *48*(8), 721–731. DOI: 10.1016/S0006-3223(00)00964-1 PMID: 11063969

McEwen, B. S. (2007). Physiology and neurobiology of stress and adaptation: Central role of the brain. *Physiological Reviews*, *87*(3), 873–904. DOI: 10.1152/physrev.00041.2006 PMID: 17615391

O'Boyle, E. H., Humphrey, R. H., Pollack, J. M., Hawver, T. H., & Story, P. A. (2011). The relation between emotional intelligence and job performance: A meta-analysis. *Journal of Organizational Behavior, 32*(5), 788–818. DOI: 10.1002/job.714

PMC. (2025). Employee–AI collaboration and emotional fatigue: Risks of over-reliance on automation. *Frontiers in Psychology, 16*, 1–12. https://pmc.ncbi.nlm.nih.gov/articles/PMC12108590/

Reivich, K., Seligman, M. E. P., & McBride, S. (2011). Master resilience training in the US Army. *The American Psychologist, 66*(1), 25–34. DOI: 10.1037/a0021897 PMID: 21219045

Rock, D. (2008). SCARF: A brain-based model for collaborating with and influencing others. *NeuroLeadership Journal, 1*(1), 1–9.

Rock, D., & Cox, C. (2012). SCARF in 2012: Updating the social neuroscience of collaborating with others. *NeuroLeadership Journal, 4*, 1–14.

Ryan, R. M., & Deci, E. L. (2000). Self-determination theory and the facilitation of intrinsic motivation, social development, and well-being. *The American Psychologist, 55*(1), 68–78. DOI: 10.1037/0003-066X.55.1.68 PMID: 11392867

Seligman, M. E. P. (2011). *Flourish: A visionary new understanding of happiness and well-being*. Free Press.

Seligman, M. E. P., Steen, T. A., Park, N., & Peterson, C. (2005). Positive psychology progress: Empirical validation of interventions. *The American Psychologist, 60*(5), 410–421. DOI: 10.1037/0003-066X.60.5.410 PMID: 16045394

Steger, M. F., Dik, B. J., & Duffy, R. D. (2006). Measuring meaningful work: The work and meaning inventory (WAMI). *Journal of Career Assessment, 20*(3), 322–337. DOI: 10.1177/1069072711436160

Tan, C. M. (2012). *Search inside yourself: The unexpected path to achieving success, happiness (and world peace)*. HarperOne.

Tang, Y. Y., Ma, Y., Wang, J., Fan, Y., Feng, S., Lu, Q., & Posner, M. I. (2009). Central and autonomic nervous system interaction is altered by short-term meditation. *Proceedings of the National Academy of Sciences of the United States of America, 106*(22), 8865–8870. DOI: 10.1073/pnas.0904031106 PMID: 19451642

Thayer, J. F., & Lane, R. D. (2000). A model of neurovisceral integration in emotion regulation and dysregulation. *Journal of Affective Disorders, 61*(3), 201–216. DOI: 10.1016/S0165-0327(00)00338-4 PMID: 11163422

Thibault, R. T., Lifshitz, M., & Raz, A. (2020). Neurofeedback: The promise of self-regulation through brain imaging. *Annual Review of Neuroscience, 43*, 433–450. DOI: 10.1146/annurev-neuro-092019-022831

Tugade, M. M., & Fredrickson, B. L. (2004). Resilient individuals use positive emotions to bounce back from negative emotional experiences. *Journal of Personality and Social Psychology, 86*(2), 320–333. DOI: 10.1037/0022-3514.86.2.320 PMID: 14769087

Urquhart, L., Laffer, A., & Miranda, D. (2022). Public perceptions of affective computing in workplace surveillance. Proceedings of the ACM on Human-Computer Interaction, 6(CSCW2), 1–28. https://doi.org/DOI: 10.1145/3555178

Wang, Y., & Zhang, Z. (2023). Artificial intelligence coaching in the workplace: Opportunities and challenges for employee learning and development. *Computers in Human Behavior, 138*, 107470. DOI: 10.1016/j.chb.2022.107470

KEY TERMS AND DEFINITIONS

Emotional Intelligence: A multifaceted set of abilities that includes the capacity to recognize, understand, manage, and express emotions effectively across a range of contexts in oneself and others, which underlies better decision-making, effective communication, and interpersonal relationships.

Feedback Intervention Theory (FIT): A theory describing how feedback influences performance based on where it draws a person's attention—on the task, self-assessment, or self-image—with task-related feedback tending to work best.

Heart Rate Variability (HRV): A quantification of the fluctuation in time between heartbeats that is employed as a marker of emotional regulation and nervous system flexibility, typically correlated with stress resilience and adaptive functioning.c5z

Neuroplasticity: The capacity of the brain to modify its function and structure throughout the lifespan by creating new neural networks in reaction to experience, learning, environmental pressures, or stress recovery.

PERMA Model: A well-being model that defines five key elements—Positive Emotion, Engagement, Relationships, Meaning, and Accomplishment—each of which contributes to growth, emotional well-being, and resilience.

Resilience: The psychological and physiological ability to adjust favorably to stress, setbacks, and adversity, allowing people to sustain or recover wellness and performance under duress.

SCARF Model: A neuroscientific framework that defines five major social drivers—Status, Certainty, Autonomy, Relatedness, and Fairness—that shape how individuals react to interactions, especially in workplaces and leadership.

Social Neuroscience: An area of inquiry that studies how brain systems and processes give rise to social behavior, such as empathy, trust, cooperation, and emotional communication.

Chapter 4
Neuroplasticity, Stress, and Resilience in the Mordern Workplace

Jossy P. George
Christ University, India

Peter Varghese Muttungal
https://orcid.org/0000-0002-1953-3830
Christ University, India

Benny Godwin J. Davidson
https://orcid.org/0000-0001-5755-2920
University of the Fraser Valley, Canada

Nishi Priya
https://orcid.org/0009-0009-2727-5497
Christ University, India

ABSTRACT

Workplace stress profoundly impacts well-being, leading to burnout and low productivity despite traditional interventions. This chapter explores how neuroplasticity offers a solution for enhanced resilience. Chronic stress alters brain regions like the amygdala and prefrontal cortex, impairing decision-making and memory. Fortunately, processes like synaptic reorganization and neurogenesis enable brain recovery. Evidence-based interventions mindfulness, exercise, social support, and sleep hygiene—strengthen stress tolerance. Workplace-level changes, including resilient leadership and ergonomics, also foster adaptability. Integrating neuroscience,

DOI: 10.4018/979-8-3373-2292-6.ch004

psychology, and organizational behavior, this framework highlights neuroplasticity's role in building individual and organizational resilience, creating more adaptive and humane workplaces

1. INTRODUCTION

Workplace stress presents a pressing global challenge directly influencing mental and physical health. While traditional organizational interventions like counselling and leave policies offer some relief, many employees continue to experience burnout, anxiety, and low productivity. This chapter delves into the concept of neuroplasticity which is the brain's inherent capacity for structural adaptation and highlights its potential as a powerful tool for bolstering individual resilience, particularly when navigating demanding, high-stress environments. Chronic workplace stress significantly alters brain structures like the amygdala, hippocampus, and prefrontal cortex, impairing decision-making, memory, and emotional regulation. However, neuroplastic mechanisms such as synaptic remodeling, neurogenesis, and functional reassignment help the brain adapt and recover. The chapter also emphasizes evidence-based strategies like mindfulness, physical activity, social support, and sleep hygiene, all of which strengthen neural pathways related to coping and stress tolerance. In addition, workplace-level interventions such as resilient leadership, organizational culture, and ergonomically designed workspaces foster systemic support for brain-based adaptability. By integrating neuroscience, psychology, and organizational behaviour, this chapter presents a holistic model for understanding and enhancing workplace resilience through neuroplasticity. Ultimately, it argues that cultivating neuroplastic resilience benefits not just individual employees but also fosters more adaptive, productive, and humane workplaces.

According to WHO, the incidence of mental health disorders in working Individuals are estimated around 15 percent (World Health Organization, 2024). More eighty three percent workers in the USA experience stressors at work which also impact their home life. A recent poll in Europe revealed that 51% of employees there identify stress as a common issue in their workplaces (Muñoz et al., 2022).

Workplace stressor has become a global concern, where prolonged exposure is causing detrimental effects on the functions of the brain. Studies on chronic stress revealed that it can increased production of cortisol that impacts the prefrontal cortex which is responsible for activities like decision making, attention and emotional regulation.In the workplace, stressors have been associated with absenteeism, job performance and productivity (Manning & Osland, 1989). In terms of health, workplace stressors can impact four negative outcomes including poor assessment of mental and physical health of self, diagnostic conditions like cardiovascular issues and death (Kivimäki & Kawachi, 2015).

Some critical determinants of workplace stressors include poor communication, financial deficiencies, ostracism, longer work hours and leadership styles (Tülay Çivici & Gulden Gumusburun Ayalp, 2024; Yao et al., 2014). The recent COVID-19 Pandemic has impacted the individual perception of work and workplace, with the shift from in person to hybrid offices. It has further exacerbated stressors causing significant work life conflict and increased techno- stress (Chan et al., 2022). Employers in the current times have realised the importance of providing effective interventions at workplace for managing stressors. Introduction of employee counselling services, mental health leaves, triaging interventions like stress management workshops etc have become a core part of extensive activities conducted for employee wellbeing. Despite these interventional supports, stressors at workplace grows exponentially in the current times.

One question to ponder on is how do people still cope up with these stressors if the effects of interventions are low. Are there any regulatory mechanisms within us which aids us through situational challenges and stressors at workplace. One answer to this question lies in the phenomenal discovery of neuronal plasticity coined by Santiago who viewed it as modifications in adult brain not related to any pathology. The brain has unique capacity to moderate stressful situations, adapting itself and retaining homeostasis (Hermans et al., 2024). The brain's capacity for neuroplasticity can help individuals develop resilience in the face of workplace stressors. This intrinsic ability of the brain to change and reorganize itself allows us to rethink challenging situations and strengthen the positive brain circuits that support healthy coping mechanisms and emotional control.

Cramer et al. (2011) defines neuroplasticity as nervous system's capacity to change its structure, function, and connections in response to internal or external stimuli. Looking into the origins of neuroplasticity, the basic idea was introduced by William James in the 1890. The term was further coined by Jerzy Konorski in 1948 and popularised by Donald Hebb in 1949. It was earlier studied in terms of understanding how the brain was able to process sensory information post injury and impairment.

The chapter aims to comprehensively review the mechanisms of stress, models of stress and different stressors at workplace and fundamentals of neuroplasticity. It also clearly articulates the distinct definitions of neuroplasticity and resilience. It further highlights how chronic workplace stress can lead to specific maladaptive changes in brain structures like the prefrontal cortex and amygdala, affecting cognitive functions and emotional regulation relevant to work performance. It also focuses on understanding different interventions including mindfulness, cognitive restructuring can promote positive neuroplastic changes that strengthen neural circuits associated with stress management and coping mechanisms in a work setting. We also aim to understand and explore how individual's pre-existing neurobiological vulnerabilities and coping styles interact with workplace factors like job control and social support to either exacerbate stress or foster resilience.

2. THE NEUROSCIENCE OF STRESS: HOW THE WORKPLACE SHAPES OUR BRAINS

A key feature of the stress response is the activation of the autonomic nervous system and the hypothalamic-pituitary-adrenal (HPA) axis. This surge is commonly understood as the fight-or-flight response, preparing the body and mind to react to threats like predators, muggings, accidents, or natural disasters. While a regulated amount of stress is important for surviving, excess amount of the same stress can lead to challenges. During the activation of HPA axis, three key structures of brain are predominantly active and get affected the most.

1. Amygdala – It is a cluster of nuclei in the medial temporal lobe of the brain which is shaped like almond. It plays a significant role in implicit emotional memories, which are often unconscious and linked to feelings. Research consistently shows that the amygdala in primates is involved in the emotional processing of social cues, such as the faces and vocalizations of their own species (Olucha-Bordonau et al., 2015). Studies show that amygdala plays an important role in in instrumental performance driven by secondary rather than primary rewards (Balleine & Killcross, 2006).
2. Hippocampus- The widely studied and famous case of H.M. in psychology profoundly demonstrated the critical role of the hippocampus in memory formation; due to its surgical removal, he permanently lost the ability to form new, declarative memories. Post this, the Hippocampus became the most thoroughly investigated brain structures (Knierim, 2015). The hippocampus connects bidirectionally with virtually all cortical association regions of the CNS. This indicates its role as a multi-modal sensory integration system within the CNS. The hippocampus is vital for learning, memory, and spatial navigation.
3. Prefrontal Cortexor PFC is located in the anterior frontal lobes is important when top-down processing is required (Harms & Pollak, 2023). The PFC becomes critical in situations where the connections between sensory input, thoughts, and actions are either weak or rapidly changing (Miller & Cohen, 2001). PFC Constitutes approximately one-third of the human brain's volume and regulates cognitive functions, behaviours, emotions, and bodily reactions in ways that

are virtually opposite to those initiated during moments of stress (Gold et al., 2002).

Stress impacts these regions of the brain through stress response system. This a complex network of psychological and physiological responses that will be delved in below.

2.1 Stress response system in humans

Imagine you are strolling in the park whilst enjoying the serene nature around you. Suddenly you perceive a slithering object around your legs which your brain perceives to be of a snake. When this occurs, your amygdala will detect it as a threat and send a warning signal to the hypothalamus, which then activates the sympathetic nervous system

The autonomic nervous system provides the body with a quick-acting control system for essential organs and functions, like regulating cardiovascular, respiratory, gastrointestinal, renal, and endocrine activity. It does this through the sympathetic nervous system, the parasympathetic nervous system, or a combination of both (Ioannis Kyrou et al., 2017). The sympathetic nervous system earned its name because its various actions usually work in unison. For example, during stressful situations, it orchestrates a unified response where your heart rate speeds up, blood vessels in your skin narrow, and digestion slows down. This integrated response is partly due to the specific anatomical arrangement of the sympathetic outflow (Watson et al., 2010).

When that signal reaches the adrenal glands via autonomic nerves, the hormone epinephrine (adrenaline) is released which triggers many physiological changes. During times of stress, your body undergoes a rapid series of changes to prepare for action. Your heart rate speeds up, efficiently pumping blood to your most crucial areas like your organs, muscles, and even the heart itself. This leads to a noticeable increase in your pulse and blood pressure. Simultaneously, your breathing quickens as your airways widen, allowing you to take in the maximum amount of oxygen. This rush of oxygen to your brain makes you more alert and sharpens your senses, like sight and hearing. At the same time, epinephrine triggers the release of stored glucose and fats, providing an immediate burst of energy throughout your entire body.

After the initial rush, the hypothalamus activates the hypothalamic-pituitary-adrenal (HPA) axis which is a network of hypothalamus, the pituitary gland, and the adrenal glands. The HPA axis is fundamental to maintaining basic bodily balance and orchestrating the stress response. Its primary pathway culminates in the production and release of cortisol. The hypothalamus initiates this process by secreting corticotrophin-releasing hormone (CRH). This is driven by both a circadian rhythm of neural input and stress-induced signals. CRH then stimulates the anterior pituitary gland to release adrenocorticotropic hormone (ACTH). ACTH travels to the adrenal cortex, prompting the secretion of cortisol into the bloodstream in the final stage (Orbell et al., 2013). HPA axis is also important in governing the production of versatile glucocorticoid hormone (Sunstrum & Inoue, 2019). These hormones impact peripheral functions like energy homeostasis and brain functions such as autonomic, emotional, and cognitive processes.

After the intense activation of HPA, the body needs to go back to its original state or attain homeostasis. This is fulfilled by Parasympathetic Nervous System (PNS), often referred to as the rest and digest system. The parasympathetic nervous system is predominantly organized around cranial and sacral spinal nerves. Its preganglionic neurons emerge from the brain or sacral spinal cord, and notably, they connect with only a small number of postganglionic neurons. These postganglionic neurons are characteristically located either directly in or immediately adjacent to the effector organs (muscles or glands) they influence. Activation of these systems leads to Pupil constriction, Reduced heart rate and lowered blood pressure, Narrowing of bronchial airways, Enhanced digestive activity, Increased production of saliva, mucus and greater urine secretion (Noyes & Barber-Westin, 2017). Acetylcholine serves as the primary neurotransmitter for both sympathetic and parasympathetic preganglionic fibers, facilitating communication (Rea, 2014).

Having explored the fundamental mechanisms of the body's stress response, we can now apply this the workplace. It's crucial to differentiate between acute and chronic workplace stress to understand how it impacts the neurobiology of the brain

2.2 Acute vs. Chronic Workplace Stress and Brain Changes

Recall the example where the brain triggers a fight-or-flight response upon perceiving a snake, often bringing anxiety and helplessness. this illustrates an episode of acute stress. While its effects can persist and contribute to chronic stress, acute stress represents the initial, short-term biophysical and psychological consequences of an experience. The brain's response to acute stress includes heightened alertness and enhanced learning and memory for the context surrounding the stressor (Segan et al., 2022). These changes are largely attributed to a region-specific increase in glutamate transmission, influenced by stress hormone exposure. Brain areas showing increased neuronal excitability after acute stress include limbic regions and the medial prefrontal cortex (van Campen et al., 2014).

Chronic stress stems from the cumulative burden of ceaseless everyday pressures, such as professional exhaustion. It's connected to unfavorable mental states and detrimental lifestyle choices such as inadequate nourishment, physical inactivity, insufficient sleep, and reliance on substances. Chronic stress also leads to consequences like inflammation and heightened coagulation. When experiencing chronic stress, a person's arousal state never fully deactivates, leading to persistently elevated cortisol levels. This can be detected in various bodily fluids and tissues, such as saliva, hair etc. Over time, these sustained changes result in subtle yet harmful physiological shifts. For instance, the increase in blood sugar caused by chronic stress has been linked to the development of type 2 diabetes (Buckle, 2015). Cultural norms profoundly shape how individuals perceive stress and engage with resilience interventions. In collectivist cultures, where the group's harmony and well-being are prioritized, stress may stem from failing to meet community expectations or causing shame to the group. Accordingly, workplace wellness programs in these settings often focus on group activities and shared goals to build collective support and resilience. In contrast, in individualistic cultures, stress is more often tied to personal failure or the pressure to succeed independently. Therefore, resilience interventions here are typically centered on individual coping strategies, personal achievement, and self-reliance.

2.3 Observable Impacts of Chronic Stress in the Workplace

Psychiatric disorders such as depression, generalized anxiety disorder (GAD), obsessive-compulsive disorder (OCD), post-traumatic stress disorder (PTSD), and addictive behaviours often involve cognitive impairments linked to chronic stress (Marin et al., 2011). Sustained periods of intense chronic stress have been shown to cause a reduction in brain volume and a decrease in its overall mass. Such alterations in brain structure modify how an individual reacts to stressors, their thought processes and recall abilities (Yaribeygi et al., 2017).

Chronic stress can lead to suppressing or ruminating of emotions. The prefrontal cortex becomes active, which in turn quiets down areas like the amygdala in the limbic system. While both suppression and rumination use similar brain regions, they show distinct patterns of neural activity (Ragen et al., 2016).

Chronic stress influences behavioural changes like procrastination and inability to self-regulate. In workplace excessive amount of stress diminishes productivity (Sirois et al., 2023). This negative impact is particularly evident when work encroaches upon employees' personal lives (Halkos & Bousinakis, 2010). Chronic stress from work appears to be the primary explanation for the poorer health observed in men with lower socioeconomic status. This stressful dynamic is further characterized by depression and a sense of despair (Kopp et al., 2007). Continued exposure towards chronic stressors could also lead to higher rates of absence and organizational impairment (Colligan & Higgins, 2008).

Chronic stress can physically manifest as sleep disturbances and weakened immune systems. A study conducted by Han et al. (2012) found that Sustained periods of stress are associated with increased levels of distal corticosteroids and disrupted sleep patterns. It also weakens the immune system by reducing B and T cell counts and diminishing NK cell activity.

Because stress alters brain structure, understanding neuroplasticity helps us see how recovery might occur

3. NEUROPLASTICITY: THE BRAIN'S CAPACITY FOR CHANGE AND ADAPTATION

In his 1890 work, "The Principles of Psychology," William James introduced the concept of brain and nervous system plasticity in his chapter on habits. While he couldn't experimentally investigate this phenomenon, his idea was rooted in contemporary biological and physiological research findings (Alcover & Rodríguez Mazo, 2012). Spanish neurologist Santiago Ramón y Cajal, using the Golgi staining technique revealed that individual neurons are the brain's fundamental building blocks. This discovery fundamentally reshaped neuroscience and changed how the brain develops (Foojan Zeine et al., 2024). During the 1890s, Meyer observed that emerging experimental evidence, indicating that nerve cells were not physically joined, substantiated James's idea of biological plasticity (Lamb, 2019). The term "neural plasticity" was later coined by Jerzy Konorski in 1948 (Puderbaugh & Emmady, 2023). In his 1949 book, The Organization of Behavior, Canadian psychologist Donald Hebb put forth the idea that experience could reshape neuronal connections. This concept describes a fundamental mechanism of synaptic plasticity (Zhao & Willing, 2018). Canadian psychiatrist and author Norman Doidge significantly popularized neuroplasticity through his book, "The Brain That Changes Itself." He underscored its potential for treating neurological conditions and enhancing cognitive function.

Now let us delve into understanding the mechanisms of neuroplasticity

3.1 Mechanisms of Neuroplasticity

Neuroplasticity can be understood via Synaptic plasticity, structural plasticity and functional plasticity

Synaptic plasticity is a fundamental mechanism driving many observable changes at the system level and within dendrites. Adaptive alterations at the synapse arise from a complex interaction of neurotransmitter release, the quantity and type of postsynaptic receptors, and the simultaneous activation of nearby structures, ultimately leading to stronger or weaker synaptic connections (Weishaupt, 2017). Over recent decades, synaptic plasticity has become recognized as a key mechanism for reshaping the brain's reward and motivation pathways. For instance, both stress and addictive substances disrupt synaptic plasticity within the mesocorticolimbic system.

This disruption then shapes the pathological learning seen in anxiety, depression, and addictive behaviors (Simmons et al., 2023). Donald Hebb's impactful cell assembly theory of synaptic plasticity is famously encapsulated by the phrase, "Cells that fire together, wire together". In contemporary terms, this theory explains how repeated and synchronized activity between a presynaptic input and postsynaptic depolarization leads to stronger synaptic connections (Zhang et al., 2024). While synaptic plasticity focuses on the connections between the neurons, the brains connection exists beyond this. Another important factor of the change in the brain is structural plasticity.

Structural plasticity encompasses various anatomical changes within neural tissue. These modifications include alterations in the number, location, and size of dendritic spines and filopodia, as well as changes in the branching patterns of dendrites and axons (Bozelos & Poirazi, 2017). Beyond its other functions, structural plasticity serves as the vital mechanism through which the brain can mend itself, a role that is arguably even more significant (Gage, 2004).

Functional plasticity refers to the brain's remarkable capacity to transfer functions from injured or lost tissue to undamaged areas. It can thereby adapt to neurological impairments. The functional Plasticity is further dependent on various factors such as individual's health and age (*APA Dictionary of Psychology*, n.d.).

According to Grafman (2000) At least four main types of functional neuroplasticity can be observed in humans: homologous area adaptation, cross-modal reassignment, map expansion, and compensatory masquerade.

- Homologous area adaptation is when a corresponding region in the opposite brain hemisphere takes over a specific cognitive function.
- Cross-modal reassignment happens when brain structures that used to process one type of sensory information begin to accept input from a different sense.
- Map expansion involves a functional brain region growing in size as a result of improved performance.
- Compensatory masquerade is the novel assignment of a specific cognitive process to complete a task.

Another important phenomenon that proves neuroplasticity is neurogenesis which is the formation of new neurons, beginning with stem and progenitor cell (Jahanshahi et al., 2011). For the majority of brain regions, this new neuron formation is confined to a brief developmental window, largely concluding shortly after birth (Apple et al., 2017). While often associated with the hippocampus and subventricular zone, adult neurogenesis also takes place in other specific brain regions like the hypothalamus (Cavallucci et al., 2020).

Positive influences such as aerobic exercises or resistance exercise like muscle building, calorie deficiency, ketogenic diet, supplements like omega -3 fatty acid increases the neurogenetic processes (Pickersgill et al., 2022). Studies have shown that mindfulness-based practices increase gray matter leading to structural neurogenesis (Hölzel et al., 2011). Negative influences of neurogenesis include lack of sleep as it affects the formation of new neurons in the hippocampus because of elements active during wakefulness (Mueller et al., 2015). social isolation and social defeat are linked to a reduction in newly generated neurons, whereas social dominance and an enriched social environment promote their increase (García-Gómez et al., 2022). Drug abuse alters neurophysiological functions at various cellular and molecular levels in the addicted brain, resulting in cognitive deficits (Ameneh Rezayof et al., 2023).

3.2 Neuroplasticity as a Buffer Against Stress

When introduced to a stressor, our brain copes in two ways either via active or passive coping. Active coping involves goal directed actions that diminish if the target is not achieved whereas passive coping is learned through extinction. Sometimes individuals also cope via actively avoiding the source of stress. Neuroplasticity occurs during all of these coping phases (Cabib et al., 2020). Via neurogenesis, individual's stress responsiveness can be increased and can reduce its susceptibility to stress. Evidence based treatment that reverse the behavioural effects of stress like presence of stimulating environments or removing adrenal glands that prevents stress hormones depends on neurogenesis (Levone et al., 2015). functional involvement of Adult Hippocampal Neurogenesis (AHN) specifically focuses on how the newly generated granule neurons (abGNs) in the dentate gyrus impact various aspects of the stress response. AHN helps make sense of information from the environment more accurately and within context. This leads to more appropriate responses to actual stressful situations, preventing "over-generalized" stress responses that might happen less frequently. This could explain why more new neurons reduce the impact of chronic stress (Surget & Belzung, 2021).

4. BUILDING RESILIENCE: LEVERAGING NEUROPLASTICITY IN THE WORKPLACE

Mrs. A is a fresh graduate and landed her first job as a marketing assistant at a fast-paced tech startup. She was excited to apply her theoretical knowledge, but the reality of the workplace quickly set in. Her manager was often demanding and provided little direct guidance, expecting Mrs. A to figure things out independently.

The workload was intense, deadlines were tight, and she often felt overwhelmed by the sheer volume of tasks.

One particularly tough week, Mrs. A was asked to do a client presentation which she has never done before. Despite her best efforts, she made several formatting errors and missed a key data point. Her manager was frustrated and pointed every mistake in front of the team.

Mrs. A stood at a crossroads at this point. She could choose to become **resilient**. This path would involve acknowledging her mistakes, seeking feedback constructively, and actively looking for resources or mentors within the company to help her improve. She might start by dedicating extra time to master the presentation software, asking a trusted senior colleague for tips, or even politely requesting more structured guidance from her manager. By reframing the setback as a learning opportunity, she could develop coping mechanisms like time management strategies, stress-reduction techniques, and a stronger belief in her ability to overcome future obstacles. This would allow her to adapt, grow, and ultimately thrive in the challenging environment.

Alternatively, Mrs. A could find other ways to cope. She might start to internalize the criticism, lose confidence, and become disengaged from her work. This could manifest as avoiding new tasks, procrastination, or even calling in sick frequently. She might complain constantly to friends and family, dwelling on the negative aspects of her job without actively seeking solutions.

In the field of positive organizational behaviour, **resilience** stands out as a fundamental element, defined as the ability to adapt positively when facing challenges. With the current work environment being increasingly disruptive and adverse, there's been a notable rise in scholarly and practical attention to workplace resilience in recent years (Hartmann et al., 2019).

Initially, resiliency was seen as a personality trait, a built-in coping style that helped certain children navigate tough situations and stay on a positive path. These kids were often labeled hardy or stress resistant. Despite this early focus, there's no single agreed-upon definition for the resiliency trait (Kuldas & Foody, 2021).

By 2015, researchers proposed that resiliency could be understood through three ecological systems: engineering, ecological, and adaptive. These weren't tied to one specific psychological theory or measurement model, but they offered a way to assess resiliency as a trait.

4.1 Theories of resilience

A key concept behind STAIR is Hobfoll's Conservation of Resources (COR) theory. This model views resources as either internal (like personal strength) or external (like social support) elements that help us cope and build resilience (Ortigo et al., 2020). When we face stressful events, especially trauma, these resources get used up. More stress on an already depleted system can lead to a breakdown in our ability to cope. Hobfoll and colleagues applied this idea to trauma, discovering that repeated traumatic experiences can cause a “spiral of loss” in resources, which in turn worsens PTSD symptoms. COR theory further posits, and research confirms, that environments lacking resources severely undermine resilience. The accumulation of resource losses is more impactful and rapid than gaining an equivalent amount (Hobfoll et al., 2015). This means events causing significant resource depletion will greatly hinder the development of resilience. Losing resources early on makes it much harder to build resilience, as it creates a domino effect where further losses become more likely. This is especially true when an individual hasn't yet established a strong reserve of protective resources to fall back on.

The Broaden-and-Build theory is fundamentally linked to resilience by explaining how positive emotions contribute to it (Fredrickson, 2004). At its heart, the concept suggests that pleasant feelings like happiness or curiosity expand a person's thinking and behaviors. This shifts them away from the limited, survival-driven reactions (such as confronting danger or fleeing) that often arise from difficult emotions. Such an expanded perspective fosters discovery, inventiveness, and relationships, which in turn helps accumulate personal assets. These assets aren't fleeting; they encompass physical, mental, social, and emotional capabilities. Crucially, these accumulated resources function as reserves that can be drawn upon during challenging times. By fostering these resources, positive emotions enhance one's ability to cope effectively with adversity and “bounce back” from setbacks. In essence, the theory suggests that cultivating positive emotions directly fuels psychological resilience, leading to greater well-being and an increased capacity to navigate future difficulties (Fredrickson, 2001).

Psychological Capital (PsyCap) is a comprehensive idea made up of several positive inner strengths: a sense of hope, belief in one's own abilities, the capacity to bounce back from setbacks, and a positive outlook (Youssef-Morgan, 2024). Research has clearly shown that having high PsyCap leads to desirable outcomes across the board, including better job performance, more positive attitudes, constructive behaviors, and improved overall physical and mental well-being.

The systems theory of resilience views resilience not as an individual trait, but as a characteristic of an entire system. The systems theory of resilience explains how entire systems like individuals, teams, or organizations adapt and maintain their purpose when facing challenges (Wright et al., 2012). It looks at how interconnected parts work together, not just individual strengths, to not only “bounce back” but also to adapt and transform in the face of adversity. This holistic view emphasizes continuous adjustment and the dynamic interplay of all system components

The stress response quickly frees energy. It enables a swift brain response, restoring body balance. Crucially, it regulates brain plasticity. Understanding this healthy stress response is vital. This helps grasp stress resilience. That's maintaining mental health during adversity. It's also quickly recovering afterward. Research separates resilience factors from mechanisms. Factors are stable variables. They predict resilient outcomes. Mechanisms are how these factors work (Hermans et al., 2024).

5. INDIVIDUAL STRATEGIES FOR CULTIVATING NEUROPLASTIC RESILIENCE

A clear link exists between trait mindfulness and resilience, confirmed by various studies. The mindfulness model proposes that mindfulness helps individuals detach from their thoughts (Zhang et al., 2023). Instead of being consumed by inner experiences, people can observe them objectively. This shift is called reperceiving and it involves stepping back to simply witness one's internal narrative. This process builds a greater tolerance for uncomfortable feelings. This in turn enhances one's capacity to handle stress and foster higher levels of resilience. The neuroscience of mindfulness reveals its link to emotion regulation, activating the anterior cingulate, insular, and prefrontal cortices. This also leads to deactivating the amygdala. Individuals with higher mindfulness show similar brain activity patterns (Wheeler et al., 2017). These functional and structural brain changes correlate with improved attentional skills and psychological well-being. An experimental study confirmed the benefits of online resilience training in workers (Joyce et al., 2019). High-risk workers like firefighters participated in a mindfulness-based program. Results showed significant improvements in adaptive resilience for the intervention group, particularly for those completing most sessions. This suggests internet-delivered mindfulness training is an effective and scalable way to boost resilience in demanding workplaces.

In order to build brain resilience and ward off neurological diseases, physical activity or exercise is important. In both human and animal studies, exercise improves cognitive function and increased stress resilience (Arida & Teixeira-Machado, 2021). According to Lancaster and Callaghan (2022) Physical activity strengthens our ability to bounce back from challenges, right down to our core biology. It encourages the brain to release chemicals like neurotransmitters and endorphins. This creates a feeling of intense well-being. That positive feeling helps to reduce unhelpful ways of thinking. It also acts as a helpful distraction from stressful events. The Brazilian Health Ministry recently underlined the crucial role of physical activity in the workplace. About 100,000 workers in Brazil are absent from their jobs due to problems with their muscles and bones linked to work. This issue also drains company finances (Matheus & Martinez, 2021). Companies like Microsoft and Google are recognising this and providing onsite fitness facilities for the employee's wellbeing (Focht, 2020). For example, Microsoft offers "$800

Stay Fit reimbursement for gym memberships". Google provides nutritious food options through cafes and micro-kitchens. According to de Sousa Fernandes et al. (2020) Brain flexibility, or neuroplasticity, happens when neurotrophins increase, changing how brain cells grow and communicate. Key neurotrophins include GDNF, NGF, NT3, NT4, and BDNF. A study showed that older women who did aerobic and resistance exercise for 12 weeks improved their thinking skills and had higher BDNF levels. This is likely because exercise makes muscles activate pathways that signal the brain, showing a strong connection between muscle and brain activity.

Our bodies' natural stress response, managed by the neuroendocrine system, can become harmful if it stays active too long. Resilience and stress are closely linked, much like two sides of a coin, and both significantly impact sleep. Lower resilience often means higher stress, which then contributes to more sleep issues (Viviana Lo Martire et al., 2024). New findings indicate that a key function of the brain during sleep involves clearing away harmful substances. This occurs through the activation of its internal cleansing mechanism. An important part of this system involves specific vessels found in the brain's protective layers (Oxana Semyachkina-Glushkovskaya et al., 2023). Studies also showed that resilience can influence the connection between mental health and sleep quality in patients with sleep issues (Allan et al., 2023). Sleep is crucial for memory consolidation and managing emotions. When sleep quality is poor, the amygdala shows a 60% increase in activity when faced with emotional triggers(Poon et al., 2024). This heightened response ultimately reduces resilience. According to Meerlo et al. (2009) recent research explores how sleep impacts the creation of new cells and their development into neurons. Briefly losing sleep for less than a day seems to have minimal impact on basic cell production. However, extended sleep restriction can reduce hippocampal cell proliferation, cell survival, and neurogenesis due to cumulative effects.

The systematic study of gratitude in medical research is quite recent. Gratitude has been defined as recognizing kindness from others or the environment, and responding to that perception with positive feelings (Hahn et al., 2024). A study conducted by Kumar et al. (2022) explored gratitude as a personal strength during the COVID-19 pandemic's early stages. The results indicated that gratitude was also linked to more positive outlooks and experience and fostered resilience.

Social support has a critical role for both physical and psychological well-being. The effects of inadequate social connections and mental health issues are well studied (John Victor Kainama et al., 2023). Social support may buffer against genetic and environmental risks, enhancing resilience to stress. This occurs through its influence on key biological systems, including the HPA axis, the noradrenergic system, and central oxytocin pathways. Recent studies in neuroplasticity and psychosocial genomics strongly support this view. They show how social and psychological factors actually change our brain's biology (Garland & Howard, 2009).

Self-control is a vital skill that helps individuals navigate challenges and thrive. Traditionally, it's been linked to reducing negative behaviors. For example, students with stronger self-control are less likely to misinterpret others' actions as hostile and exhibit aggressive behavior. Beyond curbing maladaptive outcomes, self-control also enhances positive ones(Ronen, 2021). Research shows that students with high self-control report greater perceived social support, experience more positive emotions, and have a higher sense of happiness. Some even suggest that positive emotions contribute to self-control, providing a strong foundation for goal achievement and increased resilience.

6. ORGANIZATIONAL STRATEGIES FOR FOSTERING NEUROPLASTIC RESILIENCE

Beyond the buzzwords, how do effective leaders forge resilience organizations? Let's start by understanding what we mean by a resilient organization: it's fundamentally a system's innate capacity to maintain or swiftly regain equilibrium. A critical element is a strong leadership quotient. This signifies an environment where the vast majority of team members feel empowered to contribute ideas and step into leadership capacities within their respective domains (Southwick et al., 2017). Previous research indicates that resilient organizations move away from hierarchical management. Instead, they embrace decentralization, self-organization, and shared decision-making. This approach allows for greater flexibility and adaptation, which are characteristics of complex adaptive systems (Karreinen et al., 2023).

A study was conducted on 410 Australian middle managers to understand how organizational culture impacts organizational resilience. The study found that respect for people and work, led to higher resilience in the workplace. A culture of respect for people/teamwork consistently enhances resilience in birth, growth, and revival stages of the organization. Outcome orientation, which is another core component of culture positively impacts resilience and managerial efficiency. An innovative culture consistently boosts resilience by promoting adaptability and experimentation.

A resilient organization prioritizes safety by supporting its workers in three key areas: foresight, coping, and recovery(Rangachari & L. Woods, 2020). This support is provided across individual, team, and organizational levels. The goal is to anticipate potential failures, adapt effectively when failures occur, and restore safe conditions swiftly after any incident

According to (Zhang, Xie, et al., 2024) At the organizational level, resilient leadership contributes positively to various aspects, including organizational culture, commitment, justice, performance, and overall organizational resilience. It has even been linked to increased outward foreign direct investment.At the employee level, resilient leadership fosters positive outcomes such as increased employee positivity, psychological capital, job performance, and job engagement. It also plays a role in enhancing employee resilience while simultaneously reducing turnover intention. In essence, resilient leadership promotes positive results, cultivating resilience, adaptability, and effective recovery during challenging times

Bernuzzi et al. (2022) mentions striking a healthy work-life balance is increasingly challenging due to factors like the rise of dual-earner couples and more women entering the workforce. This often leads to difficulties in meeting family needs. One key aspect of this challenge is work-life conflict, which arises when individuals perceive an incompatibility between their work and other life domains. For instance, work demands might interfere with family responsibilities, and vice versa. Such conflicts can negatively impact an individual's well-being, family functioning, job satisfaction, and performance. The lines between work and family life frequently blur, causing behaviors and emotions to spill over from one domain to the other. Resilience, however, has shown a positive correlation with work-life balance, specifically work-family balance. For example, in a study of family medicine program directors, higher resilience was linked to having adequate personal time, a healthy work-life balance, and the ability to disengage from work-related thoughts.

According to Klatt et al. (2015) Mindfulness interventions boost attention and reshape brain structures. Mindfulness is a profound discipline of consciousness, teaching us to fully inhabit our minds, hearts, and bodies in each passing moment (Baer, 2014). While often linked to meditation, mindfulness transcends a mere technique; it's a fundamental way of being. It cultivates a deep, open awareness, allowing us to perceive and experience life as it genuinely unfolds, moment by moment. Mindfulness also positively affects the parasympathetic nervous system. This is vital as our body systems respond to stress. It also reduces anxiety and builds resilience amongst workers.

Resilience coaching is gaining significant traction in both research and practice, proving effective in workplace challenges (Sipondo & Terblanche, 2024). This approach functions as a forward-thinking method to enhance mental fortitude. Research indicates that this method outperforms typical instructional techniques when it comes to developing resilience. It provides distinct benefits: a private setting for addressing struggles, the early cultivation of abilities, and candid discussions about obstacles. This equips individuals to steer their work lives and make well-informed vocational decisions.

Constant change compels companies and their employees to adapt, ensuring customer satisfaction and global market success (Beuss et al., 2021). These high and shifting demands profoundly impact the modern workplace, increasing employee stress. Consequently, work incapacity days are rising. The current understanding of workplace stress largely overlooks the physical environment where work occurs.

While research increasingly shows that the physical setting influences both job performance and satisfaction, most studies on work-related stress concentrate on psychosocial factors (Vischer, 2007). This focus often neglects the expanding field of environmental psychology, which specifically examines the impact of workspace design. Providing optimal conditions like ergonomics, natural light and green spaces can reduce stress and enhance resilience at workplace.

Workplaces often cause stress, but they also provide self-worth, daily purpose, and access to wellness programs. Many, especially in countries like the United States, also gain health insurance through their jobs. Employee Assistance Programs exemplify workplace counselling and triage services (Langlieb et al., 2022). EAPs hold immense potential to address individual needs, both during normal times and crises. By examining their current structure and leveraging new technologies, we can develop an improved EAP model for our changing global landscape. For EAPs to succeed and scale in today's competitive market, we must first conduct a bottom-up review of their care processes, supported by meaningful outcome data.

7. MEASURING AND MONITORING RESILIENCE IN THE WORKPLACE

The Brief Resilience Scale or BRS was developed to measure an individual's ability to bounce back from stress (Smith et al., 2008). Researchers tested its reliability across diverse groups, including students and patients with cardiac conditions or chronic pain. The BRS proved to be a reliable tool, measuring a single concept of resilience. Its scores consistently correlated with various personal characteristics, social relationships, coping mechanisms, and overall health in all tested groups.

The Connor-Davidson Resilience Scale (CD-RISC) is a 25-item assessment. Each item uses a 5-point rating, where higher scores indicate greater resilience (Connor & Davidson, 2003). Investigators applied this metric across various populations, including local residents, individuals receiving basic medical attention, those attending mental health clinics, and subjects in studies for widespread worry and trauma.

The RAW Scale, developed and validated in Australia in 2013, drew its initial participants from diverse sectors including health, education, commerce, information technology, finance, and manufacturing (Greco et al., 2022). Currently, several countries are utilizing the English version of the RAW Scale in their research. It is 25-item and 20-item versions (Winwood et al., 2013).

Biofeedback devices help employees by providing **real-time, measurable feedback** on their body's physiological responses to stress. For example, a device might track an employee's heart rate variability (HRV) or muscle tension and translate that data into a visual or audio signal. This allows the employee to see or hear how their body is reacting to stress and learn to consciously alter those responses.

Virtual reality (VR) offers a unique way to practice resilience skills in a controlled, safe environment. VR simulations can be designed to recreate stressful workplace scenarios, such as giving a presentation or managing a conflict.

Identifying appropriate resilience scales presents several challenges. While an overview of selected tools reveals some commonalities, significant differences exist among them. An analytical framework, developed from literature, highlights six crucial criteria for evaluating these assessment tools: encompassing multiple resilience dimensions, considering cross-scale relationships, capturing temporal dynamism, addressing uncertainties, employing participatory approaches, and facilitating action plan development (Sharifi, 2016). However, current tools have achieved only limited success in meeting these comprehensive criteria. For instance, the environmental dimension, despite its critical role in building community resilience, receives comparatively less attention in existing assessments.

8. CHALLENGES, CONSIDERATIONS AND CONCLUSION

Neuroplasticity offers immense promise for enhancing human adaptability. However, its application in workplace resilience presents significant challenges. We must carefully consider these limitations and ethical dilemmas. Effective implementation also demands attention to diverse organizational contexts. Finally, future research directions will shape evolving best practices.

Neuroplasticity is certainly not a universal remedy (Brown & Weaver, 2012). Genetic predispositions can introduce deeper, inherent challenges. Severe emotional trauma also creates more profound obstacles. Prolonged chronic stress similarly impacts the brain's capacity for change. Achieving lasting change requires consistent, dedicated effort and substantial time investment. It is a process demanding persistent engagement from individuals. Numerous promising neuroplasticity-based interventions exist or are being studied, yet many questions persist. As mentioned, challenges related to clinical trial methodology remain (Cramer et al., 2011). Key questions about the best treatment settings and who benefits most from therapy still need to be answered. Addressing these will probably require much more collaboration among research teams working across multiple locations. Future studies might concentrate on customizing these brain-adapting treatments to fit each patient's unique biological details.

A critical concern involves the risk of "resilience shaming." This occurs when organizations place undue burden on employees. They often expect individuals to cope without addressing root systemic issues. Resilience training should genuinely complement broader organizational efforts (Kristensen et al., 2023). It must not replace necessary initiatives to reduce workplace stressors. Systemic problems demand systemic solutions, not merely individual adaptation.

Ethical considerations also emerge clearly in this discussion. A balance between employer responsibility and individual accountability is crucial. Organizations hold a fundamental duty to create healthy environments. Simultaneously, employees retain agency over their personal well-being. Future advancements, like brain-monitoring technologies, raise new privacy concerns. Companies must navigate these sensitive areas with utmost transparency.

Implementing neuroplasticity-based interventions varies greatly by context. Small businesses face different resource constraints than large corporations. Distinct industries also possess unique workplace stressors. A factory environment differs significantly from a creative agency. Cultural differences further influence approaches to stress and well-being. Solutions must be adaptable and culturally sensitive to succeed.

Future directions in research hold considerable promise. Personalized neuroplasticity interventions may become commonplace. These tailored approaches would target individual needs precisely. Technology-assisted resilience building, like biofeedback and virtual reality, shows potential. Longitudinal studies are essential to evaluate long-term workplace intervention effectiveness. These will provide vital data for informed, ethical, and impactful strategies. Addressing these challenges will foster truly resilient and thriving workforces.

Workplace stress is an ongoing and multifaceted problem, with implications for employee. Neuroplasticity offers both a scientific framework and a practical guide for constructing resilient workplace strength. Through the adaptation of the brain's natural ability to modify, individuals and organizations can break free from mere survival and embark on enduring flourishing.

REFERENCES

Alcover, C. M., & Rodríguez Mazo, F. (2012). Plasticidad Cerebral y Hábito en William James: Un Antecedente para la Neurociencia Social. *Psychologia Latina, 3*(1). Advance online publication. DOI: 10.5209/rev_psla.2012.v3.n1.38737

Allan, A. C., Gamaldo, A. A., Gamaldo, C. E., Gunia, B. C., & Al, M., Edoghogho Ighodaro, & Marie, R. (2023). The promotion of sleep wellness: Resilience as a protective factor. *PubMed, 2*. Advance online publication. DOI: 10.3389/frsle.2023.1133347 PMID: 37427086

APA Dictionary of Psychology. (n.d.). Dictionary.apa.org. https://dictionary.apa.org/functional-plasticity

Apple, D. M., Fonseca, R. S., & Kokovay, E. (2017). The role of adult neurogenesis in psychiatric and cognitive disorders. *Brain Research, 1655*, 270–276. DOI: 10.1016/j.brainres.2016.01.023 PMID: 26801829

Arida, R. M., & Teixeira-Machado, L. (2021). The Contribution of Physical Exercise to Brain Resilience. *App.dimensions.ai, 14*(626769). https://doi.org/DOI: 10.3389/fnbeh.2020.626769

Baer, R. A. (2014). *Mindfulness-based treatment approaches: clinician's guide to evidence base and applications*. Elsevier/Academic Press.

Balleine, B. W., & Killcross, S. (2006). Parallel incentive processing: An integrated view of amygdala function. *Trends in Neurosciences, 29*(5), 272–279. DOI: 10.1016/j.tins.2006.03.002 PMID: 16545468

Bernuzzi, C., Sommovigo, V., & Setti, I. (2022). The role of resilience in the work-life interface: A systematic review. *Work (Reading, Mass.), 73*(4), 1–19. DOI: 10.3233/wor-205023 PMID: 36093652

Beuss, F., Sender, J., & Fluegge, W. (2021). Resilience-enhancing Workplace Design - An Approach for Workplaces in the Manual Assembly of large-scaled One-off Products. *Procedia CIRP, 104*, 313–317. DOI: 10.1016/j.procir.2021.11.053

Bozelos, P., & Poirazi, P. (2017). Impact of Structural Plasticity on Memory Capacity. *The Rewiring Brain*, 319–341. https://doi.org/DOI: 10.1016/b978-0-12-803784-3.00015-9

Brown, A., & Weaver, L. C. (2012). The dark side of neuroplasticity. *Experimental Neurology, 235*(1), 133–141. DOI: 10.1016/j.expneurol.2011.11.004 PMID: 22116043

Buckle, J. (2015). *Stress and Well-Being*. Churchill Livingstone., DOI: 10.1016/B978-0-7020-5440-2.00011-5

Cabib, S., Campus, P., Conversi, D., Orsini, C., & Puglisi-Allegra, S. (2020). Functional and Dysfunctional Neuroplasticity in Learning to Cope with Stress. *Brain Sciences, 10*(2), 127. DOI: 10.3390/brainsci10020127 PMID: 32102272

Cavallucci, V., Fidaleo, M., & Pani, G. (2020). Nutrients and neurogenesis: The emerging role of autophagy and gut microbiota. *Current Opinion in Pharmacology, 50*, 46–52. DOI: 10.1016/j.coph.2019.11.004 PMID: 31869664

Chan, X. W., Shang, S., Brough, P., Wilkinson, A., & Lu, C. (2022). Work, life and COVID-19: A rapid review and practical recommendations for the post-pandemic workplace. *Asia Pacific Journal of Human Resources, 61*(2). Advance online publication. DOI: 10.1111/1744-7941.12355

Çivici, T., & Ayalp, G.G. (2024). Modelling the critical determinants of architects' work-related stress. *International Journal of Construction Management/˜the œInternational Journal of Construction Management*, 1–14. https://doi.org/DOI: 10.1080/15623599.2024.2313817

Colligan, T. W., & Higgins, E. M. (2008). Workplace Stress: Etiology and Consequences. *Journal of Workplace Behavioral Health, 21*(2), 89–97. DOI: 10.1300/j490v21n02_07

Connor, K. M., & Davidson, J. R. T. (2003). Development of a new resilience scale: The Connor-Davidson resilience scale (CD-RISC). *Depression and Anxiety, 18*(2), 76–82. DOI: 10.1002/da.10113 PMID: 12964174

Cramer, S. C., Sur, M., Dobkin, B. H., O'Brien, C., Sanger, T. D., Trojanowski, J. Q., Rumsey, J. M., Hicks, R., Cameron, J., Chen, D., Chen, W. G., Cohen, L. G., deCharms, C., Duffy, C. J., Eden, G. F., Fetz, E. E., Filart, R., Freund, M., Grant, S. J., & Haber, S. (2011). Harnessing neuroplasticity for clinical applications. *Brain, 134*(6), 1591–1609. DOI: 10.1093/brain/awr039 PMID: 21482550

de Sousa Fernandes, M. S., Ordônio, T. F., Santos, G. C. J., Santos, L. E. R., Calazans, C. T., Gomes, D. A., & Santos, T. M. (2020). Effects of Physical Exercise on Neuroplasticity and Brain Function: A Systematic Review in Human and Animal Studies. *Neural Plasticity, 2020*(1), 1–21. DOI: 10.1155/2020/8856621 PMID: 33414823

Focht, J. (2020, February 7). *7 Companies That Make Health & Fitness a Priority | Scouted*. Mediabistro. https://www.mediabistro.com/candidates/7-companies-that-make-health-fitness-a-priority/

Fredrickson, B. L. (2001). The role of positive emotions in positive psychology: The broaden-and-build theory of positive emotions. *The American Psychologist*, *56*(3), 218–226. DOI: 10.1037/0003-066x.56.3.218 PMID: 11315248

Fredrickson, B. L. (2004). The broaden–and–build theory of positive emotions. *Philosophical Transactions of the Royal Society of London. Series B, Biological Sciences*, *359*(1449), 1367–1377. DOI: 10.1098/rstb.2004.1512 PMID: 15347528

Gage, F. H. (2004). Neuroplasticity. *Phlebolymphology*, *6*(2). Advance online publication. DOI: 10.31887/dcns.2004.6.2

García-Gómez, L., Castillo-Fernández, I., & Perez-Villalba, A. (2022). In the pursuit of new social neurons. Neurogenesis and social behavior in mice: A systematic review. *Frontiers in Cell and Developmental Biology*, *10*. Advance online publication. DOI: 10.3389/fcell.2022.1011657 PMID: 36407114

Garland, E. L., & Howard, M. O. (2009). Neuroplasticity, Psychosocial Genomics, and the Biopsychosocial Paradigm in the 21st Century. *Health & Social Work*, *34*(3), 191–199. DOI: 10.1093/hsw/34.3.191 PMID: 19728478

Gold, P. W., & Gabry, K. Eddie., Yasuda, M. R., & Chrousos, G. P. (2002). Divergent endocrine abnormalities in melancholic and atypical depression: Clinical and pathophysiologic implications. *Endocrinology and Metabolism Clinics of North America*, *31*(1), 37–62. DOI: 10.1016/s0889-8529(01)00022-6 PMID: 12055990

Grafman, J. (2000). Conceptualizing functional neuroplasticity. *Journal of Communication Disorders*, *33*(4), 345–356. DOI: 10.1016/s0021-9924(00)00030-7 PMID: 11001161

Greco, P. B. T., Ongaro, J. D., Seidel, E. J., Machado, W. de L., McEwen, K., & Magnago, T. S. B. de S. (2022). Psychometric evaluation of the Resilience at Work Scale (RAW Scale - Brazil). *Revista Brasileira de Enfermagem*, *75*(3). Advance online publication. DOI: 10.1590/0034-7167-2021-0241 PMID: 35239826

Hahn, N., Brzoska, P., & Kiessling, C. (2024). On the correlation between gratitude and resilience in medical students. *GMS Journal for Medical Education*, *41*(1), Doc8. DOI: 10.3205/zma001663 PMID: 38504862

Halkos, G., & Bousinakis, D. (2010). The effect of stress and satisfaction on productivity. *International Journal of Productivity and Performance Management*, *59*(5), 415–431. DOI: 10.1108/17410401011052869

Han, K. S., Kim, L., & Shim, I. (2012). Stress and Sleep Disorder. *Experimental Neurobiology*, *21*(4), 141. DOI: 10.5607/en.2012.21.4.141 PMID: 23319874

Harms, M. B., & Pollak, S. D. (2023). Emotion regulation. *Elsevier EBooks*. https://doi.org/DOI: 10.1016/b978-0-323-96023-6.00036-1

Hartmann, S., Weiss, M., Newman, A., & Hoegl, M. (2019). Resilience in the workplace: A multilevel review and synthesis. *Applied Psychology*, *69*(3), 913–959. DOI: 10.1111/apps.12191

Hermans, E. J., Hendler, T., & Kalisch, R. (2024). Building Resilience: The Stress Response as a Driving Force for Neuroplasticity and Adaptation. *Biological Psychiatry*, *97*(4). Advance online publication. DOI: 10.1016/j.biopsych.2024.10.016 PMID: 39448004

Hobfoll, S. E., Stevens, N. R., & Zalta, A. K. (2015). Expanding the Science of Resilience: Conserving Resources in the Aid of Adaptation. *Psychological Inquiry*, *26*(2), 174–180. DOI: 10.1080/1047840x.2015.1002377 PMID: 27069335

Hölzel, B. K., Carmody, J., Vangel, M., Congleton, C., Yerramsetti, S. M., Gard, T., & Lazar, S. W. (2011). Mindfulness practice leads to increases in regional brain gray matter density. *Psychiatry Research: Neuroimaging*, *191*(1), 36–43. DOI: 10.1016/j.pscychresns.2010.08.006 PMID: 21071182

Jahanshahi, A., Temel, Y., Lim, L. W., Hoogland, G., & Steinbusch, H. W. M. (2011). Close communication between the subependymal serotonergic plexus and the neurogenic subventricular zone. *Journal of Chemical Neuroanatomy*, *42*(4), 297–303. DOI: 10.1016/j.jchemneu.2011.09.001 PMID: 21924347

Joyce, S., Shand, F., Lal, T. J., Mott, B., Bryant, R. A., & Harvey, S. B. (2019). Resilience@Work Mindfulness Program: Results From a Cluster Randomized Controlled Trial With First Responders. *Journal of Medical Internet Research*, *21*(2), e12894. DOI: 10.2196/12894 PMID: 30777846

Kainama, J. V., Riasnugrahani, M., & Rohinsa, M. (2023). The Effect of Social Support and Religiosity Against Bereavement. *Al Qalam*, *17*(4), 2763–2763. DOI: 10.35931/aq.v17i4.2379

Karreinen, S., Paananen, H., Kihlström, L., Janhonen, K., Huhtakangas, M., Viitaaho, M., & Tynkkynen, L.-K. (2023). Living through uncertainty: A qualitative study on leadership and resilience in primary healthcare during COVID-19. *BMC Health Services Research*, *23*(1), 1–13. DOI: 10.1186/s12913-023-09223-y PMID: 36894990

Kivimäki, M., & Kawachi, I. (2015). Work Stress as a Risk Factor for Cardiovascular Disease. *Current Cardiology Reports*, *17*(9). Advance online publication. DOI: 10.1007/s11886-015-0630-8 PMID: 26238744

Klatt, M., Steinberg, B., & Duchemin, A.-M. (2015). Mindfulness in Motion (MIM): An Onsite Mindfulness Based Intervention (MBI) for Chronically High Stress Work Environments to Increase Resiliency and Work Engagement. *Journal of Visualized Experiments, 101*. Advance online publication. DOI: 10.3791/52359 PMID: 26168365

Knierim, J. J. (2015). The hippocampus. *Current Biology, 25*(23), 1116–1121. DOI: 10.1016/j.cub.2015.10.049 PMID: 26654366

KOPP, M. S., SKRABSKI, A., SZEKELY, A., STAUDER, A., & WILLIAMS, R. (2007). Chronic Stress and Social Changes: Socioeconomic Determination of Chronic Stress. *Annals of the New York Academy of Sciences, 1113*(1), 325–338. DOI: 10.1196/annals.1391.006 PMID: 17483208

Kristensen, A. K., Kristensen, M. L., & Mikkelsen, E. G. (2023). Workplace hazing in nursing: An integrative literature review. *Health Sciences Review (Oxford, England), 9*, 100120. DOI: 10.1016/j.hsr.2023.100120

Kuldas, S., & Foody, M. (2021). Neither Resiliency-Trait nor Resilience-State: Transactional Resiliency/e. *Youth & Society, 54*(8), 0044118X2110293. https://doi.org/DOI: 10.1177/0044118x211029309

Kumar, S. A., Edwards, M. E., Grandgenett, H. M., Scherer, L. L., DiLillo, D., & Jaffe, A. E. (2022). Does Gratitude Promote Resilience During a Pandemic? An Examination of Mental Health and Positivity at the Onset of COVID-19. *Journal of Happiness Studies, 23*(7), 3463–3483. DOI: 10.1007/s10902-022-00554-x PMID: 35855779

Kyrou, I., Randeva, H. S., & Tsigos, C. (2017). *Stress, Insulin Resistance, and Type 2 Diabetes*. 351–358. https://doi.org/DOI: 10.1016/b978-0-12-802175-0.00035-8

Lamb, S. (2019). Neuroplasticity: A century-old idea championed by Adolf Meyer. *Canadian Medical Association Journal, 191*(49), E1359–E1361. DOI: 10.1503/cmaj.191099 PMID: 31818930

Lancaster, M. R., & Callaghan, P. (2022). The effect of exercise on resilience, its mediators and moderators, in a general population during the UK COVID-19 pandemic in 2020: A cross-sectional online study. *BMC Public Health, 22*(1). Advance online publication. DOI: 10.1186/s12889-022-13070-7 PMID: 35468747

Langlieb, A. M., Langlieb, M. E., & Xiong, W. (2022). EAP 2.0: Reimagining the role of the employee assistance program in the new workplace. *International Review of Psychiatry (Abingdon, England), 33*(8), 1–12. DOI: 10.1080/09540261.2021.2013172 PMID: 35412422

Levone, B. R., Cryan, J. F., & O'Leary, O. F. (2015). Role of adult hippocampal neurogenesis in stress resilience. *Neurobiology of Stress, 1*, 147–155. DOI: 10.1016/j.ynstr.2014.11.003 PMID: 27589664

Lo Martire, V., Berteotti, C., Zoccoli, G., & Bastianini, S. (2024). Improving sleep to improve stress resilience. *Current Sleep Medicine Reports, 10*(1), 23–33.

Manning, M. R., & Osland, J. S. (1989). The relationship between absenteeism and stress. *Work and Stress, 3*(3), 223–235. DOI: 10.1080/02678378908251558

Marin, M.-F., Lord, C., Andrews, J., Juster, R.-P., Sindi, S., Arsenault-Lapierre, G., Fiocco, A. J., & Lupien, S. J. (2011). Chronic stress, cognitive functioning and mental health. *Neurobiology of Learning and Memory, 96*(4), 583–595. DOI: 10.1016/j.nlm.2011.02.016 PMID: 21376129

Matheus, V., & Martinez, L. (2021). The importance of workplace exercise A importância da ginástica laboral. *Revista Brasileira de Medicina do Trabalho: Publicacao Oficial da Associacao Nacional de Medicina do Trabalho-ANAMT, 19*(4), 523–528. DOI: 10.47626/1679-4435-2021-666 PMID: 35733538

Meerlo, P., Mistlberger, R. E., Jacobs, B. L., Craig Heller, H., & McGinty, D. (2009). New neurons in the adult brain: The role of sleep and consequences of sleep loss. *Sleep Medicine Reviews, 13*(3), 187–194. DOI: 10.1016/j.smrv.2008.07.004 PMID: 18848476

Miller, E. K., & Cohen, J. D. (2001). An Integrative Theory of Prefrontal Cortex Function. *Annual Review of Neuroscience, 24*(1), 167–202. https://www.annualreviews.org/content/journals/10.1146/annurev.neuro.24.1.167 PMID: 11283309

Mueller, A. D., Meerlo, P., McGinty, D., & Mistlberger, R. E. (2015). Sleep and adult neurogenesis: Implications for cognition and mood. *Current Topics in Behavioral Neurosciences, 25*, 151–181. DOI: 10.1007/7854_2013_251 PMID: 24218292

Muñoz, S., Iglesias, C. Á., Mayora, O., & Osmani, V. (2022). Prediction of stress levels in the workplace using surrounding stress. *Information Processing & Management, 59*(6), 103064. DOI: 10.1016/j.ipm.2022.103064

Noyes, F. R., & Barber-Westin, S. D. (2017). Diagnosis and Treatment of Complex Regional Pain Syndrome. *Noyes' Knee Disorders: Surgery, Rehabilitation, Clinical Outcomes*, 1122–1160. https://doi.org/DOI: 10.1016/b978-0-323-32903-3.00040-8

Olucha-Bordonau, F. E., Fortes-Marco, L., Otero-García, M., Lanuza, E., & Martínez-García, F. (2015). Amygdala. *The Rat Nervous System*, 441–490. https://doi.org/DOI: 10.1016/b978-0-12-374245-2.00018-8

Orbell, S., Schneider, H., Esbitt, S., Gonzalez, J. S., Gonzalez, J. S., Shreck, E., Batchelder, A., Gidron, Y., Pressman, S. D., Hooker, E. D., Wiebe, D. J., Rinehart, D., Hayman, L. L., Meneghini, L., Kikuchi, H., Kikuchi, H., Desouky, T. F., McAndrew, L. M., Mora, P. A., & Bruce, B. (2013). Hypothalamic-Pituitary-Adrenal Axis. *Encyclopedia of Behavioral Medicine*, 1017–1018. https://doi.org/DOI: 10.1007/978-1-4419-1005-9_460

Ortigo, K. M., Bauer, A., & Cloitre, M. (2020). Skills Training in Affective and Interpersonal Regulation (STAIR) Narrative Therapy: Making meaning while learning skills. *Emotion in Posttraumatic Stress Disorder*, 513–543. https://doi.org/DOI: 10.1016/b978-0-12-816022-0.00018-1

Pickersgill, J. W., Turco, C. V., Ramdeo, K., Rehsi, R. S., Foglia, S. D., & Nelson, A. J. (2022). The Combined Influences of Exercise, Diet and Sleep on Neuroplasticity. *Frontiers in Psychology*, *13*, 831819. DOI: 10.3389/fpsyg.2022.831819 PMID: 35558719

Poon, K., Ho, M. S. H., Tai, A. P. L., Leung, M.-K., Chan, M. C. M., & Lau, W. K. W. (2024). Mediating roles of sleep quality and resilience in the relationships between chronotypes and mental health symptoms. *Scientific Reports*, *14*(1), 5874. DOI: 10.1038/s41598-024-56688-w PMID: 38467740

Puderbaugh, M., & Emmady, P. D. (2023, May 1). *Neuroplasticity*. National Library of Medicine; StatPearls Publishing. https://www.ncbi.nlm.nih.gov/books/NBK557811/

Ragen, B. J., Roach, A., & Chollak, C. (2016). Chronic Stress, Regulation of Emotion, and Functional Activity of the Brain. *Elsevier EBooks*, 241–249. https://doi.org/DOI: 10.1016/b978-0-12-800951-2.00029-7

Rangachari, P., & Woods, L., J. (2020). Preserving organizational resilience, patient safety, and staff retention during COVID-19 requires a holistic consideration of the psychological safety of healthcare workers. *International Journal of Environmental Research and Public Health*, *17*(12), 4267. DOI: 10.3390/ijerph17124267 PMID: 32549273

Rea, P. (2014). Introduction to the Nervous System. *Clinical Anatomy of the Cranial Nerves*, *xv–xxix*. Advance online publication. DOI: 10.1016/b978-0-12-800898-0.00019-1

Rezayof, A., Ghasemzadeh, Z., & Sahafi, O. H. (2023). Addictive drugs modify neurogenesis, synaptogenesis and synaptic plasticity to impair memory formation through neurotransmitter imbalances and signaling dysfunction. *Neurochemistry International*, *169*, 105572–105572. DOI: 10.1016/j.neuint.2023.105572 PMID: 37423274

Ronen, T. (2021). The role of coping skills for developing resilience among children and adolescents. *The Palgrave Handbook of Positive Education*, 345–368. https://doi.org/DOI: 10.1007/978-3-030-64537-3_14

Segan, L., Prabhu, S., Kalman, J. M., & Kistler, P. M. (2022). Atrial Fibrillation and Stress. *JACC Clinical Electrophysiology*, *8*(8), 1051–1059. DOI: 10.1016/j.jacep.2021.12.008 PMID: 35981797

Semyachkina-Glushkovskaya, O., Fedosov, I., Penzel, T., Li, D., Yu, T., Telnova, V., & Zhu, D. (2023). Brain waste removal system and sleep: Photobiomodulation as an innovative strategy for night therapy of brain diseases. *International Journal of Molecular Sciences*, *24*(4), 3221.

Sharifi, A. (2016). A critical review of selected tools for assessing community resilience. *Ecological Indicators*, *69*, 629–647. DOI: 10.1016/j.ecolind.2016.05.023

Simmons, S. C., Grecco, G. G., Atwood, B. K., & Nugent, F. S. (2023). *Effects of prenatal opioid exposure on synaptic adaptations and behaviors across development*. *222*, 109312–109312. https://doi.org/DOI: 10.1016/j.neuropharm.2022.109312

Sipondo, A., & Terblanche, N. (2024). Organisational coaching to improve workplace resilience: A scoping review and agenda for future research. *Frontiers in Psychology*, *15*. Advance online publication. DOI: 10.3389/fpsyg.2024.1484222 PMID: 39569094

Sirois, F. M., Stride, C. B., & Pychyl, T. A. (2023). Procrastination and health: A longitudinal test of the roles of stress and health behaviours. *British Journal of Health Psychology*, *28*(3). Advance online publication. DOI: 10.1111/bjhp.12658 PMID: 36919887

Smith, B. W., Dalen, J., Wiggins, K., Tooley, E., Christopher, P., & Bernard, J. (2008). The Brief Resilience scale: Assessing the Ability to Bounce Back. *International Journal of Behavioral Medicine*, *15*(3), 194–200. DOI: 10.1080/10705500802222972 PMID: 18696313

Southwick, F. S., Martini, B. L., Charney, D. S., & Southwick, S. M. (2017). Leadership and resilience. *Springer Texts in Business and Economics*, 315–333. springer. https://doi.org/DOI: 10.1007/978-3-319-31036-7_18

Sunstrum, J. K., & Inoue, W. (2019). Heterosynaptic modulation in the paraventricular nucleus of the hypothalamus. *Neuropharmacology*, *154*, 87–95. DOI: 10.1016/j.neuropharm.2018.11.004 PMID: 30408488

Surget, A., & Belzung, C. (2021). Adult hippocampal neurogenesis shapes adaptation and improves stress response: A mechanistic and integrative perspective. *Molecular Psychiatry*. Advance online publication. DOI: 10.1038/s41380-021-01136-8 PMID: 33990771

van Campen, J. S., Jansen, F. E., de Graan, P. N. E., Braun, K. P. J., & Joels, M. (2014). Early life stress in epilepsy: A seizure precipitant and risk factor for epileptogenesis. *Epilepsy & Behavior, 38*, 160–171. DOI: 10.1016/j.yebeh.2013.09.029 PMID: 24144618

Vischer, J. C. (2007). The effects of the physical environment on job performance: Towards a theoretical model of workspace stress. *Stress and Health, 23*(3), 175–184. DOI: 10.1002/smi.1134

Watson, C., Kirkcaldie, M., & Paxinos, G. (2010). Peripheral nerves. *Brain*, **•••**, 43–54. DOI: 10.1016/b978-0-12-373889-9.50004-8

Weishaupt, N. (2017). Cortical Plasticity in Response to Injury and Disease. *Elsevier EBooks*, 37–56. https://doi.org/DOI: 10.1016/b978-0-12-801942-9.00002-1

Wheeler, M. S., Arnkoff, D. B., & Glass, C. R. (2017). The Neuroscience of Mindfulness: How Mindfulness Alters the Brain and Facilitates Emotion Regulation. *Mindfulness, 8*(6), 1471–1487. DOI: 10.1007/s12671-017-0742-x

Winwood, P. C., Colon, R., & McEwen, K. (2013). A Practical Measure of Workplace Resilience. *Journal of Occupational and Environmental Medicine, 55*(10), 1205–1212. DOI: 10.1097/jom.0b013e3182a2a60a PMID: 24064782

World Health Organization. (2024, September 2). *Mental health at work*. World Health Organization. https://www.who.int/news-room/fact-sheets/detail/mental-health-at-work

Wright, C., Kiparoglou, V., Williams, M., & Hilton, J. (2012). A Framework for Resilience Thinking. *Procedia Computer Science, 8*, 45–52. DOI: 10.1016/j.procs.2012.01.012

Yao, Y.-H., Fan, Y.-Y., Guo, Y.-X., & Li, Y. (2014). Leadership, work stress and employee behavior. *Chinese Management Studies, 8*(1), 109–126. DOI: 10.1108/cms-04-2014-0089

Yaribeygi, H., Panahi, Y., Sahraei, H., Johnston, T. P., & Sahebkar, A. (2017). The impact of stress on body function: A review. *EXCLI Journal, 2017*, 1057–1072. DOI: 10.17179/excli2017-480 PMID: 28900385

Youssef-Morgan, C. M. (2024). Psychological capital and mental health: Twenty-five years of progress. *Organizational Dynamics*, *53*(4), 101081–101081. DOI: 10.1016/j.orgdyn.2024.101081

Zeine, F., Jafari, N., Nami, M., & Blum, K. (2024). Awareness integration theory A Psychological and genetic path to self-directed Neuroplasticity. *Health Sciences Review (Oxford, England)*, *11*, 100169.

Zhang, J., Mao, Y., Wang, Y., & Zhang, Y. (2023). The relationship between trait mindfulness and resilience: A meta-analysis. *Personality and Mental Health*. Advance online publication. DOI: 10.1002/pmh.1581 PMID: 36958861

Zhang, J., & Xie, C., & Songshan (Sam) Huang. (2024). The effect of resilient leadership on employee resilience during a crisis in tourism & hospitality firms: A self-determination perspective. *International Journal of Hospitality Management*, *122*, 103886–103886. DOI: 10.1016/j.ijhm.2024.103886

Zhang, T., Dolga, A. M., Eisel, U. L. M., & Schmidt, M. (2024). Novel cross-talk mechanisms between GluA3 and Epac2 in synaptic plasticity and memory in Alzheimer's disease. *Neurobiology of Disease*, *191*, 106389. DOI: 10.1016/j.nbd.2023.106389 PMID: 38142840

Zhao, L.-R., & Willing, A. (2018). Enhancing endogenous capacity to repair a stroke-damaged brain: An evolving field for stroke research. *Progress in Neurobiology*, *163-164*, 5–26. DOI: 10.1016/j.pneurobio.2018.01.004 PMID: 29476785

Chapter 5

Organizational Characteristics of "Happy Organizations"

Evangelina Rekha Joseph
https://orcid.org/0009-0008-3819-2454
Christ University, India

Vinita Seshadri
https://orcid.org/0000-0002-0552-1943
Christ University, India

ABSTRACT

This chapter undertakes a comprehensive exploration of the specific organisational characteristics that are foundational to the cultivation and sustained presence of happiness and well-being within the workplace. Moving beyond the well-trodden path of individual-level determinants of employee satisfaction, this contribution meticulously dissects the systemic elements, deeply ingrained cultural norms, and thoughtfully designed structural frameworks that collectively contribute to the emergence of a positive and flourishing work environment. By examining existing scholarly research and synthesizing diverse theoretical perspectives, this chapter identifies the core organisational attributes that not only nurture employee happiness and a profound sense of well-being but also exert a demonstrably positive influence on critical organisational outcomes.

DOI: 10.4018/979-8-3373-2292-6.ch005

INTRODUCTION

The contemporary workplace is experiencing profound transformation and unprecedented changes driven by increasing complexity, technological advancements, competitive pressures, shifting workforce demographics, evolving societal expectations and rising awareness of the significance of employee well-being in organisational success (Lim, 2023). Employee well-being has ascended to the forefront of strategic organisational thinking from being viewed as a mere peripheral concern of the management. Managers and business leaders increasingly recognise that employee happiness is not just a moral imperative, but a key source for competitive advantage (Galiano-Coronil et al., 2024).

Employee well-being has a become an integral aspect of human sustainability, a term introduced by Deloitte's 2023 Global Human Capital report, referring to "the degree to which an organization creates value for people as human beings, leaving them with greater health and well-being, stronger skills and greater employability, good jobs, opportunities for advancement, progress toward equity, increased belonging, and heightened connection to purpose." Studies establish the need for organisations to recognise the significance of employee well-being, which is not just a luxury but an essential factor for long-term sustainability. As such, the corporate leaders must understand how they can create a thriving workplace by prioritising well-being rather than assuming that employees' needs are being met (Makinde, 2024).

The traditional work paradigm, characterised by rigid hierarchical structure, command and control management styles, and a clear distinction between professional and personal life, gives way to a more fluid, collaborative, and holistic approach to organisational design. This shift indicates the need for understanding and acknowledging that the employees are not merely the resources to be optimised but rather complex human beings who have differential emotional, psychological, and social needs that significantly impact organisational performance, creativity, and commitment to organisational goals.

This chapter addresses this critical transformation by shifting the focus from the individual-centric approaches to examining the fundamental organisational characteristics that serve as a foundation for a happy and thriving workforce. While individual resilience and coping mechanisms are indisputably important, this chapter theorises that a sustainable approach to well-being requires a deep understanding and deliberate cultivation of organisational factors that foster happiness from the foundational level. The imperative for this organisational focus becomes more evident when considering individual-based interventions' limitations. Mindfulness programs, stress management workshops, and wellness initiatives are valuable, but often fall short of their intended impact when employees return to potentially toxic work environments and unsupportive management structures. It is analogous to treating patients' symptoms while ignoring the underlying disease- an approach that might provide temporary relief but fails to address the root causes of workplace happiness. This chapter aims to offer a comprehensive framework for understanding these crucial organisational underpinnings, and offer both theoretical insights and practical guidance to organisational leaders, human resource professionals, and organisational development practitioners who seek to create workplaces where happiness flourishes organically rather than being imposed as superficial level interventions.

Defining "Happy Organisations"

The concept of "happy organisations" goes beyond the simple notions of fleeting employee satisfaction or the mere absence of explicit dissatisfaction. To develop a meaningful framework for organisational happiness, we must first establish a more nuanced and multifaceted understanding of what constitutes a genuinely happy organisation, recognising that the happiness in an organisational context is complex and profound than fleeting positive emotions.

Drawing from various established theories of well-being, positive psychology and organisational behaviour, we have explored frameworks that can offer a multidimensional perspective on human flourishing in the workplace. Well-being is not only a combination of feeling good and functioning well; it is a state of experiencing positive emotions like happiness and contentment and having a sense of purpose and control over life (Ruggeri et al., 2020). The hedonic perspective on well-being focuses on happiness and defines well-being in terms of pleasure attainment and pain avoidance; the eudaimonic perspective focuses on meaning and self-actualisation and defines well-being in terms of the degree to which a person fully functions (Ryan & Deci, 2001). In organisational terms, hedonic well-being can translate to job satisfaction, employee engagement, and reduced workplace stress and burnout. The eudaimonic well-being manifests as employees finding deeper meaning in their work, experiencing opportunities for personal and professional development, and developing a sense that their contribution matters more than they perceive. This dimension of happiness is vital in knowledge-intensive environments where employees seek financial compensation and greater psychological fulfilment.

PERMA Model (Positive emotions, Engagement, Relationships, Meaning, and Achievement), introduced by Seligman in 2011, aims at increasing and measuring well-being, and is another practical framework for understanding organisational happiness (Kovich et al., 2022). Positive Emotions in the workplace include happiness, joy, hope, and satisfaction. Engagement refers to the individual's focus, interest, and absorption in work tasks. Relationships include peers' social connections, connectedness and a sense of community in the organisation. Meaning involves the sense of a larger purpose and achievement that transcends the work goals.

A truly happy organisation, therefore, would be the one where multiple dimensions of well-being are employed, supported, and nurtured through deliberate organisational choices. The absence of challenges does not characterise happy organisations, but rather by supportive structure, leadership, relationships, and culture that fosters the employee's sense of well-being and enables them to thrive in the face of uncertainty and complexity. To understand what constitutes a happy organisation, we must critically scrutinise the differences between individual happiness experienced within the organisational context and the concept of having a "happy" organisational culture. While an individual employee might experience moments of happiness and satisfaction, a happy organisation is one where the well-being of the employees is deeply embedded in its values and becomes a pervasive characteristic of the work environment. This distinction is critical because it emphasises the difference between organisations that sporadically produce happy employees and those fundamentally designed to foster happiness as a natural outcome. Happy organisations are characterised by what the researchers term as "positive organisational energy," a collective sense of enthusiasm, vitality, and engagement that permeates the workplace.

KEY ORGANIZATIONAL CHARACTERISTICS OF HAPPY ORGANIZATIONS

This chapter section methodically examines the key organisational attributes that differentiate a happy and high-performing workplace from the rest. Each of these characteristics discussed in the chapter represents a critical dimension of the organisational design, which contributes to sustainable workplace happiness when meticulously implemented and maintained. Keeping in mind that none of these characteristics exist in isolation but are part of a larger interconnected web of mutually reinforcing factors that collectively contribute to creating conditions for human happiness and flourishing in organisational contexts.

Positive and Trustworthy Organisational Culture

Every organisation has its own culture, its set of values, beliefs, assumptions, behaviours, norms, and attitudes shared among the employees. The foundation to any happy organisation lies in this culture that guides behaviour and decision making throughout the organisation (Tsai, 2021). Culture plays a significant role in determining whether or not the organisation is a happy and healthy workplace (Kane-Urrabazo, 2006). A positive and trustworthy organisational culture is a foundation upon which all the other characteristics depend, thereby creating the psychologically safe foundation the employees need to engage genuinely and contribute through their best efforts.

Psychological safety, perhaps, is the most critical element of a positive organisational culture. It refers to taking interpersonal risks without fearing negative consequences, and there is a sense of being safe (Patil et al., 2023). Individuals with psychological safety are more likely to communicate openly, ask questions, share ideas, and accept errors without fearing adverse outcomes in their career or interpersonal relationships. Having higher levels of psychological safety creates a sense of self-worth, where employees feel secure to take risks, experiment with new ideas and engage in collaborative interactions and work that foster learning and innovation at the workplace.

Research across diverse organisational and industrial contexts demonstrates that psychological safety is a prerequisite for high creativity, performance, and employee well-being. It induces a sense of vitality, leading to higher involvement in creative work (Kark & Carmeli, 2009). When employees fear repercussions for speaking up, they withhold creative ideas and valuable information. They are reluctant to take necessary risks, and their energy is spent on self-protection rather than contributing to the organisation. Conversely, when psychological safety is present, employees acknowledge their errors and share their learning with colleagues (Smeets et al., 2021). Interestingly, positive psychological safety triggers learning behaviour at individual and team levels and is a dominant driver of learning from errors (Frazier et al., 2017).

Building psychological safety requires deliberate actions and norm-setting from the leaders at all levels of the organisation. Leaders exhibit vulnerabilities by owning up to their own mistakes and insecurities, actively seeking feedback from team members, positively handling any constructive feedback or challenging perspective and consistently demonstrating acceptance of diverse opinions in a team. Behaviourally, organisations with strong psychological safety implement mechanisms to motivate employees to communicate openly, voice their concerns, and seek feedback to improve continuously (Newman et al., 2017).

Trust is the second pillar of positive organisational culture, which extends beyond psychological safety to encompass broader beliefs about the integrity, competence, and munificence of organisational leaders and systems. Kaplan et al (2020; p.1150) define trust as a critical factor imperative for properly functioning relationships between a human and another entity. Trust encompasses three fundamental elements: 1) a trustor, who is doing the trusting and is vulnerable to harm from another individual. 2) the trustee, who is the one being trusted, and who is the individual capable of harming the trustor; and, 3) a context within which the trustee's actions are capable of causing harm or benefit to the trustor." In context of organisation, organisational trust is defined as "positive expectations individuals have about the intent and behaviors of multiple organizational members based on organizational roles, relationships, experiences, and interdependencies" (Joo et al., 2022; p.6). Trust in organisations can manifest in multiple dimensions, including trust in leadership, colleagues, organisation processes and systems, and organisational commitment to employee well-being. High-trust organisations are characterised by open communication, consistent commitment, equity, fair employee treatment, and an alignment between their stated values and actual practices. Furthermore, high-trust organisations, when compared to low-trust ones, are more fun, supportive, motivating, challenging, open, productive, transparent, and comfortable (Hurley, 2012). Building organisational trust needs sustained efforts over time, as trust is earned through recurrent positive communication and can be quickly eroded by inconsistency. Organisations can foster trust through transparent decision-making, continuous communication, fair and consistent application of organisational policies and procedures, and visible commitment to employees' development and advancement.

Inclusivity and respect are additional pillars of positive organisational culture. With growing diversity in work organisations, organisational leaders recognise the significance of inclusivity. Inclusive organisations value diversity in all forms and are not limited to only demographic diversity; they extend to the diversity of individuals' experiences, thoughts, perspectives, and working styles. Inclusivity here goes beyond complying with the equal opportunities regulations, instead creating a work environment where all employees feel valued, welcomed, and their authentic selves (Vohra et al., 2015). Creating an inclusive culture needs deliberate efforts in various organisational processes and practices, including recruitment and promotion practices, ensuring equitable access to career advancement, and fostering inclusive behaviour at all levels of the organisation.

Supportive and Empowering Leadership

Leadership is pivotal in creating and sustaining a happy organisation, going beyond traditional notions of task delegation and performance management to creating an environment where employees can thrive. Leadership effectiveness indicates self-sacrificial work behaviour focused on bringing greater benefits to the organisation and its people. Employees can be inspired to shadow leadership behaviour and have meanings associated with their actions to shape their goals. Effective leadership behaviour significantly impacts their followers and ultimately the organisational systems. It helps create high commitment and motivation and drives their desire to work in a high-performance culture. This leadership effectiveness can create a vision for the future of the workforce and the organisations to ensure that both are aligned for long-term sustainability (Cakir & Adiguzel, 2020).

Supportive and empowering leadership is representative of a fundamental shift from command-and-control approaches to a more collaborative, development and humanistic approach to leadership that recognises and acknowledges the full potential of every employee. Emotional intelligence stands as a cornerstone of supportive leadership. A leader can recognise, evaluate, and control their own and other people's emotions (Drigas et al., 2023). Drigas & Papoutsi (2021, p.123) define emotional intelligence as "a set of abilities and skills that a person must train and develop gradually and hierarchically to reach emotional self-realisation. It is the response to emotional stimuli, the recognition—the expression of emotions, the full awareness and management of our own emotions but also the emotions of others, the social skills for better intrapersonal, interpersonal, and working relationships, the empathy and compassion, the accurate discrimination of emotions with the ultimate aim of the emotional development of our potential, self–actualization, transcendence and finally the unity of emotions because humans are part of a united world". Leaders with high emotional intelligence possess self-consciousness, self-regulation, empathy, motivation, and social abilities to create a more positive work environment to support both employees and management navigate challenging times and resonate with the followers (Dutta et al., 2025).

Empathy, a key component of emotional intelligence, enables leaders to understand and share the emotions and feelings of employees, leading to more compassionate and effective leadership. Landry (2019) suggests that leaders who excel in social awareness practice empathy. They constantly strive to understand their peers' feelings and perspectives, which enables them to collaborate and communicate effectively. Interestingly, the Global leadership development firm DDI ranks empathy as the topmost leadership skill. Empathetic leaders can recognise situations where employees are struggling, support them with appropriate resources and create a psychologically safe environment where everyone feels valued and understood. This empathetic approach allows the leaders to establish a link leading to higher employee engagement levels, job satisfaction, and organisational commitment.

Supervisor's active-empathetic listening is an important antecedent of work engagement. Active listening is listening to another person to understand the message and context better. It is a skill linked with higher emotional intelligence and actively responding to the other person (Jonsdottir & Kristinsson, 2020). Active listening goes beyond simply hearing words; it is about fully engaging with the other person, seeking to understand their perspective, asking for clarifications and demonstrating genuine interest in their opinions and thoughts. Leaders engaging in active listening create environments where employees feel valued and heard, resulting in improved communication, stronger relationships, and more efficient problem-solving.

Constructive feedback is a critical component of mentorship in supportive leadership, helping mentees identify and understand their strengths and weaknesses, develop plans for addressing these weaknesses, and improve capabilities over time (Kincaid, 2025). Effective feedback is timely and specific rather than personal, focusing on growth and development rather than being judgmental and criticising the individual. Leaders and supervisors who excel at providing constructive feedback create a learning environment where employees feel supported and confident to take on new challenges.

The most transformative aspects of supportive leadership are empowerment and autonomy. Autonomy support, basic psychological needs satisfaction, autonomous motivation, and perceived leadership support are vital for leading followers towards creative performance (Hocine & Zhang, 2014), rather than micromanaging the employees' decisions and tasks, empowering leaders extending authority, resources, and necessary support to the team members to make decisions and take ownership and accountability of their work. Employee empowerment satisfies their basic needs of autonomy and competency, whilst also allowing them to leverage diverse skills and perspectives of the team to achieve better outcomes. Studies have constantly demonstrated that employees who experience a higher level of autonomy in their work report greater engagement and job satisfaction, in addition to improved motivation and performance (Slemp et al., 2015). Autonomy does not mean abandonment; instead, it means empowering leaders by providing clear expectations, appropriate resources, and ongoing support while allowing employees to determine how best to achieve their objectives (Mclean, 2010).

Coaching and development orientation distinguishes supportive leaders from those focusing on achieving immediate performance outcomes. Leaders with a coaching orientation invest time and energy in helping their team members develop new skills, expand their capabilities, and advance in their careers. This developmental approach creates loyalty and engagement while building organisational capability over time. Effective coaching involves asking powerful questions that promote reflection and learning, providing opportunities for skill development and growth, offering stretch assignments that challenge employees appropriately, and creating clear pathways for career advancement (Zuberbuhler et al., 2020). Leaders who excel at coaching report higher levels of team performance, employee engagement and retention and simultaneously, the team members feel invested, valued, and supported by their organisations.

Supportive leadership involves leaders actively acknowledging and celebrating the achievements of their team members to boost employee engagement. When these acknowledgements and appreciations are delivered authentically, they can profoundly impact employee motivation and overall well-being (Jo & Shin, 2025). Supportive leaders understand that recognition is not a one-size-fits-all approach but requires understanding individual preferences and motivations. For instance, some employees value private acknowledgement whilst others prefer public recognition. Some employees are motivated by simple emails involving gratitude and appreciation, while others are formally motivated by rewards and ceremonies. Leaders who excel at recognising the achievements of their peers, ensuring their contributions are channelled, and contributing towards meaningful outcomes.

3.3 Meaningful Work and Job Design

The work design represents one of the most powerful levers of creating organisational happiness, primarily because the day-to-day experiences at work shape the employee's well-being, work engagement and job satisfaction. Experiencing meaning and purpose in life is fundamentally important for human existence and quality of life. Meaningful work is a subject of work experience regarding how significant and valuable the work is for the people (Martela & Pessi, 2018). It is argued that understanding the key sources that make the work meaningful is vital not only for the organisation's bottom line but also from the point of supporting employee motivation, commitment and well-being (Martela et al., 2021).

Job Characteristics Theory offers a fundamental lens for understanding how work design shapes employee motivation and satisfaction. Hackman and Oldham's (1974) model suggests five core dimensions that affect specific personal and work-related outcomes, including job dimensions. These core dimensions include "autonomy, feedback, skill variety, task identity, and task significance" (Ali et al., 2014). The models suggest that when the jobs have these characteristics, they foster greater intrinsic motivation, job satisfaction and work performance.

- Autonomy refers to the degree to which independence and discretion are granted to employees in scheduling their work and determining the processes and systems used to carry out their tasks.
- Feedback concerns the degree to which the job offers clear, specific, and actionable information to improve efficiency and job performance. Receiving regular and constructive feedback from the employees can help understand how they are performing and the areas of improvement.
- Skill variety is the degree to which the job demands skills and different talents to avoid experiencing monotony, and thereby encourages employees to develop and use their diverse capabilities.
- Task identity involves completing the assigned tasks and allowing the individuals to witness the task's outcome.
- Task significance is how the job affects others within the organisation and the broader community.

Contemporary organisations increasingly acknowledge the value of job design, which enables employees to link their tasks with meaningful outcomes (Hall & Heras, 2010). This includes restricting the roles that can involve workflow, creating new opportunities for direct communication and explicitly linking individual contributions, organisational mission, and long-term impact. Meaningful work, however, extends beyond job structure; it emphasises the "Why" of the work and why it matters. Batuchina et al. (2025) suggest that meaningful work allows employees to align their work with their values and passions, creating a pathway for greater fulfilment and commitment. When the work is meaningful, the employees develop greater resilience during challenging times and view the challenges as opportunities rather than obstacles. Employees who view their efforts as contributing to the greater good report higher engagement and satisfaction (van der Deijl, 2022). Organisations that create a sense of meaning among the workforce implement practices such as regular communication, sharing updates on organisational impacts and successes, creating opportunities for employees to interact with the beneficiaries of their work, celebrating individual contributions and ensuring the organisational values resonate with the employees' personal beliefs and values.

As identified, providing challenging work and growth opportunities is critical to meaningful work. It consists of community, contribution and challenge. Jobs that are either too simple or excessively demanding can lead to disengagement, boredom (low-demand jobs), or burnout and stress (high-demand jobs). Drawing from the job characteristics theory, it is suggested that when one experiences work as important and meaningful, they are always motivated and perform better (Isaksen, 2000; Kawada et al., 2023). In contrast, the job roles that offer the right level of complexity drive the employees to use their skills and abilities and foster creative thinking and higher engagement.

Career progression and development are associated with meaningful work, too. Employees who see a clear path of career advancement in the organisation are more likely to stay committed and engaged. Having transparent career pathways ensures that employees view their potential growth opportunities and work on developing the skills and experiences required to progress. Effective career development requires mentoring, coaching, leadership programs, and engagement in cross-functional projects. Implementing these initiatives demonstrates that organisations are willing to invest in employee growth while also strengthening talent pipelines. Flexibility and personalisation in the job design are vital for addressing how different employees find meaning in the work in different ways. While some employees thrive in a team-based work environment, others prefer to work independently. While some get energised with variety and change in job design, others seek fulfilment. Organisations that can deliver meaningful work through job design empower employees and ensure their jobs are a best fit for their strengths, passions and career objectives.

3.4 Fairness and Equity

Fairness and equity serve as a fundamental pillar of organisational happiness, addressing the deep needs of justice, respect and right to equal treatment (Jha et al., 2023). When the employees perceive their organisations as fair and equitable, they are more likely to trust the management and leadership, be engaged in their work, and maintain positive relationships with their colleagues. Conversely, if the employees develop the perception of unfairness, their morale, trust, and commitment can erode quickly and this is regardless of presence of other positive organisational characteristics.

Organisational justice not only affects the organisational performance but it also a key determinant of economic well-being of the employees. The organisational justice is the employees' perception of the fairness in the organisation. A happy organisation is reflective of the employees trust and willingness to believe in the values and practices of the organisation (Yean & Yusof, 2016). Organisational justice theory offers a comprehensive framework for understanding the different dimensions of fairness and justice at workplace through distributive, procedural, and interactional justice. Distributive justice is about fairness of outcomes whether it is rewards, recognition and opportunities among the employees. Procedural justice emphasises on the fairness of processes used to make decisions and resources allocations. Finally, the interactional justice explores the quality of interpersonal treatment received by the employees during the implementation of distributive and procedural outcomes.

Distributive justice requires careful consideration towards the rewards and compensation system, the key drive of employee's happiness to ensure that the pay, promotions, bonuses and other benefits are allocated in an apparent and dependent on the job-criteria rather than any irrelevant factors and favouritism (Nameda et al., 2022). Having a substantial distributive justice requires development of transparent compensation frameworks that clearly articulate ow the payments are being made, how regularly compensation practices are being reviewed and ensuring the systems are addressing potential inequities to ensure consistent application of rewards and recognition across all the employees. Organisational commitment to fairness contributes to creating a positive and happy organisational work environment. Procedural justice focuses on ensuring the organisational processes are fair, transparent, and consistently applied across the organisations. This included unbiased decision making and based on accurate information, creating opportunities for employees to voice their opinions especially in decisions that affects them, in addition to how policies and practices are applied across different employees and situations (Baldwin, 2006).

Fairness and equity serve as cornerstones for creating thriving workplace cultures where employees feel valued, motivated, and genuinely happy (Edmondson, 2019). When organizations prioritize these principles, they unlock significant benefits that extend far beyond basic compliance requirements. Fair treatment establishes psychological safety, allowing employees to express ideas without fear of unfair judgment or retaliation (Kahn, 1990). When workers trust that decisions are made transparently and consistently, they invest more emotionally in their roles, translating into higher engagement levels where employees willingly exceed their basic responsibilities because they believe their contributions matter and will be recognized appropriately (Cropanzano et al., 2007). Equitable practices significantly reduce workplace tension and interpersonal conflicts by creating clear promotion criteria, transparent compensation structures, and fairly distributed opportunities, allowing employees to focus productive energy on work rather than worrying about unfair treatment (Colquitt, 2001). Organizations known for fairness become magnets for high-quality candidates who seek workplaces aligned with their values, while employees in equitable environments demonstrate dramatically lower turnover rates, reducing recruitment costs and preserving institutional knowledge (Hausknecht et al., 2009). When diverse perspectives are valued equally and all voices are heard, organizations benefit from enhanced creativity and problem-solving capabilities, as fair treatment encourages risk-taking and innovation through objective evaluation processes (Page, 2007). Companies that embed fairness and equity into their organizational DNA experience sustainable growth, stronger customer relationships, and enhanced reputation, as happy employees deliver superior customer service, leading to increased loyalty and business success (Harter et al., 2002). Ultimately, fairness and equity create a positive reinforcement cycle where organizational success strengthens employee satisfaction, driving continued excellence and sustainable growth.

3.5 Strong Social Connections and Teamwork

Strong social connections and effective teamwork form the foundation of employee well-being and organizational happiness, creating environments where individuals thrive personally and professionally. Research consistently demonstrates that humans are inherently social beings who derive significant meaning, support, and satisfaction from positive workplace relationships (Baumeister & Leary, 1995). When organizations prioritize building strong social connections among employees, they create robust networks of support that enhance individual resilience, reduce stress, and promote psychological well-being (Cohen & Wills, 1985). These connections are emotional buffers during challenging times, providing employees with the social resources to navigate workplace pressures and maintain mental health.

Effective teamwork amplifies the benefits of social connections by creating shared purpose and collective achievement experiences that boost individual and group morale (Katzenbach & Smith, 1993). When teams function cohesively, members experience a sense of belonging and contribution that directly enhances job satisfaction and organizational commitment (Hackman, 2002). The collaborative nature of well-functioning teams allows individuals to leverage diverse strengths, learn from colleagues, and achieve outcomes that would be impossible through individual effort alone, creating a sense of accomplishment and pride that contributes significantly to workplace happiness (Kozlowski & Ilgen, 2006).

Social connections in the workplace also facilitate knowledge sharing and learning opportunities that enhance employee development and career satisfaction (Cross & Parker, 2004). When employees feel comfortable approaching colleagues for advice, sharing experiences, and collaborating on projects, they develop professionally while building meaningful relationships that extend beyond task completion. These learning networks create environments where employees feel supported in their growth, leading to increased engagement and reduced turnover intentions (Reagans & McEvily, 2003). The informal mentoring and peer support that emerge from strong social connections provide continuous development opportunities that formal training programs often cannot replicate.

Furthermore, strong workplace relationships contribute to psychological safety, enabling employees to take risks, share innovative ideas, and admit mistakes without fear of social rejection or professional retaliation (Edmondson, 1999). This psychological safety is crucial for both individual well-being and organizational innovation, as employees in socially connected environments are more likely to engage in creative problem-solving and contribute to continuous improvement initiatives. The trust that develops through positive social interactions allows teams to experiment, learn from failures, and adapt quickly to changing circumstances, creating more resilient and agile organizations.

The impact of social connections extends beyond immediate work relationships to influence overall life satisfaction and physical health outcomes (House et al., 1988). Employees who report strong workplace friendships demonstrate lower levels of stress hormones, reduced absenteeism, and higher overall life satisfaction compared to those who feel socially isolated at work (Rook, 1987). These health benefits translate into reduced healthcare costs for organizations while contributing to more positive workplace cultures where employees genuinely enjoy coming to work.

Organizations that invest in team-building activities, collaborative workspaces, and social interaction opportunities create environments where natural connections can flourish, leading to self-reinforcing cycles of positive relationships and organizational success (Putnam, 2000). The collective happiness that emerges from strong social connections and effective teamwork creates competitive advantages through improved customer service, increased innovation, and enhanced organizational reputation. Ultimately, prioritizing social connections and teamwork represents a strategic investment in both human well-being and organizational performance, creating workplaces where employees thrive and organizations achieve sustainable success through the power of positive relationships and collaborative achievement.

3.6 Opportunities for Growth and Development

Opportunities for growth and development serve as critical catalysts for employee well-being and organizational happiness, fulfilling fundamental human needs for competence, autonomy, and purpose that drive intrinsic motivation and life satisfaction (Deci & Ryan, 2000). When organizations invest in comprehensive development programs, they create environments where employees experience continuous learning, skill enhancement, and career progression, leading to increased engagement, reduced turnover, and enhanced psychological well-being (Maurer et al., 2003). The provision of growth opportunities demonstrates organizational commitment to employee potential, fostering reciprocal loyalty and emotional investment that translates into higher performance and job satisfaction.

Professional development opportunities directly address the human need for mastery and competence, which are essential components of psychological well-being and intrinsic motivation (Pink, 2009). When employees acquire new skills, expand their knowledge base, and take on challenging assignments, they experience a sense of accomplishment and self-efficacy that enhances their overall life satisfaction and workplace engagement (Bandura, 1997). This continuous learning process not only improves job performance but also builds confidence and resilience, enabling employees to adapt to changing workplace demands while maintaining positive mental health and professional identity.

Career advancement opportunities create clear pathways for employee progression, providing hope and direction that significantly impact psychological well-being and organizational commitment (Super, 1990). When employees can envision their future within the organization and understand the steps required for advancement, they develop stronger emotional connections to their work and employer, leading to increased motivation and reduced anxiety about career stagnation (Hall, 2002). The anticipation of future growth creates positive emotional states that enhance current job satisfaction while building long-term organizational loyalty and retention.

Mentoring and coaching programs exemplify how development opportunities create dual benefits for both mentors and mentees, fostering intergenerational knowledge transfer while building meaningful workplace relationships (Kram, 1985). These programs provide personalized guidance that accelerates learning while creating social connections that enhance workplace belonging and support networks. The mentoring relationship often extends beyond professional development to encompass personal growth, creating holistic development experiences that contribute to overall employee well-being and organizational culture strength.

Cross-functional training and job rotation initiatives expose employees to diverse organizational aspects, broadening their perspectives while increasing their value to the organization and marketability in their field (Campion et al., 1994). This variety prevents job monotony and burnout while building versatile skill sets that enhance employee confidence and career security. The exposure to different departments and functions also creates organizational understanding that improves collaboration and reduces silos, contributing to more cohesive and compelling workplace environments.

Investment in employee development signals organizational values and commitment to human capital, creating cultures of learning and growth that attract high-performing individuals while retaining existing talent (Noe et al., 2010). Organizations known for robust development opportunities become employers of choice, attracting candidates who value continuous learning and career progression. This reputation creates competitive advantages in talent acquisition while building internal cultures where growth mindset and continuous improvement become embedded organizational characteristics.

The financial returns on development investments are substantial, with organizations reporting improved performance, innovation, and customer satisfaction when employees feel supported in their growth journey (Bartel, 2000). Employees who receive development opportunities demonstrate higher levels of engagement, creativity, and problem-solving capabilities, contributing to organizational adaptability and competitive advantage. The enhanced skills and knowledge gained through development programs enable organizations to respond more effectively to market changes while building internal capacity for future challenges.

Furthermore, development opportunities contribute to employee resilience and adaptability, essential qualities in rapidly changing work environments (London & Smither, 1999). When employees continuously update their skills and expand their capabilities, they become more confident in their ability to handle new challenges and career transitions, reducing anxiety and stress associated with workplace uncertainty. This resilience benefits both individual well-being and organizational stability, creating workforces capable of thriving amid change and uncertainty while maintaining positive attitudes and performance levels.

3.7 Work-Life Balance and Flexibility

Work-life balance and flexibility have emerged as fundamental pillars of employee well-being and organizational happiness, addressing the complex interplay between professional responsibilities and personal life demands that significantly impact individual satisfaction and organizational effectiveness (Greenhaus & Powell, 2006). When organizations prioritize flexible work arrangements and support employees in achieving sustainable work-life integration, they create environments where individuals can thrive holistically, leading to reduced stress, improved mental health, and enhanced job satisfaction that translates into higher performance and organizational commitment (Allen et al., 2013). The recognition that employees are whole individuals with diverse responsibilities and interests beyond work represents a paradigm shift toward more humanistic management approaches that acknowledge the interconnected nature of personal and professional well-being.

Flexible work arrangements, including remote work options, flexible scheduling, and compressed workweeks, provide employees with greater autonomy and control over their work environment and time allocation, directly addressing fundamental psychological needs for self-determination and work-life harmony (Gajendran & Harrison, 2007). This autonomy reduces role conflict and time-based stress that often arise when rigid work structures clash with personal responsibilities such as childcare, eldercare, or educational pursuits (Byron, 2005). When employees can adjust their work schedules to accommodate personal needs without compromising professional responsibilities, they experience reduced anxiety and guilt associated with competing demands, leading to improved psychological well-being and increased engagement in both work and personal domains.

Implementing work-life balance initiatives demonstrates organizational care for employee welfare, fostering psychological contracts built on mutual respect and understanding that enhance emotional attachment and organizational loyalty (Guest, 2017). Employees who perceive their organizations as supportive of their personal lives report higher organizational commitment, job satisfaction, and intention to remain with their employer long-term (Thompson et al., 1999). This emotional connection creates positive workplace cultures where employees feel valued as individuals rather than as productive resources, leading to increased discretionary effort and organizational citizenship behaviors that benefit overall organizational performance.

Remote work and flexible scheduling options have proven effective in reducing commuting stress, workplace distractions, and time pressures that traditionally contribute to employee burnout and dissatisfaction (Golden et al., 2006). Eliminating or reducing daily commutes provides employees additional time for personal activities, family engagement, and self-care practices that contribute to overall life satisfaction and stress reduction. Furthermore, the ability to work in environments that match individual preferences and peak productivity periods often results in improved work quality and efficiency, creating win-win scenarios where both employee well-being and organizational outcomes benefit from flexible arrangements.

Work-life balance initiatives also address diverse employee needs across life stages and personal circumstances, promoting inclusivity and equity in organizational policies (Clark, 2000). Parents with young children, employees pursuing educational goals, individuals caring for aging family members, and those managing health conditions benefit from flexible arrangements that accommodate their unique circumstances without penalizing their career advancement or professional standing. This inclusiveness contributes to organizational diversity and creates cultures where all employees can succeed regardless of personal situations, leading to more innovative and adaptive organizational capabilities.

The business case for work-life balance and flexibility is compelling, with research demonstrating reduced turnover, lower absenteeism, decreased healthcare costs, and improved productivity among organizations that prioritize these initiatives (Kelly et al., 2008). The cost savings associated with reduced recruitment and training and enhanced performance and employee satisfaction provide substantial returns on investment in flexibility programs. Additionally, organizations known for supporting work-life balance attract top talent who increasingly prioritize these benefits when making career decisions, creating competitive advantages in talent acquisition and retention that contribute to long-term organizational success and sustainability in an evolving employment landscape.

3.8 Effective Communication and Feedback Mechanisms

Effective communication and feedback mechanisms serve as the organizational circulatory system that sustains employee well-being and creates thriving workplace cultures where individuals feel heard, valued, and continuously supported in their professional development (Tourish & Hargie, 2004). When organizations establish robust communication channels and systematic feedback processes, they create environments characterized by transparency, trust, and mutual understanding that enhance psychological safety and job satisfaction (Klauss & Bass, 1982). The quality of organizational communication significantly influences employee engagement, stress levels, and overall workplace happiness, making it a critical determinant of individual well-being and organizational success.

Clear and consistent communication from leadership provides employees with essential information about organizational direction, expectations, and changes, reducing uncertainty and anxiety that often arise from ambiguous workplace situations (Lewis, 2011). When employees understand their roles, organizational goals, and how their contributions align with broader objectives, they experience greater sense of purpose and meaning in their work, leading to increased motivation and job satisfaction (Mayfield & Mayfield, 2002). This clarity eliminates the stress associated with role ambiguity and conflicting expectations, creating more stable psychological environments where employees can focus their energy on productive activities rather than navigating unclear directives or organizational politics.

Regular feedback mechanisms, including performance reviews, one-on-one meetings, and informal check-ins, provide employees with valuable information about their performance while demonstrating organisational investment in their growth and development (London, 2003). Constructive feedback helps employees understand their strengths and areas for improvement, enabling them to make necessary adjustments and experience continuous professional growth that enhances self-efficacy and career satisfaction (Kluger & DeNisi, 1996). The feedback process also creates opportunities for two-way communication where employees can express concerns, share ideas, and contribute to organisational improvements, fostering a sense of ownership and engagement that contributes to overall workplace happiness.

Open communication channels, encouraging employee voice and participation in decision-making processes, significantly enhance psychological well-being by addressing fundamental human needs for autonomy and influence (Morrison, 2011). When employees feel comfortable expressing opinions, raising concerns, and contributing to workplace discussions, they experience a greater sense of agency and value, increasing job satisfaction and organisational commitment (Van Dyne et al., 2003). This participative communication culture also benefits organisations from diverse perspectives and innovative ideas. It creates more effective problem-solving processes and adaptive capabilities that contribute to organisational success and employee pride in their workplace.

Effective communication also plays a crucial role in conflict resolution and relationship management, preventing minor misunderstandings from escalating into major interpersonal issues that can negatively impact team dynamics and individual well-being (Deutsch, 1973). When organisations establish clear communication protocols and conflict resolution mechanisms, they create safer psychological environments where employees can address concerns constructively without fear of retaliation or relationship damage. This proactive approach to communication management reduces workplace stress and maintains positive team cohesion that supports individual mental health and collective performance.

Technology-enabled communication platforms and feedback systems have enhanced organisations' ability to maintain consistent, timely, and accessible communication across diverse workforces, including remote and hybrid teams (Dennis & Valacich, 1999). Digital tools enable real-time feedback, peer recognition, and collaborative communication that bridges geographical and temporal barriers while creating more inclusive communication environments where all team members can participate effectively. These technological solutions also provide data and analytics that help organisations continuously identify communication gaps and improve their feedback processes.

The organisational benefits of effective communication and feedback mechanisms extend beyond employee satisfaction, including improved performance, reduced turnover, enhanced innovation, and stronger customer relationships (Clampitt & Downs, 1993). Employees who receive regular, high-quality communication and feedback demonstrate higher engagement, productivity, and loyalty. In contrast, organisations with strong communication cultures experience better coordination, faster problem-solving, and more successful change management initiatives. Ultimately, investing in communication and feedback systems creates positive reinforcement cycles. Improved relationships and performance outcomes enhance organisational culture and employee well-being, establishing sustainable competitive advantages built on human connection and continuous improvement.

3.9 Recognition and Appreciation

In today's talent war, organisations and leaders are looking for strategies to attract and retain top performers while boosting the organic employee productivity (Gall, 2024). Recognition and appreciation are vital for satisfying the fundamental needs of humans for acceptance, validation, and belonging, whilst reinforcing behaviour and achievements that contribute to organisational success. Happy organisations ensure a comprehensive recognition system that celebrates individual and collective contributions and acknowledges their efforts towards organisational goals. Workplace recognition motivates employees, offers them a sense of accomplishment, and makes them feel valued. It is argued that recognition not only boosts an individual employee's engagement but also boosts their loyalty, leading to higher retention (Gallup, 2024).

Formal recognition system offers a structured approach to acknowledging the employees' achievements and contributions. These might include employee of the month or year programs that celebrate outstanding performance, service awards that recognise tenure and loyalty, achievement awards that celebrate specific accomplishments or milestones, peer nomination programs that enable employees to recognise each other, and team awards that celebrate collective achievements and collaboration. (Amoatemaa & Kyeremeh, 2016).

Effective formal recognition programs are based on clear criteria that align with organisational values and objectives, provide meaningful rewards that recipients value, include diverse recognition opportunities that accommodate different types of contributions, ensure fair and equitable recognition across all employee groups, and maintain consistency and regularity that make recognition predictable and reliable.

Informal recognition and appreciation often impact more than formal programs because they are more personal, timely, and specific to individual contributions and circumstances. Informal recognition might involve verbal praise and appreciation expressed in the moment, written notes or emails that acknowledge specific contributions, public acknowledgement in team meetings or organisational communications, and small tokens of appreciation that demonstrate thoughtfulness and gratitude (Werquin, 2010). Building cultures of informal recognition requires leadership modelling of appreciation and gratitude, as well as training for managers on recognising and appreciating employee contributions. These cultural norms encourage peer-to-peer recognition, appreciation and awareness of individual preferences for recognition style and frequency.

Personalised recognition recognises that different employees are motivated and appreciated through different types of recognition. For example, some employees prefer public recognition and celebration, while others value more private acknowledgement and appreciation. Some employees are motivated by formal award ceremonies, and some like a simple expression of gratitude and thank you. Some employees value monetary rewards while others seek development opportunities, time off from work, or other non-monetary benefits. Effective personalised recognition, however, requires understanding the individual employees' preferences and sources of motivation, in addition to flexibility in recognition approaches and methods based on the ongoing feedback and changing needs of the workforce to ensure effectiveness of the rewards and recognition system and drive employees' satisfaction (Ramya & Vanithamani, 2023).

Interestingly, receiving career development opportunities can also serve as a powerful form of recognition by demonstrating organisational confidence in their employees' capabilities, ensuring that meaningful rewards for outstanding performance are provided simultaneously. However, this form of recognition would require establishing clear connections between performance and career advancement opportunities to ensure all employees have equitable access to opportunities and development resources.

Celebrating and acknowledging collective and individual achievements reinforces teamwork and collaboration whilst building organisational culture and pride. This building of celebrative culture requires leadership commitment towards collective acknowledgement of the achievement, identifying high performers regularly, and communication and celebrating team and organisational success, thereby allowing an integration between celebration and recognition within organisational practices.

3.10 Physical Environment and Well-being Resources

According to the Individual theory, people in specific environments have a dynamic relationship with their social, physiological, and physical environments. The theory posits that the workplace environment is interrelated with job settings and effectively affects workplace activities (Zhenjing et al., 2022). The physical work environment significantly impacts the employee's well-being, productivity, and job satisfaction, thereby assuming the role of a tangible manifestation of organisational values and priorities. Happy organisations design and maintain the physical spaces that support employees' health, comfort, and productivity while providing resources that directly support their well-being.

The physical surroundings where the employees spend their work lives are an important source of job satisfaction and significantly impact their work outcomes (Page & Tolmie, 2024). Workplace design and ergonomics can address the fundamental need for physical comfort required during work. As Klitzman & Stellman (1989) note, several organisational and environmental psychology studies reported a strong association between physical environment characteristics, including temperature, ventilation, lighting, work station design, and employees' attitudes, behaviour, satisfaction, and turnover. Effective workplace design requires understanding the regular assessments of workplace requirements involving assessing employee comfort and ergonomic needs, thereby allowing the management to make quality investments in equipment that supports the long-term health and continued comfort.

An organisation's physical environment and its design and layout can influence employee behaviour in the workplace. It is estimated that improvements in the physical design of the workplace can result in a 5-10% increase in employee productivity. Owing to this, increasingly organisations are designing the physical layout of their workplace around the employees' needs to maximise productivity and satisfaction (McGuire & McLaren, 2009). Stellman et al. (1987) found that all-day visual display terminal users experienced higher levels of job and physical stressors, thereby suggesting that work design needs to be aligned with the employees and work to ensure it contributes to employee well-being.

Collaborative and flexible spaces form the well-being resources that enable different work activities whilst promoting team collaboration and building (Scott & Manning, 2022). Well-being resources demonstrate organisational commitment to employee health and well-being whilst providing practical support for maintaining the physical and mental health of the workforce. These resources include fitness facilities, healthy food options, nutrition programs, spaces for meditation and relaxation and onsite health and wellness services. Implementing well-being resources requires evaluation of employee health and wellness needs and interests.

Safety and security measures ensure that employees feel physically and psychologically safer in their work environment. This includes ensuring physical security measures that protect the workplace from external threats, workplace safety protocols to avoid accidents and injuries, emergency response procedures and additional psychological safety measures that protect against harassment, discrimination, and other interpersonal risks (Dollard & Bakker, 2010). However, maintaining a safe and secure environment requires regular assessment of the facilities and practices to identify potential risks and threats, ensuring preventive measures are being followed, and training employees on the established safety and security procedures within the organisation. Lastly, having robust technological and infrastructure supports the organisational systems and processes and aids in boosting employee productivity and collaboration whilst ensuring a flexible work environment and effective communication. Providing effective technology requires a comprehensive understanding of diverse work preferences and requirements to ensure employees can keep pace with the changing needs and capabilities of the organisation in the market.

CURRENT CHALLENGES TO ORGANISATIONAL HAPPINESS

The pursuit of organisational happiness in contemporary workplaces faces unprecedented challenges that stem from rapid technological evolution, shifting societal expectations, and fundamental changes in work structures. These challenges have created a complex landscape where traditional approaches to employee well-being are increasingly inadequate, necessitating a deeper understanding of the barriers that prevent organisations from achieving genuine workplace happiness.

Digital transformation has introduced a pervasive challenge known as "technostress," where employees experience psychological strain from constant connectivity and information overload (Tarafdar et al., 2019). While enhancing productivity, the proliferation of digital communication tools has blurred the boundaries between work and personal life, leading to "always-on" cultures that undermine psychological recovery (Boswell & Olson-Buchanan, 2007). Research by Gallup (2023) indicates that 76% of employees experience burnout occasionally, with 28% reporting being burned out "frequently" or "always," attributed mainly to digital overwhelm and unrealistic expectations of constant availability.

The modern workforce grapples with increasing job insecurity and employment precarity, particularly following global economic disruptions and the rise of gig economy structures (Kalleberg, 2018). This uncertainty creates chronic stress that fundamentally undermines happiness, as employees struggle with anxiety about job stability, inadequate benefits, and limited career progression opportunities. The psychological impact of precarious employment extends beyond individual well-being, affecting organisational commitment and collective morale (Shoss, 2017).

Generational diversity in the workplace presents both opportunities and challenges for organisational happiness. While diversity brings varied perspectives and skills, it also creates potential friction in values, communication styles, and expectations regarding work-life integration (Twenge, 2010). Younger employees, particularly Generation Z and millennials, prioritise purpose-driven work and social impact, often clashing with traditional organisational structures that emphasise hierarchy and conventional success metrics (Deloitte, 2023). This misalignment can lead to disengagement and turnover when organisations fail to adapt their cultures to accommodate diverse generational expectations.

The mental health crisis in workplaces has reached critical proportions, with the World Health Organisation (2022) estimating that depression and anxiety disorders cost the global economy over $1 trillion annually in lost productivity. Stigma surrounding mental health continues to prevent open discussions and adequate support systems, creating environments where employees suffer in silence rather than seeking help. The COVID-19 pandemic exacerbated these issues, with remote work isolation, increased caregiving responsibilities, and economic uncertainty contributing to widespread psychological distress (Kniffin et al., 2021).

Inequality and discrimination remain persistent challenges that undermine organisational happiness for marginalised groups. Pay gaps, limited advancement opportunities, and microaggressions create hostile environments that prevent authentic engagement and well-being (Hekman et al., 2017). Despite increased awareness of diversity, equity, and inclusion initiatives, many organisations struggle with implementation, leading to performative rather than substantive change that fails to address systemic barriers to employee happiness.

The intensification of work pace and productivity demands has created what scholars term "acceleration culture," where organisations continuously increase performance expectations without corresponding increases in resources or support (Rosa, 2013). This creates unsustainable pressure, leading to chronic stress, reduced creativity, and burnout. The phenomenon is particularly pronounced in knowledge work, where cognitive demands are high, but recovery time is limited.

Finally, the erosion of social connections and community within organisations poses a significant threat to happiness. As organisations become more geographically dispersed and digitally mediated, the informal social networks traditionally providing emotional support and belonging weaken (Putnam, 2000). This social fragmentation undermines the relational aspects of work that are fundamental to human flourishing and organisational happiness.

From a neuroplasticity perspective, these contemporary challenges create particularly concerning patterns because chronic stress and negative workplace experiences can rewire the brain in ways that perpetuate unhappiness. Neuroscientific research demonstrates that prolonged exposure to workplace stressors can lead to structural changes in the brain, including reduced prefrontal cortex function and enlarged amygdala responses, creating heightened stress reactivity and decreased emotional regulation capacity (McEwen, 2017). This neurobiological reality means organisations facing these challenges must understand that employee unhappiness is temporary and can become neurologically embedded, requiring intentional interventions to reverse negative neural patterns and promote positive brain plasticity.

KEY ORGANISATIONAL BARRIERS TO EMPLOYEE HAPPINESS

Understanding the organisational barriers that impede employee happiness is crucial for leaders seeking to create genuinely supportive work environments. These barriers often manifest as systemic issues embedded in organisational structures, processes, and cultures that, despite well-intentioned efforts, continue to undermine employee well-being and engagement.

Fear-based management represents one of the most pervasive barriers to organisational happiness. This approach, characterised by authoritarian leadership styles, punitive feedback mechanisms, and cultures of blame, creates psychological environments where employees operate from a state of chronic stress rather than empowerment (Kish-Gephart et al., 2009). Research by Edmondson (2019) demonstrates that fear-based cultures significantly reduce psychological safety, leading to decreased innovation, increased turnover, and reduced overall well-being. Employees who fear retribution for mistakes or honest feedback engage in self-protective behaviours that limit their authentic participation in organisational life. Data from the Society for Human Resource Management (2023) indicates that 58% of employees report leaving jobs due to poor management relationships, with fear-based leadership being a primary factor.

Inadequate recognition and reward systems constitute another significant barrier to happiness. Many organisations operate under outdated assumptions about motivation, relying heavily on financial incentives while neglecting the intrinsic motivators that drive long-term satisfaction (Pink, 2009). When recognition is sporadic, generic, or disconnected from meaningful contributions, employees experience a sense of invisibility that undermines their sense of value and belonging. Research by Gallup (2022) found that employees who do not feel adequately recognised are twice as likely to quit within the next year, and organisations with effective recognition programs experience 31% lower voluntary turnover rates.

Communication failures and information opacity create significant barriers to happiness by fostering uncertainty, mistrust, and disengagement. When organisations fail to provide transparent communication about decisions, changes, or expectations, employees fill information voids with assumptions that are often negative and anxiety-provoking (Tourish, 2010). The absence of clear, consistent, and honest communication creates environments where rumours flourish, trust erodes, and employees feel disconnected from organisational purpose and direction. Studies indicate that organisations with highly effective communication practices are 3.5 times more likely to outperform their peers in employee engagement metrics (Watson Wyatt, 2009).

Excessive bureaucracy and rigid organisational structures impede happiness by limiting autonomy, slowing decision-making, and creating frustration with inefficient processes. When employees encounter layers of approval for routine decisions, complex procedures for simple tasks, or inflexible policies that ignore individual circumstances, they experience a sense of powerlessness that undermines intrinsic motivation (Deci & Ryan, 2000). The resulting frustration and wasted time contribute to stress and disengagement, particularly among high-performing employees who value efficiency and impact.

Work-life integration challenges have intensified as organisational boundaries have become increasingly permeable. While technology enables flexibility, it also creates expectations of constant availability that undermine recovery and personal time (Boswell & Olson-Buchanan, 2007). Organisations that fail to establish clear boundaries or respect employees' time create unsustainable conditions that lead to burnout and resentment. Research by the American Psychological Association (2023) indicates that 60% of employees report that work-related stress affects their personal relationships and family life.

Inadequate professional development and career advancement opportunities represent significant barriers to long-term happiness and engagement. Employees who perceive limited growth potential, skills stagnation, or unclear career pathways experience reduced motivation and increased likelihood of disengagement (Maslach & Leiter, 2016). This is particularly problematic for younger workers who prioritise learning and development as key factors in job satisfaction. Organisations that fail to invest in employee growth often experience increased turnover and reduced innovation capacity.

Finally, toxic workplace cultures characterised by gossip, politics, favouritism, and interpersonal conflict create environments antithetical to happiness. These cultures develop when organisations fail to establish clear values, enforce behavioural standards, or address problematic behaviours promptly and effectively (Schein, 2010). The resulting psychological contamination affects all employees, creating stress, reducing collaboration, and undermining the trust necessary for genuine organisational happiness.

The neuroplasticity implications of these organisational barriers are profound and warrant serious consideration. Neuroscientific research reveals that negative workplace experiences trigger neurobiological stress responses that, when sustained, can create lasting changes in brain structure and function (Lupien et al., 2018). Fear-based management, for instance, activates the amygdala's threat detection system, leading to chronic stress hormone release that impairs prefrontal cortex functioning responsible for executive decision-making, creativity, and emotional regulation (Arnsten, 2009). Over time, these neural patterns become entrenched, making it increasingly difficult for employees to experience positive emotions or engage authentically, even when organisational conditions improve. Conversely, understanding neuroplasticity offers hope: positive workplace interventions can rewire the brain toward greater happiness, resilience, and well-being, but only when organisations commit to sustained, evidence-based approaches that allow sufficient time for neural change to occur (Davidson & Lutz, 2008).

MEASURING AND ASSESSING ORGANISATIONAL HAPPINESS

The systematic measurement and assessment of organisational happiness represents a critical yet often overlooked component of effective workplace well-being initiatives. Without robust measurement frameworks, organisations cannot accurately diagnose happiness levels, track progress over time, or evaluate the effectiveness of interventions designed to enhance employee well-being and engagement.

Contemporary approaches to measuring organisational happiness have evolved beyond simple satisfaction surveys to encompass multidimensional frameworks that capture both hedonic and eudaimonic aspects of workplace well-being. The Workplace PERMA Profiler, adapted from Seligman's original PERMA model by Butler and Kern (2015), provides a comprehensive framework for assessing positive emotions, engagement, relationships, meaning, and accomplishment within organisational contexts. This instrument measures 15 core items across five PERMA dimensions, plus additional items for overall wellbeing, negative emotion, health, and loneliness at work, offering organisations a scientifically validated tool for understanding the complexity of workplace happiness. Research applications of the Workplace PERMA Profiler have demonstrated strong correlations with performance metrics, absenteeism rates, and turnover intentions, establishing its validity as a diagnostic and predictive tool.

Pulse surveys have emerged as particularly effective mechanisms for continuous happiness monitoring, offering organisations real-time insights into employee sentiment and well-being trends. Unlike traditional annual engagement surveys, pulse surveys typically consist of 5-10 questions administered weekly, bi-weekly, or monthly, allowing organisations to quickly detect happiness levels and respond proactively (Maylett & Wride, 2017). Companies like Microsoft and Adobe have successfully implemented pulse survey systems that track happiness indicators alongside performance metrics, enabling data-driven decision-making about workplace interventions.

Biometric and physiological indicators are increasingly incorporated into comprehensive happiness assessment strategies. Wearable technology can monitor stress indicators such as heart rate variability, sleep quality, and physical activity levels, providing objective data about employee well-being that complements self-reported measures (Genin et al., 2018). At the same time, privacy concerns require careful consideration; organisations implementing voluntary biometric monitoring report valuable insights into the relationship between workplace conditions and physiological stress responses.

360-degree feedback systems, when properly designed, can provide valuable insights into relational aspects of organisational happiness. These systems gather feedback from supervisors, peers, and subordinates about interpersonal experiences, communication effectiveness, and collaborative satisfaction (Bracken et al., 2016). When aggregated across organisational units, 360-degree feedback can reveal patterns in relationship quality that significantly impact collective happiness and psychological safety.

Advanced analytics and machine learning approaches are revolutionising happiness measurement by identifying subtle patterns in communication, collaboration, and behavioural data. Email communication patterns, meeting frequency and quality, internal social network analysis, and workspace utilisation data can provide insights into organisational dynamics influencing happiness (Waber, 2013). Companies like Humanyze have developed people analytics platforms that correlate communication patterns with engagement and performance outcomes, offering organisations unprecedented visibility into the relational infrastructure that supports or undermines happiness.

Qualitative assessment methods remain essential for understanding the contextual factors that quantitative measures cannot capture. Focus groups, ethnographic observations, and narrative interviews provide rich insights into employee experiences, cultural dynamics, and systemic issues that affect happiness (Schein, 2010). When combined with quantitative measures, qualitative assessments offer a comprehensive understanding of both what is happening and why it is occurring.

Implementation of effective happiness measurement requires careful attention to several critical factors. First, employees must perceive measurement systems as legitimate and valuable, requiring transparent communication about purposes, data usage, and improvement commitments. Second, measurement frequency must balance the need for timely insights with survey fatigue concerns. Third, organisations must demonstrate clear connections between measurement results and concrete actions, as measurement without response undermines trust and engagement.

Benchmarking happiness metrics against industry standards and best practices provides organisations with contextual understanding of their performance relative to peers. Organisations like Great Place to Work and Gallup provide industry benchmarks for comparative analysis and goal setting. However, internal benchmarking over time often provides more actionable insights than external comparisons, as organisational contexts vary significantly.

Finally, effective happiness measurement requires integration with broader organisational performance metrics to demonstrate business impact and secure ongoing investment. Research consistently demonstrates positive correlations between happiness measures and outcomes such as productivity, innovation, customer satisfaction, and financial performance (Lyubomirsky et al., 2005). Organisations that successfully integrate happiness metrics into their performance dashboards are more likely to sustain long-term commitment to employee well-being initiatives.

Emerging neuroscience research is beginning to inform happiness measurement through neuroplasticity indicators that provide objective evidence of well-being changes at the brain level. Neuroimaging studies demonstrate that workplace interventions designed to enhance happiness can produce measurable changes in brain structure and function within 8-12 weeks (Lutz et al., 2004). Organisations pioneering this approach explore biomarkers such as cortisol patterns, heart rate variability, and EEG measures of prefrontal cortex activity as objective indicators of neurological well-being (Thayer & Lane, 2009). While still in early stages, these neuroplasticity-informed measurement approaches offer the potential for more precise and objective assessment of happiness interventions. Furthermore, understanding neuroplasticity principles suggests that happiness measurement should track not just current states but also trajectories of change over time, as neural rewiring requires sustained positive experiences to create lasting improvements in well-being (Vago & Silbersweig, 2012). This neurobiological perspective emphasises the importance of longitudinal measurement strategies that can capture the gradual but profound changes as positive workplace experiences reshape employees' neural architecture toward greater happiness and resilience.

EMPLOYEE HAPPINESS CASE STUDIES

ORGANIZATIONS THAT STRUGGLED DUE TO POOR EMPLOYEE HAPPINESS

1. **Amazon Warehouses (2010s-2020s)**

Amazon's fulfilment centres exemplify how poor employee conditions can create systemic organisational problems. The company maintains one of the highest turnover rates globally at 150% annually, meaning more people quit or are fired each year than are employed. The New York Times reported turnover rates of approximately 3% per week in 2021, creating constant workforce instability.

In New Jersey alone, Amazon warehouse turnover reached 124%, almost double the rate of non-Amazon warehouses in the state. The demanding work environment, with excessive productivity targets and limited break times, resulted in counties with Amazon warehouses experiencing 117% average turnover compared to 72% in counties without Amazon facilities. This high churn model required continuous recruitment and training, creating significant operational costs. The poor working conditions also led to unionisation attempts, worker walkouts, and negative publicity that damaged Amazon's employer brand, forcing costly interventions and policy changes to address workforce retention issues.

2. **Wells Fargo Account Fraud Scandal (2016)**

Wells Fargo's toxic sales culture demonstrates how employee unhappiness can trigger massive corporate scandals. Unrealistic sales targets and intense pressure created an environment where employees opened as many as 2 million unauthorised accounts over five years. The bank fired 5,300 employees related to the fraudulent behaviour, indicating widespread systemic issues rather than isolated incidents.

The financial consequences were devastating: Wells Fargo paid $3 billion in criminal and civil settlements, representing approximately 15% of their annual profits. Additional penalties included $100 million to the CFPB and $5 million for customer compensation. The scandal led to billions in total fines, toppled two CEOs, resulted in guilty pleas from executives, and triggered a Federal Reserve asset cap that remains in place. The reputational damage required extensive rebuilding efforts, regulatory oversight intensified, and customer trust erosion persisted for years, demonstrating how employee dissatisfaction can cascade into enterprise-threatening crises.

3. Uber Culture Crisis (2017)

Uber's aggressive workplace culture created a toxic environment that nearly destroyed the company. Reports of sexual harassment, discrimination, and hostile management practices led to mass resignations throughout 2017, including CEO Travis Kalanick. The cultural problems resulted in over 20 senior executive departures within months, creating a leadership vacuum and operational instability.

The crisis delayed Uber's IPO by approximately two years, costing billions in potential valuation. Regulatory scrutiny intensified globally, with several cities banning or restricting Uber operations. Legal settlements for harassment and discrimination cases reached millions of dollars. The company's growth stagnated as talent acquisition became extremely difficult due to reputational damage. Market share erosion to competitors like Lyft accelerated during this period. Recovery required comprehensive cultural transformation initiatives, new leadership, and extensive policy overhauls, demonstrating how poor employee treatment can threaten organisational survival and require massive resources to rectify.

ORGANISATIONS THAT SUCCEEDED THROUGH EMPLOYEE HAPPINESS FOCUS

1. Salesforce

Salesforce's commitment to employee well-being through its "Ohana" (family) culture has yielded measurable results. 84% of Salesforce employees rate it as a great place to work, significantly above industry averages. A 2022 report showed that 80% of employees felt the company prioritised their mental health, reflecting successful wellness initiatives.

The company's focus on equality, including comprehensive pay equity audits and inclusive policies, has created a sustained competitive advantage. Marc Benioff's emphasis on mindfulness programs, work-life balance, and social responsibility alignment has resulted in consistent "Best Places to Work" rankings across multiple surveys. This employee satisfaction directly correlates with business performance: Salesforce grew from a startup to over $30 billion in annual revenue, with its market capitalisation exceeding $200 billion. The strong culture enables superior talent retention, reduces recruitment costs, and maintains institutional knowledge. High employee engagement scores correlate with innovation rates, customer satisfaction metrics, and market expansion success, demonstrating clear ROI on employee happiness investments.

2. Patagonia

Patagonia's employee-centric approach has created exceptional organisational loyalty and performance. The company maintains turnover rates of approximately 4%, significantly below the retail industry average of 13%. Their comprehensive benefits include on-site childcare, flexible work arrangements, and environmental mission alignment that resonates with employee values.

The company has received over 100 "Best Places to Work" awards across various publications and regions. Employee satisfaction surveys consistently show 90%+ approval ratings for leadership and workplace culture. This happiness translates into business success: Patagonia generates over $1 billion in annual revenue with remarkable profit margins for the outdoor apparel sector. Their engaged workforce drives innovation, with employees contributing ideas for sustainable practices and product development. Low turnover reduces hiring and training costs by an estimated $2-3 million annually. High employee advocacy creates authentic brand marketing, with staff as genuine ambassadors for environmental values and product quality.

3. Microsoft (Satya Nadella Era)

Microsoft's cultural transformation under CEO Satya Nadella exemplifies how employee happiness drives business performance. The shift from a competitive internal culture to a collaborative, growth-mindset approach yielded dramatic results. Employee satisfaction scores improved from below-average to consistently ranking in top tech employer lists.

The cultural change contributed to Microsoft's market capitalisation growing from approximately $300 billion in 2014 to over $2 trillion by 2023, making it one of the world's most valuable companies. Revenue growth accelerated, with the cloud computing division (Azure) becoming a market leader primarily due to employee innovation and customer focus. Glassdoor ratings improved significantly, with CEO approval ratings exceeding 95%. The empathy-driven culture enabled successful pivots to cloud computing, collaboration tools, and AI development. Employee retention in critical roles improved measurably, reducing brain drain to competitors. The transformation demonstrates how leadership commitment to employee wellbeing and market leadership can drive unprecedented business success.

4. Google/Alphabet

Google's comprehensive employee happiness strategy has maintained its position as a top employer while driving extraordinary business growth. The company consistently achieves over 90% employee satisfaction in internal surveys, with over 90% stating they would recommend Google as an employer. Their innovative perks, 20% time for passion projects, and psychological safety initiatives have become industry benchmarks.

Focusing on employee well-being has enabled Google to attract and retain top talent in highly competitive markets. Their parent company, Alphabet, maintains a market capitalisation exceeding $1.5 trillion, primarily attributed to innovation driven by engaged employees. The 20%-time policy has generated products like Gmail and AdSense, contributing billions in revenue. Employee Net Promoter Scores consistently rank among the highest in the technology sector. Low turnover in engineering and product roles has maintained competitive advantages in search, advertising, and cloud computing. The company's ability to attract PhD-level talent and maintain engagement demonstrates how employee happiness correlates with innovation capacity and market leadership in technology sectors.

CONCLUSION

This chapter explored the fundamental organisational characteristics that differentiate happy organisations from their competitors, providing a comprehensive framework for understanding how workplace well-being should be part of the deliberate organisational design rather than working in silos and adopting it to adapt to the market trends. The evidence discussed in this chapter demonstrates that creating a positive and flourishing workplace requires systematic emphasis on diverse interconnected factors, including leadership, culture, job design, interpersonal relationships, equity, development, communication, career development, recognition, and physical environment.

The journey towards organisational happiness is neither quick nor simple; it requires sustained commitment, strategic thinking and willingness to address fundamental aspects of how the organisations operate rather than developing and implementing superficial programs and strategies that offer short-term benefits but fail to create a sustained change. Organisations that can successfully create cultures of well-being and high performance understand that employees' happiness is not a goal to be reached, but rather it is a continuous process of evaluation, improvement, and adaptation that aligns with the changing needs and conditions in the business environment. The business case for happy organisations has become increasingly compelling, with studies consistently demonstrating that organisations prioritising employee well-being outperform their business peers across multiple dimensions, including innovation, productivity, customer satisfaction, retention, and financial performance. These benefits create a cycle of mutual benefits where the organisational success drives continued investment in the employee well-being, boosting organisational competitive advantage.

Considering the current trends, it is suggested that the significance of organisational happiness will continue to grow. Changing generational expectations from organisations, with younger employees increasingly prioritising meaningful work and positive cultures over compensation and career advancement opportunities, organisations will be pressured to create more engaging and fulfilling work experiences. Rapid technological advancements will continue to automate tasks, increasing the importance of creativity, innovation, and collaboration capabilities critical for flourishing positive organisational outcomes. Additionally, the continued evolution of work arrangements, accelerated by global events and technological capabilities, will push organisations to reimagine how they create culture, connection, and engagement across diverse work settings. The principles discussed in this chapter, including supportive leadership, psychological safety, meaningful work, strong interpersonal relationships, and others, will remain relevant irrespective of how and where the work is performed. However, it will push the organisation to implement and adopt new approaches and strategies aligned with the workforce requirements and business conditions. Furthermore, the growing awareness of the connection between the employee's well-being and broader societal outcomes indicates that creating happy organisations is not just a business imperative but also a social responsibility. Organisations prioritising employee well-being can generate positive externalities, including contribution to community health, work-life balance, and societal sustainability by ensuring the employees' daily sense of happiness and fulfilment rather than experiencing stress and energy depletion from work. The theoretical frameworks discussed in the chapter guide a systematic approach for organisations to initiate employee well-being interventions, whilst also offering strategic guidance for optimising their current policies and practices for the workforce.

In future, empirical investigations must examine the complex relationships among organisational variables delineated from this chapter and explore how these constructs interact in diverse sectors, cultural, and institutional contexts. Additionally, longitudinal studies examining long-term sustainability of well-being interventions and their impact on organisational resilience during adversity would contribute through evidence-based insights for organisations/practitioners committed to sustainable organisational change. Conducting such research would contribute towards the growing body of literature on positive organisational culture and evidence-based management practices.

This chapter advocates for a fundamental theoretical reorientation in the conceptualisation of organisational management of workforce and setting- it calls for a transition from treating the employee as mere human capital to adopting more humanistic approaches to position employees' well-being as a core organisational imperative. Organisations that can operationalise this paradigm transformation and systematically implement the characteristics discussed in this chapter can demonstrate superior outcomes in the form of enhanced employee well-being and experiences and boosted financial performance, thereby validating the business case for human-centred organisational design. Through strategic focus on the fundamental organisational characteristics, organisational leaders can foster human thriving, and a work environment characterised by high employee engagement and intrinsic motivation, is the foundation to a strong workforce that is organically aligned and committed to the shared organisational purpose.

REFERENCES

Ali, S. A. M., Said, N. A., Yunus, N. M., Kader, S. F. A., Latif, D. S. A., & Munap, R. (2014). Hackman and Oldham's Job Characteristics Model to Job Satisfaction. *Procedia - Social and Behavioral Sciences, 129*(2014), 46–52. Science direct. https://doi.org/DOI: 10.1016/j.sbspro.2014.03.646

Allen, T. D., Johnson, R. C., Kiburz, K. M., & Shockley, K. M. (2013). Work-family conflict and flexible work arrangements: Deconstructing flexibility. *Personnel Psychology*, *66*(2), 345–376. DOI: 10.1111/peps.12012

American Psychological Association. (2023). *Stress in America: Work and well-being*. APA Press.

Amoatemaa, A. S., & Kyeremeh, D. D. (2016). Making Employee Recognition a Tool for Achieving Improved Performance: Implication for Ghanaian Universities. *Journal of Education and Practice*, *7*(34), 46–52.

Arnsten, A. F. T. (2009). Stress signalling pathways that impair prefrontal cortex structure and function. *Nature Reviews. Neuroscience*, *10*(6), 410–422.

Baldwin, S. (2006). *Organisational Justice*. Institute for employment Studies. https://www.employment-studies.co.uk/system/files/resources/files/mp73.pdf

Bandura, A. (1997). *Self-efficacy: The exercise of control*. W. H. Freeman.

Bartel, A. P. (2000). Measuring the employer's return on investments in training: Evidence from the literature. *Industrial Relations*, *39*(3), 502–524. DOI: 10.1111/0019-8676.00177

Baumeister, R. F., & Leary, M. R. (1995). The need to belong: Desire for interpersonal attachments as a fundamental human motivation. *Psychological Bulletin*, *117*(3), 497–529. DOI: 10.1037/0033-2909.117.3.497

Boswell, W. R., & Olson-Buchanan, J. B. (2007). The use of communication technologies after hours: The role of work attitudes and work-life conflict. *Journal of Management*, *33*(4), 592–610.

Bracken, D. W., Timmreck, C. W., & Church, A. H. (2016). *The handbook of multisource feedback: The comprehensive resource for designing and implementing MSF processes*. Jossey-Bass.

Butler, J., & Kern, M. L. (2015). The PERMA-Profiler: A brief multidimensional measure of flourishing. *International Journal of Wellbeing*, *6*(2), 1–48.

Byron, K. (2005). A meta-analytic review of work-family conflict and its antecedents. *Journal of Vocational Behavior*, *67*(2), 169–198. DOI: 10.1016/j.jvb.2004.08.009

Cakir, F. S., & Adiguzel, Z. (2020). Analysis of Leader Effectiveness in Organization and Knowledge Sharing Behavior on Employees and Organization. *SAGE Open*, *9*(4), 1–14. Sagepub. https://doi.org/DOI: 10.1177/2158244020914634

Campion, M. A., Cheraskin, L., & Stevens, M. J. (1994). Career-related antecedents and outcomes of job rotation. *Academy of Management Journal*, *37*(6), 1518–1542. DOI: 10.2307/256797

Clampitt, P. G., & Downs, C. W. (1993). Employee perceptions of the relationship between communication and productivity: A field study. *Journal of Business Communication*, *30*(1), 5–28. DOI: 10.1177/002194369303000101

Clark, S. C. (2000). Work/family border theory: A new work/family balance theory. *Human Relations*, *53*(6), 747–770. DOI: 10.1177/0018726700536001

Cohen, S., & Wills, T. A. (1985). Stress, social support, and the buffering hypothesis. *Psychological Bulletin*, *98*(2), 310–357. DOI: 10.1037/0033-2909.98.2.310

Colquitt, J. A. (2001). On the dimensionality of organizational justice: A construct validation of a measure. *The Journal of Applied Psychology*, *86*(3), 386–400. DOI: 10.1037/0021-9010.86.3.386

Cropanzano, R., Bowen, D. E., & Gilliland, S. W. (2007). The management of organizational justice. *The Academy of Management Perspectives*, *21*(4), 34–48. DOI: 10.5465/amp.2007.27895338

Cross, R., & Parker, A. (2004). *The hidden power of social networks: Understanding how work really gets done in organizations*. Harvard Business Review Press.

Davidson, R. J., & Lutz, A. (2008). Buddha's brain: Neuroplasticity and meditation. *IEEE Signal Processing Magazine*, *25*(6), 176–188.

Deci, E. L., & Ryan, R. M. (2000). The "what" and "why" of goal pursuits: Human needs and the self-determination of behavior. *Psychological Inquiry*, *11*(4), 227–268. DOI: 10.1207/S15327965PLI1104_01

Deloitte. (2023). *Global Human Capital Trends 2023: Boundaryless organization*. Deloitte Insights.

Dennis, A. R., & Valacich, J. S. (1999). Rethinking media richness: Towards a theory of media synchronicity. *Proceedings of the 32nd Annual Hawaii International Conference on Systems Sciences*, 1-10. https://doi.org/DOI: 10.1109/HICSS.1999.772701

Deutsch, M. (1973). *The resolution of conflict: Constructive and destructive processes.* Yale University Press.

Dollard, M. F., & Bakker, A. B. (2010). Psychosocial safety climate as a precursor to conducive work environments, psychological health problems, and employee engagement. *Journal of Occupational and Organizational Psychology, 83*(3), 579–599.

Drigas, A., & Papoutsi, C. (2021). Nine Layer Pyramid Model Questionnaire for Emotional Intelligence. [IJOE]. *International Journal of Online and Biomedical Engineering, 17*(07), 123. DOI: 10.3991/ijoe.v17i07.22765

Drigas, A., Papoutsi, C., & Skianis, C. (2023). Being an Emotionally Intelligent Leader through the Nine-Layer Model of Emotional Intelligence—The Supporting Role of New Technologies. *Sustainability, 15*(10), 8103. MDPI. https://www.mdpi.com/2071-1050/15/10/8103

Dutta, P. K., Gupta, S., Kashyap, S., Gehlot, A., Karmakar, R., & Bhattacharya, P. (2025). *Emotional Intelligence in the Digital Era.* CRC Press.

Edmondson, A. (1999). Psychological safety and learning behavior in work teams. *Administrative Science Quarterly, 44*(2), 350–383. DOI: 10.2307/2666999

Edmondson, A. C. (2019). *The fearless organization: Creating psychological safety for learning, innovation, and growth.* Wiley.

Frazier, M. L., Fainshmidt, S., Klinger, R. L., Pezeshkan, A., & Vracheva, V. (2017). Psychological Safety: A Meta-Analytic Review and Extension. *Personnel Psychology, 70*(1), 113–165. DOI: 10.1111/peps.12183

Gajendran, R. S., & Harrison, D. A. (2007). The good, the bad, and the unknown about telecommuting: Meta-analysis of psychological mediators and individual consequences. *The Journal of Applied Psychology, 92*(6), 1524–1541. DOI: 10.1037/0021-9010.92.6.1524

Galiano-Coronil, A., Blanco-Moreno, S., & Tobar-Pesantez, L. B. (2024). Social marketing and happiness in employment. Evidences from Glassdoor. *BMC Psychology, 12*(1). Advance online publication. DOI: 10.1186/s40359-024-01882-8

Gallup. (2022). *State of the global workplace: 2022 report.* Gallup Press.

Gallup. (2023). *Employee burnout: Causes and cures.* Gallup Press.

Gallup. (2024, January 12). *Employee Recognition: Low Cost, High Impact.* Gallup. https://www.gallup.com/workplace/236441/employee-recognition-low-cost-high-impact.aspx

Genin, P. M., Degoutte, F., Finaud, J., Pereira, B., Thivel, D., & Duclos, M. (2018). Effect of a 5-month worksite physical activity program on tertiary employees overall health and fitness. *Journal of Occupational and Environmental Medicine*, *60*(12), e643–e649.

Golden, T. D., Veiga, J. F., & Simsek, Z. (2006). Telecommuting's differential impact on work-family conflict: Is there no place like home? *The Journal of Applied Psychology*, *91*(6), 1340–1350. DOI: 10.1037/0021-9010.91.6.1340

Greenhaus, J. H., & Powell, G. N. (2006). When work and family are allies: A theory of work-family enrichment. *Academy of Management Review*, *31*(1), 72–92. DOI: 10.5465/amr.2006.19379625

Guest, D. E. (2017). Human resource management and employee well-being: Towards a new analytic framework. *Human Resource Management Journal*, *27*(1), 22–38. DOI: 10.1111/1748-8583.12139

Hackman, J. R. (2002). *Leading teams: Setting the stage for great performances*. Harvard Business Review Press.

Hall, D. T. (2002). *Careers in and out of organizations*. Sage Publications.

Hall, D. T. T., & Heras, M. L. (2010). Reintegrating job design and career theory: Creating not just good jobs but smart jobs. *Journal of Organizational Behavior*, *31*(2-3), 448–462. DOI: 10.1002/job.613

Harter, J. K., Schmidt, F. L., & Hayes, T. L. (2002). Business-unit-level relationship between employee satisfaction, employee engagement, and business outcomes: A meta-analysis. *The Journal of Applied Psychology*, *87*(2), 268–279. DOI: 10.1037/0021-9010.87.2.268

Hausknecht, J. P., Rodda, J., & Howard, M. J. (2009). Targeted employee retention: Performance-based and job-related differences in reported reasons for staying. *Human Resource Management*, *48*(2), 269–288. DOI: 10.1002/hrm.20279

Hekman, D. R., Johnson, S. K., Foo, M. D., & Yang, W. (2017). Does diversity-valuing behavior result in diminished performance ratings for non-white and female leaders? *Academy of Management Journal*, *60*(2), 771–797.

Hocine, Z., & Zhang, J. (2014). Autonomy supportive leadership: A new framework for understanding effective leadership through self-determination theory. *International Journal of Information Systems and Change Management*, *7*(2), 135. DOI: 10.1504/ijiscm.2014.069397

House, J. S., Landis, K. R., & Umberson, D. (1988). Social relationships and health. *Science*, *241*(4865), 540–545. DOI: 10.1126/science.3399889

Hurley, R. F. (2012). *The decision to trust: how leaders create high-trust organizations*. Jossey-Bass C.

Isaksen, J. (2000). Constructing Meaning Despite the Drudgery of Repetitive Work. *Journal of Humanistic Psychology*, *40*(3), 84–107. DOI: 10.1177/0022167800403008

Jha, I. N., Pal, D., & Sarkar, S. (2023). Unlocking the secret to happiness at work: The power of inclusive leadership, organizational justice and workplace inclusion. *Journal of Management Development*, *43*(2). Advance online publication. DOI: 10.1108/jmd-04-2023-0136

Jo, H., & Shin, D. (2025). The impact of recognition, fairness, and leadership on employee outcomes: A large-scale multi-group analysis. *PLoS One*, *20*(1), e0312951. DOI: 10.1371/journal.pone.0312951

Jonsdottir, I. J., & Kristinsson, K. (2020). Supervisors' active-empathetic listening as an important antecedent of work engagement. *International Journal of Environmental Research and Public Health*, *17*(21). Advance online publication. DOI: 10.3390/ijerph17217976

Joo, B.-K. (Brian), Yoon, S. K., & Galbraith, D. (2022). The effects of organizational trust and empowering leadership on group conflict: Psychological safety as a mediator. *Organizational Management Journal*, *20*(1), 4–16. DOI: 10.1108/omj-07-2021-1308

Kahn, W. A. (1990). Psychological conditions of personal engagement and disengagement at work. *Academy of Management Journal*, *33*(4), 692–724. DOI: 10.2307/256287

Kalleberg, A. L. (2018). *Precarious lives: Job insecurity and well-being in rich democracies*. Polity Press.

Kane-Urrabazo, C. (2006). Management's role in shaping organizational culture. *Journal of Nursing Management*, *14*(3), 188–194. https://onlinelibrary.wiley.com/doi/full/10.1111/j.1365-2934.2006.00590.x?saml_referrer

Kaplan, A. D., Kessler, T. T., & Hancock, P. A. (2020). How Trust is Defined and its use in Human-Human and Human-Machine Interaction. *Proceedings of the Human Factors and Ergonomics Society Annual Meeting*, *64*(1), 1150–1154. DOI: 10.1177/1071181320641275

Kark, R., & Carmeli, A. (2009). Alive and Creating: The Mediating Role of Vitality and Aliveness in the Relationship between Psychological Safety and Creative Work Involvement. *Journal of Organizational Behavior, 30*(6), 785–804. DOI: 10.1002/job.571

Kark, R., & Carmeli, A. (2009). Alive and creating: The mediating role of vitality and aliveness in the relationship between psychological safety and creative work involvement. *Journal of Organizational Behavior, 30*(6), 785–804.

Katzenbach, J. R., & Smith, D. K. (1993). *The wisdom of teams: Creating the high-performance organization.* Harvard Business Review Press.

Kawada, M., & Shimazu, A. Daisuke Miyanaka, Tokita, M., Sakakibara, K., Mori, N., Fuad Hamsyah, Yuheng, L., Kojiro Shojima, & Schaufeli, W. B. (2023). Boredom and engagement at work: Do they have different antecedents and consequences? *Industrial Health.* Advance online publication. DOI: 10.2486/indhealth.2023-0028

Kelly, E. L., Kossek, E. E., Hammer, L. B., Durham, M., Bray, J., Chermack, K., Murphy, L. A., & Kaskubar, D. (2008). Getting there from here: Research on the effects of work-family initiatives on work-family conflict and business outcomes. *The Academy of Management Annals, 2*(1), 305–349. DOI: 10.1080/19416520802211610

Kincaid, R. (2025). *First Leadership.*

Kish-Gephart, J. J., Detert, J. R., Treviño, L. K., & Edmondson, A. C. (2009). Silenced by fear: The nature, sources, and consequences of fear at work. *Research in Organizational Behavior, 29*, 163–193.

Klauss, R., & Bass, B. M. (1982). *Interpersonal communication in organizations.* Academic Press.

Klitzman, S., & Stellman, J. M. (1989). The impact of the physical environment on the psychological well-being of office workers. *Social Science & Medicine, 29*(6), 733–742. DOI: 10.1016/0277-9536(89)90153-6

Kluger, A. N., & DeNisi, A. (1996). The effects of feedback interventions on performance: A historical review, a meta-analysis, and a preliminary feedback intervention theory. *Psychological Bulletin, 119*(2), 254–284. DOI: 10.1037/0033-2909.119.2.254

Kniffin, K. M., Narayanan, J., Anseel, F., Antonakis, J., Ashford, S. P., Bakker, A. B., & Vugt, M. V. (2021). COVID-19 and the workplace: Implications, issues, and insights for future research and action. *The American Psychologist, 76*(1), 63–77.

Kovich, M. K., Simpson, V. L., Foli, K. J., Hass, Z., & Phillips, R. G. (2022). Application of the PERMA Model of Well-being in Undergraduate Students. *International Journal of Community Well-being*, *6*(1), 1–20. DOI: 10.1007/s42413-022-00184-4

Kozlowski, S. W. J., & Ilgen, D. R. (2006). Enhancing the effectiveness of work groups and teams. *Psychological Science in the Public Interest*, *7*(3), 77–124. DOI: 10.1111/j.1529-1006.2006.00030.x

Kram, K. E. (1985). *Mentoring at work: Developmental relationships in organizational life*. Scott, Foresman.

Landry, L. (2019, April 3). *Why emotional intelligence is important in leadership*. Harvard Business School Online. https://online.hbs.edu/blog/post/emotional-intelligence-in-leadership

Lewis, L. K. (2011). *Organizational change: Creating change through strategic communication*. Wiley-Blackwell.

Lim, W. M. (2023). The Workforce revolution: Reimagining work, workers, and Workplaces for the Future. *Global Business and Organizational Excellence*, *42*(4). Advance online publication. DOI: 10.1002/joe.22218

London, M. (2003). *Job feedback: Giving, seeking, and using feedback for performance improvement* (2nd ed.). Lawrence Erlbaum Associates.

London, M., & Smither, J. W. (1999). Empowered self-development and continuous learning. *Human Resource Management*, *38*(1), 3–15. <3:AID-HRM2>3.0.CO;2-MDOI: 10.1002/(SICI)1099-050X(199921)38:1

Lupien, S. J., Juster, R. P., Raymond, C., & Marin, M. F. (2018). The effects of chronic stress on the human brain: From neurotoxicity, to vulnerability, to opportunity. *Frontiers in Neuroendocrinology*, *49*, 91–105.

Lutz, A., Greischar, L. L., Rawlings, N. B., Ricard, M., & Davidson, R. J. (2004). Long-term meditators self-induce high-amplitude gamma synchrony during mental practice. *Proceedings of the National Academy of Sciences of the United States of America*, *101*(46), 16369–16373.

Lyubomirsky, S., King, L., & Diener, E. (2005). The benefits of frequent positive affect: Does happiness lead to success? *Psychological Bulletin*, *131*(6), 803–855.

Makinde, O. (2024). Employee Well-Being Matters In Corporate Strategy. *Forbes*. https://www.forbes.com/sites/oludolapomakinde/2024/06/30/employee-well-being-matters-in-corporate-strategy/

Makinde, O. A. (2024). Digital transformation and employee well-being: A systematic review. *Journal of Business Research, 156*, 113–127.

Martela, F., Gómez, M., Unanue, W., Araya, S., Bravo, D., & Espejo, A. (2021). What makes work meaningful? Longitudinal evidence for the importance of autonomy and beneficence for meaningful work. *Journal of Vocational Behavior, 131*(103631), 103631. DOI: 10.1016/j.jvb.2021.103631

Martela, F., & Pessi, A. B. (2018). Significant Work Is About Self-Realization and Broader Purpose: Defining the Key Dimensions of Meaningful Work. *Frontiers in Psychology, 9*(9). Advance online publication. DOI: 10.3389/fpsyg.2018.00363

Maslach, C., & Leiter, M. P. (2016). *The burnout challenge: Managing people's relationships with their jobs*. Harvard Business Review Press.

Maurer, T. J., Weiss, E. M., & Barbeite, F. G. (2003). A model of involvement in work-related learning and development activity: The effects of individual, situational, motivational, and age variables. *The Journal of Applied Psychology, 88*(4), 707–724. DOI: 10.1037/0021-9010.88.4.707

Mayfield, J., & Mayfield, M. (2002). Leader communication strategies critical paths to improving employee commitment. *American Business Review, 20*(2), 89–94.

Maylett, T., & Wride, M. (2017). *The employee experience: How to attract talent, retain top performers, and drive results*. Wiley.

McEwen, B. S. (2017). Neurobiological and systemic effects of chronic stress. *Chronic Stress (Thousand Oaks, Calif.), 1*, 1–11.

McGuire, D., & McLaren, L. (2009). The impact of physical environment on employee commitment in call centres: The mediating role of employee well-being. *Team Performance Management, 15*(1/2), 35–48.

Mclean, S. (2010). *Autonomy, consent and the law*. Routledge-Cavendish.

Morrison, E. W. (2011). Employee voice behavior: Integration and directions for future research. *The Academy of Management Annals, 5*(1), 373–412. DOI: 10.1080/19416520.2011.574506

Nameda, A., Kumaki, Y., Hashimoto, Y., & Toda, Y. (2022). Distributive justice and development: Toward an integrative understanding of reward and responsibility distributions. *Journal of Moral Education, 52*(4), 1–18. DOI: 10.1080/03057240.2022.2134095

Newman, A., Donohue, R., & Eva, N. (2017). Psychological safety: A systematic review of the literature. *Human Resource Management Review*, *27*(3), 521–535. DOI: 10.1016/j.hrmr.2017.01.001

Noe, R. A., Hollenbeck, J. R., Gerhart, B., & Wright, P. M. (2010). *Human resource management: Gaining a competitive advantage* (7th ed.). McGraw-Hill/Irwin.

Page, G., & Tolmie, J. (2024). Physical working environments: How they affect our wellbeing and performance. *In Practice*, *46*(3), 152–156. DOI: 10.1002/inpr.420

Page, S. E. (2007). *The difference: How the power of diversity creates better groups, firms, schools, and societies*. Princeton University Press.

Patil, R., Raheja, D. K., Nair, L., Deshpande, A., & Mittal, A. (2023). The Power of Psychological Safety: Investigating its Impact on Team Learning, Team Efficacy, and Team Productivity. *The Open Psychology Journal*, *16*(1). Advance online publication. DOI: 10.2174/18743501-v16-230727-2023-36

Pink, D. H. (2009). *Drive: The surprising truth about what motivates us*. Riverhead Books.

Putnam, R. D. (2000). *Bowling alone: The collapse and revival of American community*. Simon & Schuster.

Ramya, S. R., & Vanithamani, M. R. (2023). The power of employee recognition: Building a culture of appreciation in the workplace. *Journal of Teacher Education*, *109*, 109.

Reagans, R., & McEvily, B. (2003). Network structure and knowledge transfer: The effects of cohesion and range. *Administrative Science Quarterly*, *48*(2), 240–267. DOI: 10.2307/3556658

Rook, K. S. (1987). Social support versus companionship: Effects on life stress, loneliness, and evaluations by others. *Journal of Personality and Social Psychology*, *52*(6), 1132–1147. DOI: 10.1037/0022-3514.52.6.1132

Rosa, H. (2013). *Social acceleration: A new theory of modernity*. Columbia University Press.

Ruggeri, K., Garcia-Garzon, E., Maguire, Á., Matz, S., & Huppert, F. A. (2020). Well-being is more than happiness and life satisfaction: A multidimensional analysis of 21 countries. *Health and Quality of Life Outcomes*, *18*(192), 1–16. DOI: 10.1186/s12955-020-01423-y

Ryan, R., & Deci, E. (2001). On Happiness and Human potentials: A Review of Research on Hedonic and Eudaimonic well-being. *Annual Review of Psychology, 52*(1), 141–166. DOI: 10.1146/annurev.psych.52.1.141

Schein, E. H. (2010). *Organizational culture and leadership* (4th ed.). Jossey-Bass.

Scott, B. A. B., & Manning, M. R. (2022). Designing the collaborative organization: A framework for how collaborative work, relationships, and behaviors generate collaborative capacity. *The Journal of Applied Behavioral Science, 60*(1), 149–193. sagepub. https://doi.org/DOI: 10.1177/00218863221106245

Shoss, M. K. (2017). Job insecurity: An integrative review and agenda for future research. *Journal of Management, 43*(6), 1911–1939.

Slemp, G. R., Kern, M. L., & Vella-Brodrick, D. A. (2015). Workplace Well-Being: The Role of Job Crafting and Autonomy Support. *Psychology of Well-Being, 5*(1). Advance online publication. DOI: 10.1186/s13612-015-0034-y

Smeets, L., Gijselaers, W. H., Meuwissen, R. H. G., & Grohnert, T. (2021). Beyond Psychological Safety – the Role of Direct Supervisor Behavior in Fostering Learning from Errors at the Workplace. *Vocations and Learning, 14*. Advance online publication. DOI: 10.1007/s12186-021-09272-6

Society for Human Resource Management. (2023). *Employee job satisfaction and engagement: The doors of opportunity are open*. SHRM Press.

Stellman, J. M., Klitzman, S., Gordon, G., & Snow, B. J. (1987). Work environment and the well-being of clerical and VDT workers. *Journal of Occupational Behaviour, 8*(2), 95–114. DOI: 10.1002/job.4030080202

Super, D. E. (1990). A life-span, life-space approach to career development. In Brown, D., & Brooks, L. (Eds.), *Career choice and development* (pp. 197–261). Jossey-Bass.

Tarafdar, M., Cooper, C. L., & Stich, J. F. (2019). The technostress trifecta-techno eustress, techno distress and design: Theoretical directions and an agenda for research. *Information Systems Journal, 29*(1), 6–42.

Thayer, J. F., & Lane, R. D. (2009). Claude Bernard and the heart–brain connection: Further elaboration of a model of neurovisceral integration. *Neuroscience and Biobehavioral Reviews, 33*(2), 81–88.

Thompson, C. A., Beauvais, L. L., & Lyness, K. S. (1999). When work-family benefits are not enough: The influence of work-family culture on benefit utilization, organizational attachment, and work-family conflict. *Journal of Vocational Behavior, 54*(3), 392–415. DOI: 10.1006/jvbe.1998.1681

Tourish, D. (2010). *Auditing organizational communication: A handbook of research, theory and practice*. Routledge.

Tourish, D., & Hargie, O. (2004). *Key issues in organizational communication*. Routledge.

Tsai, Y. (2021). Relationship between Organizational culture, Leadership Behavior and Job Satisfaction. *BMC Health Services Research, 11*(1). BMC. https://doi.org/DOI: 10.1186/1472-6963-11-98

Twenge, J. M. (2010). A review of the empirical evidence on generational differences in work attitudes. *Journal of Business and Psychology, 25*(2), 201–210.

Vago, D. R., & Silbersweig, D. A. (2012). Self-awareness, self-regulation, and self-transcendence (S-ART): A framework for understanding the neurobiological mechanisms of mindfulness. *Frontiers in Human Neuroscience, 6*, 296.

van der Deijl, W. (2022). Two Concepts of Meaningful Work. *Journal of Applied Philosophy*. Advance online publication. DOI: 10.1111/japp.12614

Van Dyne, L., Ang, S., & Botero, I. C. (2003). Conceptualizing employee silence and employee voice as multidimensional constructs. *Journal of Management Studies, 40*(6), 1359–1392. DOI: 10.1111/1467-6486.00384

Vohra, N., Chari, V., Mathur, P., Sudarshan, P., Verma, N., Mathur, N., Thakur, P., Chopra, T., Srivastava, Y., Gupta, S., Dasmahapatra, V., Fonia, S., & Gandhi, H. K. (2015). Inclusive Workplaces: Lessons from Theory and Practice. *Vikalpa: The Journal for Decision Makers, 40*(3), 324–362. sagepub. https://doi.org/DOI: 10.1177/0256090915601515

Waber, B. (2013). *People analytics: How social sensing technology will transform business and what it tells us about the future of work*. FT Press.

Watson Wyatt. (2009). *Effective communication: A leading indicator of financial performance*. Watson Wyatt Worldwide.

Werquin, P. (2010). Recognition of non-formal and informal learning: Country practices. *Organisation de coopération et de développement économiques OCDE. Paris. Disonible en.*

World Health Organization. (2022). *Mental health at work: Policy brief.* WHO Press.

Yean, T. F., & Yusof, A. A. (2016). Organizational Justice: A Conceptual Discussion. *Procedia: Social and Behavioral Sciences*, *219*(1), 798–803. DOI: 10.1016/j.sbspro.2016.05.082

Zhenjing, G., Chupradit, S., Ku, K. Y., Nassani, A. A., & Haffar, M. (2022). Impact of employees' Workplace Environment on employees' performance: a multi-mediation Model. *Frontiers in Public Health, 10*(890400). NCBI. https://doi.org/ DOI: 10.3389/fpubh.2022.890400

Zuberbuhler, M. J. P., Salanova, M., & Martínez, I. M. (2020). Coaching-Based Leadership Intervention Program: A Controlled Trial Study. *Frontiers in Psychology*, *10*(3066), 1–22. DOI: 10.3389/fpsyg.2019.03066

Chapter 6
Navigating the Rapid:
Organizational Change and Adaptation in the 21st Century

Evangelina Rekha Joseph
https://orcid.org/0009-0008-3819-2454
Christ University, India

ABSTRACT

In an era defined by volatility, uncertainty, complexity, and ambiguity (VUCA), organisational change is no longer episodic—it is a continuous necessity. This chapter explores how organisations can navigate constant change in a volatile, uncertain, complex, and ambiguous (VUCA) world. It examines key drivers of change technological, demographic, environmental, and socio-political and integrates classical and contemporary theories, including Lewin's model, Kotter's framework, and sensemaking. Emphasis is placed on strategic foresight, organisational agility, innovation culture and role of neuroplasticity in change implementation. Real-world case studies (e.g., Microsoft, Patagonia, GM) demonstrate practical applications from organisational and neuroplasticity perspective. The chapter highlights the importance of adaptive resilience, aligning formal and informal systems, empowering middle managers, and fostering psychological safety to enable sustainable transformation and continuous learning in rapidly evolving environments.

INTRODUCTION

The 21st century has steered an era of rapid and persistent transformation across industries, driven by global trends including technological innovation, changing workforce dynamics such as demographics, rising political instability, environmental challenges and increased digitalisation. These changes have made volatility,

DOI: 10.4018/979-8-3373-2292-6.ch006

uncertainty, complexity, and ambiguity (VUCA) in the external environment not only buzzwords but a lived organisational reality (Cernega et al., 2024). In such a turbulent environment, adapting and navigating through the changes is no longer a strategic advantage but a fundamental necessity for businesses.

Organisations today face challenges that require swift, informed and often audacious decision-making. Disruptive technologies like artificial intelligence and blockchain, the surge of platform-based economies and changing consumer behaviour are constantly reshaping the markets and business value chains (Verhoef et al., 2021). Concurrently, external shocks like COVID-19, geopolitical conflicts and climate-related disasters have further exposed the fragility of many traditional business models. These shifts emphasised the need for organisations to become more resilient, adaptive and forward-thinking. Remarkably, adaptation is not solely about survival. It is also about leveraging change as an opportunity for renewal, innovation and sustainable growth. Organisations that embrace change and adapt to it can unlock new business models, enter untapped markets and efficiently meet the evolving needs of their stakeholders. Adaptability fosters competitive advantage, enabling companies to stay ahead of their competitors and industry disruptions and maintain their relevance in the marketplace (Reeves & Deimler, 2011).

Agility is the organisational backbone, especially considering the ever-increasing ambiguity and the need for change. It acts as a springboard for the organisation; an anchor point that should remain constant despite the functional fluctuations and change. It is defined as the "ability of organisations to quickly respond to market needs by sensing, renewing, adapting and succeeding in a turbulent market" (Motwani & Kataria, 2024). Agility is critical for the organisation, as it is the ability to move quickly and efficiently rather than constantly implementing new technologies. It requires a fundamental shift in the mindset, leadership styles, processes and values.

While strategic foresight is a central competency for businesses in this environment, it is often underused and undervalued. Moqaddamerad & Ali (2024) argue that strategic sight is an imperative requirement that significantly affects organisational outcomes. Furthermore, environmental instability triggered by rapid technological shifts creates managerial challenges in designing, restructuring, and managing resources. Therefore, organisations must develop foresight to cope with these challenges (Schoemaker & Day, 2021). It is important for the leaders to cultivate the ability to anticipate the possible future, recognise the early signs of potential change, and devise flexible strategies for adapting to the changes. Proactive change management as such is required for initiating change rather than reacting defensively, and ensuing transformation is an ongoing process and not a one-time event.

This chapter makes a distinctive contribution to the organisational change literature by presenting an integrated framework that synthesises diagnostic assessment, strategic design, transformational leadership, and sustainability mechanisms into a cohesive approach to organisational adaptation. Unlike existing fragmented approaches that address these elements in isolation, this chapter provides a holistic methodology that enables organisations to simultaneously assess their change readiness while designing and implementing adaptive strategies. The framework uniquely bridges the gap between theoretical understanding and practical application by offering evidence-based tools and techniques that can be directly applied across diverse organisational contexts.

The chapter is structured to provide theoretical foundations and practical guidance through six key sections: First, we establish the theoretical underpinnings of organisational change and adaptation, examining the core concepts and contemporary challenges. Second, we present a comprehensive diagnostic framework for assessing organisational readiness for change, including assessment tools and readiness indicators. Third, we explore strategic design principles for effective change initiatives, covering planning methodologies and implementation strategies. Fourth, we examine transformational leadership approaches that drive successful change, analysing leadership competencies and change facilitation techniques. Fifth, we address sustainability mechanisms that ensure long-term transformation success, including monitoring systems and continuous improvement processes. Finally, we explore organisational change from a neuroplasticity lens, examining how insights from neuroscience can inform adaptive organisational practices.

This comprehensive approach aims to prepare managers, leaders, business consultants and students with the necessary knowledge and tools to navigate the evolving landscape of modern organisations. The chapter underscores that successful adaptation is not limited to surviving a disruption, but rather about thriving through transformation.

THEORETICAL FOUNDATIONS OF ORGANISATIONAL CHANGE

Understanding and managing organisational change requires a multifaceted approach grounded in established theoretical frameworks. These theories offer a structured approach to understanding the change processes, supporting the leaders and practitioners to navigate complex dynamics inherent to organisational transformation (Burnes, 2020). These theories serve as a diagnostic tool aiding in identifying challenges, predicting stakeholder responses and choosing/implementing the most appropriate change interventions. In the absence of robust theoretical foundations, change initiatives are at the risk of becoming reactive, fragmented or even detached from the broader organisational context (By, 2005). This chapter examines some of the most influential theoretical models and frameworks shaping contemporary understanding and implementation of organisational change.

Classical Change Models

Lewin's three-step model of change represents one of the earliest and most foundational theories in change management (Lewin, 1947). The model was developed in the 1940s, and it conceptualises change in a three-stage sequential process involving "Unfreeze, Change, and Refreeze" (Figure 1).

Figure 1. Lewin's three-step model of change

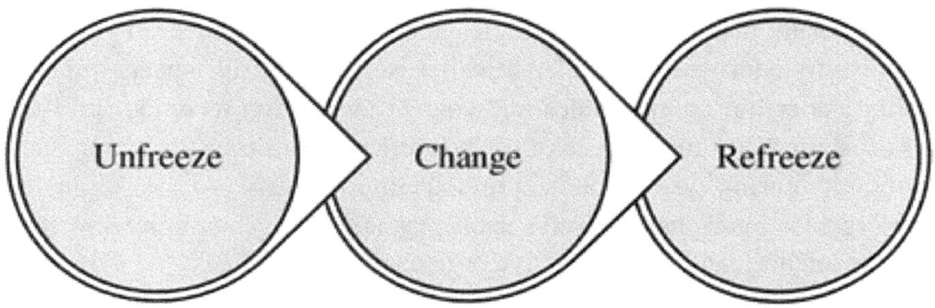

During the first-unfreezing stage, the organisations recognise the need for change and prepare to forsake their existing methods and practices (Schein & Schein, 2017). The change stage involves implementing identified new processes, structures, practices, and behaviours. The refreezing stage focuses on embedding these changes into the organisational system and culture to ensure stability and permanence (Cummings et al., 2016). Despite the model's simplistic approach, enhancing its application, accessibility, and communication value, critics argue that its linear and static nature is inadequate in today's dynamic business environment (Burnes, 2020). Nonetheless, the model remains valuable and continues to serve as a foundational concept in the change management theories.

Contemporary Leadership-Driven Models

Kotter's 8-step process offers a more comprehensive roadmap for leading organisational change and transformation (Kotter, 1996) (Figure 2). The model involves sequential steps involving "creating a sense of urgency, building a guiding coalition, forming a strategic vision for change, communicating the vision and enlisting a volunteer army, empowering others towards a broad-based action, enabling and promoting short-term wins, consolidating gains and producing more change, and anchoring new approaches in the culture" (Kotter, 2012). Kotter emphasises the importance of leadership and emotional commitment, particularly in creating urgency and maintaining momentum throughout the change process (Appelbaum et al., 2012). While the model's structured approach facilitates planning and execution, its sequential format may require adaptation in fast-paced or emergent change situations (Pollack & Pollack, 2015).

Figure 2. Kotter's 8-step change management process

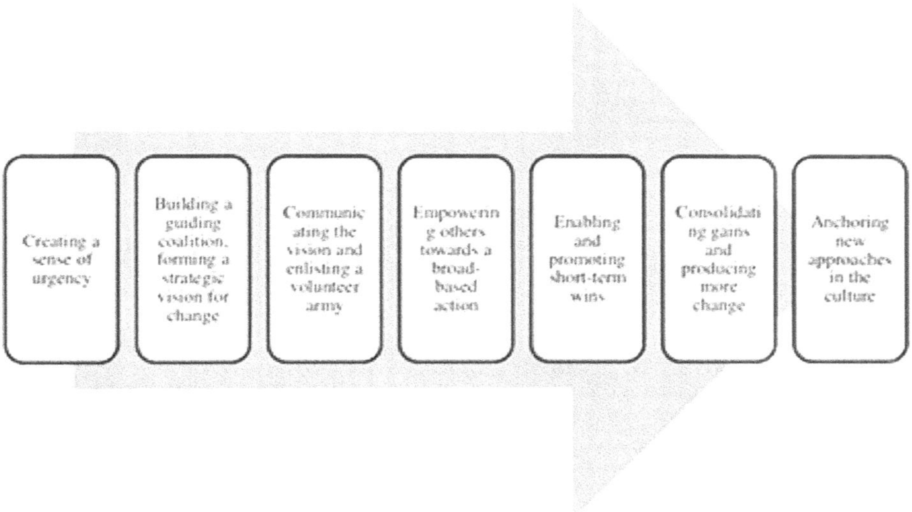

Individual-Focused Change Models

The ADKAR Model, developed by Prosci, centres on the individual as the fundamental unit of change (Hiatt, 2006). The acronym represents five essential building blocks: Awareness of the need for change, Desire to participate and support change, Knowledge of how to change, Ability to implement required skills and behaviours, and Reinforcement to sustain the change (Prosci, 2018) (Figure 3). Unlike broader organisational models, ADKAR proves particularly effective for managing personal transitions, coaching employees, and aligning communication and training strategies (Creasey, 2009). Its primary strength lies in recognising that organisational change succeeds only when individuals are adequately engaged and supported throughout the process (Hornstein, 2015).

Figure 3. ADKAR Model

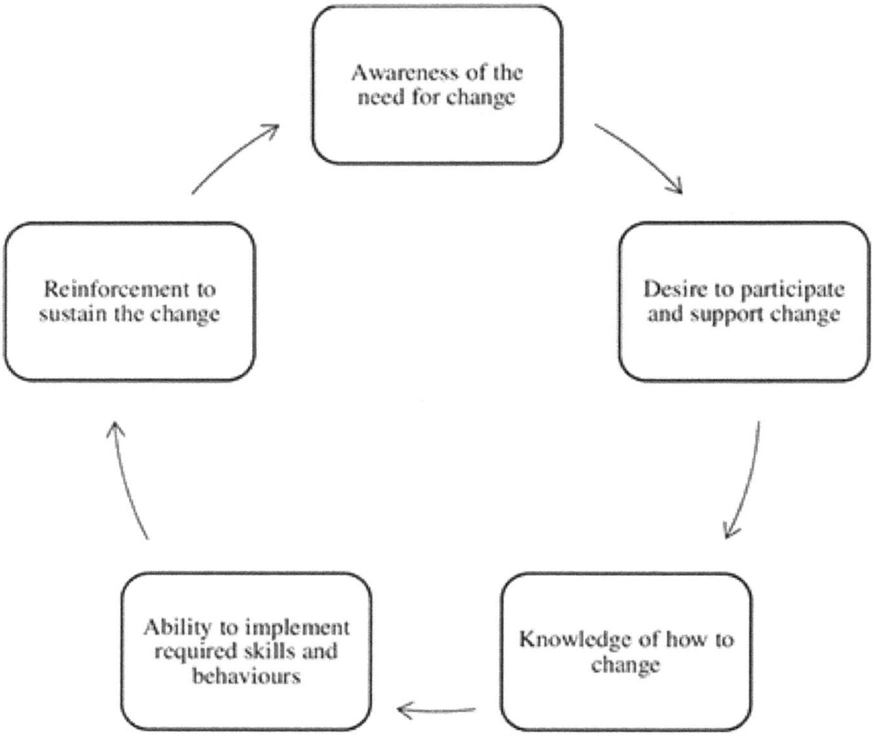

Leadership Theory Applications

Transformational Leadership Theory is pivotal in shaping change leadership approaches (Bass & Riggio, 2006). Transformational leaders inspire and motivate followers by articulating compelling visions, demonstrating genuine concern for individual needs, and challenging them to exceed performance expectations (Avolio & Bass, 2004). These leaders are change champions who cultivate trust, enthusiasm, and commitment while modelling expected behaviours and fostering innovation-supportive environments (Northouse, 2021). The relevance of transformational leadership intensifies in uncertain and ambiguous contexts, where employees seek direction and reassurance from their leaders (Eisenbach et al., 1999).

Systems and Complexity Perspectives

Systems theory and Complexity theory offer non-linear perspectives on organisational change (Senge, 2006). Systems theory conceptualises organisations as interconnected elements where changes in one component affect the entire system (Anderson, 1999). Complexity theory recognises organisations as adaptive systems with emergent properties, with unpredictable outcomes (Stacey, 2011). These perspectives caution against over-reliance on top-down, mechanistic approaches and instead advocate for flexible, iterative, and context-sensitive interventions (Marion & Uhl-Bien, 2001). Leaders who apply these theories focus on understanding the patterns, fostering collaborations and enabling self-organisational processes (Uhl-Bien et al., 2007).

Cognitive and Interpretive Approaches

Sensemaking theory, introduced by Karl Weick, emphasises the process by which individuals and groups interpret the meaning of the change events (Weick, 1995). According to Weick, it is the "process through which people assign meaning to issues or events that cause the current state of the world to be different from the expected state. Since its first clear conceptualisation" (Cristofaro, 2021). The theory posits that people's interpretation of change, through narratives, symbols and social interactions, is shaped by their reactions and behaviours (Weick et al., 2005). Leaders are critical in preparing and guiding through such interpretations, particularly during ambiguity and dissonant notes (Gioia & Chittipeddi, 1991). Sensemaking theory highlights the importance of efficient communication strategies, storytelling, and psychological safety in supporting employees navigate through the change processes (Maitlis & Christianson, 2014).

Agility and Adaptive Frameworks

Organisational agility models focus on developing organisational capacity to respond promptly and efficiently to the changing environmental conditions (Doz & Kosonen, 2008). It has become a critical capability for organisations in achieving and sustaining competitive advantage (Asghar et al., 2025). Agility is supported by the core principles, including iterative development, continuous feedback, and cross-functional collaboration, empowering organisations to react and adapt quickly to changes in the market (Teece et al., 2016). While the agile frameworks were initially developed within the software industry, they have been increasingly applied in the context of organisational design and change management (Rigby et al., 2016). These frameworks offer the tools and mindset essential for managing organisational and environmental complexity, fostering innovation, and aligning strategic intent with the operational requirements (Felipe et al., 2017).

These theoretical models provide diverse analytical perspectives for understanding and managing organisational change. From Lewin's foundational insights on the change to the contemporary theories of agility and complexity, these theoretical frameworks offer unique applicability to a specific organisational context (Todnem By, 2005). Effective change leaders avoid being inflexible and rigid to single models; instead, they integrate multiple theoretical perspectives to address their organisations' specific change needs and dynamics (Burnes, 2004). These theoretical frameworks offer practical applications that equip practitioners and managers to navigate the uncertainty and opportunity inherent in organisational transformation with greater intentionality, insight, and professional skill.

FORCES SHAPING ORGANISATIONAL CHANGE

Change is a constant process in response to problems, challenges, and opportunities. Organisations also experience changes that can be transitional, transformational, temporary, or permanent (Chalofsky et al., 2014). The response to the impact of external and internal forces disrupts the status quo and demands that organisations adopt new ways of thinking, functioning, and competing in the marketplace. Understanding these forces is vital for leaders to develop responsive and future-ready organisations. Several key drivers shape the context for organisational change, and this section explores some of these factors (Figure 4).

Figure 4. Factors shaping organisational change

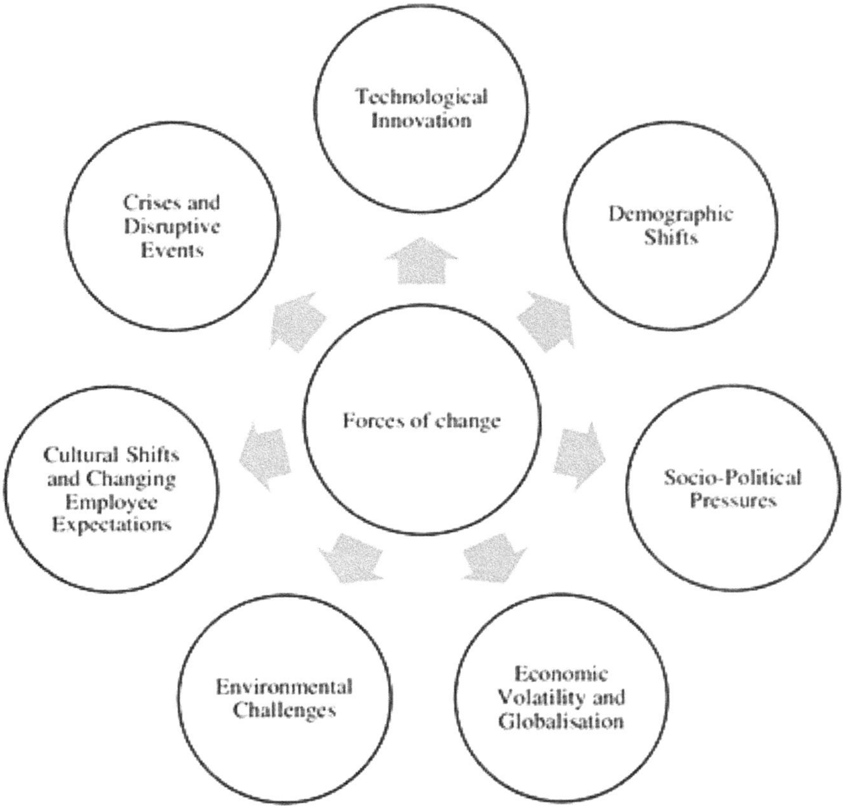

Technological innovation today, perhaps, is the most visible and powerful drive of change. The advent of artificial intelligence, machine learning, blockchain, and the Internet of Things (IoT) is increasingly shaping diverse fields and altering how organisations operate (Rejeb et al., 2023). Furthermore, automation is reshaping the job roles and workflows, whilst data analytics enables more accurate and predictive decision-making in the organisation. Drake (1994) noted that information technology has been driving adaptive changes that are both mere responses and fundamental. Digital transformation initiatives are disrupting traditional business models and forcing companies to reconsider the value creation and service delivery. For example, the rise in platform-based businesses like Uber, Amazon, and Airbnb demonstrates how technology can redefine entire industries. As such, it is critical for the organisation to constantly monitor and assess the business landscape to identify the technological trends and examine the degree of impact they can have on its operations, customers, employees, and the competitive advantage.

Demographic shifts are transforming the workforce and consumer base in the marketplace. An ageing population in many developed countries, like Australia, has created labour shortages and increasing demand for healthcare and retirement services (Earl et al., 2017). According to Tubussum et al. (2024), demographic shifts driven by labour shortages and the need for robust talent management strategies have been evident across industries, and companies are under immense pressure to develop and implement strategies that can help to attract and retain qualified employees. Companies confronting a diverse generational structure in their workforce require strategic change implementation to develop a strong collaborative workforce that adds to business advantage and supports the company's success in the long run. The generational shifts require companies to employ new approaches to talent management, organisational culture, and leadership, especially given that Gen Z and Millennials value different work aspects and expectations. This demographic shift drives organisational change in workforce planning and staffing (Simmons, 2024).

Socio-political pressure is another key factor influencing organisational change. (Rocque, 2017). Social, political, and economic pressure can drive drastic organisational changes. For instance, the rising awareness of social justice, environmental sustainability, and corporate ethics is pushing organisations to rethink their roles and responsibilities and think beyond profits. Every corporation is a place of power where many stakeholders and rationales come into conflict with each other. In the event of change, it can be postulated that organisational change is not primarily a technical problem, but rather a socio-political problem (Soenen, 2010). Movements like #MeToo, Black Lives Matter, and Climate activism have heightened the expectations for accountability, transparency, and ethical leadership. Government and regulators are introducing stringent compliance requirements aligned with ESG (environmental, social, and governance) standards. Additionally, organisations that fail to respond to these external pressures risk reputational damage, loss of public trust, and legal consequences. In contrast, those organisations that proactively embrace social responsibility can experience boosted brand value and strengthened stakeholder relationships.

Globalisation has made the world economy highly volatile (Mendoza & Quadrini, 2023). This can contribute to an unpredictable business environment. While globalisation brings many benefits for businesses, such as access to new markets, consumer base, lowering costs of products and diversification, it also brings challenges like increased competition, navigating cultural differences, and pressure on quality and price. Additionally, fluctuating exchange rates, inflation, trade wars, and supply chain disruptions have become standard features of the global economy. COVID-19 forced many organisations to close, leading to an unprecedented disruption of the commerce and supply chain in most business sectors, with significant effects on the supply chain, consumer demands, cash flow, sales, and marketing (Donthu &

Gustafsson, 2020). It revealed vulnerabilities in the systems and processes of companies globally. Companies must develop more resilience and prepare their people to stay resilient in periods of uncertainty. There is a need for business models that can enable them to withstand shocks and adapt to the changing market dynamics.

Another critical factor driving organisational change in the contemporary era is the **environmental challenges**, which have led to businesses rethinking their entire organisational practices. As society continues to grow increasingly aware of the state of the environment owing to rising concerns over climate change, resource scarcity, and environmental degradation, there is pressure on the businesses to reduce the effects of their business operations on environment through efficient resource consumption, ethical sourcing, and emission of pollutants among others (Impact Day, 2023). Consumers, investors, and stakeholders increasingly prioritise environmental stewardship, prompting businesses to shift towards more circular economy models. Sustainable practices are no longer a moral imperative but a source of competitive differentiation and should be integrated into the core business strategy.

Cultural shifts and changing employee expectations significantly influence how organisations attract, retain, and motivate talent. Employees' expectations are changing, and so should their experience, especially as people are changing, social structures are altering, and what employees value at work and in their personal lives has shifted as well. Employees today seek more than just tangible financial rewards; they value purpose, flexibility, inclusion, and personal growth while working. Pannerselvam & Balaram (2022) argue that knowing what the employees expect in the workplace can help organisations understand the talent needs and preferences and then initiate action on those expectations. Additionally, with the rise of remote working, the gig economy, and hybrid working arrangements, the traditional employment norms are being challenged and continue to redefine the workplace experience. This requires organisations to be flexible and adaptive by offering more personalised and meaningful work environments, investing in employee well-being and fostering a greater sense of belonging. Organisations can foster a sense of belonging by creating an organisational culture that fosters psychological safety, allowing individuals to contribute and feel included and valuable, while also being comfortable expressing their concerns and vulnerabilities (Waller, 2021). Furthermore, there is a need to evolve the psychological contract between employer and employee to ensure that organisations are recognising and reconfiguring their value propositions.

One cannot disregard the impact of **crisis and disruptive events** such as wars, natural disasters, pandemics, and cyberattacks, which catalyse the introduction of both urgent and radical changes. These unforeseen events often test the organisational resilience and emphasise the significance of crisis preparedness, management, continuity planning and adaptive leadership. For example, the COVID-19 pandemic accelerated the digital transformation, uncovered the gaps in healthcare and logistics, and forced global experimentation in remote work. While these crises reveal systemic weaknesses, they also offer opportunities for reinvention and innovation. However, this requires organisations to ensure a high level of trust across the organisational roles and levels, a clear vision and objectives, capitalisation on previous experiences dealing with minor and major crises and crisis management among the employees (Cedergren & Hassel, 2024).

In summary, the forces shaping the organisational changes are multifaceted and interrelated—from demographic to technological, social, economic, environmental and geopolitical. Successful organisations do not just react to these forces but anticipate them in advance and strategically respond to them. By understanding these factors, organisations can look beyond the challenges and leverage the opportunities to lead the change.

DESIGNING EFFECTIVE CHANGE STRATEGIES

The contemporary business landscape demands a fundamental reassessment of how organisations approach change management, moving beyond reactive measures toward strategic, integrated frameworks that can withstand the complexities of modern disruption. While the traditional change management models offer a foundational understanding of change, the accelerators of technological advancement, market volatility, and global interconnectedness would require more sophisticated and robust approaches that can balance organisational agility and strategic foresight (Kotter, 2012; Burnes, 2020). This chapter examines the key dimensions of effective change strategies, arguing for a holistic integration of various factors for designing effective change strategies.

The most important principle from contemporary change management literature is that strategic planning precedes ad-hoc reactions. Kotter (2012) explains how organisations without a clear strategic direction for change initiatives tend to spread their resources too thin across similar, competing, and fragmented efforts, ultimately failing to achieve a sustainable transformation. This is also supported by Higgs & Rowland's (2005) research, which shows that organisations with defined change strategies are 67% more likely to achieve their transformation goals than those with reactive approaches. On the other hand, critics of overly structured change planning, such as Weick and Quinn (1999), argue that too much focus on blueprints can result in inflexible rigidity that dampens adaptive responses to new challenges. This tension highlights a critical paradox in change management: While a strategic direction offers vital coherence, organisations must navigate unpredictable realities and retain some strategic chaos. The solution is not choosing between planning and adaptability, but adopting what Eisenhardt and Martin (2000) call "dynamic capabilities", the ability to integrate, build, and reconfigure competencies to respond to a rapidly changing environment.

Strategic foresight using scenario planning is vital to traditional strategic planning methods. It provides a way to manage ambiguity while avoiding the pitfalls of overly rigid forecasting. Van der Heijden (2005) regards scenario planning as a learning tool that fosters cognitive adaptability within organisations and helps challenge constraining mental models. Its strength is not in predictive accuracy but in the ability to think beyond linear extrapolations, evidenced by Shell's successful navigation of oil crises via scenario-based preparedness (Wack, 1985). Regardless, scenario planning suffers from a critical lack of scrutiny. Schoemaker (1995) points out that scenario exercises are plagued with cognitive biases like anchoring and confirmation bias, which limit the effectiveness of expanding perspective. Also, the thoroughness and resource demands of comprehensive scenario planning may prove impractical for smaller firms or those needing immediate resolution. Even with these restrictions, the complex nature of business dealings today makes some systematic forward-looking planning even more essential, even if organisations have to modify scenario planning.

The shift towards more agile organisational designs marks an important change in business strategic focal points. However, these changes bring to light critical issues around the applicability or durability of genuinely flexible organisational models. Worley et al. (2014) advocate for agile organisational structures. They argue that traditional hierarchical organisations are poorly aligned with modern change demands because of their slow decision-making and limited collaboration across functions. The success of technology companies like Spotify and Netflix in implementing agile organisational principles provides compelling evidence for the potential of flexible structures (Kniberg & Ivarsson, 2012). Still, not every organisation may benefit from alternate structures as suggested by the agile movement, which focuses less on hierarchy and distributed decision making. As noted by Gulati and Puranam (2009), some industries marked by deep technical complexity and strict regulatory scrutiny may do better with more traditional forms characterised by greater division of labour and specialisation, along with defined accountability models. In addition, many firms underestimate the significant cultural change needed to embrace agile frameworks, resulting in surface-level agile adoption where practices are changed without fundamental shifts in worldview.

The cultivation of innovation culture emerges as both a prerequisite for and consequence of effective change management, creating what Senge (1990) describes as a "learning organisation" capable of continuous adaptation. Research by Ahmed (1998) demonstrates that organisations with strong innovation cultures are significantly more resilient during periods of change, as their emphasis on experimentation and learning reduces the psychological and operational barriers to transformation.

Establishing dedicated innovation labs, idea management systems, and failure-tolerant environments represents concrete mechanisms for fostering innovative capacity (Chesbrough, 2003). However, the innovation culture paradigm faces several critical challenges that limit its universal applicability. First, emphasising experimentation and risk-taking may conflict with operational efficiency requirements, particularly in cost-sensitive industries or during financial constraints (March, 1991). Second, the cultural transformation required to shift from risk-averse to innovation-oriented mindsets often takes years to achieve and may face significant resistance from employees and managers comfortable with established practices (Christensen, 1997). Third, measuring and evaluating innovation efforts remain problematic, as traditional performance metrics may not capture the long-term value of failed experiments or incremental improvements.

Integrating big data and analytics into change management decision-making represents a paradigm shift toward evidence-based transformation strategies, yet this data-driven approach introduces new complexities and potential limitations. McAfee and Brynjolfsson (2012) argue that organisations leveraging data analytics for change initiatives achieve superior outcomes by identifying patterns and relationships that would be invisible through traditional analysis methods. The ability to analyse employee engagement data, customer behaviour patterns, and market trends provides unprecedented insight into change requirements and potential resistance factors (Davenport & Harris, 2007). However, the data-driven approach to change management faces several critical limitations that warrant careful consideration. First, data availability does not guarantee its relevance or accuracy for change management purposes, and organisations may suffer from what Kahneman (2011) describes as "data illusion" – the false confidence that comes from having access to large quantities of information. Second, emphasising quantitative analysis may overlook important qualitative factors such as organisational culture, employee emotions, and informal networks that significantly influence change outcomes (Pfeffer & Sutton, 2006). Third, the complexity of organisational systems means that data analytics may identify correlations without revealing causal relationships, potentially leading to misguided change interventions.

The alignment of vision, structure, and culture represents the ultimate integration challenge in change management, requiring organisations to achieve coherence across multiple organisational dimensions simultaneously. Burke and Litwin's (1992) organisational performance model demonstrates the interconnectedness of these elements, arguing that sustainable change requires systematic attention to their alignment. This holistic perspective finds support in Beer and Nohria's (2000) research on change approaches, which shows that successful transformations address both "hard" structural elements and "soft" cultural factors. However, the alignment paradigm faces practical implementation challenges that may limit its effectiveness in real-world settings. The complexity of achieving simultaneous change across multiple organisational dimensions often exceeds managerial capacity and organisational resources, leading to incomplete or superficial alignment efforts (Sirkin et al., 2005). Additionally, the dynamic nature of organisational environments means that alignment achieved at one point may quickly become obsolete as external conditions change, requiring continuous realignment efforts that may exhaust organisational energy and attention.

While contemporary change management approaches offer sophisticated frameworks for navigating organisational transformation, their effectiveness depends on thoughtful adaptation to specific organisational contexts and realistic assessment of implementation constraints. Integrating strategic planning, scenario thinking, structural agility, innovation culture, data-driven decision-making, and systemic alignment represents an ideal that few organisations can fully achieve. However, pursuing these capabilities remains essential for long-term organisational survival and success. Future research should focus on developing a more nuanced understanding of how these change management dimensions interact in different organisational contexts and how practitioners can prioritise and sequence their implementation for maximum impact.

LEADING CHANGE IN TURBULENT TIMES

Contemporary organisations operate within what Bennis and Nanus (1985) first characterised as environments of volatility, uncertainty, complexity, and ambiguity (VUCA). In such contexts, organisational change has transitioned from episodic to perpetual organisational reality (Burnes, 2004). Leadership quality is central to successful organisational transformation, which functions not merely as a hierarchical position but as the primary determinant of change initiative outcomes (Kotter, 2012).

Effective change leaders do not simply administer change processes; they actively construct and direct transformation efforts while providing strategic guidance through periods of organisational disruption (Higgs & Rowland, 2005). Empirical studies indicate that leadership effectiveness accounts for approximately 70% of the variance observed in change implementation success rates, establishing it as the most significant variable in transformational endeavours (Beer & Nohria, 2000). Given contemporary challenges, including technological disruption and global economic volatility, the ability to lead change effectively has evolved from an advantageous competency to an organisational survival requirement (Yukl, 2013).

Transformational organisational change requires what Bass (1985) termed visionary leadership as its fundamental element. Successful change leaders initiate transformation by developing compelling organisational visions beyond incremental modifications to envision inspirational future organisational states (Bennis & Nanus, 1985). These visions function as navigational guides, providing directional clarity during periods when organisational turbulence may create disorientation (Conger & Kanungo, 1998).

Compelling change visions balance ambitious aspirations and realistic attainability, creating future scenarios that challenge and maintain credibility (Kouzes & Posner, 2017). The effectiveness of such visions derives from their capacity to generate what Seligman (2011) identifies as "positive tension" a constructive dissatisfaction with current conditions combined with enthusiasm for future possibilities.

Kelleher's leadership at Southwest Airlines during industry upheaval exemplifies this approach; rather than focusing solely on cost reduction measures, he articulated a vision of democratised air transportation that sustained employee engagement throughout difficult transitions (Gittell, 2003). Similarly, Nadella's transformation of Microsoft involved communicating a vision of cloud-first, mobile-first operations that provided the conceptual architecture for aligning diverse organisational divisions around unified transformation objectives (Shaw, 2017). Visionary leaders transcend future prediction to create compelling rationales for collective future construction.

Sophisticated change leaders recognise that organisational transformation is driven primarily by compelling narratives rather than analytical documents and strategic frameworks (Denning, 2007). Storytelling is a powerful mechanism for converting abstract visions into emotionally meaningful experiences that generate authentic organisational engagement (Heath & Heath, 2007).

Successful change narratives address rational and emotional dimensions, acknowledging present realities while building confidence in future organisational potential (Kotter & Cohen, 2002). Effective change stories incorporate several essential components: acknowledgement of existing challenges, articulation of purpose beyond financial objectives, celebration of incremental achievements, and demonstration of authentic leadership commitment (Denning, 2007).

IBM's (International Business Machines) transformation from hardware manufacturing to solutions provision illustrates this narrative approach; leadership developed a compelling story arc that honoured organisational heritage while generating excitement about evolutionary direction (Gerstner, 2002). The most impactful change narratives do not minimise difficulties but position them as essential elements of a worthwhile organisational journey. Through narratives that balance transparency with optimism, leaders transform change from an imposed process into a collaborative achievement (Shaw, 2017).

Sustainable transformational change rarely succeeds through hierarchical mandates alone (Kotter, 1996). Skilled change leaders develop comprehensive coalitions that extend beyond formal organisational authority structures, strategically empowering employees throughout organisational levels (Higgs & Rowland, 2005). By identifying and developing informal influencers individuals whose perspectives carry significant weight regardless of formal position leaders establish essential networks of change advocates who amplify transformation efforts (Armenakis & Harris, 2009).

These coalitions serve as tangible representations of change visions, translating broad initiatives into contextually relevant meaning across diverse organisational segments (Burnes, 2004). Central to this approach is what Spillane (2006) conceptualises as "distributed leadership" intentionally transferring decision-making authority to individuals most proximate to implementation challenges.

During Best Buy's organisational turnaround, Joly's leadership demonstrates this principle; a critical success factor was his decision to grant store managers unprecedented autonomy to respond to local market conditions (Joly, 2021). Similarly, Bertolini's transformation of Aetna involved establishing cross-functional teams with substantial authority to drive innovation from multiple organisational perspectives (Bertolini, 2019). Through coalition building across hierarchical boundaries and employee empowerment in implementation processes, leaders convert theoretical change into a practical organisational reality.

Within environments characterised by significant disruption, emotional intelligence and resilience distinguish effective change leaders (Goleman, 1995). The turbulence accompanying substantial transformation generates intense emotional responses including anxiety, resistance, confusion, and fatigue that logical arguments cannot adequately address (Bridges, 2003). Leaders possessing high emotional intelligence navigate this complex human landscape by demonstrating three essential capabilities: self-awareness regarding their emotional responses to change; empathy toward diverse employee reactions; and the ability to regulate organisational emotional climate (Bar-On, 2006).

This emotional competence enables leaders to respond strategically rather than impulsively to inevitable setbacks accompanying ambitious change efforts (Boyatzis & McKee, 2005). Resilient change leaders model what Peterson (2000) terms "realistic optimism" acknowledging difficulties while maintaining confidence in eventual success.

Mulcahy's leadership of Xerox through its existential crisis exemplifies emotional steadiness during intense pressure, creating psychological safety that enabled necessary but difficult changes (Mulcahy, 2014). Similarly, Mulally's guidance of Ford's transformation without government assistance was characterised by a consistent emotional tone—captured in his philosophy of "skilful optimism"—providing stability amid automotive industry chaos (Hoffman, 2012). Through emotionally intelligent and resilient leadership, transformational leaders establish psychological foundations enabling organisations to adapt rather than fracture during turbulent change.

Fear and uncertainty constitute the most significant obstacles to organisational transformation, creating cognitive and emotional resistance capable of undermining well-designed change initiatives (Armenakis & Harris, 2009). Effective change leaders overcome these barriers through communication strategies characterised by transparency, consistency, and bidirectional engagement (Lewis, 2007).

Transparency involves sharing positive developments, challenges, and setbacks, building trust through honesty rather than false assurance (Mishra & Spreitzer, 1998). Consistency requires aligning messages across communication channels and periods, avoiding contradictions that generate cynicism (Armenakis & Harris, 2009). Most importantly, bidirectional communication transforms information dissemination into authentic dialogue, creating feedback mechanisms that enable leaders to identify and address concerns in real time (Lewis, 2007).

During significant turbulence periods, successful change leaders substantially increase communication frequency and accessibility, recognising that information vacuums rapidly fill with rumours and anxiety (Kotter, 1996). Nadella's leadership of Microsoft's cultural transformation included regular town halls where challenging questions were encouraged rather than discouraged (Shaw, 2017). Similarly, during Starbucks' restructuring period, Schultz conducted unprecedented open forums where employee concerns could be expressed directly to leadership (Schultz & Gordon, 2011). Through communication, acknowledging uncertainty while clarifying processes and priorities, leaders transform fear from an impediment into motivation for meaningful change.

Leading organisational change in turbulent environments requires sophisticated integration of vision, narrative, coalition-building, emotional intelligence, and strategic communication (Yukl, 2013). Leaders who successfully transform organisations amid disruption understand that change fundamentally constitutes a human process rather than merely a technical one (Bridges, 2003). They recognise that their primary function is not to possess all solutions but to create conditions that enable collective wisdom to emerge and develop (Senge, 2006).

As organisations confront increasingly complex challenges, the capacity for effective change leadership will continue to distinguish those that merely survive from those that genuinely flourish (Kotter, 2012). Ultimately, successful change leadership rests upon a paradox: providing sufficient stability to make transformation possible (Heifetz & Linsky, 2002).

NEUROSCIENCE AND ORGANISATIONAL CHANGE: THE ROLE OF NEUROPLASTICITY

Neuroscience has reshaped how we understand and manage organisational change in recent years. Neuroplasticity is at the heart of this emerging perspective the brain's ability to rewire and reorganise itself in response to experience, learning, and environmental stimuli (Doidge, 2007). This biological reality has profound implications for change leadership, as it suggests that transformation is not only a structural or strategic endeavour but also a psychological and neurological one. Organisational change efforts often falter not due to poor strategy but because they overlook how deeply habits, behaviours, and thought patterns are ingrained in the human brain.

Neuroplasticity reveals why change is difficult. Most daily behaviours in organisations how meetings are run, how people interact, how decisions are made are governed by automatic neural patterns that have been reinforced over time through repetition and familiarity. These patterns, once entrenched, create resistance to change not necessarily out of defiance but because the brain prefers efficiency and familiarity over the uncertainty of learning something new (Rock & Schwartz, 2006). Thus, real and lasting organisational change must account for this neurological reality by creating conditions that encourage the unlearning of old behaviours and the reinforcement of new ones.

One of the most critical enablers of neuroplasticity in organisational settings is psychological safety a shared belief that it is safe to take risks, ask questions, and admit mistakes without fear of punishment (Edmondson, 1999). When employees feel threatened or judged, the amygdala the brain's fear centre activates, impairing learning and creativity. Conversely, psychologically safe environments dampen the threat response and enable the prefrontal cortex, responsible for problem-solving and innovation, to engage fully. Leaders play a crucial role in fostering this safety through empathy, transparency, and open communication.

Another cornerstone of neuroplastic change is repetition with reward. Neuroscience shows that when paired with positive emotional experiences, repeated behaviours trigger the release of dopamine, a neurotransmitter that strengthens the neural circuits associated with the new behaviour (Duhigg, 2012). This insight has been widely adopted by organisations such as Nationwide and Microsoft, which use regular reinforcement mechanisms like recognition rituals, storytelling, and employee showcases to celebrate incremental progress and establish new cultural norms. Over time, these repeated, emotionally rewarding experiences help hardwire desired behaviours into organisational life.

Equally powerful is the role of narrative and storytelling in rewiring organisational behaviour. The human brain is hardwired for stories—they engage emotion, aid memory retention, and help individuals make sense of change (Heath & Heath,

2007). Effective change leaders use narrative to connect abstract visions of change with concrete, personal meaning for employees. For example, Satya Nadella's "learn-it-all" vision at Microsoft became more than a slogan it was embodied through personal stories, leadership behaviour, and systems that encouraged curiosity and experimentation. These narratives created emotional connections that made the change memorable and meaningful, forming new mental associations.

Social learning is another vital mechanism through which neuroplasticity supports change. By activating mirror neurons, individuals model the behaviours of those around them, especially those they perceive as influential or trustworthy (Boyatzis & McKee, 2005). This underscores the importance of identifying and empowering informal leaders and change champions who can act as behavioural models. By making desired behaviours visible and socially contagious, these peer influencers accelerate the adoption of new habits across teams and departments.

Furthermore, structured experimentation and reflection enhance cognitive flexibility a key element of neuroplasticity. Organisations that institutionalise feedback loops, retrospective reviews, and safe-to-fail pilots create environments where employees learn iteratively and adaptively. This approach mirrors the brain's trial-and-error learning mechanisms. Companies like Toyota and IBM have used structured learning practices (e.g., Kaizen huddles, "stop-the-line" protocols, and transformation units) to systematically cultivate new neural and organisational pathways for dealing with complexity and uncertainty.

Critically, neuroplasticity also emphasises the importance of environment design, or what behavioural economists refer to as choice architecture (Thaler & Sunstein, 2008). Leaders can make the desired behaviours more likely and effortless by intentionally shaping the physical, digital, and procedural environments in which employees operate. For instance, redesigning workspaces to encourage collaboration or modifying onboarding processes to embed new cultural values from day one nudges people toward change without conscious resistance.

In sum, neuroscience reminds us that successful organisational change requires more than a change in systems it demands a change in the human brain. Neuroplasticity offers a hopeful message: people can change, teams can grow, and organisations can evolve but only when change is managed in a way that supports the biological processes underlying learning, emotion, and habit formation. Leaders who understand and apply neuroscience-informed strategies empathy, repetition, recognition, narrative, social learning, and environmental nudging are far more likely to embed change not just into plans and policies, but into their people's minds and daily practices. Building organisations that can adapt, rewire, and grow will be a fundamental source of sustainable competitive advantage as we continue to navigate complex and disruptive environments.

CHANGE OUTCOMES

We have all been there. A new initiative launches with great fanfare, and everyone seems excited. However, for a few weeks or months, things seem different. However, then, almost imperceptibly, old habits creep back in. Meetings return to their familiar patterns, processes revert to how they have always been done, and before you know it, it is as if the change never happened.

This is not just a feeling complex data backs it. Research consistently shows that approximately 70% of organisational change initiatives fail to achieve their intended outcomes (Kotter, 2012; Beer & Nohria, 2000). This sobering reality reflects what many of us have experienced firsthand: initial enthusiasm gradually gives way to what experts call "change decay" the slow drift back to old ways of doing things. Why does this happen so predictably? The answer lies in how we fundamentally misunderstand what change requires. Many organisations treat transformation like an event something you announce, implement, and then move on from. Successful change is like learning to play an instrument: it requires consistent practice, patience, and a deep understanding that mastery takes time (Higgs & Rowland, 2005).

Think about the most successful changes you have witnessed in your career. Chances are, they were not just about new policies or procedures they fundamentally altered how the organisation operated at its core. This is what researchers call "embedding" change, and it requires rewiring both the formal and informal systems that govern daily work life (Schein & Schein, 2017). At the formal level, this means aligning three critical elements: your processes (how work gets done), your policies (the rules that guide behaviour), and your performance metrics (what you measure and reward). When Paul O'Neill became CEO of Alcoa, he did not just talk about safety he completely restructured reporting systems, compensation packages, and promotion criteria around safety metrics (O'Neill & Knox, 2001). This made it virtually impossible for the organisation to return to old safety practices without deliberately dismantling these new systems.

However, formal changes only tell half the story. The informal aspects of organisational culture the unwritten rules, the casual conversations, the way meetings work often have even more influence on daily behaviour. When Satya Nadella led Microsoft's cultural transformation, the company did not just change policies; it redesigned meeting formats, collaboration tools, and even office spaces to reinforce a growth mindset in every interaction (Nadella, 2017). The magic happens when these formal and informal changes work together, creating what systems theorists call "coherent alignment." Instead of fighting against each other, every element of the organisation starts pulling in the same direction, making new behaviours feel natural rather than forced.

One of the organisation's most significant mistakes is treating change like a straight line from point A to point B. Successful transformation looks more like a spiral you are generally moving in the right direction but constantly course-correcting based on what you learn (Senge, 1990). The best change initiatives build this learning directly into their design through continuous feedback loops. This is not just about quarterly surveys or annual reviews it is about creating multiple ways to sense how things are going. Toyota's legendary continuous improvement culture provides a perfect example: daily team huddles, visual management boards, and their famous "andon cord" system that allows employees to stop the production line when they spot a problem (Liker, 2004). These mechanisms create what researchers call "strategic sensitivity" the ability to detect weak signals before they become significant problems.

When IBM underwent its massive service transformation, it established dedicated transformation units that monitored implementation challenges and rapidly tested solutions (Palmisano, 2004). Perhaps most importantly, organisations that sustain change create psychological safety environments where people feel comfortable sharing when things are not working without fear of blame or punishment (Edmondson, 1999). When feedback is seen as intelligence rather than criticism, organisations can adapt and evolve their change efforts rather than letting them slowly decay.

While CEO vision statements get most of the attention in change stories, sustaining transformation often happens in the organisation's middle. Middle managers and informal peer networks are the unsung heroes determining whether change sticks (Balogun & Johnson, 2004). Middle managers occupy what researchers call "the critical intersection" close enough to the front lines to understand operational realities, but senior enough to influence how strategies get translated into action. During GE's Six Sigma transformation, the company invested heavily in developing middle managers as certified "Black Belts" and "Green Belts," giving them both the skills and authority to drive change in their areas (Harry & Schroeder, 2000).

Equally important are the informal influence networks—the respected colleagues people listen to when deciding whether a change is worth embracing. Progressive Insurance understood this dynamic when they identified and developed "change champions" throughout the organization people who advocated for transformation not because of their job titles, but because of their social influence (Leonard-Barton, 1988). These groups function like organisational immune systems. They either reject change as a foreign intrusion or help it become part of the organisation's natural defence mechanisms. Innovative organisations deliberately cultivate these critical cohorts as change amplifiers rather than leaving them to become implicit sources of resistance.

At its heart, organisational change is about changing human behaviour, which requires understanding how our brains work. Neuroscience research shows us that celebrating progress is lovely and neurologically necessary for making new behaviours stick (Duhigg, 2012). When we celebrate wins, our brains release dopamine, strengthening the neural pathways associated with new behaviours. Over time, this transforms effortful change into an automatic habit. Nationwide Insurance understood this when they implemented weekly recognition rituals specifically highlighting customer-centric behaviours during their "On Your Side" transformation, creating regular neurological reinforcement for the changes they wanted to see (Nationwide, 2015).

Beyond formal recognition, sustainable change requires what behavioural economists call "choice architecture" designing environments that make desired behaviours easier and more natural than alternatives (Thaler & Sunstein, 2008). Zappos did not just train employees on customer service; they embedded their values into hiring processes, office layouts, and daily workflows that consistently reinforced their customer-first approach (Hsieh, 2010). Perhaps most powerfully, organisations that sustain change regularly share stories about how new behaviours create meaningful outcomes. These narratives transform abstract values into emotionally resonant experiences that motivate people through the inevitable implementation challenges (Heath & Heath, 2007).

Sustainable organisational change is not about having the perfect plan or the most charismatic leader. It is about understanding that transformation is a human process that requires patience, systems thinking, and a deep respect for how people change. By embedding new approaches into organisational infrastructure, creating continuous learning loops, empowering middle managers and peer networks, and supporting the neurological process of habit formation, organisations can move beyond the 70% failure rate to create changes that truly last. The key insight is simple but profound: change is not something you do to an organization it is something you build into how the organisation thinks, learns, and evolves. Real, lasting change becomes possible when transformation becomes part of your organisational DNA rather than an external imposition.

CASE INSIGHTS: LEARNING FROM REAL-WORLD CHANGE

Case Study 1: Microsoft's Cloud Transformation Under Satya Nadella

Microsoft faced an existential crisis until Satya Nadella became the company's CEO in 2014. The tech giant was losing its market relevance in the cloud-first, mobile-first segment. With its Windows-centric strategy becoming obsolete, competitors like Google and Amazon dominated cloud services (Denning, 2021). With the stock deteriorating over time, the company's internal competition had characterised its culture. Creativity and innovation were stifled with everyone having a "know-it-all" attitude (Tabrizi, 2023). Rather than attempting a complete organisational overhaul, Satya Nadella employed a dual transformation strategy involving maintaining the profitable legacy businesses while building on cloud capabilities through Azure (Somasegar, 2024). Employing this structural ambidexterity enabled Microsoft to continue serving its existing customer base while allowing heavy investments in future growth areas.

More significantly, Nadella successfully shifted the company's culture from being "know-it-all" to "learn-it-all," fostering creativity, innovation, and collaboration across siloed divisions (Tabrizi, 2023). While this transformation was necessary, it faced significant internal resistance, especially from the leaders of the Windows division, who believed that the shift to cloud services could cannibalise their core business (Denning, 2021). Additionally, when Office 365 was made available on Android and iOS instead of Windows phones, many traditional employees believed this to be a betrayal of the company's platform strategy (Somasegar, 2024). Understanding and acknowledging this resistance, Nadella addressed the workforce by effectively communicating the vision behind the changes being implemented, and he established that cloud services would boost the company's core strengths. He demonstrated this with evidence of early cloud adoption boosting Microsoft product usage, contrary to the existing customer base's assumptions of replacement or decreased use (Denning, 2021).

The key breakthrough came when the company positioned itself as a productivity and platform company rather than a Windows company. The change in value proposition and market positioning exhibits its refined vision to help customers achieve more rather than focusing on its products (Tabrizi, 2023). Furthermore, Microsoft created a space for innovation that transcended traditional boundaries. Within five years, the company became the second-largest public company based on market capitalisation. Azure became one of the biggest competitors to Amazon Web Services, and Office 365 completely transformed into a comprehensive cloud productivity platform (Somasegar, 2024).

Microsoft's transformation under Satya Nadella exemplifies how neuroplasticity manifests in organisational change. Nadella's decision to shift the cultural identity from a "know-it-all" to a "learn-it-all" organisation was not merely rhetorical it was designed to create repeated behavioural experiences that could rewire how employees thought, collaborated, and innovated (Tabrizi, 2023). By introducing structures such as growth mindset workshops, agile experimentation protocols, and performance metrics that rewarded learning over perfection, Microsoft reinforced new neural pathways among employees at every level. Psychological safety was enhanced through leadership humility and transparency, helping suppress fear-based responses and enabling the prefrontal cortex responsible for strategic thinking and innovation to engage fully. Additionally, storytelling was heavily employed. Nadella used personal anecdotes and success narratives to anchor the transformation emotionally, increasing cognitive resonance and memory retention. Informal leaders were identified and empowered to model the desired behaviours, activating mirror neurons in their teams. Over time, this cascading effect turned new behaviours into ingrained habits. From a neuroscience perspective, Microsoft's change succeeded because it aligned structural, emotional, and cognitive stimuli to reinforce desired behaviours. By embedding learning into the neural and cultural fabric of the organisation, Microsoft used neuroplasticity to drive lasting, sustainable change (Tabrizi, 2023).

Case Study 2: Patagonia's Stakeholder Capitalism Evolution

Another case of understanding the need for change and implementing successful change would be Patagonia and its journey toward radical environmental activism and stakeholder capitalism. This approach intensified after founder Yvon Chouinard transferred company ownership to a trust and nonprofit dedicated to fighting climate change (Gelles, 2022). However, this transformation was strategically built on decades of gradual cultural change that began in the 1980s, when Patagonia started integrating environmental considerations into its business decisions even at the expense of short-term profits (Chouinard, 2005).

Patagonia's evolution challenged traditional assumptions about corporate purpose and consumer behaviour. For instance, its 2011 "Don't Buy This Jacket" Black Friday ad defied conventional marketing logic but ultimately strengthened brand loyalty among environmentally conscious consumers (Patagonia, 2011). This counterintuitive strategy attracted new customers who valued authenticity and transparency over traditional advertising (Whelan & Fink, 2016).

Despite its success, Patagonia faced scepticism from industry peers who questioned whether its activism would alienate customers or hurt profitability (Whelan & Fink, 2016). Instead of retreating, the company expanded initiatives like Worn Wear, 1% for the Planet, and Action Works to deepen its environmental impact (Patagonia, n.d.-a). These programs reinforced Patagonia's belief that environmental responsibility can enhance business performance.

From a neuroscience perspective, Patagonia's transformation exemplifies organisational neuroplasticity. The company created emotionally resonant experiences such as linking daily work to global environmental outcomes that helped rewire employee behaviour (Rock & Schwartz, 2006). Campaigns like "Don't Buy This Jacket" represented symbolic rituals, reinforcing a shared narrative. Over time, Patagonia institutionalised its values through hiring, peer recognition, and leadership modelling, triggering social learning and embedding environmental commitment into its cultural fabric (Whelan & Fink, 2016).

Case Study 3: General Motors' Electric Vehicle Transformation

Another case of understanding how change is driven by market and internal forces is General Motors' (GM) EV transformation. GM's commitment to becoming carbon neutral by 2040 and eliminating tailpipe emissions from light-duty vehicles represents one of the most ambitious transformations in the automobile industry (General Motors, n.d.). Interestingly, the company that once discontinued its EV programs and lobbied against emission standards had to fundamentally reimagine its operations, identity, and relationship with consumers and regulators (Vlasic, 2021).

A key driver of this change was CEO Mary Barra, who managed multiple dimensions of transformation—technical, operational, manufacturing, cultural, and market-facing (Gelles, 2021). GM adopted a dual operating system, maintaining profitable internal combustion engine (ICE) production while investing heavily in EV development through dedicated teams and resources (Lee, 2024). This approach helped the company navigate internal resistance from engineers and executives whose careers were built on ICE expertise. Many questioned whether consumers would accept EVs and whether GM could compete with Tesla's technology (Vlasic, 2021).

To address these concerns, GM acquired talent from tech companies, partnered with battery manufacturers like LG Energy Solution, and developed the Ultium platform to rival competitors (General Motors, n.d.; Lee, 2024). This led to the launch of vehicles like the GMC Hummer EV and Cadillac Lyriq, showcasing that EVs could deliver luxury-level performance and innovation (Gelles, 2021). By reframing EVs as superior in acceleration, cost-efficiency, and technology, GM shifted its narrative from reluctant compliance to competitive leadership.

GM's transformation also illustrates how neuroplasticity can enable strategic agility. GM's routines and decision-making were wired around ICE technologies for decades, creating cognitive lock-in. Under Barra's leadership, the dual operating system allowed gradual exposure to new paradigms, reducing threat responses and enabling cognitive flexibility (Rock & Schwartz, 2006). Cross-functional teams, immersive EV labs, and external hires helped rewire mental models. Recognition of early adopters and narrative reframing reinforced new neural pathways, enabling GM to repattern collective thinking—a hallmark of neuroplastic change.

Lessons Across Cases

The cases discussed have been drawn from different industries, collectively demonstrating real-world organisation transformation rooted in strategically recognising the need for change and implementing neuroscience-aligned strategies. A unique characteristic identified across these cases is the deliberate rejection of the binary thinking between legacy systems and emerging paradigms. These organisations adopted a dual operating mindset, balancing business continuity and sustainability with innovation. For instance, Microsoft retained its legacy products while building a strong foothold in the cloud ecosystem; Patagonia fused profitability with environmental activism; and General Motors focused on internal combustion engines and EV development. These approaches reflect an understanding that change does not demand the abandonment of past activities, but instead calls for integrating current practices with future-oriented capabilities.

Notably, all these organisations experienced initial resistance, whether from internal stakeholders, cultural inertia, or market scepticism. Nonetheless, rather than viewing this resistance as a barrier to change, these organisations reframed it as a source of insight and vigour and, aligning with the neuroplasticity theory which emphasises the importance of repeated exposure, emotional safety and gradual rewiring for mental models. Microsoft used its learning culture to retrain the cognitive habits among the staff; Patagonia relied on emotional storytelling to embed new values, and GM cultivated new thinking through experiential innovation.

All these cases underscore that sustainable transformation emerges not merely from having visionary strategies, but also from the ability of an organisation and its people to continuously adapt, learn, and embed changes even at behavioural and neurological levels. By responding strategically to external pressures and cultivating internal agility, each organisation built long-term change capacity, anchored in systems and how their people think, feel, and act.

CONCLUSION

We have demonstrated throughout this investigation that change is a fundamental constant of our contemporary world, rather than just an occasional disruption. The people and organisations that prosper in this setting recognise that flexibility is a fundamental skill that needs to be intentionally developed. As the evidence makes clear (as seen in the case of IBM, Patagonia and GM), those who embrace uncertainty, create resilient yet adaptable systems, and cultivate collective intelligence routinely outperform their more inflexible counterparts. Rethinking resilience as an active adaptive process instead of a passive capacity to tolerate stress opens up new avenues for development even under the most trying conditions. This change is nothing less than a progression in how we think about organisational capability, acknowledging that strength comes from adapting to change, growing from it, and eventually using it.

Importantly, this adaptive capacity is deeply connected to neuroplasticity the brain's inherent ability to reorganise itself by forming new neural pathways in response to experience, learning, and environmental demands. Just as individuals rewire habits through repetition, reflection, and reinforcement, organisations must build processes that enable employees to unlearn outdated practices and embrace new ones. Embedding change at the neurological level through emotionally resonant experiences, social learning, and supportive environments ensures that transformation becomes sustained and internalised.

Moving forward necessitates a fundamental mental shift from reacting to disruption and proactively influencing change. This entails establishing settings that value experimentation, acknowledge failure as a helpful learning opportunity, and actively seek out and incorporate a variety of viewpoints. It entails creating mechanisms to identify subtle signs of change before they become overwhelming forces and strengthening organisational muscle memory to react quickly rather than rigidly. It necessitates developing psychological adaptability, lifelong learning, and individually accepting uncertainty. At the group level, it necessitates open communication, decentralised decision-making, and shared accountability for results. Adaptive resilience encourages us to proceed deliberately in the face of uncertainty, changing direction as new information becomes available and new difficulties appear instead of waiting for ideal information or specific results.

Adaptive resilience will distinguish those who survive from those who truly thrive as we navigate a world that is becoming more complex and interconnected. The leaders who can inspire action while allowing for reflection, drive performance while fostering wellbeing, and articulate a compelling vision while remaining humble enough to revise it will have the most significant impact. This well-rounded approach acknowledges that resilience is about addressing problems in a sustainable way that respects our human limitations and makes use of our extraordinary growth potential, not about superhuman strength or emotionless endurance. Incorporating these ideas into our personal and corporate cultures fosters creativity, fulfilling work, and long-lasting influence even during difficult times. The future belongs to those who can adapt wisely, empathise with others, and persevere with purpose, not to the strongest or fastest. By understanding and applying the principles of neuroplasticity in how we design, communicate, and lead change, we actively shape the future, not merely prepare for it.

RECOMMENDATIONS FOR RESEARCHERS AND PRACTITIONERS

To build on the insights presented in this chapter, several key recommendations are offered to guide future research and enhance practical implementation in organisational change management. For researchers, five critical areas warrant attention. First, exploring the neurobiological foundations of change through further interdisciplinary studies is needed to deepen the understanding of how neuroplasticity influences individual and collective change in organisational settings, with longitudinal research examining neural and behavioural shifts during transformation processes offering critical insights. Second, developing context-sensitive change frameworks requires researchers to design adaptive change models tailored to specific industries, organisational sizes, and cultural contexts, with comparative studies exploring how change frameworks perform across different sectors, enriching practical application. Third, investigating the role of informal networks is essential, as the influence of peer networks and informal leadership on sustaining change warrants deeper empirical exploration, with future studies assessing how these networks can be strategically cultivated to accelerate or inhibit change adoption. Fourth, evaluating dual operating systems requires more scholarly work to examine the efficacy and trade-offs of maintaining legacy operations alongside innovative business models, with research exploring how ambidexterity can be optimised in resource-constrained settings. Finally, measuring long-term change outcomes is crucial, as while many studies focus on short-term change implementation, future research should examine mechanisms that enable long-term sustainability and prevent change decay.

For practitioners, five key areas demand focused attention to ensure successful change implementation. First, embedding change through culture and systems requires practitioners to ensure alignment between formal systems, such as policies, metrics, and processes and informal dynamics, including stories, rituals, and symbols, to embed change effectively. Second, leveraging neuroplasticity in change design involves applying neuroscience principles such as repetition, positive reinforcement, psychological safety, and social learning to reinforce new behaviours and facilitate habit formation. Third, cultivating adaptive leadership necessitates investment in leadership development that emphasises emotional intelligence, storytelling, and the ability to lead through ambiguity, with leaders modelling the behaviours they seek to institutionalise. Fourth, building resilience through feedback loops requires implementing continuous learning mechanisms such as retrospectives, real-time feedback, and pilot experiments that allow for course correction and ongoing adaptation. Finally, fostering innovation-friendly environments involves encouraging experimentation, tolerating intelligent failure, and designing work environments that promote collaboration and reflective learning, as these elements are essential to building a learning organisation.

REFERENCES

Ahmed, P. K. (1998). Culture and climate for innovation. *European Journal of Innovation Management*, 1(1), 30-43Anderson, P. (1999). Complexity theory and organisation science. *Organization Science*, *10*(3), 216–232.

Appelbaum, S. H., Habashy, S., Malo, J. L., & Shafiq, H. (2012). Back to the future: Revisiting Kotter's 1996 change model. *Journal of Management Development*, *31*(8), 764–782.

Armenakis, A. A., & Harris, S. G. (2009). Reflections: Our journey in organizational change research and practice. *Journal of Change Management*, *9*(2), 127–142.

Asghar, J., Kanbach, D. K., & Kraus, S. (2025). *Toward a multidimensional concept of organizational agility: a systematic literature review*. Management Review Quarterly., DOI: 10.1007/s11301-025-00497-6

Avolio, B. J., & Bass, B. M. (2004). *Multifactor leadership questionnaire: Manual leader form, rater, and scoring key for MLQ (Form 5x-Short)*. Mind Garden.

Balogun, J., & Johnson, G. (2004). Organizational restructuring and middle manager sensemaking. *Academy of Management Journal*, *47*(4), 523–549.

Bar-On, R. (2006). The Bar-On model of emotional-social intelligence (ESI). *Psicothema*, *18*, 13–25. PMID: 17295953

Bass, B. M. (1985). *Leadership and performance beyond expectations*. Free Press.

Bass, B. M., & Riggio, R. E. (2006). *Transformational leadership* (2nd ed.). Lawrence Erlbaum Associates.

Beer, M., & Nohria, N. (2000). Cracking the code of change. *Harvard Business Review*, *78*(3), 133–141. PMID: 11183975

Bennis, W., & Nanus, B. (1985). *Leaders: The strategies for taking charge*. Harper & Row.

Bertolini, M. (2019). *Mission-driven leadership: My journey as a radical capitalist*. Currency.

Boyatzis, R., & McKee, A. (2005). *Resonant leadership: Renewing yourself and connecting with others through mindfulness, hope, and compassion*. Harvard Business Review Press.

Bridges, W. (2003). *Managing transitions: Making the most of change*. Da Capo Press.

Burke, W. W., & Litwin, G. H. (1992). A causal model of organizational performance and change. *Journal of Management, 18*(3), 523–545.

Burnes, B. (2004). Managing change: A strategic approach to organisational dynamics. *Financial Times Prentice Hall.*

Burnes, B. (2020). The origins of Lewin's three-step model of change. *The Journal of Applied Behavioral Science, 56*(1), 32–59.

By, R. T. (2005). Organisational change management: A critical review. *Journal of Change Management, 5*(4), 369–380.

Cedergren, A., & Hassel, H. (2024). Building organisational adaptive capacity in the face of crisis: Lessons from a public sector case study. *International Journal of Disaster Risk Reduction, 100.* Advance online publication. DOI: 10.1016/j.ijdrr.2023.104235

Cernega, A., Nicolescu, D. N., Mele canu Imre, M., Ripszky Totan, A., Arsene, A. L., erban, R. S., Perpelea, A.-C., & Nedea, M.-I. (Ilie), & Pi uru, S.-M. (2024). Volatility, Uncertainty, Complexity, and Ambiguity (VUCA) in Healthcare. *Health Care, 12*(7), 773. DOI: 10.3390/healthcare12070773 PMID: 38610195

Chalofsky, N., Rocco, T. S., & Morris, M. L. (2014). *Handbook of human resource development.* Wiley.

Chesbrough, H. W. (2003). *Open Innovation.* Harvard Business School Press.

Chouinard, Y. (2005). *Let My People Go Surfing: The Education of a Reluctant Businessman.* Penguin Books.

Christensen, C. M. (1997). *The Innovator's Dilemma.* Harvard Business School Press.

Conger, J. A., & Kanungo, R. N. (1998). *Charismatic leadership in organizations.* Sage Publications.

Creasey, T. (2009). *Defining change management.* Prosci.

Cristofaro, M. (2021). Organizational sensemaking: A systematic review and a co-evolutionary model. *European Management Journal, 40*(3). Advance online publication. DOI: 10.1016/j.emj.2021.07.003

Cummings, S., Bridgman, T., & Brown, K. G. (2016). Unfreezing change as three steps: Rethinking Kurt Lewin's legacy for change management. *Human Relations, 69*(1), 33–60.

Davenport, T. H., & Harris, J. G. (2007). *Competing on Analytics.* Harvard Business School Press.

Day, I. (2023, July 2). *Environmental Impact in Business.* Impact Day. https://impactday.eu/blog/environmental-impact-in-business/

Denning, S. (2007). *The secret language of leadership: How leaders inspire action through narrative.* Jossey-Bass.

Denning, S. (2021, June 20). *How Microsoft's digital transformation created a trillion dollar gain.* Forbes. https://www.forbes.com/sites/stevedenning/2021/06/20/how-microsofts-digital-transformation-created-a-trillion-dollar-gain

Doidge, N. (2007). *The brain that changes itself: Stories of personal triumph from the frontiers of brain science.* Viking.

Donthu, N., & Gustafsson, A. (2020). Effects of COVID-19 on Business and Research. *Journal of Business Research, 117*(1), 284–289. NCBI. https://www.sciencedirect.com/science/article/pii/S0148296320303830

Doz, Y. L., & Kosonen, M. (2008). *Fast strategy: How strategic agility will help you stay ahead of the game.* Pearson Education.

Drake, M. A. (1994). Technological Innovation and Organizational Change. *Journal of Library Administration, 19*(3-4), 39–53. DOI: 10.1300/j111v19n03_04

Duhigg, C. (2012). *The power of habit: Why we do what we do in life and business.* Random House.

Earl, C., Taylor, P. R., Roberts, C., Huynh, P., & Davis, S. J. (2017). *The Workforce Demographic Shift and the Changing Nature of Work: Implications for Policy, Productivity, and Participation.* Advanced Series in Management., DOI: 10.1108/s1877-636120170000017002

Edmondson, A. C. (1999). Psychological safety and learning behavior in work teams. *Administrative Science Quarterly, 44*(2), 350–383. DOI: 10.2307/2666999

Eisenbach, R., Watson, K., & Pillai, R. (1999). Transformational leadership in the context of organizational change. *Journal of Organizational Change Management, 12*(2), 80–89.

Eisenhardt, K. M., & Martin, J. A. (2000). Dynamic capabilities: What are they? *Strategic Management Journal, 21*(10-11), 1105–1121.

Felipe, C. M., Roldán, J. L., & Leal-Rodríguez, A. L. (2017). Impact of organizational culture values on organizational agility. *Sustainability, 9*(12), 2354.

Gelles, D. (2021, November 17). *Biden Praises G.M. for Electric Vehicle Push. The New York Times.* https://www.nytimes.com/2021/11/17/business/gm-electric-vehicles-biden.html

Gelles, D. (2022, September 14). *Billionaire No More: Patagonia Founder Gives Away the Company. The New York Times.* https://www.nytimes.com/2022/09/14/climate/patagonia-climate-philanthropy-chouinard.html

Gerstner, L. V. (2002). *Who says elephants can't dance? Inside IBM's historic turnaround.* HarperBusiness.

Gioia, D. A., & Chittipeddi, K. (1991). Sensemaking and sensegiving in strategic change initiation. *Strategic Management Journal, 12*(6), 433–448.

Gittell, J. H. (2003). *The Southwest Airlines way: Using the power of relationships to achieve high performance.* McGraw-Hill.

Goleman, D. (1995). *Emotional intelligence: Why it matters more than IQ.* Bantam Books.

Gulati, R., & Puranam, P. (2009). Renewal through reorganization: The value of inconsistencies between formal and informal organization. *Organization Science, 20*(2), 422–440.

Harry, M., & Schroeder, R. (2000). *Six Sigma: The breakthrough management strategy revolutionizing the world's top corporations.* Currency.

Heath, C., & Heath, D. (2007). *Made to stick: Why some ideas survive and others die.* Random House.

Heifetz, R., & Linsky, M. (2002). *Leadership on the line: Staying alive through the dangers of leading.* Harvard Business Review Press.

Hiatt, J. (2006). *ADKAR: A model for change in business, government, and our community.* Prosci Learning Center Publications.

Higgs, M., & Rowland, D. (2005). All changes great and small: Exploring approaches to change and its leadership. *Journal of Change Management, 5*(2), 121–151.

Hoffman, B. G. (2012). *American icon: Alan Mulally and the fight to save Ford Motor Company.* Crown Business.

Hornstein, H. A. (2015). The integration of project management and organizational change management is now a necessity. *International Journal of Project Management, 33*(2), 291–298.

Hsieh, T. (2010). *Delivering happiness: A path to profits, passion, and purpose*. Grand Central Publishing.

Joly, H. (2021). *The heart of business: Leadership principles for the next era of capitalism*. Harvard Business Review Press.

Kahneman, D. (2011). *Thinking, Fast and Slow*. Farrar, Straus and Giroux.

Kniberg, H., & Ivarsson, A. (2012). *Scaling Agile @ Spotify with Tribes, Squads, Chapters & Guilds*. Spotify Engineering Culture.

Kotter, J. P. (1996). *Leading change*. Harvard Business Review Press.

Kotter, J. P. (2012). *Leading change*. Harvard Business Review Press.

Kotter, J. P., & Cohen, D. S. (2002). *The heart of change: Real-life stories of how people change their organizations*. Harvard Business Review Press.

Kouzes, J. M., & Posner, B. Z. (2017). *The leadership challenge: How to make extraordinary things happen in organizations*. Jossey-Bass.

Lee, M. (2024, October 3). *How GM's EV Strategy Just Went From Setbacks to Success. EV News Daily*. https://www.evnewsdaily.com/2024/10/03/how-gms-ev-strategy-just-went-from-setbacks-to-success

Leonard-Barton, D. (1988). Implementation as mutual adaptation of technology and organization. *Research Policy*, *17*(5), 251–267.

Lewin, K. (1947). Frontiers in group dynamics: Concept, method and reality in social science. *Human Relations*, *1*(1), 5–41.

Lewis, L. K. (2007). An organizational stakeholder model of change implementation communication. *Communication Theory*, *17*(2), 176–204.

Liker, J. K. (2004). *The Toyota way: 14 management principles from the world's greatest manufacturer*. McGraw-Hill.

Maitlis, S., & Christianson, M. (2014). Sensemaking in organizations: Taking stock and moving forward. *The Academy of Management Annals*, *8*(1), 57–125.

March, J. G. (1991). Exploration and exploitation in organizational learning. *Organization Science*, *2*(1), 71–87.

Marion, R., & Uhl-Bien, M. (2001). Leadership in complex organizations. *The Leadership Quarterly*, *12*(4), 389–418.

McAfee, A., & Brynjolfsson, E. (2012). Big data: The management revolution. *Harvard Business Review*, *90*(10), 60–68. PMID: 23074865

Mendoza, E. G., & Quadrini, V. (2023, January 1). *Unstable Prosperity: How Globalization Made the World Economy More Volatile*. National Bureau of Economic Research. https://doi.org/DOI: 10.3386/w30832

Mishra, A. K., & Spreitzer, G. M. (1998). Explaining how survivors respond to downsizing: The roles of trust, empowerment, justice, and work redesign. *Academy of Management Review*, *23*(3), 567–588.

Moqaddamerad, S., & Ali, M. (2024). Strategic foresight and business model innovation: The sequential mediating role of sensemaking and learning. *Technological Forecasting and Social Change*, *200*, 123095–123095. DOI: 10.1016/j.techfore.2023.123095

Motors, G. (n.d.). *Electric Vehicles*. https://www.gm.com/innovation/electric-vehicles

Motwani, J., & Kataria, A. (2024). Organization agility: A literature review and research agenda. *International Journal of Productivity and Performance Management*. Advance online publication. DOI: 10.1108/ijppm-07-2023-0383

Mulcahy, A. (2014). Lessons from Xerox's near-death experience. *Harvard Business Review*, *92*(5), 78–86.

Nadella, S. (2017). *Hit refresh: The quest to rediscover Microsoft's soul and imagine a better future for everyone*. HarperBusiness.

Nationwide. (2015). *Annual report: On your side*. Nationwide Mutual Insurance Company.

Northouse, P. G. (2021). *Leadership: Theory and practice* (9th ed.). SAGE Publications.

O'Neill, P., & Knox, R. (2001). *The power of safety: How Paul O'Neill transformed Alcoa and created a model for corporate transformation*. Portfolio.

Palmisano, S. J. (2004). Leading change when business is good: An interview with Samuel J. Palmisano. *Harvard Business Review*, *82*(12), 60–70. PMID: 15605566

Patagonia. (2011, November 25). *Don't Buy This Jacket, Black Friday and the New York Times*. https://www.patagonia.com/stories/dont-buy-this-jacket-black-friday-and-the-new-york-times/story-18615.html

Patagonia. (n.d.-a). *Environmental & Social Footprint*. https://www.patagonia.com/our-footprint/

Peterson, C. (2000). The future of optimism. *The American Psychologist*, *55*(1), 44–55. PMID: 11392864

Pfeffer, J., & Sutton, R. I. (2006). *Hard Facts, Dangerous Half-Truths, and Total Nonsense*. Harvard Business School Press.

Pollack, J., & Pollack, R. (2015). Using Kotter's eight stage process to manage an organisational change program: Presentation and practice. *Systemic Practice and Action Research*, *28*(1), 51–66.

Prosci. (2018). *Best practices in change management - 2018 edition*. Prosci Inc.

Reeves, M., & Deimler, M. (2011, August). *Adaptability: The New Competitive Advantage*. Harvard Business Review. https://hbr.org/2011/07/adaptability-the-new-competitive-advantage

Rejeb, A., Rejeb, K., Appolloni, A., Jagtap, S., Iranmanesh, M., Alghamdi, S., Alhasawi, Y., & Kayikci, Y. (2023). Unleashing the power of internet of things and blockchain: A comprehensive analysis and future directions. *Internet of Things and Cyber-Physical Systems*, *4*. Advance online publication. DOI: 10.1016/j.iotcps.2023.06.003

Rigby, D. K., Sutherland, J., & Takeuchi, H. (2016). Embracing agile. *Harvard Business Review*, *94*(5), 40–50.

Rock, D., & Schwartz, J. (2006). The neuroscience of leadership. *Strategy & Business*, (43), 1–10. https://www.strategy-business.com/media/file/sb43_06207.pdf

Rocque, R. (2017). *Organizational Change*. Ijanm.com. https://ijanm.com/HTML_Papers/International%20Journal%20of%20Advances%20in%20Nursing%20Management__PID__2017-5-4-21.html

Schein, E. H., & Schein, P. (2017). *Organizational culture and leadership* (5th ed.). Jossey-Bass.

Schoemaker, P. J. (1995). Scenario planning: A tool for strategic thinking. *Sloan Management Review*, *36*(2), 25–40.

Schoemaker, P. J. H., & Day, G. (2021). Preparing Organizations for Greater Turbulence. *California Management Review*, *63*(4), 000812562110220. DOI: 10.1177/00081256211022039

Schultz, H., & Gordon, J. (2011). *Onward: How Starbucks fought for its life without losing its soul*. Rodale Books.

Seligman, M. E. P. (2011). *Flourish: A visionary new understanding of happiness and well-being*. Free Press. *discipline: The art and practice of the learning organization*. Doubleday.

Senge, P. M. (1990). *The Fifth Discipline*. Doubleday.

Senge, P. M. (2006). *The fifth discipline: The art and practice of the learning organization* (Revised ed.). Doubleday.

Shaw, G. (2017). *Making software teams effective: How agile practices lead to great outcomes*. O'Reilly Media.

Simmons, M. (2024, January 22). *The impact of demographic shifts on workforce planning and staffing*. Staffing Industry Analysts. https://www.staffingindustry.com/editorial/staffing-stream/impact-demographic-shifts-workforce-planning-and-staffing

Sirkin, H. L., Keenan, P., & Jackson, A. (2005). The hard side of change management. *Harvard Business Review*, *83*(10), 108–118. PMID: 16250629

Soenen, G. (2010). *A socio-political approach to organisational change, In: Change: Moving from intend to implementation* (pp. 84–90). Celerant Consulting; Houston.

Somasegar, S. (2024, February 1). *10 years into Nadella's run, Microsoft is a company transformed.* GeekWire. https://www.geekwire.com/2024/microsofts-resurgence-reflecting-on-satya-nadellas-leadership-a-decade-after-he-became-ceo

Spillane, J. P. (2006). *Distributed leadership*. Jossey-Bass.

Stacey, R. D. (2011). *Strategic management and organisational dynamics: The challenge of complexity* (6th ed.). Pearson Education.

Tabrizi, B. (2023, February 20). *How Microsoft became innovative again*. Harvard Business Review. https://hbr.org/2023/02/how-microsoft-became-innovative-again

Teece, D., Peteraf, M., & Leih, S. (2016). Dynamic capabilities and organizational agility: Risk, uncertainty, and strategy in the innovation economy. *California Management Review*, *58*(4), 13–35.

Thaler, R. H., & Sunstein, C. R. (2008). *Nudge: Improving decisions about health, wealth, and happiness*. Yale University Press.

Todnem By, R. (2005). Organisational change management: A critical review. *Journal of Change Management*, *5*(4), 369–380.

Tubussum, H., Hashim, H. M., Javaid, M. U., & Marimuthu, M. (2024). Impact of demographic variables on interplay of workplace bullying, organizational culture, and organizational performance: SEM-based multi group analysis. *Work (Reading, Mass.)*. Advance online publication. DOI: 10.1177/10519815241289761 PMID: 40297878

Uhl-Bien, M., Marion, R., & McKelvey, B. (2007). Complexity leadership theory: Shifting leadership from the industrial age to the knowledge era. *The Leadership Quarterly*, *18*(4), 298–318.

Van der Heijden, K. (2005). *Scenarios: The Art of Strategic Conversation*. John Wiley & Sons.

Verhoef, P. C., Broekhuizen, T., Bart, Y., Bhattacharya, A., Qi Dong, J., Fabian, N., & Haenlein, M. (2021). Digital transformation: a Multidisciplinary Reflection and Research Agenda. *Journal of Business Research, 122*(122), 889–901. Sciencedirect. https://doi.org/DOI: 10.1016/j.jbusres.2019.09.022

Vlasic, B. (2021, October 6). *How Mary Barra Is Remaking GM's Culture for the EV Era. Axios*. https://www.axios.com/2021/10/06/mary-barra-gm-electric-vehicles-culture

Wack, P. (1985). Scenarios: Uncharted waters ahead. *Harvard Business Review*, *63*(5), 73–89.

Waller, L. (2021). Fostering a Sense of Belonging in the Workplace: Enhancing Well-Being and a Positive and Coherent Sense of Self. *The Palgrave Handbook of Workplace Well-Being*, 341–367. https://doi.org/DOI: 10.1007/978-3-030-30025-8_83

Weick, K. E. (1995). *Sensemaking in organizations*. SAGE Publications.

Weick, K. E., & Quinn, R. E. (1999). Organizational change and development. *Annual Review of Psychology*, *50*(1), 361–386. PMID: 15012461

Weick, K. E., Sutcliffe, K. M., & Obstfeld, D. (2005). Organizing and the process of sensemaking. *Organization Science*, *16*(4), 409–421.

Whelan, T., & Fink, C. (2016). *The Comprehensive Business Case for Sustainability. Harvard Business Review*. https://hbr.org/2016/10/the-comprehensive-business-case-for-sustainability

Worley, C. G., Williams, T., & Lawler, E. E.III. (2014). *The Agility Factor*. Jossey-Bass.

Yukl, G. (2013). *Leadership in organizations*. Pearson.

Chapter 7
Measuring the Immeasurable?
Quantifying Neuroplasticity's Impact on Organizational Performance and Workforce Productivity

Joanne Vázquez de Príncipe
https://orcid.org/0000-0001-6510-0086
Purposeful Advising & Coaching LLC, USA & University of Phoenix, USA

ABSTRACT

This chapter examines one of the most critical yet underexplored aspects of neuroplasticity in organizational contexts: the rigorous measurement of its impact on quantifiable metrics. Building on this publication's core idea that neuroplasticity offers deep insights into individual and team capabilities, this contribution to the body of knowledge addresses the challenge of transforming ideas into tangible evidence. It reviews existing research, points out gaps in current studies, and proposes new frameworks and methods to evaluate how fostering neuroplasticity can enhance effectiveness, productivity, innovation, retention, and business excellence in a competitive global environment. By examining different levels—individual, team, and organizational—and identifying relevant indicators, this chapter aims to provide researchers, practitioners, and students with a clear plan for future research and practical application, transforming the "so what?" question into meaningful insights that support the enablement of neuroplastic, healthy organizations and thriving workforces globally.

DOI: 10.4018/979-8-3373-2292-6.ch007

INTRODUCTION

While the concept of neuroplasticity's benefits is gaining traction, rigorously measuring its direct impact on quantifiable organizational performance metrics, such as return on investment (ROI), productivity, innovation rates, and employee retention, remains a significant challenge and a substantial area for future research. This is where the rubber meets the road for businesses and where scientific rigor is most needed. This is the ultimate "so what?" question for organizations. This chapter directly addresses the objectives of this publication by moving beyond the "how" and "why" of neuroplasticity in organizational dynamics to the crucial "what" and "how much" of its impact. By proposing frameworks and methodologies for measurement, this chapter provides the empirical foundation necessary to validate the claims made throughout the book regarding the influence of neuroplasticity on organizational effectiveness and workforce productivity.

This chapter aims to transform a fascinating biological concept of neuroplasticity into an actionable business imperative, offering industry thought leaders, practitioners, and academics the tools to assess the return on investment for fostering a neuroplastic workforce as an empowerment tool to counteract the resistance to change associated with integrating artificial intelligence technologies into the modern workplace (Ooi, 2025). Without measurement, the full strategic value of this concept cannot be fully realized or justified in a business context (Workhuman, 2025). Hence, the research herewith endeavors to catalyze neuroplasticity, empowering future generations for the evolutionary work ahead (Harlianto et al., 2025; Ooi, 2025; Rajagopalan & Jacob, 2025; Vázquez de Príncipe, 2024a). The background section presents a discussion of the potential impact of neuroplasticity on organizational effectiveness and workforce productivity, drawing on evidence, mechanisms, and strategic value derived from a comprehensive literature review. This multidisciplinary chapter offers a comprehensive examination of the application and measurable outcomes of cultivating a neuroplastic organization for the rapidly evolving future of work, as the integration of AI continues to become more prevalent and embedded in the fabric of organizations, becoming business as usual.

BACKGROUND

Neuroplasticity, the brain's ability to adapt and reorganize, has emerged as a powerful lens for understanding and enhancing organizational effectiveness and workforce productivity. Recent research demonstrates that neuroplasticity-based interventions, neuroscience-informed leadership, and the integration of neurodiversity can drive measurable improvements in employee engagement, adaptability, creativity, and overall business outcomes (Akşahin et al., 2023; Badi, 2019; Cullen et al., 2014; Gee, 2025a; Kaur, 2024; Kiss et al., 2019; Lee et al., 2024; LeFevre-Levy et al., 2023; McCreedy, 2024; Ooi, 2025). Case studies and empirical models show that organizations prioritizing psychological safety, well-being, and adaptive leadership can achieve up to 70% greater employee flexibility and 30% higher team productivity (Gee, 2025a; Ooi, 2025). The application of neuroscience in organizational settings, including neuroergonomics and neuroleadership, has provided actionable frameworks for optimizing performance, resilience, and innovation (Akşahin et al., 2023; Becker & Cropanzano, 2010; Becker et al., 2011; Cucino et al., 2021; Dehais et al., 2020; Guarnier & Chimenti, 2023; McCreedy, 2024; Stasielowicz, 2020; Waldman et al., 2019). However, the field faces challenges related to measurement, ethical considerations, and the need for more robust, longitudinal evidence (Ashkanasy et al., 2014; Dehais et al., 2020; Healey & Hodgkinson, 2014; Silver et al., 2023; Waldman et al., 2019). This literature review synthesizes the latest interdisciplinary research, including case studies, measurement indicators, and strategic frameworks, to assess the scope and magnitude of neuroplasticity's influence on healthy organizational outcomes (Bell-Ellis, 2013).

METHODS

A comprehensive search was conducted across over 170 million research papers in Consensus, encompassing Semantic Scholar, PubMed, and additional databases. The search strategy included targeted queries on neuroplasticity, organizational effectiveness, workforce productivity, measurement indicators, case studies, leadership, artificial intelligence, neurodiversity, and related interdisciplinary themes. In total, 1,044 research study papers were identified, 621 were screened, 369 were deemed eligible, and the top 50 most relevant and high-quality papers were included in this literature review. In Table 1, the matrix below indicates the process flow of the search and selection process for the included studies in the literature review for this book chapter.

Table 1. Literature Review Process Flow Matrix.

Identification	Screening	Eligibility	Included
1044	621	369	50

Eight unique search groups were executed, systematically covering foundational concepts, measurement, business context, ethical considerations, interdisciplinary links, and case studies.

RESULTS

Neuroplasticity and Organizational Effectiveness: Empirical Evidence

Multiple studies and case analyses demonstrate that neuroplasticity-based interventions—such as cognitive flexibility training, neuroleadership development, and resilience programs—can significantly enhance organizational effectiveness. For example, longitudinal interventions with leaders and teams have shown improvements in creativity (up to 60%), emotional regulation (70%), and team productivity (30%) (Akşahin et al., 2023; Gee, 2025a; Lee et al., 2024; McCreedy, 2024; Ooi, 2025). Organizations that foster psychological safety and well-being report up to 70% more employee flexibility and outperform traditional governance models (Akşahin et al., 2023; Cullen et al., 2014; Gee, 2025a; Ooi, 2025).

Measurement Indicators and Methodological Advances

Advances in neuroimaging and neuroergonomics have enabled the practical assessment of brain function in workplace settings, providing new measurement indicators for engagement, adaptability, and performance (Becker et al., 2011; Dehais et al., 2020; Waldman et al., 2019). Neurophysiological markers (e.g., prefrontal cortex activity) and behavioral metrics (e.g., adaptability, work engagement) are increasingly used to evaluate the impact of neuroplasticity-based interventions (Dehais et al., 2020; Lee et al., 2024; McCreedy, 2024; Waldman et al., 2019). However, methodological challenges remain, including the need for standardized protocols and longitudinal data (Ashkanasy et al., 2014; Dehais et al., 2020; Waldman et al., 2019).

Strategic Business Value, Leadership, and Neurodiversity

Neuroscience-informed leadership and the integration of neurodiversity are linked to enhanced innovation, adaptability, and business performance (Akşahin et al., 2023; Gee, 2025a; Guarnier & Chimenti, 2023; Kaur, 2024; Kiss et al., 2019; LeFevre-Levy et al., 2023; Nowak, 2021; Ooi, 2025). Empowering leadership styles, psychological capital, and holistic-driven inclusive practices for neurodivergent employees contribute to inspiring innovation that fosters higher engagement, reduced turnover, and improved organizational reputation (Akşahin et al., 2023; Branicki et al., 2024; Burhan & Khan, 2023; LeFevre-Levy et al., 2023; Niswaty et al., 2021; Rollnik-Sadowska & Grabińska, 2024; Silver et al., 2023; Szulc, 2022; Taghikilani-damavandi, 2024). Case studies from leading organizations (e.g., Microsoft, Unilever, Tata) illustrate the scalability and ROI of regenerative, neuroplasticity-driven business models (Gee, 2025a; Ooi, 2025).

Artificial Intelligence, Ethical Considerations, and Organizational Constraints

AI-driven tools and neurodata analytics are increasingly used to optimize decision-making, monitor employee well-being, and personalize interventions (Ateeq et al., 2025; De Fátima Soares Borges et al., 2020; Gee, 2025a; Ooi, 2025; Pereira et al., 2021; Shaikh et al., 2023). While these technologies can boost productivity and resilience, they also raise ethical concerns regarding privacy, transparency, and the commodification of neurodiverse talent (Ashkanasy et al., 2014; Ateeq et al., 2025; Guarnier & Chimenti, 2023; Healey & Hodgkinson, 2014; Kietzman, 2024; Silver et al., 2023). Organizational constraints include the need for cultural change, leadership buy-in, and the development of universally accessible frameworks (Branicki et al., 2024; Silver et al., 2023; Szulc, 2022). In Table 2, the comparison matrix below presents the key literature review studies that indicate the influence of neuroplasticity on organizational effectiveness and productivity. Furthermore, in Table 3, the matrix below indicates the top contributing authors and journals that appeared most frequently in the included studies of the literature review.

Table 2. Key Neuroplasticity Studies Comparison Matrix.

Key Study	Methodology	Sample/Context	Key Results	Strategic Focus
(Ooi, 2025)	Mixed-method, longitudinal case studies	100+ year intergenerational enterprises, global corporations	70% more employee flexibility, 40% better governance, long-term growth	Regenerative organizational development, AI, neuroplasticity
(Gee, 2025a)	Longitudinal intervention, empirical analysis	200 leaders (CEOs, execs, youth)	60% increased creativity, 70% more emotional regulation, 30% better team productivity	Neuroplasticity, AI, systemic leadership
(Waldman et al., 2019)	Review, methodological analysis	Organizational research settings	Practical brain imaging, new measurement indicators	Neuroscience methods in organizational research
(McCreedy, 2024)	Literature review, case studies	Organizational transformation programs	Improved employee experience, sustainable behavior change	Neuro-informed transformation frameworks
(Akşahin et al., 2023)	Qualitative, content analysis	School administrators, teachers	Neuroleadership increased participation, organizational culture, strategic success	Neuroleadership, behavior change

Table 3. Top Research Contributors Matrix.

Type	Top Contributors	Research Studies
Author	Rachel Ooi	(Gee, 2025a; Ooi, 2025)
Author	W. Becker	(Becker & Cropanzano, 2010; Becker et al., 2011)
Author	D. Waldman	(Ashkanasy et al., 2014; Waldman et al., 2019)
Journal	*Sustainability*	(Akşahin et al., 2023; Jelača et al., 2022; Rollnik-Sadowska & Grabińska, 2024)
Journal	*Journal of Organizational Behavior*	(Ashkanasy et al., 2014; Becker & Cropanzano, 2010)
Journal	*International Journal of Psychiatry*	(Gee, 2025a)

LITERATURE REVIEW DISCUSSION

The research consensus indicates that neuroplasticity has a substantial and positive influence on organizational effectiveness and workforce productivity, especially when integrated with adaptive leadership, inclusive practices, and AI-driven tools (Akşahin et al., 2023; Cullen et al., 2014; Dehais et al., 2020; Gee, 2025a; Kaur, 2024; Kiss et al., 2019; Lee et al., 2024; LeFevre-Levy et al., 2023; McCreedy, 2024; Ooi, 2025). The most substantial evidence comes from longitudinal case studies and empirical models, which demonstrate measurable gains in creativity, adaptability, and team performance (Akşahin et al., 2023; Gee, 2025a; McCreedy, 2024; Ooi, 2025). The application of neuroscience in organizational settings has moved from theoretical promise to actionable frameworks, with neuroleadership and neuroergonomics providing practical tools for leaders and HR professionals (Akşahin et al., 2023; Becker & Cropanzano, 2010; Becker et al., 2011; Cucino et al., 2021; Dehais et al., 2020; Guarnier & Chimenti, 2023; McCreedy, 2024; Stasielowicz, 2020; Waldman et al., 2019).

However, the field is not without limitations. Measurement challenges, ethical concerns, and the risk of over-reliance on the "business case" for neurodiversity highlight the need for more robust, interdisciplinary, and ethically grounded research (Ashkanasy et al., 2014; Ateeq et al., 2025; Dehais et al., 2020; Healey & Hodgkinson, 2014; Silver et al., 2023; Waldman et al., 2019). There is also a need for more longitudinal, industry-specific studies and the development of universally accessible frameworks that benefit both neurodiverse and neurotypical employees (Branicki et al., 2024; Guarnier & Chimenti, 2023; Kietzman, 2024; Silver et al., 2023; Szulc, 2022). Table 4 below presents a matrix with the key claims and supporting evidence identified in the literature review studies.

Table 4. Claims and Evidence Strength Matrix.

Claim	Evidence Strength Level	Reasoning	Studies
Neuroplasticity-based interventions improve organizational effectiveness and productivity.	**Evidence strength: Strong (9/10)**	Multiple longitudinal and case studies show significant gains in flexibility, creativity, and team performance.	(Akşahin et al., 2023; Cullen et al., 2014; Gee, 2025a; Kiss et al., 2019; Lee et al., 2024; McCreedy, 2024; Ooi, 2025)
Neuroscience-informed leadership enhances engagement, adaptability, organizational ambidexterity, and business outcomes.	**Evidence strength: Strong (8/10)**	Empirical and qualitative studies have linked neuroleadership to increased participation, a more positive culture, and greater success.	(Akşahin et al., 2023; Gee, 2025a; Guarnier & Chimenti, 2023; Kaur, 2024; Kiss et al., 2019; LeFevre-Levy et al., 2023; Nowak, 2021; Ooi, 2025)
Neurodiversity and holistic inclusive practices drive innovation and reduce turnover.	**Evidence strength: Moderate (7/10)**	Reviews and large-sample-size studies demonstrate unique talents and improved outcomes with proper holistic management strategies.	(Branicki et al., 2024; LeFevre-Levy et al., 2023; Rollnik-Sadowska & Grabińska, 2024; Silver et al., 2023; Szulc, 2022; Taghikilanidamavandi, 2024)
AI and neurodata analytics can optimize productivity, but they also raise ethical concerns.	**Evidence strength: Moderate (6/10)**	Systematic reviews and case studies highlight both benefits and risks of AI in workplace settings.	(Ateeq et al., 2025; De Fátima Soares Borges et al., 2020; Gee, 2025a; Ooi, 2025; Pereira et al., 2021; Shaikh et al., 2023)
Measurement of neuroplasticity in organizations is advancing but lacks standardization.	**Evidence strength: Moderate (5/10)**	Methodological reviews note progress in neuroimaging and behavioral metrics but call for more robust protocols.	(Becker et al., 2011; Dehais et al., 2020; Waldman et al., 2019)
Over-reliance on the business case for neurodiversity may undermine strategic optimization goals for inclusion.	**Evidence strength: Weak (3/10)**	Critical reviews warn of risks in commodifying neurodiverse talent while neglecting universal holistic design, which can erode their sense of belonging and make them feel like a tool or a means to an end, rather than a valued member of the team.	(Silver et al., 2023)

Research Gaps

Despite promising evidence, gaps remain in the standardization of measurement indicators, longitudinal industry-specific studies, and the development of universally accessible frameworks. There is also a need for more research on the ethical implications of neurodata and AI, as well as the intersection of neuroplasticity with organizational culture and policy. The matrix in Table 5 below depicts the research topics and study attributes in the literature review, as well as highlights the key areas with limited research coverage.

Table 5. Research Gaps Matrix.

Topic/Attribute	Leadership Interventions	AI/ Neurodata Tools	Inclusive Practices	Measurement Studies	Longitudinal Case Studies
Creativity/ Productivity	7	4	5	3	6
	6	3	4	2	5
Neurodiversity Outcomes	2	**1**	7	**1**	2
Ethical/Policy Issues	**1**	2	2	**GAP**	**1**
Organizational Culture	3	**1**	3	**1**	2

Literature Review Summary

The literature review strongly supports the view that neuroplasticity, when harnessed through holistic leadership, inclusive practices, and technology, can significantly enhance organizational effectiveness and workforce productivity. However, realizing its full potential requires addressing measurement, ethical, and implementation challenges. Furthermore, there are clear gaps in literature, which are an opportunity to elevate neuroplasticity as a 'must-have' in today's workplace, as opposed to considering it as a 'nice to have' or nonexistent, as the research indicates.

MAIN FOCUS OF THE CHAPTER

To begin understanding the foundational importance of neuroplasticity, let us imagine a world in which we are unable to learn from our mistakes, we never start a new hobby or learn a new skill, and we cannot learn to navigate a new city. This is a world without neuroplasticity. Neuroplasticity is an intriguing brain function that people critically depend on in their daily lives. Next, the discussion will delve into what neuroplasticity is and how it functions.

What Is Neuroplasticity?

Neuroplasticity is our brain's ability to adapt in response to experience. In other words, it is our ability to learn new things or develop new skills. As we navigate life, we must develop behaviors that suit our context and help us not only survive but also thrive. The brain accomplishes this by creating new connections between neurons, thereby altering the way distinct parts of the brain communicate with each other, which supports the new behavior or skill we seek to learn (Kleim & Jones, 2008).

Neuroplasticity might be one of the brain's most crucial functions. If the neural circuitry that is created throughout our early development were fixed, we would not be able to adapt to new contexts or environments, the functional effects of brain damage would be permanent, and we would not be able to experience the joy of learning. We would be stuck as the person we were when our brains stopped developing. All humans would remain cognitively 20-somethings forever.

How Does Neuroplasticity Work?

Current research shows that neuroplasticity involves various forms of changes that occur in the brain. One kind of change is in the actual structure of brain cells. For instance, Sagi et al. (2012) conducted a neuroimaging study that found structural changes in brain areas associated with learning and memory following a task that engaged these processes.

Another type of structural change observed in brain cells is dendritic growth. Dendrites are branch-like projections that extend from one cell and connect to another one to allow for communication between them. Hence, this communication comes in the form of a synapse, which is the flow of information from one neuron to another. Neurons can have multiple dendrites that branch out in all directions and synapse on many other neighboring cells.

Therefore, more dendrites mean a single neuron can talk to more neurons at one time. Dendritic growth is the growth and branching of a neuron's dendrites, which leads to changes in the way distinct parts of the brain can communicate with each other. Ploughman et al. (2015) exclaimed that these changes in communication can lead to changes in neuronal networks, or an assembly of neurons that work together to support functions. Furthermore, it is these changes that people observe in work-life dynamic behaviors, demonstrating a growth mindset, as noted by neuroplasticity (Ploughman et al., 2015; Vázquez de Príncipe, 2023).

Neuroplasticity Theoretical Principles

Literature identified ten distinct principles of neuroplasticity. These theoretical principles incorporate the following: (1) use it or lose it; (2) use it and improve it; (3) specificity; (4) repetition matters; (5) intensity matters; (6) time matters; (7) salience matters; (8) age matters; (9) transference or generalization; and (10) interference (Harlianto et al., 2025; Kleim & Jones, 2008; Rajagopalan & Jacob, 2025; Vázquez de Príncipe, 2024a). Next, the author emphasizes the implications of three of these neuroplasticity principles.

1. **Use It or Lose It:** This is a phrase that you are already familiar with, but it is a good encapsulation of one aspect of neuroplasticity. As we saw in the examples of neuroplasticity in blind people and deaf people, the brain is not a fan of letting areas go unused. The 'use it or lose it' principle refers to this same process. However, unlike in these examples, the neural networks that are not being used are not necessarily repurposed for something better. Luckily, even if we do not use it and subsequently lose it, we can regain it through dedicated training and consistency.
2. **Intensity Matters:** The intensity principle can be thought of as the dosage required to achieve a result. This will depend heavily on what you are trying to do and may not necessarily apply to your goals. For example, let us say you are trying to quit a bad habit. There is no way for you to intensely not do something. Nevertheless, if you are learning to play piano, sitting down for five minutes and playing one of those songs you can play with two fingers will not improve your piano-playing ability. Sitting down for an hour and challenging yourself with complicated pieces, on the other hand, will improve your ability to play the piano.
3. **Age Matters:** This principle refers to the fact that our brains are much more plastic (or adaptable) when we are young than when we are old. Though an aging brain is still more than capable of learning new things, it just might not come as quickly or as easily as it would to a younger person.

Neuroplasticity is one brain function that humanity can all appreciate. This brain function is the basis for so much of what makes life rich and enjoyable. Today, researchers are still discovering the mechanisms that underlie how it works. However, it is evident in the literature that neuroplasticity plays a substantial role in our daily lives.

THE IMPERATIVE OF MEASUREMENT IN THE NEUROPLASTIC ERA

Brain plasticity, also known as neural plasticity or neuroplasticity, refers to the brain's inherent ability to restructure its neural pathways and synaptic connections in response to learning, experience, and environmental stimuli (Forget & Le Pertel, 2024; Hawkins, 2021; Pascual-Leone et al., 2005). Today, neuroplasticity is recognized as a foundational concept for organizational effectiveness and workforce productivity, as evidenced by a 2019 international conference dedicated to highlighting the practical application of brain science as a driver of achieving organizational success (Al Halyan et al., 2019). Tomlinson (2021) emphasized the importance of neuroplasticity in the workplace as a catalyzing tool for improving managerial approaches. Forget and Pertel (2024) indicated that promoting brain health at work plays a critical role in facilitating learning and memory retention, which is a key factor in optimizing organizational performance.

From Theoretical Understanding to Practical Application

The shift from a theoretical understanding of neuroplasticity to the practical need for measurable outcomes is crucial because it transforms this powerful biological idea from abstract science into actionable strategies with clear value. While the theory of neuroplasticity explains how the brain changes, measuring its results quantifies the real-world impact of those changes, especially in organizational and personal growth (Workhuman, 2025).

Firstly, measurable outcomes validate and justify investments. Organizations and individuals are increasingly funding programs designed to enhance cognitive flexibility, adaptability, and resilience. By quantifying the impact of these neuroplasticity-enhancing initiatives—such as better learning rates, lower resistance to change, or improved well-being, it becomes possible to demonstrate a clear ROI for their efforts in enhancing cognitive flexibility, adaptability, and resilience, moving neuroplasticity from an abstract concept to a strategic imperative (Barua & Vázquez de Príncipe, 2025; Psychology Today, 2025). This shifts neuroplasticity from a "nice-to-have" idea to a strategic necessity, making it a "must-have."

Secondly, it allows for the optimization and refinement of interventions. By measuring specific outcomes, practitioners can determine which strategies effectively promote desired neuroplastic changes and which do not. This data-driven approach enables continuous improvement of training programs, leadership development, and well-being initiatives, ensuring they are tailored for maximum impact (BetterUp, 2021; Neuro Business School, 2024).

Thirdly, measurable outcomes support targeted interventions and accountability. Knowing which specific aspects of neuroplasticity are affected, such as increased cognitive flexibility and improved social-emotional regulation, enables the creation of more precise and effective intervention programs (Bakhtiari, 2025; Peaceful Leaders Academy, 2025; Pollack, 2025). It also holds leaders and program designers accountable for delivering tangible results, rather than relying solely on anecdotal evidence of progress (News-Medical, 2019).

Ultimately, this transition bridges the gap between scientific research and practical application. It converts complex neurological processes into clear and actionable metrics that can guide business decisions, organizational design, learning curriculum development, HR strategies, and personal self-improvement goals, making the profound theoretical understanding of neuroplasticity practically applicable and impactful (Frontiers, 2016; Vázquez de Príncipe, 2023; 2024b). Without measurable outcomes, the deep theoretical knowledge of neuroplasticity would remain largely academic, limiting its ability to create meaningful change in areas such as workforce adaptability to modern technologies or personal resilience in the face of challenges (Workhuman, 2025; Lim et al., 2017; Ludvik, 2023).

Interconnected Hyper-Competitive Global Dynamics

Kamkankaew et al. (2022) emphasized the "increasing dynamics of the competitive environment and the need for hyper-competition among businesses" as a driving force behind the urgency for understanding the impact of neuroplasticity in the evolving workplace (pp. 9-20). Today, hyper-competitive global dynamics create an urgency for organizations to become neuroplastic (Scientific Research Archives, 2023). These dynamics underscore why merely understanding neuroplasticity theoretically is insufficient, emphasizing the imperative need for measurable outcomes, which this chapter aims to highlight.

Firstly, pervasive global hyper-competition is a primary driver of business evolution. In an intensely competitive landscape, organizations must become agile to continuously adapt, innovate, and optimize performance to maintain relevance and secure a competitive edge (DailyBot, 2024; Scientific Research Archives, 2023). Secondly, the rapidly changing work environment necessitates organizational neuroplasticity (PerspectAI, 2024). This includes the accelerating pace of technological advancements, such as the integration of artificial intelligence, which demands constant learning and adaptability from the workforce to remain career relevant (CertLibrary, 2025; Ooi, 2025).

Organizations face continuous market shifts and unforeseen disruptions, requiring them to be inherently agile in problem-solving and embracing change. A neuroplastic organization, therefore, becomes essential for rapid learning and skill acquisition, ensuring individuals and teams can quickly grasp new knowledge and apply new competencies (Paradiso Solutions, 2025). This directly contributes to cognitive flexibility, enabling the organization to demonstrate agility in navigating complex situations and generating novel solutions (Agile by Design, 2024).

Furthermore, the need for enhanced organizational resilience is paramount. In a volatile global landscape, organizations must be able to "bounce back" from market shocks and maintain performance under pressure. By fostering neuroplasticity, organizations build a collective capacity for adaptation, ensuring their survival and prosperity in dynamic conditions (Marín, 2025; Scientific Research Archives, 2023). Without this inherent adaptability, organizations risk falling behind competitors who can leverage the workforce's 'cognitive and emotional malleability' for global strategic advantage (PerspectAI, 2024).

Moreover, the urgency for neuroplastic organizations stems primarily from the rapidly evolving hyper-competitive dynamics and the demanding global landscape (Scientific Research Archives, 2023). There are five predominantly competitive global dynamics creating this urgency for cultivating neuroplastic organizations, as depicted in *Figure 1* below.

Figure 1. Urgency for Cultivating Neuroplastic Organizations Venn Graphic.

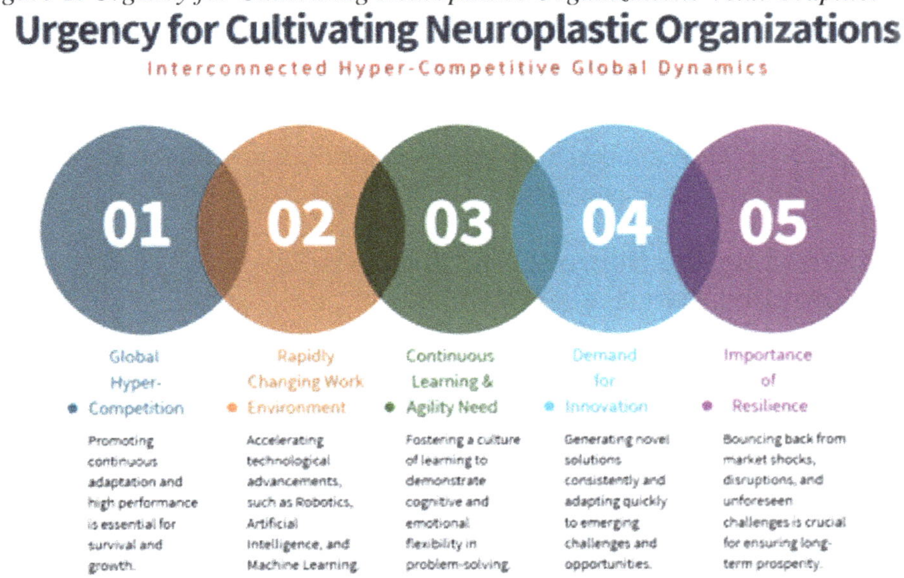

In the current state of global hyper-competition, organizations operate in an intensely competitive environment where continuous adaptation and high performance are essential for their survival and growth (Scientific Research Archives, 2023).

This relentless competition requires businesses to continually evolve in order to stay ahead, rather than playing catch-up.

The accelerating technological advancements, such as the integration of artificial intelligence, robotics, and machine learning, drive the rapidly changing work environment (Ooi, 2025). The modern workplace is in a constant state of flux, especially in the new post-pandemic normal, which incorporates various work settings for employees, including office, remote, and hybrid (Gallup, 2023). This necessitates a workforce that can rapidly acquire new skills and embrace new ways of working with minimal resistance to change (Workhuman, 2025).

To remain relevant and competitive, organizations must cultivate a culture that encourages individuals and teams to continually learn, adapt, and exhibit cognitive flexibility in problem-solving (Lim et al., 2017; Ludvik, 2023). The demand for innovation drives the dynamic nature of global markets, compelling organizations to continually develop novel solutions and respond swiftly to emerging challenges and opportunities. Lastly, the importance of resilience in an unpredictable global economy—the ability of an organization to "bounce back" from market shocks, disruptions, and unforeseen challenges—is crucial for maintaining high-performance levels and ensuring long-term prosperity (Marín, 2025).

In summary, these interconnected dynamics underscore why fostering neuroplasticity within an organization is not only advantageous but also a strategic necessity for long-term success. The goal of this chapter is to examine methods and frameworks for assessing the impact of neuroplasticity on organizational performance and workforce productivity. In the next section, the discussion will focus on integrating holistic principles of neuroplasticity to enhance organizational health and employee well-being, benefiting the entire ecosystem.

Understanding the Neuroplastic Organization: A Conceptual Overview

The concept of neuroplasticity encompasses the brain's natural malleable ability to adapt to change via the nervous system, which researchers universally refer to as "rewiring your brain" (Costandi, 2016, p. 1). Hence, this is what enables humans and all organisms with a nervous system to adapt to their environment and make informed decisions based on circumstances and lived experiences. Costandi (2016) emphasized that neuroplasticity is the "intrinsic and fundamental property of all nervous systems" (p. 2). Therefore, mechanisms such as synaptic plasticity and adult neurogenesis are inherently linked to learning, adaptation, and cognitive function, as evidenced by the occurrence of lifelong brain changes (Appelbaum & Gallagher, 2000). Intentional activities and exercises specific to "brain training" have been shown to alleviate "addiction and pain" and even improve sleep (Costandi, 2016, pp. 85, 115).

To provide contextual alignment, a Neuroplastic Organization is a conceptual framework that applies the principles of neuroplasticity—the brain's ability to change and adapt—to the collective capabilities of an organization and work-life dynamics (Vázquez de Príncipe, 2023). It is an organization designed and managed to leverage the inherent human capacity for learning, adaptation, and cognitive flexibility at individual, team, and systemic levels (Lim et al., 2017; Ludvik, 2023). Furthermore, a neuroplastic organization is one that consciously cultivates an environment where the collective "brain," similarly to a living organism, is continuously learning, adapting, improving, and evolving, mirroring the brain's capacity for intrinsic change. *Figure 2* below illustrates the author's conceptualization of six core wellness attributes of a neuroplastic organization.

Figure 2. Neuroplastic Organization Wellness Attributes.

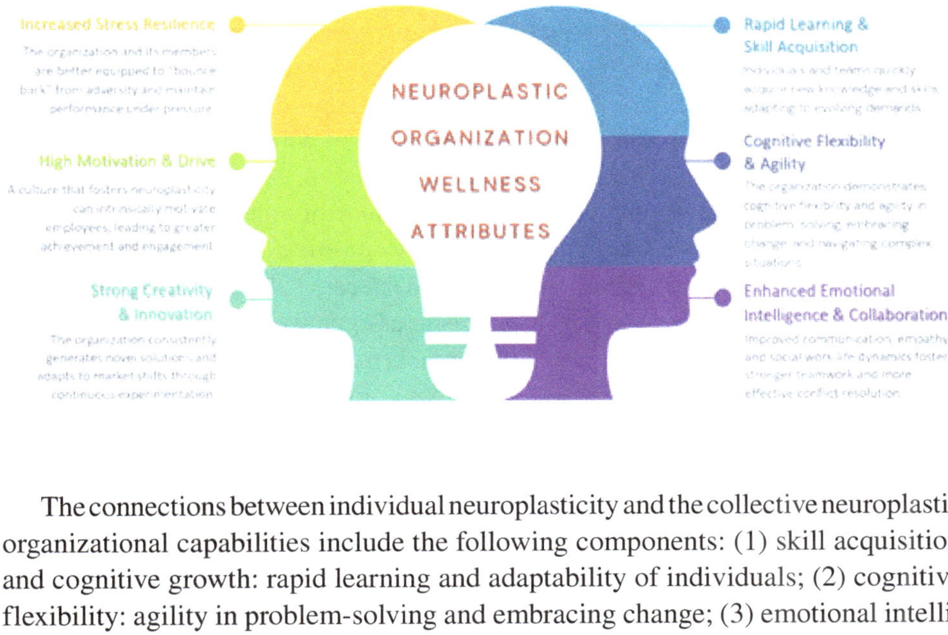

The connections between individual neuroplasticity and the collective neuroplastic organizational capabilities include the following components: (1) skill acquisition and cognitive growth: rapid learning and adaptability of individuals; (2) cognitive flexibility: agility in problem-solving and embracing change; (3) emotional intelligence and social neuroscience: improved communication, collaboration, and team dynamics; (4) stress and resilience: the capacity to bounce back from adversity; (5) motivation and drive: intrinsic motivators for achievement; and (6) creativity and innovation: generating novel solutions (Algumaei et al., 2023; Gallup, 2025).

Individual and group neuroplastic changes theoretically impact organizational performance through multiple key pathways. Firstly, enhanced individual cognitive functions, such as learning, memory, and decision-making, lead to increased productivity and adaptability, forming the bedrock of organizational output (Lim et al., 2017; Ludvik, 2023). Secondly, fostering a growth mindset and learning agility at both individual and team levels directly fuels innovation, enabling organizations to generate innovative ideas and adapt rapidly to market shifts (Appelbaum & Gallagher, 2000; Kirby & Smith, 2021; Staneiu, 2023). Thirdly, improved emotional regulation and collective emotional intelligence contribute to greater organizational resilience, reducing burnout and enhancing the capacity to navigate disruptions (Marín, 2025; Peaceful Leaders Academy, 2025). Lastly, better communication and collaboration, stemming from enhanced social cognition, streamline operations and improve problem-solving across the organization. Leaders who cultivate their neuroplasticity can foster strategic adaptability, enabling the organization to make informed decisions and effectively manage change in dynamic global environments (Faw, 2025; Neuro Business School, 2024; Lim et al., 2017; Ludvik, 2023).

By implementing these types of specific, measurable, and supportive multidisciplinary interventions, organizations can intentionally cultivate the cognitive and emotional attributes of neuroplasticity. Employing a strategic approach fosters enriched capabilities, leading to a more adaptable, resilient, and high-performing workforce that aligns with and enhances the organizational performance outcomes. For instance, the foundational Guest Model of HRM, introduced by David Guest in 1987, emphasizes the importance of aligning strategies with business objectives to improve performance and financial results (Behera et al., 2022; Khaneja & Arora, 2024; Makanur, 2025). It emphasizes the importance of consistent HR practices, including recruitment and training, which lead to positive outcomes and, in turn, influence employee behavior and organizational success (SHRM, 2025). Nonetheless, the model faces criticism for its idealistic assumptions about alignment and employee reactions (Makanur, 2025). As an alternative, *Figure 3* below presents the author's enhanced *Strategic HRM Model for Cultivating Neuroplasticity* in the modern workplace.

Figure 3. Strategic HRM Model for Cultivating Neuroplasticity.

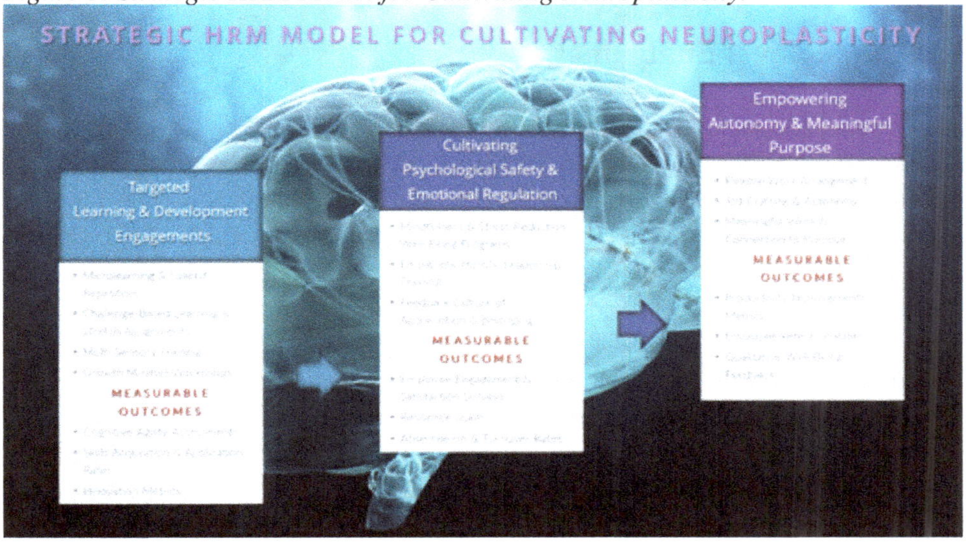

The Measurement Challenge: Why Quantifying Neuroplasticity's Impact is Complex

The challenges in measuring neuroplasticity are fivefold. Firstly, there is the Black Box Problem, which is that directly observing neuroplastic changes in the workplace is impractical and unethical. Businesses tend to rely on behavioral and performance proxies. Secondly, there is the conflict between causality and correlation, producing the difficulty in isolating neuroplasticity as the sole or primary cause of performance improvements amidst the myriad organizational factors, such as culture, leadership, strategy, and market conditions (Faw, 2025). Thirdly, the lagging indicators of neuroplastic changes are often gradual, and their organizational impact may not be immediately apparent, requiring longitudinal studies. Fourthly, defining "Neuroplasticity" in organizational terms encompasses translating a biological concept into measurable organizational constructs (e.g., what does "high organizational neuroplasticity" look like numerically?). Lastly, there is data scarcity and privacy concerns due to the lack of standardized metrics and sensitivity surrounding the collection of highly personal data related to brain function or well-being (Barua & Vázquez de Príncipe, 2025).

SOLUTIONS AND RECOMMENDATIONS

Individual Level Enablement

At the individual level, the focus must be on the changes in employees' cognitive abilities, emotional regulation, and learning capacity. The proxies or indicators can include cognitive assessments, such as pre- and post-training tests that measure problem-solving speed, memory recall, attention span, and critical thinking (Drobin et al., 2025; Ludvik, 2023). The learning and development metrics should focus on faster skill acquisition rates, higher completion rates for new certifications, and the demonstrated application of new skills in practical settings. The self-reported measurements to consider include surveys on perceived adaptability, resilience, growth mindset, and engagement, using science-backed, validated psychological scales (SHRM, 2025; Staneiu, 2023). Physiological markers must be leveraged with caution in a specific environment, considering both ethical implications and legal regulatory considerations. These factors include measuring stress hormone levels (cortisol) and heart rate variability (HRV), which serve as indicators of stress management and recovery. An alternative non-invasive experimental approach to measuring well-being incorporates pre- and post-activity self-monitoring to assess the recovery rebounding impact after prayer, meditation, reflection, nature bathing, tapping, or deep breathing exercises (Barua & Vázquez de Príncipe, 2025; Malchiodi, 2020).

A deeper, more intentional focus on cultivating the cognitive and emotional attributes of neuroplasticity through specific, measurable, and supportive multidisciplinary interventions would involve multiple key applications within a neuroplastic organization. In this section, the author recommends three potential individual-level solutions, providing details on application, measurable outcomes, and suggested empirical research to explore (Table 6). First, targeted learning and development experiential engagements by designing deep critical-thinking activities to stimulate neuroplasticity in the workplace proactively. Second, cultivating psychological safety and emotional regulation fosters an environment where employees feel secure enough to experiment, make mistakes, and openly express their emotions. Lastly, empowering autonomy and fostering a sense of meaningful purpose and well-being provide employees with a sense of control and driven focus, thereby enhancing aligned motivation and neuroplasticity, which organically improves productivity at the individual level.

Table 6. Individual Level Workplace Neuroplasticity Solutions Matrix.

Individual Solutions	Application Details	Measurable Outcomes	Empirical Research
Targeted Learning and Development Experiential Engagements	**Microlearning and spaced repetition:** Presenting content in brief sessions to enhance memory retention. **Challenge-based learning and stretch assignments:** Presenting innovative, complex problems to challenge the mind beyond its comfort zone. **Multi-sensory instructing:** Using different sensory inputs (visual, auditory, kinesthetic) to activate more areas of the brain. **Growth Mindset workshops:** Explicitly teaching the concept of neuroplasticity and the belief that abilities are developed through effort.	**Cognitive agility assessments:** Pre- and post-program assessments measuring problem-solving speed, adaptability, and cognitive flexibility. **Skill acquisition and application rates:** Monitoring how quickly new skills are learned and used in contextual scenario-based simulations and real-life experimentations. **Innovation metrics:** The number of innovative ideas, process improvements, or pilot projects launched.	De Príncipe, J. V. (2024). The Perfectionism Mindset Impacting Resilience in Multicultural and Multigenerational Leadership Work-Life Experiences [Chapter 1]. In Vázquez de Príncipe, J. (Ed.), Resilience of Multicultural and Multigenerational Leadership and Workplace Experience, 1–31. IGI Global Scientific Publishing. ISBN: 9798369318027. https://doi.org/10.4018/979-8-3693-1802-7.ch001 Staneiu, R. M. (2023). Nurturing neuroplasticity as an enabler for Growth Mindset through lifelong learning and knowledge dynamics. *Proceedings of the International Conference on Business Excellence, 17*(1), 1264–1274. https://doi.org/10.2478/picbe-2023-0113

continued on following page

Table 6. Continued

Individual Solutions	Application Details	Measurable Outcomes	Empirical Research
Cultivating Psychological Safety and Emotional Regulation	**Mindfulness and stress reduction initiatives:** Help employees regulate emotions and reduce the negative impact of chronic stress on brain plasticity. **Empathetic leadership education:** Train leaders to practice empathy, active listening, and transparent communication. **Feedback culture:** Establish continuous, constructive feedback loops focused on learning and development, such as two-way communications, and foster continuous improvement mindset and a self-leadership mindset.	**Employee engagement and satisfaction surveys:** Including questions on psychological safety and perceived support for learning from mistakes. **Resilience scales:** Using tools like the Connor-Davidson Resilience Scale (CD-RISC) to measure improvements in individual and team resilience. **Absenteeism and turnover rates:** Monitoring reductions in these metrics as indicators of improved well-being.	Connor, K. M., & Davidson, J. R. T. (2003). Development of a new resilience scale: The Connor-Davidson Resilience Scale (CD-RISC). *Depression and Anxiety, 18*(2), 76–82. https://doi.org/10.1002/da.10113 Pollack, J. (2025). *Five Tools Every Organizational Leader Must Know to Improve Their Workplace Culture Today* [eBook]. Peaceful Leaders Academy. https://peacefulleadersacademy.com/wp-content/uploads/2024/06/Five-Tools-Every-Organization-Leader-Must-Know-to-Improve-Their-Workplace-Culture-Today.pdf

continued on following page

Table 6. Continued

Individual Solutions	Application Details	Measurable Outcomes	Empirical Research
Empowering Autonomy, Meaningful Purpose, and Well-Being	**Flexible work arrangements:** Offering flexible schedules and hybrid models to reduce cognitive fatigue. Astute strategic planning and positive culture enablement are essential to ensure the sustainability of these efforts, as recent implications have highlighted potential pitfalls if work-life balance opportunities are not managed efficiently, such as not aligning structured expectations, assigning reporting structure, or measurable accountabilities, regardless of the work setting—whether remote, hybrid, in-person, or a combination thereof. **Job crafting and autonomy:** Empowering employees to shape their roles and decide how to do their jobs, as field experts, and promoting a sense of community. **Meaningful work and connection to purpose:** Clearly articulate the organization's mission and how individual roles contribute to it, then facilitate employees' purposeful connectivity through coaching, mentoring, shadowing, and career pathing activities, which help foster autonomy and employee well-being.	**Productivity metrics:** Tracking improvements in efficiency and quality of work. **Employee retention rates:** Monitoring higher retention rates, which often correlate with greater autonomy and purpose. **Qualitative feedback:** Gathering insights through interviews and focus groups on employees' sense of meaningful purpose, control, and well-being, which can lead to identifying by employees or employers untapped gifts, talents, and passions.	Barua, B., & de Príncipe, J. V. (2025). How Workplace Loneliness Impacts Workplace Wellbeing?: Leader Influence in Lowering the Impact [Chapter 11]. In Barua, B. (Ed.), *Practices, Challenges, and Deterrents in Workplace Wellbeing: Strategies for Building Resilient and Thriving Workplaces*, 263–284. IGI Global Scientific Publishing. https://doi.org/10.4018/979-8-3693-6079-8.ch011 Wang, L., & Xie, T. (2023). Double-Edged Sword Effect of Flexible Work Arrangements on Employee Innovation Performance: From the Demands–Resources–Individual Effects Perspective. *Sustainability, 15*(13), 10159. https://doi.org/10.3390/su151310159

Ludvik (2023) stated that neuroplasticity activities, such as "cognitive and psychobehavioral interventions," can promote new brain and neuro pathways in higher education. This is crucial for increasing learners' abilities in creativity, compassion, and critical thinking (p. 2005). Therefore, neuroplasticity capacity-building efforts in the workplace and education systems have the potential to foster a sense of peace, self-esteem, resilience to stress and anxiety, deep sleep, and overall well-being. These benefits can lead to strong relational connections and the development of constructive social networks. Essentially, these industry-related transformational stories encourage readers to be proactive in self-development and career planning, ensuring flexibility and preparedness for whatever the future of work may hold.

Team Level Enablement

At the team level, the focus must be on the changes in team members' cognitive abilities, emotional regulation, and learning capacity to enhance dynamic collaboration, collective problem-solving, team adaptability, and psychological safety. A Gallup (2025a) report found that hybrid work has stabilized since 2022, with about half of all workplaces maintaining a hybrid model. This trend continues even as more remote-capable employees find themselves working on teams distributed across various locations. As a result, team coordination and trust have become key management challenges in this new era of flexible work, calling for a new neuroleadership-focused approach (Aithal & Satpathy, 2024; Gallup, 2025a; Saluja et al., 2024). Christ (2025) exclaimed, "[m]anagers will need to coordinate the future of work by addressing challenges such as team coordination and trust."

Hybrid work has stabilized since 2022, with about half of all workplaces maintaining a hybrid model, and the average hybrid worker now spends 46% (~2.3 days) of their week in the office, up from 42% in 2022 (Gallup, 2025a). This trend, in conjunction with a significant shift in team composition, as the percentage of remote-capable employees collaborating with colleagues across various time zones has more than doubled, with a Gallup (2025a) study indicating an increase from 13% in 2023 to 27% in 2025. These evolving workplace experiences and work-life dynamics have made team coordination and trust-building a defining management challenge in the new hybrid era (de Príncipe, 2024; Gallup, 2023; Vázquez de Príncipe, 2023). Literature uncovered that the responsibility for hybrid work schedules is split almost equally among three groups: 34% of workers make their own schedule, 35% have their schedule decided by their manager or team, and 31% have their schedule determined by leadership (Gallup, 2025a).

The author recommends that teams collectively take ownership of scheduling to provide a structured system in which everyone contributed to the decision-making process in a neuroplastic exercise. Research indicates that when teams, rather than individuals, establish shared norms for a hybrid work business model, team members are reported to be more productive and less anxious, while feeling a sense of belonging in the workplace (Algumaei et al., 2023; Gallup, 2023; Forget & Le Pertel, 2024). It is essential to emphasize that the team leader's management style significantly impacts the team's success, which requires alignment with the organizational structure and business objectives, while maintaining an empathetic, appreciative, and trustworthy reciprocal approach (de Príncipe, 2024; Vázquez de Príncipe, 2023). Essentially, leading with empathy has the most significant potential to enhance team performance and productivity among remote and hybrid workers (Aithal & Satpathy, 2024; Gallup, 2023; Saluja et al., 2024).

Table 7. Team Level Workplace Neuroplasticity Solutions Matrix.

Team Solutions	Application Details	Measurable Outcomes	Empirical Research
Enhanced collaboration and team adaptability	**Team dynamics:** Once optimal resources have been aligned to a team, identify the relational reciprocal trust dynamics, management style, levels of productivity, and anxiety associated with cross-functional interactions, inter-team disposition, and capacity imbalances. In hybrid settings, managing coordination and trust is critical for success.	**Collaboration metrics:** Frequency and quality of interactions, successful handoffs, and a reduction in conflicts.	Algumaei, M., Hettiarachchi, I. T., Farghaly, M., & Bhatti, A. (2023). The neuroscience of team dynamics: exploring neurophysiological measures for assessing team performance. *IEEE Access, 11*, 129173–129194. https://doi.org/10.1109/ACCESS.2023.3332907 Forget, M., & Le Pertel, N. (2024). Enhancing neuroplasticity and promoting brain health at work: The role of learning and memory in workplace performance. In *Learning and Memory-From Molecules and Cells to Mind and Behavior.* IntechOpen. https://doi.org/10.5772/intechopen.1002894
	Team functioning: Leverage software, dashboards, and AI agents to monitor return on investment, business value proposition, and appreciation at work.	**Team performance metrics:** Project completion rates, adherence to deadlines, quality of deliverables, and efficiency gains in team tasks.	Gee, R. O. W. (2025b). From neurons to organizations: awakening regenerative mindsets with neuroplasticity, AI & systemic consciousness. *Journal of Biotechnology & Biomedicine, 8*, 21–58. https://doi.org/10.26502/jbb.2642-91280178 Ooi, R. W. G. D. (2025). *From neurons to organizations: Awakening regenerative mindsets through neuroplasticity, AI, and systemic consciousness.* https://doi.org/10.26502/jbb.2642-91280178

continued on following page

Table 7. Continued

Team Solutions	Application Details	Measurable Outcomes	Empirical Research
	Team composition: Leverage psychometric tools to understand multigenerational team effectiveness and efficiency in meeting internal and external stakeholders' or customers' needs.	**Team Surveys:** Assessments of psychological safety, collective efficacy, shared mental models, work setting, strengths, talents, and perceived adaptability.	Rajagopalan, J., & Jacob, A. M. (2025). Optimizing Learning Across Ages: Insights from Neuroplasticity for Corporate Reskilling and Upskilling. In Sustainable Digital Technology and Ethics in an Ever-Changing Environment: Volume 1 (pp. 675–690). Cham: Springer Nature Switzerland. https://doi.org/10.1007/978-3-031-86712-5_57 Harlianto, J., Rudi, R., & Ginting, H. C. B. G. (2025). Building Collaboration in a multigenerational workforce: Intergenerational inclusivity and mental health as key drivers of workplace productivity. In *Enhancing Workplace Productivity Through Gender Equity and Intergenerational Communication* (pp. 283–316). IGI Global Scientific Publishing. https://doi.org/10.4018/979-8-3373-2903-1.ch011
	Network Analysis: Map communication patterns and information flow to identify key "hubs" of learning and adaptation. Facilitates identifying opportunities for shadowing between teams. Gaps in resourcing and redundancy patterns can be determined.	**Ecosystem diagramming:** Visual mapping of communication patterns and information flow between collaborative teams.	Aithal, P. S., & Satpathy, C. P. D. J. (2024). Exploring neuro management: bridging science and leadership–an overview. *International Journal of Applied Engineering & Management Letters, 8*(2), 39–73. ISSN: 2581–7000. Saluja, S., Kukreja, J., & Sharma, S. (Eds.). (2024). *Building Organizational Resilience with Neuroleadership.* IGI Global Scientific Publishing. ISBN: 9798369317860.

The Global Workplace Perspective

Gallup's 2025 State of the Global Workplace Report highlights six key insights on individual and team well-being. First, there is a decline in global well-being, indicating that global life evaluations have fallen to 33% thriving (Gallup, 2025a). Older and female managers have experienced the most significant drops in well-being, with stress, sadness, and loneliness being prevalent among employees. Stress affects 40% of the global workforce (Gallup, 2025a). The impact of engagement on well-being reveals that engaged employees are more likely to thrive in life overall, reporting fewer negative emotions, such as stress and sadness. Specifically, half of engaged employees thrive, compared to only a third of disengaged employees.

Second, manager burnout is increasing, indicating managers' well-being has declined significantly, with burnout leading to absenteeism, turnover, and reduced team performance. Specifically, female managers saw a 7-point drop in well-being, and older managers experienced a 5-point decline. Third, the Gallup (2025a) study shows regional variations in individual and team well-being, with Australia and New Zealand having the highest percentage of thriving employees (56%), while South Asia has the lowest (15%), and, to no surprise, stress levels are highest in regions like the United States and Canada (50%).

Fourth, the role of training and development is rapidly evolving due to the integration of AI tools into the workplace, which varies depending on the specific industry and disciplines. Manager training and development must focus on significantly improving well-being, engagement, and productivity, leveraging scenario-based and simulation exercises to foster deep, critical, and systems thinking aligned with business objectives. Engagements must facilitate the practical application of AI tools for identified roles at risk of becoming obsolete, while providing reskilling and upskilling opportunities for individuals and teams to remain relevant in the evolving future of work. Recommended organizational development (OD) strategic initiatives: career pathing and job crafting (SHRM, 2025). Hence, Gallup (2025a) emphasizes that managers who receive training report higher levels of thriving, and their teams benefit from increased engagement.

Fifth, daily negative emotions inform toxic workplace behaviors when left unaddressed. Gallup (2025a) highlights that stress, anger, sadness, and loneliness are prevalent among employees globally, with stress being the most reported emotion. These negative emotions foster a sense of isolation and disconnection, which are associated with reduced workplace engagement and overall life satisfaction. Sixth, a deteriorating workplace environment is a common denominator for declining engagement and well-being, which impacts the ability of teams to enable neuroplasticity and negatively affects productivity (Qualtrics, 2025; SHRM, 2025). Therefore, it is imperative to intentionally address manager engagement and provide support to reverse these trends by instilling holistic leadership that nurtures a positive culture, thereby restoring teams' psychological safety and organizational health levels, which enables neuroplasticity (de Príncipe, 2024; McKinsey, 2024; Picchi, 2025; SHRM, 2025; Vázquez de Príncipe, 2023).

Trending Work-Life Dynamics: Remote and Hybrid Work Models

The *Workplace Q2 Remote Work Trends Study* conducted by Gallup (2025b) surveyed 17,660 U.S. workers to analyze remote and hybrid work trends. The key findings reveal that 54% of managers trust their teams' productivity while working remotely, highlighting confidence in remote work arrangements. These findings underscore the growing adoption of remote and hybrid work models and their influence on workplace strategies. The study also examines hybrid work preferences, revealing their increasing adoption across various industries. Data was weighted to reflect U.S. workforce demographics, ensuring reliability. These insights provide valuable guidance for organizations adapting to evolving workplace models. Overall, the Gallup (2025b) study highlights the growing acceptance and trust in remote and hybrid work models, particularly among managers, while providing reliable demographic insights into organizational trending solutions.

Organizational Level Enablement

At the organizational level, the focus must be multifaceted, addressing global, internal, and external changes that significantly impact business and workforce dynamics. In turn, it affects overall productivity, innovation outcomes, market responsiveness, employee retention, and financial performance levels (Behera et al., 2022; Khaneja & Arora, 2024). Additionally, Al Halyan et al. (2019) exclaimed that organizational success is achievable by leveraging neuroplasticity. Furthermore, Behera et al. (2022) explained that there is a "mediating factor in the relationship between cognition and neuroplasticity in investors in investment decision making" (p. 21). Moreover, Khaneja and Arora (2024) stated that there is potential for neuroscience to play a significant role in transforming business for the future of work in the modern workplace. Ultimately, the neuroplastic organizations will be the ones with staying power, no matter what comes next in the world of business.

Table 8. Organizational Level Workplace Neuroplasticity Solutions Matrix.

Organizational Solutions	Application Details	Measurable Outcomes	Empirical & Conceptual Research
Workforce productivity	A focus on increasing the output and efficiency of the workforce. This involves streamlining processes, providing necessary tools, and training, and setting clear performance goals.	Output per employee, revenue per employee, and efficiency ratios like cost per unit produced.	Forget et al., 2024; Ooi, 2025
Innovation output	Initiatives aimed at fostering a culture of creativity and the successful development and implementation of innovative ideas. This includes encouraging experimentation and providing resources for research and development.	Metrics to consider are the number of new products/ services launched, patent applications, revenue from new ventures, time-to-market for innovative ideas, and the number of implemented employee suggestions.	Al Halyan et al., 2019; Gee, 2025.
Market responsiveness	The ability of the organization to adapt to changes quickly and effectively in the market, customer demands, and competitive landscape indicates sustainable resilience. This includes strategic flexibility and operational agility.	Time to recover from market shocks and successful pivots in strategy.	Cannavacciuolo et al., 2017; Saluja et al., 2024.
Employee retention	Strategies focused on reducing voluntary turnover and retaining valuable employees. This includes competitive compensation, professional development opportunities, and a positive work environment.	Monitoring of voluntary turnover and absenteeism rates as indicators of impending mass workforce exits. Proactively leverage "stay interviews" with high-potentials and rainmakers.	Cannavacciuolo et al., 2017; Saluja et al., 2024
Talent management	The systematic process of attracting, developing, and retaining high-performing employees. This involves career pathing, succession planning, and internal mobility programs.	Retention rates of high-performing employees and success rates of internal mobility programs.	de Príncipe, 2021; Faw, 2025

continued on following page

Table 8. Continued

Organizational Solutions	Application Details	Measurable Outcomes	Empirical & Conceptual Research
Financial performance	The overall financial health and success of the organization is often an indirect result of improvements in other areas.	Scorecards and dashboards facilitate monitoring to track revenue growth, profitability, risks, and ROI. Measuring internal and external influences; leverage the PESTLE Analysis to synthesize impacts.	Khaneja & Arora, 2024; Behera et al., 2022
Engagement and culture assessments	Efforts to cultivate a positive organizational culture promote growth, learning, and employee well-being. This is often measured through surveys and feedback mechanisms.	Organization-wide perception survey results indicate a growth mindset, a learning orientation, and enhanced psychological well-being.	Barua & Vázquez de Príncipe, 2025; Staneiu, 2023

The Great Employee Detachment

Research insights emphasize the importance of addressing employee engagement, offering managerial support, and fostering a positive workplace culture to promote overall well-being (Qualtrics, 2025; SHRM, 2025). Specifically, employees feel detached due to issues in organizational culture, leadership communication, resource investment, and performance management. Gallup's (2024) study discovers that many employees feel disconnected from their workplaces, with 32% describing their environment as isolated or impersonal, 44% of Gen Z workers specifically feeling a lack of emotional connection, and 41% of remote employees report a lack of cohesion in organizational culture. Leadership issues exacerbate this sense of detachment, as 29% of employees report a lack of clear, honest, and consistent communication, highlighting a desire for transparent, two-way trust and fewer top-down directives (Gallup, 2024). Furthermore, a quarter of employees feel their organizations under-invest in people and resources, citing inadequate pay, tools, and staffing. Finally, Gallup (2024) states that a significant number of employees, 14%, believe there is a lack of constructive feedback, appreciative recognition, and career management opportunities, indicating a need for more performance management accountability and clear growth paths in their roles.

Neuroplasticity Connections with Organizational Health and Employee Well-Being

Neuroplasticity, the brain's ability to reorganize neural connections, has a profound impact on individual psychology, organizational health, and employee well-being. In psychology, it explains how learning, experiences, and therapies like cognitive behavioral therapy (CBT) can literally "rewire" the brain, foster cognitive flexibility, and aid in mental health recovery from conditions such as anxiety and post-traumatic stress disorder (PTSD) (Psychology Today, 2024, 2025). This inherent adaptability also underpins resilience, as healthy habits and stress management can enhance the brain's ability to cope with adversity (de Príncipe, 2024).

For organizational health, a neuroplastic workforce is essential for navigating dynamic environments, especially with the integration of AI (Gee, 2025b; ISG, 2025; McKinsey, 2024; Ooi, 2025). Such a workforce demonstrates greater adaptability to change, quickly acquiring new skills and adopting modern technologies, which are vital for continuous learning. This promotes innovation and problem-solving, as employees become more open to experimentation and diverse perspectives. Organizations with high collective neuroplasticity are more resilient, capable of "bouncing forward" from disruptions and finding opportunities amid chaos.

Furthermore, neuroplasticity has a direct impact on employee well-being (Barua & Vázquez de Príncipe, 2025). Practices that promote brain adaptability, such as mindfulness and adequate sleep, are essential for stress management and emotional regulation (Peaceful Leaders Academy, 2025; Pollack, 2025). When employees are empowered to learn and grow, their job performance improves, and they experience a greater sense of confidence and competence.

Ultimately, fostering neuroplasticity creates a culture of trust and psychological safety, where employees feel supported in their development, contributing to both individual flourishing and organizational success. The brain's remarkable ability to reorganize itself by forming new neural connections throughout life has profound implications for individual psychology, organizational health, and employee well-being (Barua & Vázquez de Príncipe, 2025; McKinsey, 2024). It is the biological foundation for learning, adapting, and recovering from challenges.

Neuroplasticity and Psychology

In psychology, neuroplasticity provides the scientific basis for how experiences, learning, and therapeutic interventions can literally "rewire" the brain. For example, having cognitive flexibility means being able to adapt thinking patterns, shift perspectives, and learn new skills. Engaging in novel and challenging activities enhances neuroplasticity, promoting this flexibility (Positive Psychology, 2018).

Several mental health interventions associated with neuroplasticity are foundational to the effectiveness of therapies such as Cognitive Behavioral Therapy (CBT) and Eye Movement Desensitization and Reprocessing (EMDR). By consciously changing negative thought patterns or reprocessing traumatic memories, individuals can create new, healthier neural pathways, helping to overcome conditions such as anxiety, depression, and PTSD (Psychology Today, 2024; 2025; Sunshine Coast Psychology Clinic, 2025). Resilience, the brain's capacity to adapt to adverse conditions, is linked to neuroplasticity. Healthy habits (exercise, diet, sleep) and stress management techniques can promote neuroplasticity, enhance resilience, and the ability to cope with stress (Bell-Ellis, 2013; Henry et al., 2017; Mave Health, 2024).

Neuroplasticity and Organizational Health

For organizations, a workforce with high neuroplasticity is inherently more adaptable and resilient, directly contributing to overall organizational health (Bell-Ellis, 2013; McKinsey, 2024), particularly in dynamic environments. In an era of rapid technological shifts, such as the integration of AI, organizations require employees who can quickly learn new processes, adopt new tools, and adapt to new ways of working. Neuroplasticity is the key to this continuous learning and adaptation (Gee, 2025b; ISG, 2025; Ooi, 2025). Innovation and problem-solving systems, as well as design thinking approaches, are accelerated in a neuroplastic organization, which is more open to experimentation, learning, risk-taking, critical thinking, and diverse perspectives (Drobin et al., 2025; Ludvik, 2023). This fosters a positive culture where employees are better equipped to identify challenges, propose creative solutions, and drive innovation (Forbes, 2024).

Organizational resilience is evident when the ecosystem is healthy (Bell-Ellis, 2013). Just as individuals can "bounce back" (resilience) and "bounce forward" (adaptability), organizations with neuroplastic workforces are better positioned to withstand disruptions and thrive in uncertainty (de Príncipe, 2024; Marín, 2025). This collective brain adaptability allows for quicker recovery and the ability to find opportunities amidst chaos (Hannah Brown, 2024). Subsequently, this promotes a culture that encourages neuroplasticity, thereby reducing resistance to change organically. Leaders can help employees view large-scale transformations as achievable challenges and learning opportunities, thereby reducing resistance to modern technologies and processes (ISG, 2025).

Neuroplasticity and Employee Well-being

Literature indicates that there is a link between neuroplasticity and employee well-being that is direct and significant. Building stress management skills is critical for building personal resilience, as chronic stress can negatively impact neuroplasticity. Conversely, practices that promote neuroplasticity, such as mindfulness, adequate sleep, and physical activity, can help manage stress, improve emotional regulation, and enhance overall mental well-being (Mave Health, 2024; Frontiers, 2023; Peaceful Leaders Academy, 2025; Pollack, 2025). Establishing a connection between job performance and personal growth is essential for achieving job satisfaction and self-realization.

Therefore, providing opportunities for continuous learning and skill development stimulates neuroplasticity, resulting in enhanced job performance, increased productivity, and improved career progression. Employees who can adapt and grow feel more confident and competent (Aithal & Satpathy, 2024; Mega Lifesciences, 2024). Cultivating a sense of belonging is grounded in the principles of positive psychology, promoting psychological safety in the workplace. Leading healthy organizations that foster holistic, positive cultures that empower neuroplasticity often do so by creating environments of trust, open communication, and psychological safety (Bell-Ellis, 2013). When employees feel safe to take risks, make mistakes, and learn, their brains are more receptive to new connections and learning, directly impacting their well-being (Barua & Vázquez de Príncipe, 2025; ISG, 2025; Saluja et al., 2024). Hence, neuroplasticity is not just a biological concept; it is a fundamental aspect that underpins an individual's psychological health, directly influences an organization's capacity for adaptation and innovation, and is crucial for the holistic well-being of its employees in an ever-changing world (Barua & Vázquez de Príncipe, 2025; Vázquez de Príncipe, 2024a; Qualtrics, 2025).

FUTURE RESEARCH DIRECTIONS

Practical Implications for Researchers

Five essential practical implications for researchers, anchored in neuroplasticity as a means to optimize the future of work, are emphasized herewith. First, developing standardized neuroplasticity metrics is necessary to create validated, accessible, and ethically sound tools designed explicitly for organizational settings to promote healthier work-life dynamics, workplace experiences, and overall climate wellness (de Príncipe, 2024; Vázquez de Príncipe, 2023). Second, leveraging AI and Machine Learning in data analysis for advanced analytics is needed to identify patterns in large datasets, potentially correlating behavioral proxies with subtle neuroplastic changes (Gee, 2025b; ISG, 2025; Ooi, 2025).

Third, research designed with personalized interventions is needed to explore how neuroplasticity principles can inform individualized development plans for employees, thereby optimizing their cognitive and emotional well-being (Barua & Vázquez de Príncipe, 2025; Vázquez de Príncipe, 2024b; Qualtrics, 2025). Fourth, designing and evaluating neuro-leadership training OD interventions for integration into leadership development programs that teach and coach leaders how to cultivate neuroplasticity in themselves and their teams, thereby enhancing work-life dynamics and organizational health (AACSB, 2024; de Príncipe, 2024; Long, 2006; McKinsey, 2024; Torch, 2023; Vázquez de Príncipe, 2023). Fifth, more long-term impact research studies are needed on the sustained effects of neuroplasticity-informed practices on organizational longevity and competitive advantage.

Practical Implications for Practitioners

There are three practical implications for future research directions that practitioners should consider. First, partner with organizations to galvanize leadership by starting with observable proxies and pilot programs that incorporate all workplace settings, including in-person, remote, and hybrid environments (Gallup, 2023; Neuro Business School, 2024). Second, become an ambassador or an advocate for a data-driven, holistic approach to human capital development within your organization, professional association, or communities (Vázquez de Príncipe, 2024b). Lastly, promoting organizational leadership that fosters a neuroplastic culture is a strategic investment that would boost the bottom line by increasing individual, team, and organizational efficiency and effectiveness at all levels (AACSB, 2024; Long, 2006; Neuro Business School, 2024).

Methodological Considerations for Robust Measurement

There are seven recommended methodologies for researchers and practitioners to consider for robustly measuring the impact of neuroplasticity in today's modern workplace. First, conducting longitudinal studies is an essential component for tracking neuroplastic changes and their lagged impact over time. Second, incorporating pre- and post-intervention measurements across extended periods. Third, leveraging control groups when implementing interventions, such as neuroplasticity-informed training, in experimental groups versus control groups can help to establish causality.

Fourth, a mixed-methods approach is instrumental in combining quantitative data, such as performance metrics and surveys, with qualitative insights derived from interviews, focus groups, and case studies to provide a richer understanding of the "how" and "why" factors. Fifth, the multi-source data collection methodology enables the gathering of data from various sources within the complex organizational ecosystem, including HR systems, performance reviews, project management software, employee surveys, and external market data (Aithal & Satpathy, 2024; Long,

2006; Marín, 2025; Saluja et al., 2024; Vázquez de Príncipe, 2024b). Sixth, the proxy variable selection technique emphasizes the careful selection and validation of proxy variables that are theoretically linked to neuroplastic changes and have a demonstrable impact on organizational outcomes.

Seventh, ethical considerations must always be maintained when addressing privacy, informed consent, and the responsible use of neuroscience-informed assessments in the workplace. Eighth, transparency in data collection and usage is paramount. Lastly, utilizing advanced statistical analytical techniques, such as regression analysis, structural equation modeling, and time-series analysis, should be considered to identify significant relationships and control for confounding neuroplastic variables, thereby facilitating the in-depth interpretation of research findings and outcomes.

Proposed Open Research Questions for Exploration

The author recommends that future research should focus on standardizing measurement, exploring ethical frameworks, and developing universally accessible neuroplasticity-based interventions. In *Table 6*, the matrix identifies future study opportunities that can be leveraged by addressing the open research questions below to advance the field of neuroplasticity within organizational contexts. Furthermore, the research value propositions in this matrix indicate the 'why' these research questions need to be explored for expanding the industry-related body of knowledge.

Table 9. Open Research Questions.

Question	Research Value Proposition
How can standardized, longitudinal measurement protocols for neuroplasticity-based interventions be developed and validated in diverse organizational settings?	Standardization is essential for comparing outcomes, scaling interventions, and ensuring scientific rigor across industries and cultures.
What are the long-term ethical and organizational impacts of integrating neurodata and AI-driven tools in workforce management?	Understanding these impacts is crucial to striking a balance between productivity gains and considerations of privacy, fairness, and employee well-being.
How can universally accessible frameworks for neurodiversity and neuroplasticity be designed to benefit both neurodiverse and neurotypical employees?	Inclusive culture and holistic leadership frameworks, such as Business Resource Groups (BRG) and Digital Employee Experience (DEX) strategies, can maximize innovation and well-being, ensuring that interventions do not inadvertently exclude or commodify certain groups.

CONCLUSION

Neuroplasticity presents a transformative pathway for enhancing organizational effectiveness and productivity. However, its full potential will only be realized through rigorous, inclusive, and ethically grounded research and practice. In the case of Dr. Wilms (2023), the HR Auditor at Top Employers Institute, emphasized that neuroplasticity can play a crucial role in employee development, productivity, and overall well-being (Qualtrics, 2025). This neuroplastic-centered approach was operationalized through strategic, holistic frameworks at Coca-Cola Europacific Partners, incorporating the DEX strategy and BRG practices to transform business onboarding with holistic employee life cycle systems and processes (Wilms, 2023).

The factors that have the potential to influence neuroplastic-minded approaches in today's workplace include system processes such as collaborative business value experiences, continuous learning, and positive environmental stimuli. Empirical research has shown that providing opportunities for employees to learn and develop new skills can stimulate neuroplasticity and enhance brain function triggered by intellectual stimulation, which increases self-realization, motivation, and engagement while reducing apathy, disengagement, and job dissatisfaction (SHRM, 2025; Vázquez de Príncipe, 2021; Zhelnina, 2024). As a result, it has the potential to lead to improved performance and creativity in the workplace, as demonstrated by Wilms (2023) through the integration of employees' innovative design ideas and cultural feedback into cohesive onboarding business solutions. It is important to understand that neuroplasticity refers to the brain's ability to reorganize itself by forming new neural connections throughout our lives.

Neuroplasticity Connections to Resilience, Perfectionism, and Personal Growth

Neuroplasticity, the brain's ability to reorganize itself through the formation of new neural connections, is fundamentally linked to resilience, perfectionism, and personal growth. It serves as the biological mechanism enabling our capacity for change and adaptation throughout life. In terms of resilience, neuroplasticity allows the brain to form new pathways for coping with stress and adversity, facilitating functional recovery and enhancing psychological robustness. Practices such as mindfulness and exercise actively promote this brain flexibility, strengthening an individual's ability to "bounce back."

For perfectionism, which often involves rigid thought patterns, neuroplasticity offers a pathway to reshape these ingrained habits. By consciously engaging in new behaviors and reframing expectations, individuals can weaken unhelpful neural connections and build healthier ones, moving from an "all-or-nothing" mindset to one that embraces growth. This process enables greater cognitive flexibility and reduces the maladaptive aspects of perfectionism.

Regarding personal growth, neuroplasticity is the cornerstone of continuous learning, skill acquisition, and self-improvement. Every new experience, learned skill, or positive habit adopted involves physical changes in the brain's structure and function. The more consistently new behaviors are practiced, the stronger their corresponding neural pathways become. This understanding empowers individuals to actively shape their cognitive and emotional responses actively, thereby fostering ongoing development in areas such as emotional intelligence, critical thinking, and adaptability (Drobin et al., 2025; Ludvik, 2023).

The literature on neuroplasticity highlights that our potential for transformation is lifelong, driven by our experiences and intentional efforts. Neuroplasticity, the brain's incredible ability to adapt and reorganize itself by forming and strengthening new neural connections, is deeply intertwined with resilience, perfectionism, leadership, and personal growth. It provides the biological mechanism by which our experiences, efforts, and even our mindsets can shape our brains and, consequently, our behaviors and abilities to reach our fullest potential in life.

Neuroplasticity and Resilience Alignment

Resilience is the capacity to cope with, adapt to, and recover from adversity. Neuroplasticity is the fundamental biological process that enables this process. When individuals face challenges or stress, their brains can form new neural pathways to process these experiences, develop new coping mechanisms, and learn from the situation (Journal Agent, 2024). This "rewiring" enables functional recovery and adaptation, making the brain more flexible and resilient in the face of future stressors (Wang & Xie, 2023). Practices like mindfulness, regular exercise, and learning new skills actively promote neuroplasticity, thereby enhancing psychological resilience and stress management (Aithal & Satpathy, 2024; Mave Health, 2024; Saluja et al., 2024; UCLA Health, 2016).

Neuroplasticity and Perfectionism Alignment

Perfectionism often involves rigid thought patterns, an intense fear of failure, and an inability to accept anything less than flawless outcomes. This cognitive rigidity can be understood through the lens of established, unhelpful neural pathways. However, neuroplasticity offers a pathway to overcome maladaptive perfectionism. By consciously engaging in new behaviors and thought patterns—such as practicing self-compassion, reframing expectations, accepting mistakes as learning opportunities, and embracing novelty—individuals can weaken old, rigid neural connections and form new, healthier ones (Perfectionism Rewired Podcast, 2025; Kent, 2023). This process allows for a shift from an "all-or-nothing" mindset to one that embraces growth and imperfection.

Neuroplasticity and Personal Growth Alignment

Personal growth, encompassing continuous learning, skill acquisition, and self-improvement, is directly enabled by neuroplasticity. Every new experience, every skill learned, and every habit formed involves the creation or strengthening of neural pathways (Boyd, 2015). Whether it is learning a new language, a musical instrument, or a complex professional skill, the brain physically changes to accommodate this new knowledge. The more a new behavior or skill is practiced, the stronger these neural connections become (Physiopedia, 2023). This habit formation, through neuroplasticity, explains how consistent actions become automatic responses.

By deliberately choosing and repeating positive habits, individuals can "wire" these behaviors into their brains. Hence, self-improvement is the understanding that the brain is not fixed, but rather malleable throughout life, empowering individuals to shape their cognitive and emotional responses actively. This includes improving emotional intelligence, critical thinking, and adaptability by engaging in challenging activities and reflecting on experiences (Drobin et al., 2025; Ludvik, 2023). Even experiencing struggle and difficulty during practice is crucial for neuroplastic change, as it forces the brain to adapt and find innovative solutions (Boyd, 2015). Hence, neuroplasticity is the biological engine that drives our capacity for change, making resilience achievable, offering a path to reshape perfectionistic tendencies, and serving as the foundation for lifelong personal development.

The Measurable Advantage Consideration of the Powerful Adaptive Brain

While challenging, measuring the impact of neuroplasticity on organizational performance is vital for validating its strategic importance. The author reiterates that this measurement, though indirect, provides the necessary empirical evidence to justify investments in neuroplasticity-informed organizational development initiatives. Conversely, the research reinforces the transformative potential of organizations that understand, cultivate, and measure the adaptability of the human brain, ensuring their resilience and prosperity in the dynamic future of work (Geraerts, 2018). Moreover, the author provides two practicum simulation opportunities, depicted in *Figures 4* and *5* below, as potential neuroplasticity-informed pilot case studies of scenario-based holistic culture initiatives, namely. Example One is *Measuring the*

Impact of a "Values-Driven Leadership" Initiative, and Example Two is *Measuring the Effectiveness of a "Well-Being and Psychological Safety" Initiative.*

Neuroplasticity-Informed Scenario-Based Holistic Culture Simulations

Pilot Study Example One

Figure 4. Measuring the Impact of a "Values-Driven Leadership" Initiative.

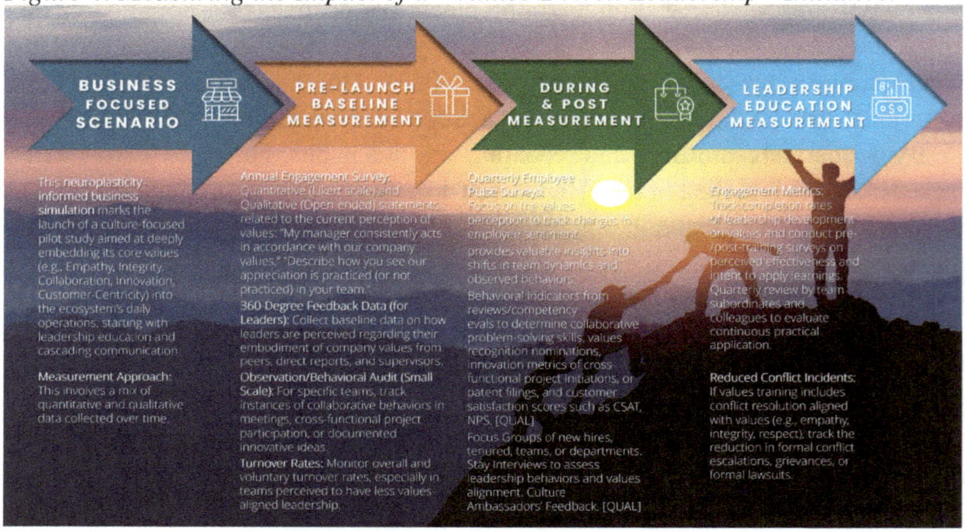

Pilot Study Example Two

Figure 5. Measuring the Effectiveness of a "Well-Being and Psychological Safety" Initiative.

Key Themes and Arguments

The research identified six key themes and arguments for measuring neuroplasticity in the workplace. First, a significant gap exists, despite increasing conceptual interest, in the literature regarding how to quantify the tangible impact of neuroplasticity initiatives on specific organizational performance metrics. Second, while this chapter's effort to establish clear theoretical pathways connecting individual brain adaptability to measurable organizational outcomes, such as productivity, innovation, and resilience, is essential, the link between biology and business impact needs case studies to validate the conceptual framework (de Príncipe, 2024). Third, using a multi-level measurement approach requires employing a comprehensive method that considers individual, team, and organizational levels to fully capture the influence of neuroplasticity on enhancing organizational performance and workforce productivity.

Fourth, leveraging a Proxy-Based Assessment acknowledges the difficulty of direct neural measurement in the workplace and proposes the use of robust, validated behavioral and performance proxies. Fifth, optimizing methodological rigor emphasizes the need for robust research designs, particularly longitudinal and mixed methods approaches, to establish credible evidence. Lastly, operationalizing a strategic imperative intervention can elevate the position for measuring the impact of neuroplasticity as a strategic imperative for organizations seeking a competitive edge in a hyper-competitive global landscape, as illustrated in the example simulations above.

The Neurodiverse Forgotten Population: Harnessing Neuroplasticity to Strengthen EF

Kietzman (2024), an educator and learning institution administrator, emphasized that although the neurodiverse population has long been overlooked, significant progress has been made in recent years to break the stigma surrounding it. Business models have emerged, providing systemic instruction and tools that develop educators, families, and professionals (Lim et al., 2017; Ludvik, 2023). Renowned organizations such as Google, Hewlett-Packard, and Caterpillar have begun focusing on empowering the neurodiverse population by leveraging these emerging business models—for example, replicating them to support the workforce's executive functions (EF) capabilities to facilitate learning within neuroplastic-enabled workplaces (Kietzman, 2024).

Neuroplasticity, the brain's natural ability to adapt by strengthening existing neural pathways and forming new ones, offers significant potential for improving executive functions vital in the workplace, including perception, memory, attention, and decision-making. For individuals, especially those who are non-neurotypical (e.g., ADHD or Autism), actively engaging in new and demanding environments encourages brain rewiring, resulting in improved cognitive flexibility, planning, and problem-solving (Neuro Frontiers, 2025; Sandoval, 2025). In particular, Neuro Frontiers (2025) emphasized that there are many neuro-myths related to ADHD, such as the brain cannot change and there is only one type; however, research shows the opposite is true. In fact, effective neuroplasticity therapies for ADHD incorporate sensory and movement-based holistic interventions that target brain health practices (Neuro Frontiers, 2025). The key is always to treat the symptoms, not the diagnosis label, which is why inclusive initiatives must be multifaceted and integrated into work-life dynamics, rather than affinity-based, as this can trigger isolation and harm individuals' well-being.

Consequently, in a workplace setting, promoting neuroplasticity helps create a more adaptable and high-performing workforce. Activities that support this include regular physical and mental exercise, which boosts neurotrophic factors vital for neuronal health and function, thereby enhancing sustained focus and coordination in demanding roles (Sandoval, 2025). Therefore, providing various workspaces as an option is highly recommended. The ability to visit other offices or build communities, such as arts/crafts or walking/running clubs, promoting volunteer community projects, and team sports competitions are effective practices, as well as team innovation challenge awards. Participating in artistic expression fosters creativity, discipline, and stress relief, all of which are valuable for innovative problem-solving and improving employee well-being (Barua & Vázquez de Príncipe, 2025; Qualtrics, 2025). Carefully selected video games can serve as engaging practice for planning and executive functions, directly applicable to complex projects (Edge Foundation, 2024).

In closing, the author emphasizes that prayer, along with other holistic practices such as mindfulness and meditation, enhances concentration and self-awareness, which are crucial for attentional control and emotional regulation in high-pressure work environments. Language learning is a powerful mental stimulant, boosting working memory, problem-solving, and task-switching skills, all of which benefit diverse roles and global teams (Psychology Today, 2024, 2025). The author also encourages readers to leverage the power of deep focus music therapy videos while studying, analyzing, working, or writing sessions, such as 14 Hz Beta, Alpha, or Gamma Brain Waves, to enhance concentration and maintain extended focus periods (Jiao, 2025). Therefore, incorporating a variety of these neuroplasticity activities into one's daily routines and work practices consistently provides the repeated brain stimulation needed to strengthen and grow neural pathways. This intentional focus on neuroplasticity can lead to significant improvements in managing cognitive demands and overall work performance within team dynamics (Algumaei et al., 2023; Gallup, 2025a, 2025b; Psychology Today, 2024, 2025). Lastly, the section below provides a guided reflection exercise to explore neuroplasticity and workplace enablement at the individual, team, and organizational levels, which can empower the neurodiverse populations and beyond.

REFLECTION QUESTIONS

Empowering The Neurodiverse Forgotten Population Together

1. How do societal norms and misperceptions contribute to the challenges faced by neurodiverse individuals in the workforce?
2. Contemplate on your workplace ecosystem and team dynamics. What strategies could be employed at your organization or team to combat these challenges?
3. In what ways could the educational system be reformed to better support the transition of neurodiverse students from K-12 education to higher education or the workforce?
4. Are there opportunities where you can collaborate with others to bring positive change?

5. Analyze the potential impacts of incorporating *disability consultancy groups* within businesses to help train employers, leaders, or managers on the strengths of neurodiverse individuals. How might this intervention transform the employment landscape for the neurodiverse community?
6. Reflect on your own educational or workplace experiences. Have you observed or been part of practices that either support or hinder the inclusion and success of neurodiverse individuals?

ACKNOWLEDGEMENT

This research received no specific grant from any funding agency in the public, commercial, or not-for-profit sectors.

REFERENCES

AACSB. (2024, February 28). *The role of quantum leadership in a complex world.* https://www.aacsb.edu/insights/articles/2024/02/the-role-of-quantum-leadership-in-a-complex-world

Agile by Design. (2024, March 8). *Organizational agility - A complete guide + best examples in 2024.* https://www.agilebydesign.com/blog/organizational-agility

Aithal, P. S., & Satpathy, C. P. D. J. (2024). Exploring neuro management: Bridging science and leadership–an overview. *International Journal of Applied Engineering & Management Letters, 8*(2), 39–73. ISSN: 2581–7000.

Akşahin, H., Dagli, G., Altınay, F., Altınay, Z., Altinay, M., Soykurt, M., Bahçelerli, N., & Adedoyin, O. (2023). Contributions of Neuroleadership to the School Administrator and Teachers for the Development of Organizational Behavior. *Sustainability.* Advance online publication. DOI: 10.3390/su152115443

Al Halyan, D. A., Al Badi, A., & Al Hosani, B. (2019, November). Applying brain science to achieve organizational success. In *Abu Dhabi International Petroleum Exhibition & Conference* (p. D031S093R004). SPE. https://doi.org/DOI: 10.2118/197532-MS

Algumaei, M., Hettiarachchi, I. T., Farghaly, M., & Bhatti, A. (2023). The neuroscience of team dynamics: Exploring neurophysiological measures for assessing team performance. *IEEE Access: Practical Innovations, Open Solutions, 11*, 129173–129194. DOI: 10.1109/ACCESS.2023.3332907

Appelbaum, S., & Gallagher, J. (2000). The competitive advantage of organizational learning. *Journal of Workplace Learning, 12*, 40–56. https://www.doi.org/10.1108/13665620010316000

Ashkanasy, N., Becker, W., & Waldman, D. (2014). Neuroscience and organizational behavior: Avoiding both neuro-euphoria and neuro-phobia. *Journal of Organizational Behavior, 35*, 909–919. DOI: 10.1002/JOB.1952

Ateeq, K., Oswal, N., Jawabri, A., Masaeid, T., Alquqa, E., Basha, S., & Alami, R. (2025). The transformative impact of Artificial Intelligence (AI) on Organizational Behaviour (OB): A study of employee engagement, performance, and ethical implications. *Journal of Posthumanism*. https://doi.org/DOI: 10.63332/joph.v5i4.1221

Badi, A. (2019). Applying Brain Science to Achieve Organizational Success. Day 3 Wed, November 13, 2019. https://doi.org/DOI: 10.2118/197532-ms

Bakhtiari, H. (2025). *Harnessing Neuroplasticity: Evidence-Based Approaches to Behavioral Modification in Contemporary Society*. https://doi.org/DOI: 10.20944/preprints202505.0235.v1

Barua, B., & Vázquez de Príncipe, J. (2025). How Workplace Loneliness Impacts Workplace Wellbeing?: Leader Influence in Lowering the Impact. [Chapter 11] In Barua, B. (Ed.), *Practices, Challenges, and Deterrents in Workplace Wellbeing: Strategies for Building Resilient and Thriving Workplaces* (pp. 263–284). IGI Global Scientific Publishing., DOI: 10.4018/979-8-3693-6079-8.ch011

Becker, W., & Cropanzano, R. (2010). Organizational neuroscience: The promise and prospects of an emerging discipline. *Journal of Organizational Behavior*, *31*, 1055–1059. DOI: 10.1002/JOB.668

Becker, W., Cropanzano, R., & Sanfey, A. (2011). Organizational Neuroscience: Taking Organizational Theory Inside the Neural Black Box. *Journal of Management*, *37*, 933–961. DOI: 10.1177/0149206311398955

Behera, Y. D. P., Nanda, S. S., Sharma, S., & Sahoo, T. R. (2022). Examining risk absorption capacity as a mediating factor in the relationship between cognition and neuroplasticity in investors in investment decision making. *International Journal of Financial Studies*, *10*(1), 21. DOI: 10.3390/ijfs10010021

BetterUp. (2021, September 29). *What is neuroplasticity and why is it important?* https://www.betterup.com/blog/what-is-neuroplasticity

Boyd, L. (2015). After watching this, your brain will not be the same [Video]. TEDx Vancouver, YouTube. https://youtu.be/LNHBMFCzznE?si=OJyB9dZWtlm0uD3z

Branicki, L., Brammer, S., Brosnan, M., Lazaro, A., Lattanzio, S., & Newnes, L. (2024). Factors shaping the employment outcomes of neurodivergent and neurotypical people: Exploring the role of flexible and homeworking practices. *Human Resource Management*. Advance online publication. DOI: 10.1002/hrm.22243

Burhan, Q., & Khan, M. (2023). Empowering leadership ripple effect: Improving employee engagement, performance, and knowledge sharing through relational energy and autonomy. *European Business Review*. Advance online publication. DOI: 10.1108/ebr-08-2023-0239

Cannavacciuolo, L., Iandoli, L., Ponsiglione, C., Maracine, V., Scarlat, E., & Nica, A. S. (2017). Mapping knowledge networks for organizational re-design in a rehabilitation clinic. *Business Process Management Journal, 23*(2), 329–348. DOI: 10.1108/BPMJ-01-2016-0028

CertLibrary. (2025). *How embracing a culture of continuous learning drives accelerated business expansion* [Blog]. https://www.certlibrary.com/blog/how-embracing-a-culture-of-continuous-learning-drives-accelerated-business-expansion/

Christ, C. (2025, September 8). Gallup: Hybrid work rates have stabilized — and flexibility is key. HR Drive. https://www.hrdive.com/news/gallup-hybrid-work-stabilized-flexibility-key-rto-remote/759477/

Connor, K. M., & Davidson, J. R. T. (2003). Development of a new resilience scale: The Connor-Davidson Resilience Scale (CD-RISC). *Depression and Anxiety*, *18*(2), 76–82. PMID: 12964174

Costandi, M. (2016). *Neuroplasticity*. Essential Knowledge Series. The MIT Press.

Cucino, V., Passarelli, M., Di Minin, A., & Cariola, A. (2021). Neuroscience approach for management and entrepreneurship: A bibliometric analysis. *European Journal of Innovation Management*. Advance online publication. DOI: 10.1108/ejim-01-2021-0015

Cullen, K., Edwards, B., Casper, W., & Gue, K. (2014). Employees' Adaptability and Perceptions of Change-Related Uncertainty: Implications for Perceived Organizational Support, Job Satisfaction, and Performance. *Journal of Business and Psychology*, *29*, 269–280. DOI: 10.1007/S10869-013-9312-Y

DailyBot. (2024, May 8). *The key to achieving organizational agility: Strategies and best practices*. https://www.dailybot.com/insights/the-key-to-achieving-organizational-agility-strategies-and-best-practices

De Fátima Soares Borges, A., Laurindo, F., Spínola, M., Gonçalves, R., & Mattos, C. (2020). The strategic use of artificial intelligence in the digital era: Systematic literature review and future research directions. *International Journal of Information Management*, *57*, 102225. DOI: 10.1016/J.IJINFOMGT.2020.102225

De Príncipe, J. V. (2024). The Perfectionism Mindset Impacting Resilience in Multicultural and Multigenerational Leadership Work-Life Experiences. [Chapter 1] In Vázquez de Príncipe, J. (Ed.), *Resilience of Multicultural and Multigenerational Leadership and Workplace Experience* (pp. 1–31). IGI Global Scientific Publishing., DOI: 10.4018/979-8-3693-1802-7.ch001

Dehais, F., Lafont, A., Roy, R., & Fairclough, S. (2020). A Neuroergonomics Approach to Mental Workload, Engagement, and Human Performance. *Frontiers in Neuroscience, 14.* Advance online publication. DOI: 10.3389/fnins.2020.00268 PMID: 32317914

Drobin, A., Kaplinsky, V., Akimova, O., Martyniuk, Z., Matiushchenko, I., & Boiarchuk, O. (2025). Developing critical thinking in schoolchildren: The neuroscientific aspect. *BRAIN. Broad Research in Artificial Intelligence and Neuroscience, 16(*1 Sup1), 221–230. http://dx.doi.org/DOI: 10.70594/brain/16.S1/18

Edge Foundation. (2024, May 24) *Edge executive function coaching: Harnessing neuroplasticity to enhance executive function* [Blog]. https://edgefoundation.us14.list-manage.com/track/click?u=656b67cd5d0fc82bccdd35283&id=a278d25d11&e=169b59af22

Faw, R. (2025). Adapting leadership styles in changing environments [Doctoral Dissertation]. Liberty University, School of Business. https://digitalcommons.liberty.edu/cgi/viewcontent.cgi?article=7861&context=doctoral

Forbes. (2024, October 31). *How the concept of neuroplasticity can apply to work.* https://www.forbes.com/councils/forbesbusinesscouncil/2024/10/31/how-the-concept-of-neuroplasticity-can-apply-to-work/

Forget, M., & Le Pertel, N. (2024). Enhancing Neuroplasticity and promoting brain health at work: The role of learning and memory in workplace performance. In *The Role of Learning and Memory-From Molecules and Cells to Mind and Behavior.* IntechOpen., DOI: 10.5772/intechopen.1002894

Frontiers. (2016, July 26). *Neuroplasticity and clinical practice: Building brain power for health.* https://pmc.ncbi.nlm.nih.gov/articles/PMC4960264/

Frontiers. (2023, May 15). Editorial: *The impact of chronic stress on neuroplasticity and abnormal behavior.* https://www.frontiersin.org/journals/behavioral-neuroscience/articles/10.3389/fnbeh.2023.1208351/full

Gallup. (2023). *Global indicator: Hybrid work.* https://www.gallup.com/401384/indicator-hybrid-work.aspx

Gallup. (2024). Why employees are detached: Four themes. https://www.gallup.com

Gallup. (2025a). *State of the global workplace: Understanding employees, informing leaders – 2025 Report.* https://www.gallup.com

Gallup (2025b). *Workplace Q2 topline remote work trends: May 7-16, 2025.* Gallup Panel. https://www.gallup.com

Gee, R. (2025a). From neurons to nations: Regenerative leadership and integrated consciousness for systemic shifts in the Anthropocene. *International Journal of Psychiatry*. Advance online publication. DOI: 10.33140/ijp.10.01.03

Gee, R. O. W. (2025b). From neurons to organizations: Awakening regenerative mindsets with neuroplasticity, AI, and systemic consciousness. *Journal of Biotechnology and Biomedicine*, *8*, 21–58. DOI: 10.26502/jbb.2642-91280178

Geraerts, E. (2018). *Better minds: How insourcing strengthens resilience and empowers your brain.* Terra-Lannoo, Uitgeverij.

Guarnier, K., & Chimenti, P. (2023). Advancing in the neuroleadership field: a systematic and integrative review. *Cadernos EBAPE.BR*. https://doi.org/DOI: 10.1590/1679-395120220184x

Hannah Brown. (2024, January 31). *Lead with learning series: Organization resiliency and adaptability.* https://hannahbrown.co/2024/01/31/lead-with-learning-series-organization-resiliency-and-adaptability/

Harlianto, J., Rudi, R., & Ginting, H. C. B. G. (2025). Building collaboration in a multigenerational workforce: Intergenerational inclusivity and mental health as key drivers of workplace productivity. In *Enhancing Workplace Productivity Through Gender Equity and Intergenerational Communication* (pp. 283–316). IGI Global Scientific Publishing., DOI: 10.4018/979-8-3373-2903-1.ch011

Hawkins, J. A. (2021). *Brain plasticity and learning*. Springer International Publishing.

Healey, M., & Hodgkinson, G. (2014). Rethinking the philosophical and theoretical foundations of organizational neuroscience: A critical realist alternative. *Human Relations*, *67*, 765–792. DOI: 10.1177/0018726714530014

Henry, M. S., Gendron, L., Tremblay, M. E., & Drolet, G. (2017). Enkephalins: Endogenous Analgesics with an emerging role in stress resilience. *Neuroplasticity, 2017*, 1–11. https://pmc.ncbi.nlm.nih.gov/articles/PMC5662798/ doi:10.1155/2017/1546125

ISG. (2025). *How to encourage neuroplasticity for a changing workplace.* https://isg-one.com/articles/how-to-encourage-neuroplasticity-for-a-changing-workplace

Jelača, M., Bjekić, R., Berber, N., Aleksić, M., Slavić, A., & Marić, S. (2022). Impact of Managers' Emotional Competencies on Organizational Performance. *Sustainability*. Advance online publication. DOI: 10.3390/su14148800

Journal Agent. (2024, September 1). *The role and importance of neuroplasticity in developing psychological resilience*. https://jag.journalagent.com/jern/pdfs/JERN_21_3_250_253.pdf

Kamkankaew, P., Phattarowas, V., Khumwongpin, S., Limpiaongkhanan, P., & Sribenjachot, S. (2022). Increasing competitive environment dynamics and the need of hyper-competition for businesses. *International Journal of Sociologies and Anthropologies Science Reviews*, *2*(5), 9–20.

Kaur, V. (2024). Mirror neurons and neuroplasticity: The dyadic neurological foundations bridging entrepreneur-level and enterprise-level capabilities. *Journal of Business Venturing Insights*. https://doi.org/DOI: 10.1016/j.jbvi.2024.e00497

Kent, J. (2023). *Perfectionism might be hurting you: Here's how to change your relationship to achievement*. Harvard Summer School. https://summer.harvard.edu/blog/perfectionism-might-be-hurting-you-heres-how-to-change-your-relationship-to-achievement/

Khaneja, S., & Arora, T. (2024). The potential of neuroscience in transforming business: A meta-analysis. *Future Business Journal*, *10*(1), 77. DOI: 10.1186/s43093-024-00369-7

Kietzman, F. (2024). *The Neurodiverse: A forgotten population* [Video]. Sage Knowledge. https://doi.org/DOI: 10.4135/9781529686302

Kirby, A., & Smith, T. (2021). *Neurodiversity at work: Drive innovation, performance, and productivity with a neurodiverse workforce*. Kogan Page Publishers.

Kiss, A., Libaers, D., Barr, P., Wang, T., & Zachary, M. (2019). CEO cognitive flexibility, information search, and organizational ambidexterity. *Strategic Management Journal*. Advance online publication. DOI: 10.5465/AMBPP.2019.12521

Kleim, J. A., & Jones, T. A. (2008). *Principles of experience-dependent neural plasticity: implications for rehabilitation after brain damage*.

Lee, L., Healy, M., Fischer, N., Tong, K., Chen, A., Sahakian, B., & Kourtzi, Z. (2024). Cognitive flexibility training for impact in real-world settings. *Current Opinion in Behavioral Sciences*, *59*. Advance online publication. DOI: 10.1016/j.cobeha.2024.101413

LeFevre-Levy, R., Melson-Silimon, A., Harmata, R., Hulett, A., & Carter, N. (2023). Neurodiversity in the workplace: Considering neuroatypicality as a form of diversity. *Industrial and Organizational Psychology: Perspectives on Science and Practice*, *16*, 1–19. DOI: 10.1017/iop.2022.86

Lim, D. H., Yoo, M. H., Kim, J., & Brickell, S. A. (2017). *Learning Agility: The Nexus Between Learning Organization, Transformative Learning, and Adaptive Performance*. Adult Education Research Conference. https://newprairiepress.org/aerc/2017/papers/28

Long, W. J. (2006). Quantum theory and neuroplasticity: Implications for social theory. *Journal of Theoretical and Philosophical Psychology*, *26*(1-2), 78.

Ludvik, M. J. B. (Ed.). (2023). *The neuroscience of learning and development: Enhancing creativity, compassion, critical thinking, and peace in higher education*. Taylor & Francis.

Makanur, S. (2025). Guest Model of HRM. Scribd. https://www.scribd.com/presentation/871712541/Guest-Model-of-HRM

Malchiodi, C. (2020). *Tapping the Healing Rhythms of the Vagal Nerve* [Blog]. Psychology Today. https://www.psychologytoday.com/us/blog/arts-and-health/202004/tapping-the-healing-rhythms-of-the-vagal-nerve

Marín, L. J. (2025, July 2). *Organizational resilience: Key to competitive companies*. Inspenet. https://inspenet.com/en/articulo/organizational-resilience-in-companies/

Mave Health. (2024, September 10). *Neuroplasticity and mental health: 9 ways to rewire your brain*. https://www.mavehealth.com/blogs/neuroplasticity-and-mental-health

McCreedy, R. (2024). Change on the Brain? The Neuroscience of Organizational Transformation. *International Journal of Applied Research in Management and Economics*. https://doi.org/DOI: 10.33422/ijarme.v7i3.1402

McKinsey & Company. (2024, February 12). *Organizational health is (still) the key to long-term performance*. https://www.mckinsey.com/capabilities/people-and-organizational-performance/our-insights/organizational-health-is-still-the-key-to-long-term-performance

Mega Lifesciences. (2024, June 11). *Importance of mental health in the workplace: Brain health tips*. https://www.megawecare.com/good-health-by-yourself/brain-health/the-importance-of-brain-health-at-the-workplace

Neuro Business School. (2024). The role of neuroplasticity in enhancing leadership skills. https://eunbs.com/the-role-of-neuroplasticity-in-enhancing-leadership-skills/

Neuro Frontiers. (2025). *Condition: ADHD.* https://www.neurofrontiers.org/condition/adhd/

News-Medical. (2019, December 12). *What is neuronal plasticity and why is it important?* https://www.news-medical.net/life-sciences/What-is-Neuronal-Plasticity-and-Why-Is-It-Important.aspx

Niswaty, R., Wirawan, H., Akib, H., Saggaf, M., & Daraba, D. (2021). Investigating the effect of authentic leadership and employees' psychological capital on work engagement: Evidence from Indonesia. *Heliyon, 7.* Advance online publication. DOI: 10.1016/j.heliyon.2021.e06992 PMID: 34027188

Nowak, R. (2021). Foundations of strategic flexibility: Focus on cognitive diversity and structural empowerment. *Management Research Review.* Advance online publication. DOI: 10.1108/mrr-02-2021-0130

Ooi, R. W. G. D. (2025). From neurons to organizations: Awakening regenerative mindsets through neuroplasticity, AI, and systemic consciousness. *Journal of Biotechnology and Biomedicine.* Advance online publication. DOI: 10.26502/jbb.2642-91280178

Paradiso Solutions. (2025, July 8). *Continuous learning: What it is & why it is important for your business.* https://www.paradisosolutions.com/blog/continuous-learning-for-business/

Pascual-Leone, A., Amedi, A., Fregni, F., & Merabet, L. B. (2005). The plastic human brain cortex. *Annual Review of Neuroscience, 28*(1), 377–401. PMID: 16022601

Peaceful Leaders Academy. (2025). Emotional regulation in the workplace: Importance & how to develop [Blog]. https://peacefulleadersacademy.com/blog/emotional-regulation-in-the-workplace/

Pereira, V., Hadjielias, E., Christofi, M., & Vrontis, D. (2021). A systematic literature review on the impact of artificial intelligence on workplace outcomes: A multi-process perspective. *Human Resource Management Review.* Advance online publication. DOI: 10.1016/j.hrmr.2021.100857

Perfectionism Rewired Podcast. (2025, March 23). *Neuroplasticity.* https://perfectionismrewired.com/episode/neuroplasticity-oa64

Perspect, A. I. (2024, March 4). How to leverage neuroplasticity at work. https://perspect.ai/blog/neuroplasticity-in-the-workplace/

Physiopedia. (2023). *Neuroplasticity*. https://www.physio-pedia.com/Neuroplasticity

Picchi, A. (2025, May). *Leading creative teams with the brain in mind: A neuroscience perspective on team safety and innovation in design leadership*. https://www.researchgate.net/publication/392161060

Ploughman, M., Austin, M. W., Glynn, L., & Corbett, D. (2015). The effects of poststroke aerobic exercise on neuroplasticity: A systematic review of animal and clinical studies. *Translational Stroke Research*, *6*, 13–28. PMID: 25023134

Pollack, J. (2025). *Five tools Every Organizational Leader Must Know to Improve Their Workplace Culture Today* [eBook]. Peaceful Leaders Academy. https://peacefulleadersacademy.com/wp-content/uploads/2024/06/Five-Tools-Every-Organization-Leader-Must-Know-to-Improve-Their-Workplace-Culture-Today.pdf

Positive Psychology. (2018, July 25). *What is neuroplasticity? A psychologist explains* [+14 tools]. https://positivepsychology.com/neuroplasticity/

Positive Psychology. (2019, September 3). *How to use the Connor Davidson Resilience Scale (CD-RISC)*. https://positivepsychology.com/connor-davidson-brief-resilience-scale/

Psychology Today. (2024, May). *Learn to rewire your ADHD brain to improve cognitive skills*. https://www.psychologytoday.com/us/blog/mythbusting-adhd/202405/learn-to-rewire-your-adhd-brain-to-improve-cognitive-skills

Psychology Today. (2025, June 18). *Neuroplasticity*. https://www.psychologytoday.com/us/basics/neuroplasticity

Qualtrics. (2025). *What is employee well-being*? https://www.qualtrics.com/experience-management/employee/well-being/

Rajagopalan, J., & Jacob, A. M. (2025). Optimizing learning across ages: Insights from neuroplasticity for corporate reskilling and upskilling. In *Sustainable Digital Technology and Ethics in an Ever-Changing Environment* (Vol. 1, pp. 675–690). Springer Nature Switzerland., DOI: 10.1007/978-3-031-86712-5_57

Rollnik-Sadowska, E., & Grabińska, V. (2024). Managing Neurodiversity in Workplaces: A Review and Future Research Agenda for Sustainable Human Resource Management. *Sustainability*. Advance online publication. DOI: 10.3390/su16156594

Sagi, Y., Tavor, I., Hofstetter, S., Tzur-Moryosef, S., Blumenfeld-Katzir, T., & Assaf, Y. (2012). Learning in the fast lane: New insights into neuroplasticity. *Neuron*, *73*(6), 1195–1203. PMID: 22445346

Saluja, S., Kukreja, J., & Sharma, S. (Eds.). (2024). *Building organizational resilience with neuroleadership*. IGI Global Scientific Publishing.

Sandoval, W. (2025). ADHD Brain Training: Can exercising your brain help with attention and focus? Healthline. https://www.healthline.com/health/adhd/adhd-brain-training

Scientific Research Archives. (2023, June 29). *Surviving the gauntlet: Navigating the hypercompetitive work culture*. https://sciresjournals.com/ijlsra/sites/default/files/IJLSRA-2023-0068.pdf

Shaikh, F., Afshan, G., Anwar, R., Abbas, Z., & Chana, K. (2023). Analyzing the impact of artificial intelligence on employee productivity: The mediating effect of knowledge sharing and well-being. *Asia Pacific Journal of Human Resources*. Advance online publication. DOI: 10.1111/1744-7941.12385

SHRM. (2025). Measuring employee engagement. https://www.shrm.org/executive/resources/articles/pages/measuring-employee-engagement.aspx

Silver, E., Nittrouer, C., & Hebl, M. (2023). Beyond the business case: Universally designing the workplace for neurodiversity and inclusion. *Industrial and Organizational Psychology: Perspectives on Science and Practice*, *16*, 45–49. DOI: 10.1017/iop.2022.99

Staneiu, R. M. (2023). Nurturing Neuroplasticity as an enabler for Growth Mindset through lifelong learning and knowledge dynamics. In *Proceedings of the International Conference on Business Excellence, 17*(1), 1264-1274. Sciendo.

Stasielowicz, L. (2020). How important is cognitive ability when adapting to changes? A meta-analysis of the performance adaptation literature. *Personality and Individual Differences*. Advance online publication. DOI: 10.1016/j.paid.2020.110178

Sunshine Coast Psychology Clinic. (2025). *Neuroplasticity and psychology as an intervention: Shaping minds for a better life* [Blog]. https://www.sunshine-coast-psychology.com.au/Blog/NeuroplasticityandPsychologyasanIntervention-645/

Szulc, J. (2022). AMO model for neuro-inclusive remote workplace. *Personnel Review*. Advance online publication. DOI: 10.1108/pr-02-2022-0085

Taghikilanidamavandi, L. (2024). Unlocking potential: Embracing neurodiversity and mental health in the workplace. *Economics, Finance and Management Review*. https://doi.org/DOI: 10.36690/2674-5208-2024-4-80-93

Tomlinson, L. S. (2021). *Neuroplasticity as a transformational tool for improving managerial approaches* [Doctoral dissertation, University of Zagreb, Faculty of Economics and Business, Department of Organization and Management]. https://urn.nsk.hr/urn:nbn:hr:148:977756

Torch. (2023). *Research Report: Leveraging coaching and mentoring to create more effective leaders*. Harvard Business School Publishing.

UCLA Health. (2016, February 8). *In pursuit of resilience: Stress, epigenetics, and brain plasticity*. https://www.uclahealth.org/sites/default/files/documents/mcewen-resilience.pdf?f=f540d7f6

Vázquez-de-Príncipe, J. (2021). *Investigating leadership and human-intelligences during public-sector organizational change: A mixed-methods study* [Doctoral Dissertation, University of Phoenix]. ProQuest Dissertations & Theses (PQDT) Global – ProQuest Publishing, 2637194604.

Vázquez de Príncipe, J. (2023). The Work-Life Dynamics cultural impact: Evolving the future of strategic holistic ecosystems post pandemic. [Chapter 7] In Tennin, K. (Ed.), *Measuring the Effectiveness of Organizational Development Strategies During Unprecedented Times* (pp. 119–155). IGI Global Scientific Publishing., DOI: 10.4018/978-1-6684-8392-3.ch007

Vázquez de Príncipe, J. (Ed.). (2024a). *Resilience of Multicultural and Multigenerational Leadership and Workplace Experience*. IGI Global Scientific Publishing., DOI: 10.4018/979-8-3693-1802-7

Vázquez de Príncipe, J. (2024b). The Evolution of human resources management in the post-pandemic world. [Chapter 1] In Even, A. M., & Christiansen, B. (Eds.), *Effective Human Resources Management in the Multigenerational Workplace* (pp. 1–22). IGI Global Scientific Publishing., DOI: 10.4018/979-8-3693-2173-7.ch001

Waldman, D., Wang, D., & Fenters, V. (2019). The Added Value of Neuroscience Methods in Organizational Research. *Organizational Research Methods*, *22*, 223–249. DOI: 10.1177/1094428116642013

Wang, L., & Xie, T. (2023). Double-Edged Sword Effect of Flexible Work Arrangements on Employee Innovation Performance: From the Demands–Resources–Individual Effects Perspective. *Sustainability*, *15*(13), 10159. DOI: 10.3390/su151310159

Wilms, Y. (2023, May 30). Neuroplasticity to Upskill Your Workforce [Video]. Top Employers Institute. https://youtu.be/uLyG_pBzkIA?si=Q0-uB81u7s7b7_uP

Workhuman. (2025, January 22). *Adapting to change in the workplace: Challenges and strategies*. https://www.workhuman.com/blog/changing-workplace/

Zhang, S., Sun, J., & Gao, X. (2020). The effect of fatigue on brain connectivity networks. *Brain Science Advances*, *6*(2), 120–131. DOI: 10.26599/BSA.2020.9050008

Zhelnina, A. (2024). 11 Apathy. *Dark Emotions: Difficult Emotional Experiences in Social and Everyday Life*.

ADDITIONAL READING

Bell-Ellis, R. (2013). Integrating spirit at work: A ripple of hope for healthy organizational cultures. In *Handbook of Faith and Spirituality in the Workplace* (pp. 333–343). Emerging Research and Practice., DOI: 10.1007/978-1-4614-5233-1_21

Gallup. (2023). *Global indicator: Hybrid work*. https://www.gallup.com/401384/indicator-hybrid-work.aspx

Gallup. (2025). *State of the global workplace: Understanding employees, informing leaders – 2025 Report*. https://www.gallup.com

Harlianto, J., Rudi, R., & Ginting, H. C. B. G. (2025). Building collaboration in a multigenerational workforce: Intergenerational inclusivity and mental health as key drivers of workplace productivity. In *Enhancing Workplace Productivity Through Gender Equity and Intergenerational Communication* (pp. 283–316). IGI Global Scientific Publishing., DOI: 10.4018/979-8-3373-2903-1.ch011

Rajagopalan, J., & Jacob, A. M. (2025). Optimizing learning across ages: Insights from neuroplasticity for corporate reskilling and upskilling. In *Sustainable Digital Technology and Ethics in an Ever-Changing Environment* (Vol. 1, pp. 675–690). Springer Nature Switzerland., DOI: 10.1007/978-3-031-86712-5_57

Saluja, S., Kukreja, J., & Sharma, S. (Eds.). (2024). *Building organizational resilience with neuroleadership*. IGI Global Scientific Publishing.

Torch. (2023). *Research Report: Leveraging coaching and mentoring to create more effective leaders*. Harvard Business School Publishing.

Vázquez de Príncipe, J. (Ed.). (2024a). *Resilience of Multicultural and Multigenerational Leadership and Workplace Experience*. IGI Global Scientific Publishing., DOI: 10.4018/979-8-3693-1802-7

Zhang, S., Sun, J., & Gao, X. (2020). The effect of fatigue on brain connectivity networks. *Brain Science Advances*, *6*(2), 120–131. DOI: 10.26599/BSA.2020.9050008

KEY TERMS AND DEFINITIONS

Alternative Non-invasive Holistic Well-being Measurements: Incorporates pre- and post-activity self-monitoring to assess the recovery rebounding impact after prayer, meditation, reflection, gratitude journaling, nature bathing, Vagus Nerve Stimulation (VNS), also known as tapping, or deep breathing practices to foster individual resilience by holistically inducing a sense of whole-person wellness, which impacts work-life dynamics.

Executive Functions: Crucially, this inherent neuroplasticity offers significant potential for improving vital executive functions in the workplace, including emotional regulation, perception, memory, attention, and decision-making abilities.

Neuroplastic Organization Wellness Attributes: A neuroplastic business exhibits the attributes of rapid learning and skill acquisition, cognitive flexibility and agility, enhanced social-emotional intelligence and collaboration, increased stress resilience, high motivation and drive, strong creativity and innovation, which builds a collaborative capacity for adaptation, ensuring its survival and prosperity in hyper-competitive global dynamic conditions, including artificial intelligence (AI) workplace dynamics.

Neuroplasticity: The brain's natural ability to reorganize its neural pathways and synaptic connections in response to learning, experience, and environmental stimuli, forming new neural networks. This adaptive capability intricately fosters one's personal resilience by overcoming perfectionism and enhancing overall growth.

Resilience: The brain's capacity to adapt to adverse conditions through practice is associated with neuroplasticity enablement in work-life dynamics, leading to healthier, more productive organizations and a happier, thriving workforce.

Strategic HRM/OD Model for Cultivating Neuroplasticity: Three measurable outcomes designated for collaborative efforts between HRM and organizational development disciplines, incorporating targeted holistic practices, such as co-creating learning and development engagements (design/systems thinking), cultivating psychological safety and emotional regulation (individual, group, and team coaching), and empowering autonomy and meaningful purpose (organizational culture ecosystem design, job crafting, career pathing).

Workforce Malleability: The organization's empowerment of inherent adaptability to leverage their workforce's cognitive and emotional malleability for strategic advantage to deter the risk of falling behind their global competitors, while consciously leveraging new technologies to streamline systems and processes, and eliminating redundancies, it is imperative to promote and facilitate human innovative creativity opportunities to strike a holistic balance in the ecosystem.

Chapter 8
Rewiring the Workplace: Harnessing Neuroplasticity for Cognitive Agility and Skill Mastery in the Age of Disruption

Rismawati Rismawati
https://orcid.org/0000-0001-8046-6764
Universitas Muhammadiyah Palopo, Indonesia

Anuradha Ranjan
https://orcid.org/0000-0003-1820-4370
Amity University, Mohali, India

Kirti Sharma
Chandigarh School of Business, India

ABSTRACT

This chapter examines the strategic application of neuroplasticity in contemporary enterprises to improve cognitive agility and cultivate future-ready talent. As disruption transforms industries, conventional paradigms of learning and performance are inadequate. This book delineates how restructuring the brain facilitates the transformation of the workplace through an integrated framework of neuroscience, organizational behavior, and human resource innovation. Topics encompass the creation of neuro-friendly workplaces, the utilization of immersive learning technology, the advancement of adaptive leadership, and the assessment of neural growth via performance indicators. By synchronizing learning ecosystems with cerebral processes, companies can transition from static skill training to dynamic, ongoing transformation. This book offers a visionary and pragmatic framework for creating neuroadaptive workplaces that excel in complexity and promote sustainable innovation

DOI: 10.4018/979-8-3373-2292-6.ch008

INTRODUCTION

Navigating Disruption Through Brain-Based Work Design

1.1 The Nature of Disruption in the 21st Century Workplace

The contemporary workplace is experiencing a profound upheaval, driven by rapid technological progress, worldwide interconnectivity, and evolving cultural expectations. Disruption has emerged as a distinguishing trait of the 21st century, transforming the ways individuals work, learn, and interact with enterprises. The essential structure of the workplace is being reinvented through automation, artificial intelligence, remote work, and digital collaboration platforms (Lim, 2023; Aroles et al., 2021; Mahmoud et al., 2024). Conventional positions are being dismantled and supplanted with fluid, dynamic functions that necessitate ongoing learning and cognitive adaptability.

In this environment of instability and unpredictability, traditional work paradigms—focused on repetitive tasks, inflexible hierarchies, and linear career advancement—are inadequate. Organizations must now function within a rapidly changing ecosystem, where adaptation and innovation are essential for survival rather than mere competitive advantages. Employees are required to continuously reskill and upskill, adopting new technologies and techniques while sustaining high productivity and mental well-being (Khurana et al., 2022; Miceli et al., 2021).

Moreover, disruption transcends technology. Demographic changes, including multi-generational workforces and increased cultural diversity, compel firms to reevaluate inclusivity, communication methods, and leadership strategies. The emergence of hybrid and remote work models has created challenges concerning engagement, cooperation, and mental health, necessitating leaders to cultivate workplaces that foster psychological safety and cognitive resilience.

Central to these issues is a crucial understanding: the human brain is not immutable. It possesses exceptional adaptability via the process of neuroplasticity. This ability to reconfigure provides a compelling metaphor and a pragmatic approach for managing disruption. As firms must undergo structural transformation, individuals must experience neurological transformation—adopting new cognitive frameworks, embracing uncertainty, and fostering learning agility (Reddy, 2025).

Comprehending the essence of disruption in this epoch necessitates a dual emphasis: one on external technical and societal influences, and another on internal cognitive and emotional competencies. The intersection of neuroscience and organizational development provides a promising foundation for redefining work, with an emphasis on fostering the brain's plasticity to create adaptive, inventive, and future-oriented workplaces. Consequently, disruption should be regarded not solely as a threat, but as a chance to fundamentally reconfigure labor structures that correspond with the advancing capabilities of the human intellect.

1.2 The Rise of Neuroplasticity as a Strategic Organizational Tool

As disruption transforms the work environment, firms are increasingly acknowledging the human brain—not merely technology or capital—as their paramount asset. In this new era, neuroplasticity has become a revolutionary notion with significant implications for organizational efficacy. Neuroplasticity, the brain's intrinsic capacity to restructure by establishing new neural connections throughout life, provides a promising and practical basis for human adaptation, skill development, and innovation (Ammirato et al., 2023; Paesano, 2023).

The emergence of neuroplasticity in organizational settings signifies a transition from mechanistic paradigms of workforce development to a more dynamic and biological comprehension of learning and performance. It recognizes that every employee have the potential for cognitive change, growth, and evolution, irrespective of age or experience (Joshua, 2022; Marzola et al., 2023). This viewpoint reinterprets conventional beliefs regarding skill and potential, highlighting that intelligence and ability are not immutable characteristics but may be developed via intentional effort, constructive feedback, and enhanced surroundings.

Organizations that strategically leverage neuroplasticity do not solely concentrate on external training programs; they reconfigure work systems, learning frameworks, and leadership methodologies to correspond with optimal brain learning processes. They cultivate environments that prioritize introspection, mindfulness, and iterative learning. They employ coaching, mentoring, and experiential learning frameworks that foster profound brain involvement. In doing so, they cultivate both competence and confidence, enabling employees to manage uncertainty and adapt to change with cognitive agility (Li & Yeo, 2024; Babüroğlu and Selsky, 2021).

Moreover, neuroplasticity corresponds with modern need for continuous education and skill enhancement. As job responsibilities transform and new technologies arise, the capacity for ongoing knowledge acquisition and application becomes an essential asset. Organizations that include neuroplastic principles into their talent development strategies are poised to cultivate resilient workforces that can thrive amidst disruption rather than simply endure it.

The strategic use of neuroplasticity also encompasses leadership. Leaders who get the neurobiology of behavior and motivation are more adept at facilitating team learning, managing change, and fostering adaptive thinking. Thus, neuroplasticity represents not merely a scientific understanding but also a leadership philosophy—perceiving human potential as malleable, learning as perpetual, and disturbance as a stimulus for development (Fagerdal et al., 2022).

1.3 From Routine Skills to Cognitive Agility: Why Rewiring Matters Now

In an era characterized by rapid disruption, the conventional emphasis on regular skills is inadequate to satisfy the requirements of modern workplaces. The skills that previously ensured stability—task repetition, memorization, and procedural execution—are progressively being mechanized, outsourced, or made obsolete by advancing technologies. Organizations currently demand not only skill repetition but also cognitive agility—the capacity for flexible thinking, rapid learning, and seamless adaptation to new conditions (Jaafar et al., 2025; Franco & Landini, 2022).

Cognitive agility signifies a shift from inflexible task-oriented skills to mental flexibility and metacognitive awareness. It includes the ability to change viewpoints, discard obsolete frameworks, and utilize knowledge in various situations. In this dynamic environment, characterized by poorly defined problems and undefined solutions, personnel must be prepared to challenge assumptions, navigate ambiguity, and integrate interdisciplinary knowledge (Nyamrunda & Freeman, 2021; Park & Park, 2021). These talents are not inherent; they are developed by deliberate mental restructuring based on the principles of neuroplasticity.

Rewiring the brain to enhance agility necessitates more than the acquisition of new information; it demands the development of novel neural pathways that promote curiosity, resilience, and innovative problem-solving. Monotonous, low-involvement activities minimally activate the brain's adaptive mechanisms. Conversely, environments that stimulate intellectual engagement, foster experimentation, and permit safe failure facilitate the development of resilient cognitive structures that can be applied across various contexts and jobs (Berrone et al., 2023).

Organizations that regard cognitive agility as a strategic asset commence by reassessing their job design, team structure, and training delivery. They progress from transactional learning to immersive, learner-centric experiences. They acknowledge that agility thrives in environments characterized by psychological safety, autonomy, and ongoing feedback. Rewiring thus transforms into a collective endeavor—an corporate pledge to advance not only employees' knowledge but also their cognitive processes.

Furthermore, the necessity to cultivate cognitive agility is time-sensitive. As disruptions escalate—stemming from digital revolution, global crises, or socio-political changes—the expense of cognitive inflexibility becomes untenable. Individuals who do not adapt risk obsolescence, whereas those who foster agility are positioned to spearhead innovation, resilience, and influence. Cognitive agility is, in essence, a tangible concept. It is a cultivable, trainable, and quantifiable competency—one that embodies the profound reality that the most adaptable organizations are those that master internal reconfiguration.

BACKGROUND

The Neuroscience of Learning and Skill Formation

2.1 Mechanisms of Neuroplasticity: How the Brain Learns and Adapts

Neuroplasticity, defined as the brain's capacity for self-modification, serves as the biological basis that allows individuals to acquire, discard, and reacquire knowledge throughout their lives. The human brain is not a static organ; it is dynamic, continually establishing, reinforcing, diminishing, or removing synaptic connections based on experience, environment, and deliberate effort. This capacity is fundamental to all types of learning and skill acquisition, rendering it essential for both individual growth and organizational change (Zeine et al., 2024; Chen and Goodwill, 2022).

Neuroplasticity at the cellular level is facilitated by neurons in the brain, which can establish new pathways or strengthen existing ones via synaptic plasticity. When an individual encounters a novel stimulus or consistently pursues a new ability, the brain modifies the efficiency of signal transmission among neurons (Joshua, 2022; Dzyubenko & Hermann, 2023). This process facilitates the acquisition of knowledge and skills while also allowing for the alteration of behaviors, habits, and emotional responses throughout time.

Neuroplastic adaptation may manifest as both structural and functional changes. Structural plasticity denotes the brain's capacity to undergo physical alterations in its architecture, such as the development of new dendrites or modifications in the density of grey matter in particular areas. Functional plasticity entails the reallocation of functions from impaired regions of the brain to intact ones, a phenomenon frequently observed in stroke therapy or trauma healing. Collectively, these processes underscore the brain's resilience and its ability for lifelong learning (Innocenti, 2022; Johnson & Cohen, 2023).

Neuroplasticity is not a passive process. It necessitates deliberate involvement. Environments that provide novelty, challenge, feedback, and emotional significance promote plastic changes more efficiently than monotonous or low-stimulation environments. Sleep, physical activity, stress management, and social interaction significantly affect the rate and permanence of brain reorganization. In the workplace, comprehending these mechanisms offers a framework for creating learning methods and surroundings that correspond with the brain's natural functioning. By connecting organizational strategies with neuroplastic principles, leaders can cultivate a workforce adept at gaining new skills and thriving in complexity, uncertainty, and continual change. This table synthesizes and contextualizes the underlying theories underpinning neuroplastic change in organizational learning environments, detailing significant theoretical contributions, their main principles, and practical applications in workplace settings.

Table 1. Comparative Theoretical Foundations of Neuroplasticity in Organizational Contexts

Theory / Study	Key Concept	Application in Workplace	Discipline Origin
Hebbian Learning (Hebb, 1949)	"Neurons that fire together wire together"	Designing collaborative team-based learning	Neuroscience
Kolb's Experiential Learning (1984)	Learning through experience cycles	Learning design via feedback loops and reflection	Educational Psychology
Doidge's Neuroplasticity (2007)	Lifelong brain change potential	Adult upskilling and digital retooling	Clinical Neuroscience
Immersive Learning (VR/AR)	Stimuli-driven engagement	Neurocognitive stimulation for training	Cognitive Tech
Cognitive Load Theory	Optimizing memory pathways	Structuring tasks for cognitive efficiency	Instructional Design

Source: Compiled by the authors based on interdisciplinary literature review

2.2 Cognitive Load and Memory Consolidation in High-Stakes Environments

In high-pressure professional settings—marked by time constraints, unpredictability, and performance demands—cognitive efficiency is essential for success. Nevertheless, such environments frequently impose significant demands on the brain's processing capabilities, resulting in what is termed cognitive load. Cognitive load denotes the extent of mental work necessary to execute a task, analyze information, and make decisions. When this cognitive load surpasses an individual's working memory capacity, performance declines, learning halts, and errors escalate (Skulmowski & Xu, 2022; Chen et al., 2023).

Cognitive load is classified into three categories: intrinsic, external, and germane. Intrinsic load pertains to the inherent complexity of the work; extraneous load encompasses distractions or ineffective instructional design; and germane load refers to the cognitive resources allocated to genuine learning and schema development. Successful learning and performance in high-stakes contexts necessitate the regulation of all three categories, aiming to reduce extraneous factors and enhance germane processing (Wang et al., 2022; Daniel, 2023).

Memory consolidation is crucial in converting short-term experiences into long-term knowledge. During consolidation, the brain stabilizes and integrates newly acquired information, frequently occurring during rest or sleep. In settings that require swift learning—such as crisis management, innovation-centric sectors, or fast-paced digital systems—providing sufficient opportunities for consolidation is crucial. In the absence of this phase, even the most rigorous training or exposure may not result in enduring competence (Xie et al., 2025).

Stress constitutes another pivotal element. Moderate stress can improve alertness and concentration, however chronic or severe stress hinders memory development by affecting hippocampus function and diminishing neuronal plasticity. High-stakes situations must combine urgency with support by ensuring psychological safety, delivering unambiguous feedback, and allowing cognitive space for contemplation (Matar et al., 2024).

Organizations that comprehend the neuroscience of cognitive load and memory can create systems that facilitate, rather than obstruct, learning. Techniques like as information chunking, spaced practice, retrieval activities, and minimizing multi-tasking can markedly enhance both instant understanding and long-term retention. Furthermore, leaders are essential in fostering settings that prioritize knowledge, respect mental capacity, and regard learning as a continuous process. By effectively controlling cognitive load and promoting memory consolidation, businesses empower their personnel to operate under pressure while cultivating profound, transportable knowledge essential for enduring growth and adaptability.

Table 2. Cognitive Load vs. Organizational Response Matrix

Cognitive Load Type	Organizational Challenge	Recommended Intervention
Intrinsic Load	Complex tasks/new skill domains	Scaffolded learning modules
Extraneous Load	Poorly structured environments	Simplified interfaces, UX training
Germane Load	Deep learning & integration	Problem-based learning, mentorship

Source: Author updated with workplace applications

2.3 The Role of Emotional States in Learning and Workplace Adaptation

Emotions are integral to cognition; they play a crucial role in how individuals learn, retain information, and adjust in the workplace. The brain handles emotional information in conjunction with cognitive systems, affecting attention, motivation, memory consolidation, and decision-making. Emotional states can significantly enhance or impede neuroplastic development, thereby influencing an employee's capacity to learn new skills and adapt to evolving work situations (Saeed et al., 2022; Li et al., 2021).

Positive emotional states, like curiosity, optimism, and engagement, augment the brain's capacity to assimilate and remember new information. These emotions engage reward circuits and elevate dopamine levels, thus promoting the development of new synaptic connections. When employees experience psychological safety and emotional support, their cognitive processes are more receptive to learning, creativity, and collaboration. Under these circumstances, adaptive actions are reinforced, and neuronal patterns associated with growth become more robust and resilient (Ekman et al., 2022; White et al., 2021).

In contrast, adverse emotional states—such as worry, dread, and persistent stress—can diminish cognitive ability and obstruct neuroplasticity. The secretion of stress hormones such as cortisol disrupts the hippocampus, the brain area crucial to memory and learning. In high-pressure circumstances characterized by emotional suppression or ambiguity, individuals may struggle with concentration, knowledge recall, or adaptability to shifting demands. With the passage of time, these situations may result in mental tiredness, disengagement, and a decline in organizational performance (Luciana & Collins, 2022; Kang et al., 2023).

Understanding the correlation between emotional well-being and learning is crucial for firms seeking to develop adaptive, future-ready workforces. Emotional intelligence, mindfulness techniques, empathic leadership, and inclusive cultures are not merely "soft" solutions; they are neurocognitive facilitators. They establish settings that facilitate the brain's ability to reorganize, recuperate from setbacks, and reinforce novel behavioral patterns (Kumprang & Suriyankietkaew, 2024; Pesqueira and Almeida, 2025).

Furthermore, emotion is crucial for adaptation throughout changes, like the adoption of new technologies, the shift to hybrid work models, or the reorientation of career routes. Emotional awareness allows individuals to identify internal resistance, reinterpret problems, and activate cognitive resources for transformation. By acknowledging the emotional aspect of learning and adaptability, companies can access greater depths of human potential. In a perpetually disruptive environment, emotional intelligence serves as the conduit between cognitive adaptability and organizational agility, facilitating harmonious evolution for both individuals and systems. The following table delineates the essential steps of a neuroadaptive learning cycle designed for contemporary workplaces, based on the aforementioned neurocognitive mechanisms.

Table 3. Five-Stage Neuroadaptive Learning Cycle in the Workplace

Stage	Description	Example of Implementation
Neuronal Stimulation	Curated stimuli to activate learning	Interactive simulations, gamification
Adaptive Learning	Microlearning with feedback	Short digital modules with quizzes
Memory Consolidation	Internalization via reflection	Reflective journaling, peer discussion
Cognitive Agility	Applying learning to new contexts	Job rotation, cross-functional assignments
Rewiring	Long-term structural brain change	Culture embedding, longitudinal coaching

Source: Authors elaborate organizational learning frameworks

ORGANIZATIONAL INTERVENTIONS TO ENHANCE NEUROPLASTICITY

3.1 Neuro-Friendly Work Environments: Designing for Cognitive Growth

A neuro-friendly work environment is not only a location for task completion, but a dynamic ecosystem purposefully crafted to promote learning, improve focus, and cultivate cognitive resilience. Within the framework of neuroplasticity, the physical and psychological attributes of the workplace play a crucial role in influencing the brain's reorganization for adaptation and proficiency. When workspaces fit with the brain's requirements, employees are more inclined to maintain focus, manage emotions, retain new information, and apply skills proficiently (Tyagi et al., 2024).

The establishment of a neuro-friendly environment commences with sensory and spatial design. Natural light, proximity to vegetation, ergonomic furnishings, and tranquil areas have been demonstrated to alleviate mental tiredness and facilitate cognitive rejuvenation. Environments that facilitate movement, diversity, and individual regulation of lighting or temperature activate the brain's attentional networks and diminish stress, hence enhancing engagement with intricate activities (Xu et al., 2024; Zhang et al., 2024).

The psychological structure of the environment is equally significant. Cognitive development thrives in environments that cultivate psychological safety, allowing individuals to inquire, question assumptions, acknowledge errors, and investigate concepts without the apprehension of criticism. A culture of trust and emotional support stimulates brain pathways associated with reward and motivation, reinforcing the basis for enduring learning and innovative thought (Singha 2024; McClintock and Fainstad, 2022).

Furthermore, the design of workflows and tasks is crucial in fostering neuroplastic involvement. Occupations that require problem-solving, teamwork, and opportunities for learning engage the brain far more than monotonous, repetitive occupations. Integrating components like job rotation, project-based assignments, or cross-functional collaboration stimulates employees to adopt novel perspectives, create new brain pathways, and cultivate transferable cognitive skills (Thiruchanuru, 2025; Marois & Lafond, 2022).

An really neuro-friendly setting recognizes the significance of rest and recuperation. Breaks, flexible timetables, and assistance for sleep hygiene are essential neurological requirements rather than just luxury. Similar to how neurons consolidate memories during periods of rest, the mind rejuvenates clarity and insight when afforded the opportunity to pause. The firms that emphasize neuro-friendly design convey a significant message: that employee potential is dynamic, and that every element of the workplace contributes to cognitive development. By creating environments that enhance and nurture cognitive function, organizations can transform into catalysts for innovation, resilience, and growth.

3.2 Learning Architectures: Microlearning, Immersive Tech, and Habitual Practice

As firms endeavor to develop adaptive talent in an era of disruption, traditional training techniques are demonstrating inadequacy. Conventional workshops and sporadic seminars sometimes do not yield enduring behavioral transformation or profound skill development. Effective learning architectures must be intentionally crafted to correspond with the brain's natural processes of encoding, storing, and retrieving knowledge, utilizing the concepts of neuroplasticity. Prominent tactics include microlearning, immersive technologies, and habitual practice (Goldstein et al., 2024; Wang et al., 2023).

Microlearning entails the provision of material in brief, concentrated chunks that can be swiftly assimilated and readily reiterated. This approach corresponds with the brain's inclination for modest cognitive loads and spaced repetition, rendering it optimal for consolidating knowledge and fostering new neural connections over time (Alias & Razak, 2025). By incorporating microlearning into daily workflows—via smartphone reminders, interactive quizzes, or concise video modules—organizations may guarantee that learning evolves into a continuous, integrated process rather than a distinct, episodic occurrence.

Immersive technologies, such as virtual reality (VR), augmented reality (AR), and interactive simulations, generate extremely engaging settings that replicate real-world difficulties. These technologies simultaneously engage various sensory channels, so augmenting attention, emotional involvement, and memory encoding

(Sung et al., 2022; Wenk et al., 2023). Immersive learning experiences offer a secure and engaging environment for experimentation and neural reconfiguration, particularly for the acquisition of intricate skills like decision-making under pressure and managing interpersonal dynamics. Students are not merely educated but fundamentally changed through experience engagement.

The role of habitual practice is equally crucial, as it transforms conscious skills into automatic habits. The brain fortifies neural networks via repetition and reinforcement, a crucial mechanism for skill acquisition. Organizations must provide regular opportunity for people to utilize new knowledge in significant contexts. This encompasses real-time feedback, coaching systems, social learning communities, and reflection tools that reinforce learning into enduring competence (Zeine et al., 2024; Dubinsky & Hamid, 2024).

By incorporating these three elements—microlearning for accessibility, immersive technology for depth, and habitual practice for sustainability—organizations establish a resilient learning framework that reflects the brain's adaptive characteristics. Instead of perceiving learning as an isolated activity, it is recognized as a continuous, neurologically enhanced process integrated into the workflow. By doing so, corporations do not merely provide information; they cultivate settings that foster intellectual growth, enhance skills, and embed innovation as a fundamental aspect of existence.

3.3 Leadership Models that Promote Brain-Based Learning Cultures

In a time characterized by complexity and swift transformation, leadership transcends mere direction-setting or performance enhancement; it increasingly involves fostering conditions conducive to continual learning and cognitive development. Leaders who adopt brain-based methodologies acknowledge that organizational advancement is intrinsically linked to individual cognitive development. By connecting leadership behaviors with neuroplasticity principles, organizations can cultivate cultures characterized by adaptability, curiosity, and innovation as integral components of their identity (Gkintoni et al., 2022; Goldberg, 2022; Ahsan, 2025).

Brain-based leadership commences with self-awareness and emotional management. Leaders that exemplify composure in high-pressure situations, exhibit empathy, and constructively address stress stimulate mirror neurons within their teams, thereby affecting emotional states and fostering psychological safety. This safety is crucial for neuroplastic development, as the brain optimally learns in contexts characterized by low threat and engaged reward pathways through trust, recognition, and autonomy (Sharma and Jindal, 2024).

Effective leaders perceive mistakes as chances for learning rather than as failures. When employees are motivated to experiment, reflect, and iterate without the apprehension of retribution, their cognitive processes are more inclined to participate in exploratory behaviors that promote the formation of new neural connections. Leaders who normalize failure as an aspect of growth empower teams to undertake intellectual risks, pose audacious inquiries, and reconfigure cognitive frameworks in beneficial manners (Benoliel & Berkovich, 2021; Walton et al., 2024).

A distinguishing feature of brain-based leadership is the capacity to create and sustain learning ecosystems. These leaders transcend mere delegation of training to HR departments; they integrate learning into team rituals, establish consistent feedback mechanisms, and advocate for collaborative problem-solving. They emphasize long-term growth rather than immediate results and allocate resources to technology and platforms that tailor learning routes to employees' cognitive types and changing requirements (Sathiyaseelan et al. 2024).

These leadership approaches are not constrained by hierarchical frameworks. Leaders, regardless of their formality, can act as catalysts for neuroplastic change by deliberately influencing, communicating, and fostering mentality. They pose thoughtful inquiries, promote metacognition, and assist individuals in deriving significance from their endeavors—essential catalysts for enduring development. By fostering a culture grounded in neurobiological principles, organizations enable their personnel to perpetually develop. Consequently, these adaptive, cognitively nimble individuals serve as the foundation of robust, future-oriented institutions—where leadership transcends mere direction and focuses on unleashing the boundless potential of the human intellect.

Table 4. Cultural & Demographic Inclusivity Considerations

Dimension	Implications for Neuroadaptive Practice	Recommended Strategies
Age Diversity	Different neuroplastic capacities	Lifespan-based learning design
Neurodivergence	Varying cognitive styles	Universal design for learning (UDL)
Gender & Culture	Diverse motivation and social cognition	Inclusive communication frameworks

Source: Constructed by the authors based on literature in inclusive design, neurodiversity, and leadership equity

SOLUTIONS AND RECOMMENDATIONS

Cognitive Agility and Skill Mastery as Drivers of Organizational Performance

4.1 Defining Cognitive Agility and Its Organizational Value

Cognitive agility denotes the capacity of individuals to alter mental frames, modify thinking habits, and respond adaptively to new and intricate situations. It is not solely a cognitive attribute but a dynamic competency that empowers people to function adeptly in contexts characterized by uncertainty, ambiguity, and swift transformation. In contemporary companies, characterized by perpetual disruption and the insufficiency of linear thinking, cognitive agility has become a vital factor influencing both individual and organizational performance (Rafi et al., 2022; Al-Omoush, 2022).

Cognitive agility prioritizes the manner of thinking over the accumulation of knowledge, in contrast to conventional assessments of intelligence or competence. It encompasses metacognition—the recognition of one's cognitive processes—and the deliberate regulation of mental tactics. Cognitively agile individuals may transition among several thinking kinds, including analytical, creative, emotive, and strategic, based on situational requirements. This capacity to alternate cognitive states improves decision-making, promotes innovation, and facilitates intricate problem-solving (Renzulli & Reis, 2021; Kumar et al., 2023).

Cognitive agility at the organizational level manifests as adaptation, resilience, and ongoing enhancement. Teams consisting of intellectually nimble individuals are more inclined to accept change, react positively to feedback, and work across divisions. They are more adept at questioning assumptions, evaluating new hypotheses, and deriving lessons from failure. This fosters a culture of learning and invention crucial for enduring competitiveness (Goraya et al., 2024; Jaafar et al., 2025).

Moreover, cognitive agility facilitates strategic alignment and implementation. In dynamic industries, the capacity to swiftly understand changing priorities, analyze intricate data, and adjust strategies is essential. Leaders possessing elevated cognitive agility can foresee change, perceive uncertainty as an opportunity, and navigate their teams through transition while maintaining focus and morale (Christofi et al., 2024; Jaafar et al., 2025).

The significance of cognitive agility is enhanced when bolstered by neuroplastic learning contexts. When firms intentionally foster environments conducive to cognitive development—such as cultures rich in feedback, reflective practices, and psychologically safe spaces—cognitive agility becomes both teachable and scalable. It transforms from an uncommon attribute into an essential competency integrated into the organization's foundation. In the era of disruption, cognitive agility is vital, not optional. It helps individuals to excel in complexity and allows organizations to progress from mere survival to strategic reinvention, harnessing the full capacity of the human intellect.

4.2 Neuroplasticity and the Future of Upskilling and Reskilling

The rapid advancement of technology and changing employment demands have rendered upskilling and reskilling essential for the survival of both individuals and companies. In this evolving setting, neuroplasticity provides a fundamental perspective for rethinking the future of workforce development. Neuroplasticity reconceptualizes abilities as dynamic and talent as adaptable, viewing human potential as flexible, expandable, and sensitive to specific interventions (Asiedu & Tenakwah, 2025; Li, 2024).

Upskilling pertains to the augmentation of current skills, whereas reskilling is the acquisition of wholly new competencies, frequently in anticipation of a transition in position or career trajectory. Both processes necessitate the brain's development of new neural pathways, assimilation of novel information, and modification of behavioral patterns. Neuroplasticity—the brain's ability to reorganize itself in response to experiences—serves as the basic mechanism enabling such transformation at every life stage.

For learning to yield enduring skill acquisition, the brain must be engaged continuously and purposefully. Passive engagement with content is inadequate. Learning settings should be structured to stimulate cognitive engagement, emotional significance, and practical application. These components enhance synaptic plasticity and facilitate the consolidation of knowledge into long-term memory. Consequently, forthcoming educational models should transcend traditional lectures and discrete modules, embracing immersive, reflective, and collaborative formats that align with optimal cognitive learning processes (Mystakidis et al., 2021; Dubovi, 2022).

Neuroplasticity research underscores the significance of personalized learning trajectories. Each brain is distinct, molded by individual experiences, strengths, and inclinations. Organizations that use adaptive technologies, individualized feedback mechanisms, and variable pacing enable learners to advance in accordance with their cognitive profiles. This not only promotes outcomes but also increases motivation and engagement.

Leaders and talent developers (Nandini et al., 2022; Malek and Almarri, 2023) must also adopt a growth-oriented mentality. When neuroplasticity is integrated into the company culture, difficulties are viewed as chances for neural advancement rather than threats to capability. Errors serve as opportunities for cognitive restructuring. Feedback serves as a catalyst for development. In these situations, the workforce becomes increasingly adaptable, robust, and adept at managing upheaval. By integrating upskilling and reskilling plans with neuroplasticity principles, organizations may transition from reactive talent management to proactive capability development, preparing individuals for both current roles and future transformations.

4.3 From Individual Brain Change to Collective Intelligence

Neuroplasticity initiates at the individual brain level, although its profound potential is realized when personal alterations amalgamate into communal intelligence. This transition in the workplace signifies a movement from solitary skill enhancement to a connected framework of collective knowledge, flexible reasoning, and cooperative problem-solving. As individuals reconfigure their cognitive processes to enhance agility, creativity, and reflection, they foster the development of a cognitively enriched organizational culture (Sulis, 2021).

Collective intelligence denotes the emerging ability of a collective to think, learn, and act in manners that exceed the aggregate of its individual components. It originates not solely from the information possessed by each individual but also from their interactions, communication, and collective responses to change. When individual neuroplastic development is fostered by a culture of openness, psychological safety, and knowledge dissemination, these connections transform into the synapses of a communal intellect—perpetually changing, learning, and innovating (Ha & Tang, 2022; Secundo et al., 2021).

Organizations that foster this evolution do more than educate people; they create an environment where thinking is a collective endeavor. Cross-functional collaboration, inclusive decision-making, and different views enhance brain diversity, mitigating cognitive rigidity and fostering complex reasoning. Teams comprised of cognitively agile individuals exhibit enhanced speed in pattern recognition, strategic adaptation, and collaborative solution development in real-time (Demir et al., 2023; Nazir et al., 2022).

The transition from individual to group cognition is not automatic. It necessitates deliberate frameworks that promote interdependence without uniformity—environments where discourse is prioritized over argument, and inquiry above assurance. Leaders should act as connections rather than mere directors, fostering cultures that encourage individuals to share ideas, challenge assumptions, and benefit from one another's intellectual experiences (Stodden et al., 2023; Schlicht & Starzak, 2019).

Digital platforms significantly enhance collective intelligence. Technologies facilitating collaborative learning, instantaneous feedback, and knowledge curation can enhance and expedite the dissemination of insights throughout an organization. In this environment, the organization functions as a dynamic neural network—constantly assimilating, integrating, and reacting to information in a manner that reflects the brain's adaptive characteristics (Secundo et al., 2021; Kretschmer et al., 2022). As individual reconfigurations proliferate across teams and departments, they provide the groundwork for a genuinely intelligent enterprise—one that adapts, evolves, and flourishes amid disruption not via top-down control, but through the collective ingenuity of its personnel. This neural network metaphor elucidates the dynamic interaction between individual neuroplastic changes and the development of collective intelligence in organizations, demonstrating how communication patterns, cognitive diversity, and feedback loops form a living, adaptive knowledge system.

Figure 1. The Neuroadaptive Learning Cycle: Stages of Brain-Based Workplace Transformation

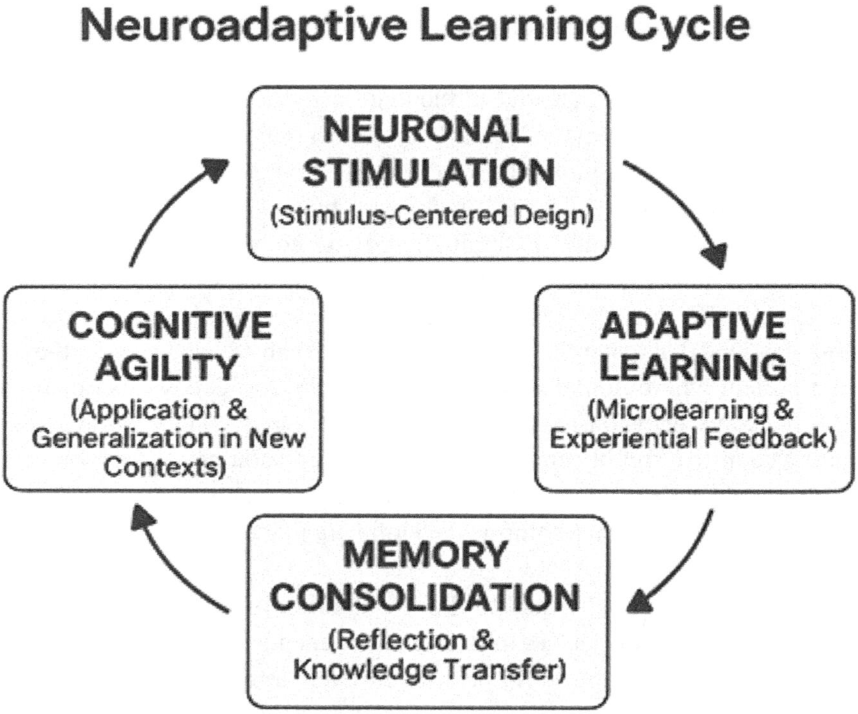

Source: Constructed by the authors experiential learning theory. Canva design

FUTURE RESEARCH DIRECTIONS

Strategic Implications and Future Directions

5.1 Policy Recommendations for Integrating Neuroscience in HR Practices

To effectively leverage neuroplasticity in developing future-ready businesses, human resource (HR) policies must be fundamentally reoriented. HR strategies can no longer depend exclusively on static competency frameworks, inflexible performance indicators, or uniform training methodologies. Policies should embody a dynamic comprehension of human learning, development, and adaptation, rooted in neuroscientific principles (Roy, 2024; Kobets et al., 2024).

The fundamental tenet of neuroscience-informed HR is that each employee is a developing system, susceptible of change given appropriate conditions. Policies should consequently transition from assessing ability based on static characteristics to fostering it through surroundings that promote neuroplastic transformation. Recruitment tactics must emphasize cognitive agility, learning potential, and emotional intelligence, rather than solely focusing on experience or technical proficiency. Assessment models must incorporate flexibility, curiosity, and growth trajectories (Schiuma et al. 2022; Leso et al., 2023).

Learning and development policies must prioritize ongoing, individualized, and context-specific education. Organizations ought to institutionalize microlearning platforms, immersive simulations, and mentorship programs aimed at enhancing brain engagement and memory consolidation. Human Resources must ensure that learning is integrated into the daily workflow, rather than confined to discrete events. This method not only corresponds with the brain's knowledge retention mechanisms but also equalizes access to developmental opportunities across various jobs and levels (Roy et al., 2025; Sharma and Kohli, 2024).

Workplace wellness policies must likewise adapt. Neuroplasticity flourishes when the brain is rested, secure, and emotionally stable. This necessitates HR to promote policies that endorse mental health, work-life integration, and psychological safety. Regular pauses, flexible scheduling, and mindfulness programs have become essential for maintaining cognitive performance and resilience (Singha2024).

Moreover, performance management systems must to be restructured to incentivize experimentation, iterative enhancement, and collaborative learning. Organizations should normalize failure as an integral component of neuronal development rather than penalize it. Feedback systems must be prompt, constructive, and conducive to cognitive processing—reinforcing positive behaviors and providing clear and considerate guidance for redirection. Neuroscience should not be limited to research;

it must be integrated into the strategic framework of the firm. By integrating neuroplastic principles into HR practices, firms demonstrate a profound commitment to realizing human potential, enhancing not only productivity but also purpose, growth, and transformative influence in an era of disruption. The subsequent table presents a phased implementation roadmap to assist corporate leaders in converting neuroplasticity concepts into practical workplace transformation tactics.

Table 5. Organizational Rewiring Flow – Implementation Plan

Step	Focus Area	Tools & Techniques
Neuro-Capability Audit	Cognitive readiness	Surveys, burnout metrics, upskilling index
Learning Architecture Design	Learning pathways	Microlearning, immersive tech, design thinking
Neurocompetent Leadership	Cognitive leadership	Neuroscience-informed coaching, EI training
KPI Alignment	Performance integration	Learning agility metrics, resilience indicators
Iteration & Evolution	Feedback & scale	Learning analytics, continuous improvement loops

Source: Developed by the authors based on strategic HR models and neuroadaptive learning design

5.2 Measuring the Impact: From Neuroplastic Change to KPIs

Translating the imperceptible process of neuroplasticity into concrete organizational results poses both a challenge and an opportunity. The internal rewiring of brain networks can and should be assessed by strategic key performance indicators (KPIs) that reflect enhancements in cognitive agility, learning capacity, and skill mastery. Organizations must advance their evaluation frameworks to encompass not just the knowledge acquired but also the manner in which individuals adapt, implement, and transfer that knowledge in practical scenarios (Singha, 2024; Reddy, 2025).

A primary consequence of neuroplasticity in the workplace is accelerated learning velocity—the rapidity and profundity with which individuals develop new skills. Key Performance Indicators (KPIs) may encompass time-to-competency for newly acquired skills, the rate of upskilling and reskilling, and the percentage of the workforce participating in continuous learning initiatives. These factors denote not merely participation, but active involvement in brain-based development activities (Marois & Lafond, 2022; Chen et al. 2021).

Cognitive flexibility is another crucial criterion, evaluated by behavioral signs including the capacity to switch tasks, adapt positively to change, and innovate under duress. Pulse surveys, adaptive performance evaluations, and 360-degree feedback mechanisms can effectively document changes in cognitive patterns, emotional control, and resilience in high-pressure contexts (van et al., 2022; Lin, 2021).

The transfer and application of knowledge further demonstrate the effects of neuroplasticity. Can staff convert fresh insights into enhanced procedures, superior decision-making, or innovative solutions? Organizations can assess this using innovation metrics, project results, and the success rates of cross-functional collaboration (Marzola et al., 2023; Prigatano et al., 2021). The team-level intelligence—the cumulative ability of organizations to address challenges and acquire knowledge collectively—can serve as an indicator of collective cognitive advancement. Indicators may encompass the frequency of multidisciplinary idea exchange, the efficacy of knowledge-sharing platforms, and the diversity of perspectives represented in strategic formulation (Tavoletti et al., 2024; Gupta et al., 2023).

Organizations must incorporate well-being and mental performance metrics into a holistic measurement plan. Neuroplastic development is unsustainable in the absence of psychological equilibrium. Monitoring burnout rates, employee engagement, cognitive tiredness, and emotional resilience provides a comprehensive understanding of the organizational climate for brain-based learning (Fraboni et al., 2023; Dias et al., 2024). By correlating neuroplastic development with pertinent KPIs, organizations can transition from anecdotal success narratives to evidence-based methodologies. This transition allows leadership to invest assuredly in settings, processes, and behaviors that realize the brain's full potential—converting imperceptible change into quantifiable excellence.

5.3 Building the Neuroadaptive Organization of Tomorrow

The future of work is reserved for firms that not only react to disruption but also adapt alongside it. The essence of this transformation is the neuroadaptive organization—an entity that methodically synchronizes its structures, cultures, and strategies with the principles of human cognitive learning, adaptation, and flourishing. These firms regard neuroplasticity not as a mere intellectual concept, but as a pragmatic framework for enduring innovation, agility, and resilience (Makanadar, 2024; Zeine et al., 2024).

A neuroadaptive organization redefines talent development as an enduring neural progression. Learning is integrated into the workflow, facilitated by immediate feedback, practical experience, and technology-driven customization. Formal training constitutes but one component of a more extensive ecosystem encompassing coaching, collaborative research, reflective intervals, and knowledge exchange. Learning is ongoing, contextual, and cumulative, rather than episodic (Dai et al., 2023).

Subsequently, these firms reevaluate leadership paradigms. Leaders are not solely decision-makers but neuro-facilitators who foster conditions that promote intellectual curiosity and emotional security. They promote exploration, accept failure as an integral aspect of development, and motivate teams to expand cognitive limits. The leader's function transitions from directing performance to unleashing potential.

Neuroadaptive organizations prioritize flexibility rather than rigidity in their structure. They transition from isolated departments and straight career trajectories, prioritizing agile teams, role adaptability, and cross-functional cooperation. These dynamics reflect the brain's intrinsic network architecture—decentralized, adaptive, and interlinked (Savitha & Kumar, 2025). Technology serves not just as a tool for automation but also as a facilitator of cognitive processes. AI, VR, and data analytics are combined not to supplant human cognition, but to enhance it—customizing development, improving decision-making, and broadening creative potential. Digital tools serve as extensions of the organizational intellect, facilitating collective intelligence and scalable adaptive learning (Ha & Tang, 2022; Järvelä et al., 2023).

These firms integrate a development mentality and neuroethical awareness into their core identity. They regard mental health, diversity of thought, and inclusivity as essential and non-negotiable pillars of advancement. All policies, practices, and platforms are formulated with consideration for cognitive processes (Cleven et al., 2023; Wexler & Specker Sullivan, 2023). Establishing the neuroadaptive organization is not a technological obstacle; it is a human endeavor. It necessitates leaders who recognize the boundless capabilities of the mind and are prepared to transform the workplace into a dynamic, educational institution. Such businesses will not merely endure disruption; they will spearhead the navigation through it.

CONCLUSION

The process of establishing neuroadaptive organizations commences with a fundamental transformation in our perception of human potential. This book argues that, unlike traditional theories which perceive talent as static and learning as a distinct phase, cognitive agility and skill mastery are ongoing processes influenced by neurobiology. Neuroplasticity provides a robust basis for this transition, demonstrating that the brain—similar to the workplace—is intrinsically adaptable and receptive to deliberate design.

By incorporating neuroplastic principles into workplace environments, educational frameworks, leadership paradigms, and human resource policies, firms can cultivate conditions in which workers do not only respond to change but actively progress through it. The means for transforming the workplace, ranging from microlearning and immersive technologies to brain-based leadership and inclusive cultures of experimentation, are now available. What is necessary is the readiness to implement them with intent and regularity.

Moreover, the correlation between individual cerebral development and collective intelligence highlights a significant realization: as individual minds are restructured, so are teams, departments, and entire companies. This systematic development allows businesses to foster agility at scale, helping them to create, collaborate, and lead in unpredictable environments. Measurement and policy alignment are crucial for maintaining these transformations. By monitoring cognitive engagement, learning velocity, emotional well-being, and team adaptability, firms can transform imperceptible brain changes into strategic advantages.

Neuroplasticity transcends a mere scientific idea; it is a strategic necessity. Organizations that comprehend and adopt this reality will not only endure the tides of disruption but will also become designers of the future workplace: adaptable, insightful, and human-centric. The future workplace is currently being shaped by the individuals we decide to empower.

REFERENCES

Ahsan, M. J. (2025). Cultivating a culture of learning: The role of leadership in fostering lifelong development. *The Learning Organization*, *32*(2), 123–138. DOI: 10.1108/TLO-03-2024-0099

Al-Omoush, K. S. (2022). Understanding the impact of intellectual capital on e-business entrepreneurial orientation and competitive agility: An empirical study. *Information Systems Frontiers*, *24*(2), 437–452. DOI: 10.1007/s10796-020-10092-7

Alias, N. F., & Razak, R. A. (2025). Revolutionizing learning in the digital age: A systematic literature review of microlearning strategies. *Interactive Learning Environments*, *33*(1), 1–21. DOI: 10.1080/10494820.2024.2331638

Ammirato, S., Felicetti, A. M., Linzalone, R., Corvello, V., & Kumar, S. (2023). Still our most important asset: A systematic review on human resource management in the midst of the fourth industrial revolution. *Journal of Innovation & Knowledge*, *8*(3), 100403. DOI: 10.1016/j.jik.2023.100403

Aroles, J., Cecez-Kecmanovic, D., Dale, K., Kingma, S. F., & Mitev, N. (2021). New ways of working (NWW): Workplace transformation in the digital age. *Information and Organization*, *31*(4), 100378. DOI: 10.1016/j.infoandorg.2021.100378

Asiedu, E., & Tenakwah, E. S. (2025). Future-proofing your workforce: Upskilling and reskilling as HR's top priorities. *Strategic HR Review*, *24*(1), 17–22. DOI: 10.1108/SHR-01-2025-0008

Babüroğlu, O. N., & Selsky, J. W. (2021). Toward reconfiguring sociotechnical systems design: Digitally infused work systems and the "Platform-STS". In Shani, A. B., & Noumair, D. A. (Eds.), *Research in Organizational Change and Development* (Vol. 29, pp. 63–87). Emerald Publishing Limited., DOI: 10.1108/S0897-301620210000029004

Benoliel, P., & Berkovich, I. (2021). Learning from intelligent failure: An organizational resource for school improvement. *Journal of Educational Administration*, *59*(4), 402–421. DOI: 10.1108/JEA-07-2020-0155

Berrone, P., Rousseau, H. E., Ricart, J. E., Brito, E., & Giuliodori, A. (2023). How can research contribute to the implementation of sustainable development goals? An interpretive review of SDG literature in management. *International Journal of Management Reviews*, *25*(2), 318–339. DOI: 10.1111/ijmr.12331

Chen, O., Paas, F., & Sweller, J. (2023). A cognitive load theory approach to defining and measuring task complexity through element interactivity. *Educational Psychology Review*, *35*, 63. DOI: 10.1007/s10648-023-09782-w

Chen, Q., Baran, T. M., Turnbull, A., Zhang, Z., Rebok, G. W., & Lin, F. V. (2021). Increased segregation of structural brain networks underpins enhanced broad cognitive abilities of cognitive training. *Human Brain Mapping*, *42*(10), 3202–3215. DOI: 10.1002/hbm.25428

Chen, S. A., & Goodwill, A. M. (2022). Neuroplasticity and adult learning. In Illeris, K. (Ed.), *Third International Handbook of Lifelong Learning* (pp. 1–19). Springer International Publishing., DOI: 10.1007/978-3-030-67930-9_43-1

Christofi, K., Chourides, P., & Papageorgiou, G. (2024). Cultivating strategic agility: An empirical investigation into best practice. *Global Business and Organizational Excellence*, *43*(3), 89–105. DOI: 10.1002/joe.22241

Cleven, A. J., Renaud, A., Larose-Pierre, M., McQuade, B., Griffin, B. L., Johnson, C., & Hughes, J. A. (2023). Associating growth mindset with emotional intelligence and why it's needed for professional identity formation. *American Journal of Pharmaceutical Education*, *87*(6), 100110. DOI: 10.1016/j.ajpe.2023.100110

Dai, D. Y., Yun, D., Dai, X., & Sun, Q. (2023). *Talent development from the perspective of developmental science*. Springer., DOI: 10.1007/978-3-031-46204-7

Demir, M., Canan, M., & Cohen, M. C. (2023). Modeling team interaction and decision-making in agile human–machine teams: Quantum and dynamical systems perspective. *IEEE Transactions on Human-Machine Systems*, *53*(4), 720–730. DOI: 10.1109/THMS.2023.3241234

Dias, S., Luís, S., & Cruz, B. (2024). Measuring well-being at organizational context: Exploring the Better Life Index as a measurement tool. *Social Responsibility Journal*, *20*(1), 45–62. DOI: 10.1108/SRJ-07-2023-0375

Dubinsky, J. M., & Hamid, A. A. (2024). The neuroscience of active learning and direct instruction. *Neuroscience and Biobehavioral Reviews*, *163*, 105737. DOI: 10.1016/j.neubiorev.2024.105737

Dubovi, I. (2022). Cognitive and emotional engagement while learning with VR: The perspective of multimodal methodology. *Computers & Education*, *184*, 104495. DOI: 10.1016/j.compedu.2022.104495

Dzyubenko, E., & Hermann, D. M. (2023). Role of glia and extracellular matrix in controlling neuroplasticity in the central nervous system. *Seminars in Immunopathology*, *45*(3), 377–387. DOI: 10.1007/s00281-023-00995-3

Ekman, R., Fletcher, A., Giota, J., Eriksson, A., Thomas, B., & Bååthe, F. (2022). A flourishing brain in the 21st century: A scoping review of the impact of developing good habits for mind, brain, well-being, and learning. *Mind, Brain and Education: the Official Journal of the International Mind, Brain, and Education Society*, *16*(1), 13–23. DOI: 10.1111/mbe.12305

Fagerdal, B., Lyng, H. B., Guise, V., Anderson, J. E., Thornam, P. L., & Wiig, S. (2022). Exploring the role of leaders in enabling adaptive capacity in hospital teams: A multiple case study. *BMC Health Services Research*, *22*, 908. DOI: 10.1186/s12913-022-08296-5

Fraboni, F., Brendel, H., & Pietrantoni, L. (2023). Evaluating organizational guidelines for enhancing psychological well-being, safety, and performance in technology integration. *Sustainability*, *15*(10), 8113. DOI: 10.3390/su15108113

Franco, C., & Landini, F. (2022). Organizational drivers of innovation: The role of workforce agility. *Research Policy*, *51*(2), 104423. DOI: 10.1016/j.respol.2021.104423

Gkintoni, E., Halkiopoulos, C., & Antonopoulou, H. (2022). Neuroleadership as an asset in educational settings: An overview. *Emerging Science Journal*, *6*(4), 893–904. DOI: 10.28991/esj-2022-06-04-016

Goldberg, H. (2022). Growing brains, nurturing minds—Neuroscience as an educational tool to support students' development as life-long learners. *Brain Sciences*, *12*(12), 1622. DOI: 10.3390/brainsci12121622

Goldstein, A., Grinstein-Dabush, A., Schain, M., Wang, H., Hong, Z., Aubrey, B., & Hasson, U. (2024). Alignment of brain embeddings and artificial contextual embeddings in natural language points to common geometric patterns. *Nature Communications*, *15*, 2768. DOI: 10.1038/s41467-024-46631-y

Goraya, M. A. S., Yaqub, M. Z., Khan, M. A., Akram, M. S., & Alofaysan, H. (2024). Transforming performance: How agility, response, resilience and support shape success in digital strategies. *Information Technology & People*. Advance online publication. DOI: 10.1108/ITP-05-2024-0592

Gupta, P., Nguyen, T. N., Gonzalez, C., & Woolley, A. W. (2023). Fostering collective intelligence in human–AI collaboration: Laying the groundwork for COHUMAIN. *Topics in Cognitive Science*, *15*(3), 564–582. DOI: 10.1111/tops.12679

Ha, D., & Tang, Y. (2022). Collective intelligence for deep learning: A survey of recent developments. *Collective Intelligence*, *1*, 26339137221114874. DOI: 10.1177/26339137221114874

Innocenti, G. M. (2022). Defining neuroplasticity. *Handbook of Clinical Neurology, 184*, 3–18. DOI: 10.1016/B978-0-12-819410-2.00001-1

Jaafar, M., Khan, K. N., & Salman, A. (2025). *A systematic review and framework for organizational agility antecedents towards Industry 4.0.* Management Review Quarterly., DOI: 10.1007/s11301-025-00489-6OUCI

Järvelä, S., Nguyen, A., & Hadwin, A. (2023). Human and artificial intelligence collaboration for socially shared regulation in learning. *British Journal of Educational Technology, 54*(5), 1057–1076. DOI: 10.1111/bjet.13325

Johnson, B. P., & Cohen, L. G. (2023). Applied strategies of neuroplasticity. *Handbook of Clinical Neurology, 196*, 599–609. DOI: 10.1016/B978-0-323-98817-9.00011-9

Joshua, A. M. (2022). Neuroplasticity. In *Physiotherapy for adult neurological conditions* (pp. 1–30). Springer Nature., DOI: 10.1007/978-981-19-0209-3_1

Kang, B., Zhao, C., Ma, J., Wang, H., Gu, X., Xu, H., & Xiao, L. (2023). Electroacupuncture alleviates pain after total knee arthroplasty through regulating neuroplasticity: A resting-state functional magnetic resonance imaging study. *Brain and Behavior, 13*(3), e2913. DOI: 10.1002/brb3.2913

Khurana, I., Dutta, D. K., & Ghura, A. S. (2022). SMEs and digital transformation during a crisis: The emergence of resilience as a second-order dynamic capability in an entrepreneurial ecosystem. *Journal of Business Research, 150*, 623–641. DOI: 10.1016/j.jbusres.2022.06.048

Kobets, K., Terentieva, N., Shkvyria, N., Lysytsia, N., & Siemak, I. (2024). Digitalization and its impact on the development of contemporary marketing strategies. *Economic Affairs, 69*(2), 1021–1040. DOI: 10.46852/0424-2513.3.2024.26

Kretschmer, T., Leiponen, A., Schilling, M., & Vasudeva, G. (2022). Platform ecosystems as meta-organizations: Implications for platform strategies. *Strategic Management Journal, 43*(3), 405–424. DOI: 10.1002/smj.3250

Kumar, P., Pillai, R., Kumar, N., & Tabash, M. I. (2023). The interplay of skills, digital financial literacy, capability, and autonomy in financial decision making and well-being. *Borsa Istanbul Review, 23*(1), 1–12. DOI: 10.1016/j.bir.2022.09.001

Kumprang, K., & Suriyankietkaew, S. (2024). Mechanisms of organizational mindfulness on employee well-being and engagement: A multi-level analysis. *Administrative Sciences, 14*(6), 121. DOI: 10.3390/admsci14060121

Leso, B. H., Cortimiglia, M. N., & Ghezzi, A. (2023). The contribution of organizational culture, structure, and leadership factors in the digital transformation of SMEs: A mixed-methods approach. *Cognition Technology and Work*, *25*(1), 1–17. DOI: 10.1007/s10111-022-00714-2

Li, J., & Yeo, R. K. (2024). Artificial intelligence and human integration: A conceptual exploration of its influence on work processes and workplace learning. *Human Resource Development International*, *27*(3), 367–387. DOI: 10.1080/13678868.2024.2348987

Li, L. (2024). Reskilling and upskilling the future-ready workforce for Industry 4.0 and beyond. *Information Systems Frontiers*, *26*(5), 1713–1732. DOI: 10.1007/s10796-022-10308-

Li, W., Huan, W., Hou, B., Tian, Y., Zhang, Z., & Song, A. (2021). Can emotion be transferred?—A review on transfer learning for EEG-based emotion recognition. *IEEE Transactions on Cognitive and Developmental Systems*, *14*(3), 833–846. DOI: 10.1109/TCDS.2021.3098842

Lim, W. M. (2023). The workforce revolution: Reimagining work, workers, and workplaces for the future. *Global Business and Organizational Excellence*, *42*(4), 5–10. DOI: 10.1002/joe.22218

Lin, C. E. (2021). Cognitive flexibility. In Volkmar, F. R. (Ed.), *Encyclopedia of Autism Spectrum Disorders* (pp. 1–3). Springer., DOI: 10.1007/978-1-4614-6435-8_102461-1

Luciana, M., & Collins, P. F. (2022). Neuroplasticity, the prefrontal cortex, and psychopathology-related deviations in cognitive control. *Annual Review of Clinical Psychology*, *18*, 443–469. DOI: 10.1146/annurev-clinpsy-081219-111203

Mahmoud, A. B., Berman, A., Reisel, W., Fuxman, L., & Hack-Polay, D. (2024). Examining generational differences as a moderator of extreme-context perception and its impact on work alienation organizational outcomes: Implications for the workplace and remote work transformation. *Scandinavian Journal of Psychology*, *65*(1), 70–85. DOI: 10.1111/sjop.12955

Makanadar, A. (2024). Neuro-adaptive architecture: Buildings and city design that respond to human emotions, cognitive states. *Research in Globalization*, *8*, 100213. DOI: 10.1016/j.resglo.2024.100213

Malek, M. A., & Almarri, K. (2023, December). The role of adaptive and transformational leadership in the successful adoption and implementation of new technologies and innovations in organizations. In *European, Mediterranean, and Middle Eastern Conference on Information Systems* (pp. 120–132). Springer Nature Switzerland. https://doi.org/DOI: 10.1007/978-3-031-56481-9_9OUCI+1

Marois, A., & Lafond, D. (2022). Augmenting cognitive work: A review of cognitive enhancement methods and applications for operational domains. *Cognition Technology and Work*, *24*(4), 729–746. DOI: 10.1007/s10111-022-00715-1

Marzola, P., Melzer, T., Pavesi, E., Gil-Mohapel, J., & Brocardo, P. S. (2023). Exploring the role of neuroplasticity in development, aging, and neurodegeneration. *Brain Sciences*, *13*(12), 1610. DOI: 10.3390/brainsci13121610

Matar, D., Serhan, A., El Bilani, S., Faraj, R. A., Hadi, B. A., & Fakhoury, M. (2024). Psychopharmacological approaches for neural plasticity and neurogenesis in major depressive disorders. In *Recent Advances and Challenges in the Treatment of Major Depressive Disorder* (pp. 27–48). Springer Nature Singapore., DOI: 10.1007/978-981-97-4402-2_2

McClintock, A. H., & Fainstad, T. (2022). Growth, engagement, and belonging in the clinical learning environment: The role of psychological safety and the work ahead. *Journal of General Internal Medicine*, *37*(9), 2291–2296. DOI: 10.1007/s11606-022-07493-6

Miceli, A., Hagen, B., Riccardi, M. P., Sotti, F., & Settembre-Blundo, D. (2021). Thriving, not just surviving in changing times: How sustainability, agility and digitalization intertwine with organizational resilience. *Sustainability*, *13*(4), 2052. DOI: 10.3390/su13042052

Mystakidis, S., Berki, E., & Valtanen, J. P. (2021). Deep and meaningful e-learning with social virtual reality environments in higher education: A systematic literature review. *Applied Sciences (Basel, Switzerland)*, *11*(5), 2412. DOI: 10.3390/app11052412

Nandini, W., Gustomo, A., & Sushandoyo, D. (2022). The mechanism of an individual's internal process of work engagement, active learning and adaptive performance. *Economies*, *10*(7), 165. DOI: 10.3390/economies10070165MDPI

Nazir, S., Price, B., Surendra, N. C., & Kopp, K. (2022). Adapting agile development practices for hyper-agile environments: Lessons learned from a COVID-19 emergency response research project. *Information Technology Management*, *23*(3), 193–211. DOI: 10.1007/s10799-022-00370-y

Nyamrunda, F. C., & Freeman, S. (2021). Strategic agility, dynamic relational capability and trust among SMEs in transitional economies. *Journal of World Business, 56*(3), 101175. DOI: 10.1016/j.jwb.2020.101175

Paesano, A. (2023). Artificial intelligence and creative activities inside organizational behavior. *The International Journal of Organizational Analysis, 31*(5), 1694–1723. DOI: 10.1108/IJOA-09-2020-2421

Park, S., & Park, S. (2021). How can employees adapt to change? Clarifying the adaptive performance concepts. *Human Resource Development Quarterly, 32*(1), E1–E15. DOI: 10.1002/hrdq.21411

Pesqueira, A., & Almeida, D. (2025). Rethinking the future of mental health: AI, bioprinting, and the ethical evolution of diagnostics for ADHD and OCB. In *Smart Healthcare, Clinical Diagnostics, and Bioprinting Solutions for Modern Medicine* (pp. 233–258). IGI Global. https://doi.org/DOI: 10.4018/979-8-3373-0659-9.ch012

Prigatano, G. P., Braga, L. W., Johnson, S. F., & Souza, L. M. (2021). Neuropsychological rehabilitation, neuroimaging and neuroplasticity: A clinical commentary. *NeuroRehabilitation, 49*(2), 255–265. DOI: 10.3233/NRE-218024PubMed

Rafi, N., Ahmed, A., Shafique, I., & Kalyar, M. N. (2022). Knowledge management capabilities and organizational agility as liaisons of business performance. *South Asian Journal of Business Studies, 11*(4), 397–417. DOI: 10.1108/SAJBS-05-2020-0145

Reddy, K. J. (2025). Neuroplasticity and technology. In *Innovations in neurocognitive rehabilitation: Harnessing technology for effective therapy* (pp. 137–169). Springer Nature Switzerland., DOI: 10.1007/978-3-031-88117-6_7

Renzulli, J. S., & Reis, S. M. (2021). The three-ring conception of giftedness: A change in direction from being gifted to the development of gifted behaviors. In S. M. Reis (Ed.), *Reflections on gifted education* (pp. 55–86). Prufrock Press.ResearchGate

Roy, P., Pandit, R., & Naskar, S. K. (2025). Innovative approaches to problem-solving through educational technology integration. In Zhang, J., & Yahaya, W. W. (Eds.), *Integrating Technology in Problem-Solving Educational Practices* (pp. 375–402). IGI Global., DOI: 10.4018/979-8-3693-6745-2.ch016

Roy, S. (2024). Cognition, human potential, and talent management. In Shukla, V. K., Kulkarni, P., Gaur, D., Lacap, P. J., & Omrane, A. (Eds.), *Industry 4.0 and People Analytics* (pp. 1–23). CRC Press., DOI: 10.1201/9781003414193-1

Saeed, I., Khan, J., Zada, M., Ullah, R., Vega-Muñoz, A., & Contreras-Barraza, N. (2022). Towards examining the link between workplace spirituality and workforce agility: Exploring higher educational institutions. *Psychology Research and Behavior Management*, *15*, 31–49. DOI: 10.2147/PRBM.S344651

Sathiyaseelan, A., Balasundaram, S., Zirkler, M., & Patangia, B. (2024). Innovative leadership and sustainable development: Exploring resource-efficient strategies among Indian managers. *Business Strategy & Development*, *7*(4), e70049. DOI: 10.1002/bsd2.70049

Savitha, M., & Kumar, S. P. (2025). Building an agile organizational structure to drive innovation and technology. *Cuestiones de Fisioterapia*, *54*(2), 2321–2334. DOI: 10.48047/qej54344

Schiuma, G., Schettini, E., Santarsiero, F., & Carlucci, D. (2022). The transformative leadership compass: Six competencies for digital transformation entrepreneurship. *International Journal of Entrepreneurial Behaviour & Research*, *28*(5), 1273–1291. DOI: 10.1108/IJEBR-01-2021-0087

Schlicht, T., & Starzak, T. (2019). Prospects of enactivist approaches to intentionality and cognition. *Synthese*, *198*(Suppl 1), 89–113. DOI: 10.1007/s11229-019-02361-z

Secundo, G., Shams, S. M. R., & Nucci, F. (2021). Digital technologies and collective intelligence for healthcare ecosystem: Optimizing Internet of Things adoption for pandemic management. *Journal of Business Research*, *131*, 563–572. DOI: 10.1016/j.jbusres.2020.10.030

Sharma, D., & Jindal, P. (2024). Role of neuroleadership through effective communication in organization and employee development. In *Neuroleadership development and effective communication in modern business* (pp. 107–118). IGI Global., DOI: 10.4018/979-8-3693-4350-0.ch008

Sharma, R. C., & Kohli, D. (2024). Embracing digital transformation for sustainable development: The synergy between learning organizations, human resource management, and service quality. *Journal of Chinese Human Resource Management*, *14*(3), 3–17. DOI: 10.47297/wspchrmWSP2040-800501.20231403

Singha, R. (2024). Stress, resilience, and brain performance. In Sharma, R. C., & Kohli, D. (Eds.), *Building Organizational Resilience With Neuroleadership* (pp. 14–29). IGI Global., DOI: 10.4018/979-8-3693-1785-3.ch002

Singha, S. (2024). Nurturing positive organizational climates to enhance work success: A positive psychology approach. In Baykal, E. (Ed.), *Fostering Organizational Sustainability with Positive Psychology* (pp. 84–107). IGI Global., DOI: 10.4018/979-8-3693-1524-8.ch007

Skulmowski, A., & Xu, K. M. (2022). Understanding cognitive load in digital and online learning: A new perspective on extraneous cognitive load. *Educational Psychology Review, 34*(1), 171–196. DOI: 10.1007/s10648-021-09624-7

Stodden, D. F., Pesce, C., Zarrett, N., Tomporowski, P., Ben-Soussan, T. D., Brian, A., & Weist, M. D. (2023). Holistic functioning from a developmental perspective: A new synthesis with a focus on a multi-tiered system support structure. *Clinical Child and Family Psychology Review, 26*(2), 343–361. DOI: 10.1007/s10567-022-00376-1

Sulis, W. (2021). Contextuality in neurobehavioural and collective intelligence systems. *Quantum Reports, 3*(4), 592–614. DOI: 10.3390/quantum3040038

Sung, E. C., Han, D. I. D., Bae, S., & Kwon, O. (2022). What drives technology-enhanced storytelling immersion? The role of digital humans. *Computers in Human Behavior, 132*, 107246. DOI: 10.1016/j.chb.2022.107246

Tavoletti, E., Şahin, F., Taras, V., Farrell, W., & Florea, L. (2024). A configurational analysis of university ranking, cognitive ability and cultural intelligence in global virtual teams: The role of academic pedigree. *International Journal of Management Education, 22*(3), 101040. DOI: 10.1016/j.ijme.2024.101040

Thiruchanuru, S. (2025). Enhance the workforce efficiency and well-being using neurotechnology: A review. In *Technological enhancements for improving employee performance, safety, and well-being* (pp. 171–184). IGI Global., DOI: 10.4018/979-8-3693-9631-5.ch009

Tyagi, S., Tyagi, M., Srivastava, A. K., & Saluja, S. (2024). Neuroscience marketing: A new age marketing. In *Building Organizational Resilience With Neuroleadership* (pp. 215–229). IGI Global., DOI: 10.4018/979-8-3693-1785-3.ch014

Walton, C. C., Purcell, R., Pilkington, V., Hall, K., Kenttä, G., Vella, S., & Rice, S. M. (2024). Psychological safety for mental health in elite sport: A theoretically informed model. *Sports Medicine (Auckland, N.Z.), 54*(3), 557–564. DOI: 10.1007/s40279-023-01912-2

Wang, A. Y., Kay, K., Naselaris, T., Tarr, M. J., & Wehbe, L. (2023). Better models of human high-level visual cortex emerge from natural language supervision with a large and diverse dataset. *Nature Machine Intelligence, 5*(12), 1415–1426. DOI: 10.1038/s42256-023-00753-y

Wang, C.-C., Cheng, P. K.-H., & Wang, T.-H. (2022). Measurement of extraneous and germane cognitive load in the mathematics addition task: An event-related potential study. *Brain Sciences*, *12*(8), 1036. DOI: 10.3390/brainsci12081036

Wenk, N., Penalver-Andres, J., Buetler, K. A., Nef, T., Müri, R. M., & Marchal-Crespo, L. (2023). Effect of immersive visualization technologies on cognitive load, motivation, usability, and embodiment. *Virtual Reality (Waltham Cross)*, *27*(1), 307–331. DOI: 10.1007/s10055-021-00565-8

Wexler, A., & Specker Sullivan, L. (2023). Translational neuroethics: A vision for a more integrated, inclusive, and impactful field. *AJOB Neuroscience*, *14*(4), 388–399. DOI: 10.1080/21507740.2021.2001078

White, A. L., Boardman, F., McNiven, A., Locock, L., & Hinton, L. (2021). Absorbing it all: A meta-ethnography of parents' unfolding experiences of newborn screening. *Social Science & Medicine*, *287*, 114367. DOI: 10.1016/j.socscimed.2021.114367

Xu, J., Qiu, B., Zhang, F., & Zhang, J. (2024). Restorative effects of pocket parks on mental fatigue among young adults: A comparative experimental study of three park types. *Forests*, *15*(2), 286. DOI: 10.3390/f15020286

Zeine, F., Jafari, N., Nami, M., & Blum, K. (2024). Awareness Integration Theory: A psychological and genetic path to self-directed neuroplasticity. *Health Sciences Review (Oxford, England)*, *11*, 100169. DOI: 10.1016/j.hsr.2024.100169

Zhang, P., Yu, Z., Hou, G., Shu, P., Bo, Y., Shi, Y., & Nie, R. (2024). Enhancing cognitive performance and physiological benefit in workspaces through patterns of biophilic design: A restorative approach. *Buildings (Basel, Switzerland)*, *14*(10), 3293. DOI: 10.3390/buildings14103293

Chapter 9
Building Resilient Organizations:
Effect on Workforce Well-Being by Mediation of Labor Productivity in Gedeo Zone, Ethiopia

Shashi Kant
https://orcid.org/0000-0003-4722-5736
Lovely Professional University, India

Dawit Jabo
https://orcid.org/0009-0002-1828-1982
Dilla University, Ethiopia

Zinabu Wolde
Dilla University, Ethiopia

Zerihun Kinde Alemu
https://orcid.org/0000-0002-8546-2882
Ethiopian Civil Service University, Ethiopia

Rakesh Kumar Gautam
https://orcid.org/0009-0001-4381-2338
Lovely Professional University, India

ABSTRACT

The chapter primary goal is to investigate the connections among productivity of labor force, well-being of workforce, and building resilient corporates. It tackles the issue of how corporate resilience might augment workforce happiness and engage-

DOI: 10.4018/979-8-3373-2292-6.ch009

ment in a dynamic workplace. In order to investigate the links among the constructs, the investigation uses a quantitative investigation approach, collecting data from workforces in different corporates through surveys. Prominent outcomes show that incubating resilient corporates has a beneficial influence on productivity of labor force, which in turn greatly augments well-being of workforce. The outcomes imply that corporates that place a high priority on resilience not only augment operational influence but also manifested a more salubrious workplace for staff members. Future investigation will look at the precise mechanisms underlying these connections as well as how individual differences may mitigate the benefits of resilience

INTRODUCTION

In the current worldwide financial environment, building resilient corporates has grown more and more important (Maru et al., 2020). According to Ayele (2019), a company's resilience is its ability to adapt, rebound, and thrive in the face of adversity, whether it be societal difficulties, technological changes, or financial crises. These global trends are micro cosmoses, a region marked by significant cultural variety and financial potential. Apart from being notable for indigenous progress, the interplay of business resilience, labor force productivity, and worker well-being also addresses bigger global issues on social stability as well as sustainable fiscal development (Sahle et al., 2019). The concept of corporate resilience is based on systems theory, management, attitude, and other fields (Fetene & Kedir, 2024). Originally studied in natural contexts, resilience was later extended to business behavior, reflecting the need for companies to keep running despite obstacles (Riché et al., 2009). According to empirical research increasingly concentrating on the link between resilience and well-being, resilient companies often encourage higher degrees of employee satisfaction and involvement (Abebe & Kjrholt, 2009). Studies show that companies that give employee well-being top priority have more resilience and lower turnover rates, which helps both companies and employees (Debisa, 2022).

Even with the wealth of research, though, implementing resilience-building strategies presents practical difficulties (Bedasa & Deksisa, 2024). Corporates in the Gedeo Zone might lack understanding about the benefits of resilience along with well-being, cultural obstacles, and resource constraints for a case study (Fekadu et al., 2024). These difficulties show clear research voids, particularly with respect to understanding how, in certain circumstances; labor efficiency changes the link between company resilience and well-being of workers. This chapter aims to fill in these gaps by providing a thorough analysis of how incubating resilient corporates might increase labor force productivity and, thus, improve the well-being of workers in the Gedeo Zone. This study will show beneficial frameworks for regional business leaders and politicians by means of the research of regional case studies and the integration of concepts from foreign literature. Apart from showing best practices, the study will provide feasible answers fit for the Gedeo Zone's specific socio-fiscal environment. "How do building resilient business entities influence well-being of workforce through the mediation of efficiency among labor force?" This study is important as it seeks to find the mechanisms linking engagement, well-being, and resilience, so providing a comprehensive knowledge that might direct future research and practical treatments. Through this study, the chapter wants to highlight a route for corporates in Gedeo and comparable conditions, finally outcompeting in a more productive and impact staff that enhances the specific well-being of society.

BACKGROUND OF INVESTIGATION

The concept of business resilience has become more prominent as companies worldwide confront an environment that is more erratic and uncertain. Resilient companies have to be able to predict change, respond fast, and recover from failures if they are to keep the involvement and well-being of their employees (Sahle et al., 2019). Especially in the Gedeo Zone, a region with a rich cultural past and significant financial potential, resilience stands out (Ayele, 2019). Problems including resource depletion, infrastructure shortcomings, and sociopolitical unrest might hinder indigenous people's ability to grow and maintain resilience (Maru et al., 2020). A indigenous agricultural cooperative in the Gedeo Zone offers a real-world case study of these difficulties (Abebe & Kørholt, 2009). This cooperative, which seeks to improve the lives of smallholder farmers, found it challenging to implement resilience-building projects because of restricted access to modern farming techniques and technology. Though there was hope for more production and improved member well-being, the cooperative found it difficult to get funds for skill development programs for its members (Riché et al., 2009). Engagement and specific well-being stalled as many farmers were unable to utilize optimum techniques that would increase their resilience to market alongside weather changes (Debisa, 2022).

A medium-sized manufacturing company in the Gedeo Zone aiming to increase staff satisfaction and involvement by using a resilience framework provides another example (Bedasa & Deksisa, 2024). The company shown that higher workplace resilience might result in more involvement. Inclusion, however, faced several challenges including administrative resistance and staff workers accustomed to more traditional ways of operating (Fekadu et al., 2024). Workforces' emotions of uncertainty and disengagement caused by a lack of a clear communication strategy as well insufficient skill development program programs undercut the expected advantages of the resilience effort.

A recent study done on a neighboring educational institution aiming to foster resilience by increasing support networks for educators as well as learners (Biadgilign, 2023) has revealed comparable problems. Though they recognized the need of fostering a supportive culture, the personnel of the institution was hindered in their ability to implement advancementful interventions by budgetary constraints and inadequate skill development initiatives (Yusuf, 2019). Burnout among instructors caused students' resilience to decline, hence showing the vital link between worker involvement, corporate resilience, and well-being (Mohammadi et al., 2023). These case studies best reflect the real-world difficulties companies in the Gedeo Zone face trying to foster resilience. A lack of resources, cultural resistance, and inadequate skill development program aggravates a cycle of low resilience and reduced well-being of workers. Indigenous people as well as politicians wishing to improve the business environment must handle these issues if they are to be addressed. Emphasizing how incubating resilient corporates might increase worker involvement and well-being in the Gedeo Zone, this study aims to more clearly show these concerns and provide ideas that could direct more specific legislative frameworks as well as regional practices.

STATEMENT OF THE PROBLEM

Especially in places like the Gedeo Zone, there has never been a greater need for resilient corporates supporting worker well-being and increasing labor force productivity (Maru et al., 2020). In a perfect world, companies would not only be able to withstand outside pressures but also foster an environment where employees could thrive, hence increasing production and satisfaction levels everywhere (Ayele, 2019). Though, despite the increased awareness of its impact, the research and practical implementation of resilience still leave much to be desired (Sahle et al., 2019). Studies provide mixed results on the direct impact of company resilience on employee well-being and involvement (Abebe & Kjørholt, 2009). While some research suggests a clear positive connection, others reveal that the relationship is more complex and shaped by mediating factors like business culture and people traits (Riché et al., 2009). This difference causes uncertainty among stakeholders about the best path forward, which leads to different results about the impact of initiatives aimed at increasing resilience.

Theoretical voids make this landscape much more difficult. A great deal of the research thus far has focused on general resilience frameworks, ignoring the particular mediating function of labor force productivity under different cultural and financial circumstances, such as those in the Gedeo Zone. The lack of a theoretical basis unique to the setting limits the relevance of the results and hinders the creation of tailored strategies that may perform better in local settings. Apart from that, there are still open issues on how several foundations, such regional financial situations and company designs, affect the relationship among well-being, involvement, and resilience (Fetene & Kedir, 2024). Little is known for a case study regarding how Gedeo Zone customs and community values could either assist or hinder the inclusion of resilience strategies. Along with these issues, actual world barriers including a lack of resources, inadequate skill development programs, and cultural hostility exacerbate the situation for businesses trying to use resilience-building techniques (Debisa, 2022).

This study aims to bridge these gaps by providing a comprehensive analysis of the relationship between labor force productivity, company resilience, and well-being of workers in the Gedeo Zone. Examining indigenous case studies and incorporating concepts from overseas literature would help this study to create beneficial frameworks for companies and governments. It will look at the mediating role of labor force productivity in order to close the theoretical and knowledge gaps that have hindered past research. The results will also indicate feasible solutions for implementation problems, guiding future researchers and decision-makers toward development resilience-building initiatives tailored to the specific financial situation of the Gedeo Zone.

RELATED THEORIES

Resilience Theory: According to this hypothesis, companies can develop the technical expertise required to anticipate, react to, and recover from disturbances (Fetene & Kedir, 2024). It shows the capacity to develop from errors, creativity, and flexibility. Resilient companies are better equipped to handle crises, hence ensuring continuity and stability (Riché et al., 2009). This idea is particularly relevant in places like the Gedeo Zone, where native companies face significant difficulties. Building a culture of resilience helps corporates to increase their capacity to change and improve employee involvement and well-being (Bedasa & Deksisa, 2024).

Human Capital Theory: Human capital theory's major emphasis is on the significance of people's skills, knowledge, and experiences as essential resources for companies. According to this theory, investing in personnel resulted in increased production and resilience (Debisa, 2022). When one considers company resilience, the requirement of investing in human capital increases. According to Maru et al. (2020), resilient companies know that their staff members are essential in conquering challenges and fostering creativity. Since it allows companies to increase general well-being and involvement by giving employees with the expressed skill development program and support, this theory is a crucial paradigm for examining the relationship between productivity of labor force alongside resilience (Fekadu et al., 2024).

Job Demands-Resources (JD-R) Model: According to the JD-R Model, the balance between job demands as well as available resources affects the well-being of the workforce (Ayele, 2019). Although too high job expectations can lead to burnout and decreased involvement, enough resources may increase engagement and resilience (Biadgilign, 2023). This approach underlines how vital company resources and support systems are to developing a resilient workforce. By addressing employment needs and offering sufficient resources, companies may foster an environment that promotes well-being of workforce alongside increases labor force output (Kassaye, 2024).

Social Exchange Theory: This concept shows the mutual character of relationships in the office and argues that when employees feel their efforts and rewards are properly traded, they are more likely to be involved and efficient (Sahle et al., 2019). Encouraging good social relationships helps resilient companies to increase workforce trust, commitment, and well-being (Abebe & Kjørholt, 2009). Resilience depends on sufficient human connections and a good business culture, according to this theory. According to Muhammadi et al. (2023), businesses may increase employee satisfaction and involvement by fostering an environment of respect and collaboration.

The main theoretical underpinning for this study was Corporate Resilience Theory. This choice is founded on the particular conditions and difficulties Gedeo Zone companies face. Emphasizing resilience helps the study to explore how businesses may create the technical knowledge's expressed to thrive in the face of adversity, so creating a direct link between resilience alongside worker involvement along with well-being (Kassaye, 2024). This theoretical framework allows for a comprehensive examination of the several fiscal foundations affecting company behavior in the Gedeo Zone, hence producing information that might direct indigenous corporate strategies as well as more specific policy efforts.

DEFINITION AND ORIGIN OF CONCEPTS

Corporate Resilience

Emphasizing flexibility, adaptability, as well as strategic foresight, this concept enables companies to overcome challenges. "Resilience" is derived from the Latin term "resilire," which means "to spring back." Originally used for materials and physical systems, the concept was changed for use in business environments in the second half of the 20th century as a result of systems theory and attitude research

(Fetene & Kedir, 2024). Studies showed that companies might become resilient in the same manner that people and communities do when confronted with suffering by incubating sufficient internal processes and fostering a culture of continuous learning and improvement (Abebe & Kørholt, 2009).

Productivity of labor force

According to Ayele (2019), it is the measure of the number of products and services generated per labor hour. It is a vital indicator of financial resilience as it reveals how power labor is applied in the manufacturing process. Especially in the writings of economists including David Ricardo as well as Adam Smith who examined the relationship between labor input and output, classical fiscals is where the concept of productivity of labor force originally appeared (Riché et al., 2009). Over time, engagement has been looked at from several viewpoints including business policies, workforce skills, and technological developments (Debisa, 2022). Often calculated in contemporary fiscals by breaking down total output by total hours worked, productivity of labor force is a simple measure of how well labor supports fiscal progress and company resilience (Bedasa & Deksisa, 2024).

Well-being of workforce

Well-being of workforce is the specific mental, emotional, and physical well-being of employees in the workplace. It includes basics like mental health, work-life balance, job satisfaction, and a sense of community inside a company (Sahle et al., 2019). The origins of the concept may be traced back to early industrial mentality, when academics began to examine how working circumstances affected employee well-being and performance (Debisa, 2022). Emphasizing the importance of interpersonal relationships and worker happiness in increasing involvement, the Hawthorne Studies done in the 1920s and 1930s set the groundwork for modern ideas of workplace well-being (Fekadu et al., 2024). According to Yusuf (2019), the well-being of workers is today seen as a vital component of business since it affects not just personal resilience but also the whole effectiveness as well culture of the company.

Corporate Culture

Corporate culture is the shared beliefs, attitudes, and practices shaping people's interactions as well as teamwork inside a company (Biadgilign, 2023). It incorporates the unspoken habits, traditions, and behaviors shaping decision-making and the involvement of business workers. Scholars like Edgar Schein and Geert Hofstede invented the term "corporate culture" in the 1980s to help clarify how culture shapes corporate power (Mohammadi et al., 2023). Schein's approach especially describes culture as a multi-layered concept made up of underlying assumptions, declared values, alongside artifacts. A good business culture may foster resilience and finally increase employee involvement and well-being by providing workers a sense of commitment and interest (Abbink, 2018).

EMPIRICAL LITERATURE REVIEW

Resilient Corporates and Well-Being

Building resilient corporates, which provide an environment where individuals can negotiate challenges and uncertainty, strongly affects worker well-being (Fetene & Kedir, 2024). According to Abebe and Kørholt (2009), resilience is the ability of a business to adapt, rebound, and thrive under adversity. Companies that build resilience foster a culture that honors worker empowerment, participation, and support (Ayele, 2019). Apart from increasing company resilience, this culture change helps employees' specific well-being by enabling them to feel valued and more equipped to handle pressure. One of the key ways resilient companies increase the well-being of their employees is by creating supportive systems and policies (Sahle et al., 2019). This means providing programs like skill development program, mental health support, and pleasant work hours. For a case study, workforces are more likely to feel competent and accomplished when companies participate in their skill development program and show processes supporting technical knowledge increase (Riché et al., 2009). Such investments result in increased motivation and job satisfaction, which are especially important for certain well-being. Resilient companies are also urged to have open communication and feedback, so allowing employees to voice their concerns and suggestions (Debisa, 2022). Emotional well-being is dependent on a feeling of community and belonging, which is fostered by this inclusive approach.

Resilient companies are also more equipped to manage crises and shocks, thereby reducing the bad effects on employees (Fekadu et al., 2024). Open and honest communication about changes coupled with preventative measures including offering mental health services helps to lower workforces' anxiety and uncertainty for a case study (Maru et al., 2020). When workforces observe their corporation actively striving to protect their interests, their faith in direction increases as well as their emotional well-being is enhanced (Biadgilign, 2023). A focus on work-life balance enhances even more the connection between well-being of workforce and organizational resilience (Yusuf, 2019). According to resilient companies, workforces who advance manage their personal and professional life are more engaged and productive. Establishing policies that promote flexibility—such as remote work options or customizable work hours—helps employers help employees better manage their responsibilities. This balance increases employees' specific well-being by reducing stress and allowing them to rest and maintain their mental health.

Apart from these benefits, resilient companies often foster a good corporate culture that appreciates shared values and interests (Mohammadi et al., 2023). A strong culture of resilience fosters workforce cooperation, creativity, and mutual support. Being part of a unified team striving toward shared objectives makes people more fulfilled and more interested (Abbink, 2018). This social support system is vital for mental and emotional wellness as it allows people the consolation that they are not alone in their battles. Ultimately, developing resilient corporates significantly and complicatedly affects employee well-being. By fostering a culture that emphasizes communication, support, and balance, corporates may increase staff engagement, satisfaction, and specific mental health (Kassaye, 2024). This positive feedback loop benefits workforces and the company as a whole, which then results in increased labor force production and business resilience. Investing in resilience is therefore crucial to nurturing an advance conscious, healthy workplace and transcends basic strategic decision-making (Chaiya, 2025).

Building Resilient Corporates and Productivity of labor force

Building resilient corporates has a significant impact on labor force productivity as resilience fosters an environment where employees may operate at their topmost level despite challenges (Ayele, 2019). Resilient corporates are those that are flexible and proactive; they create mechanisms that let employees shape their way through challenges and uncertainty (Debisa, 2022). Because employees are more involved and driven when they feel supported and ready to manage their job obligations, this empowerment results in increased involvement. One notable method that resilience increases labor force productivity is through the invention of sufficient support systems encouraging worker development and well-being (Sahle et al., 2019). Companies that engage in technical knowledge development, mentorship, and skill development initiatives not only increase staff capacities but also show workers confidence (Abebe & Kjørholt, 2009). When workers feel equipped and prepared to meet job expectations, they are more likely to work hard and take initiative in their duties (Riché et al., 2009). Eventually, engagement is increased by this proactive attitude, which also fosters a more dynamic work culture and greater production levels.

Resilient corporations are technically knowledgeable in managing change and uncertainty in today's fast-paced business environment together with that (Maru et al., 2020). Whether prompted by global crises, market changes, or technical innovations, resilient companies respond swiftly and effectively to unanticipated shocks (Fetene & Kedir, 2024). To minimize downtime and maintain organizational continuity, they use backup plans and pleasant strategies. Apart from lowering the negative effects of interruptions on production, this response increases worker confidence, hence more likely keeping them alert and engaged under disturbance. Being ability to change direction and modify assures that work processes remain impact and production levels are preserved (Bedasa & Deksisa, 2024). Along with, resilient corporates foster an environment of collaboration and open communication, both of which are essential for raising worker involvement (Biadgilign, 2023). Staff members feel at ease providing ideas and critiques, hence fostering an environment of invention and creativity. Cooperative teams can employ several points of view to solve more challenging problems and provide new ideas that increase production (Fekadu et al., 2024). Workforces cooperate to achieve shared goals, which not only improves job quality but also reduces project deadlines. The sense of togetherness and interest increases morale and involvement while decreasing turnover rates.

Along with, resilient corporates' focus on worker well-being is crucial for improving labor force productivity (Yusuf, 2019). Companies that give mental health, work-life balance, and specific workforce satisfaction top priority help to create a healthy and driven staff (Mohammadi et al., 2023). Engaged workforces that receive adequate sleep are more productive as they are better able to concentrate, be innovative, and solve issues (Abbink, 2018). Resilient companies ensure that employees can perform at their best by implementing well-being-promoting policies including pleasant work hours and wellness programs, which increases production and efficiency (Chaiya, 2025). Finally, improving company resilience offers long-term benefits that extend beyond temporary increases in involvement. Resilient companies are better placed to attract and retain excellent people as employees search more and more for environments that appreciate support and flexibility (Kassaye, 2024).

Productivity of labor force and Well-Being

Productivity of labor force shapes individual satisfaction as well as the specific health of the company, hence considerably affecting well-being of workforce (Sahle et al., 2019). Defined as the number of goods and services produced per hour of labor, productivity of labor force is one well-known indicator of fiscal efficacy and efficiency (Riché et al., 2009). Usually, a high degree of productivity provides employees a sense of accomplishment and satisfaction, which benefits their emotional and mental well-being. Many basics converge to form a culture in which well-being and involvement help one another, so this relationship is complicated (Debisa, 2022). One notable effect of increased labor force productivity on well-being is the increase of work satisfaction (Abebe & Kørholt, 2009). Workforces often have an adequateer feeling of achievement when they can fulfill their duties affect and attain their own interests (Mohammadi et al., 2023). This feeling of expertise is crucial for workforce morale since it is intimately linked to intrinsic drive. Workforces that feel valued and appreciated for their efforts are more likely to interact positively with their job. This creates a positive feedback loop in which increased involvement results in well-being, which then drives even more involvement (Bedasa & Deksisa, 2024).

Along with, more labor force productivity might enable workers feel less stressed and have less job to perform (Ayele, 2019). People in environments where engagement is maximized would be able to complete activities more rapidly, resulting in a more reasonable work-life balance (Fekadu et al., 2024). Since excessive workloads and long hours are often linked with burnout and declining mental health, this balance is vital for wellness. When companies implement policies increasing engagement—such as simplified processes or smart use of technology—workforces can have a more sustainable pace of work (Biadgilign, 2023). Apart from improving personal well-being, this stress reduction fosters a more positive company culture where employees feel supported and have the chance to succeed (Maru et al., 2020). The relationship between worker engagement and well-being is also affected by the availability of resources and support systems at work (Yusuf, 2019). Companies that value involvement highly often support staff development, wellness projects, and skill training (Abbink, 2018).

In addition, increased labor force productivity can enable companies to operate financially, therefore releasing resources for initiatives supporting the well-being of their employees (Fetene & Kedir, 2024). Companies that see an increase in involvement often have the means to improve working conditions, support workplace health projects, and increase their benefit packages. These improvements can significantly increase the well-being of the workforce by offering increased access to health services, mental health support, and friendlier work schedules (Kassaye, 2024). Workforces are more likely to develop an emotional connection with their company when they feel that their business values their well-being, hence increasing involvement and specific satisfaction (Chaiya, 2025). Finally, one must overlook how worker involvement shapes society well-being. Engagement supports financial growth, which improves public services and increases living standards (Abebe, 2024).

Resilient Corporates, Well-Being Through Productivity of labor

The dynamic relationship between labor force productivity and the development of resilient corporates as well as well-being of workforce profoundly shapes workplace results (Ayele, 2019). According to Sahle et al. (2019), resilient corporations are designed to thrive and adapt under hardship, fostering a work culture where employees feel supported and empowered. Apart from increasing business resilience, this resilience foundation is a major determinant of labor force productivity, which directly affects employee well-being. Companies that give resilience top priority create systems and policies supporting continuous development and adaptability (Maru et al., 2020). Often, this proactive approach means supporting staff development, mental health programs, and promoting workplace environments (Abebe & Kjrholt, 2009). Productive workforces tend to feel more fulfilled and successful, which enhances their specific well-being (Debisa, 2022). Resilient companies create a good feedback loop whereby output increases and employee involvement and satisfaction follow suit.

Workforce unhappiness is also caused by stress and uncertainty, which resilient companies are technically knowledgeable about managing (Riché et al., 2009). By implementing effective shift management strategies and maintaining lines of communication open, these companies may reduce the stress caused by disruptions (Fetene & Kedir, 2024). When workers believe their company is resilient, they are more likely to feel safe in their roles and be able to focus on their tasks without concern about the future (Bedasa & Deksisa, 2024). Stability in the workplace enables workforces to give their best effort and focus on their tasks. More productive workforces also say they feel less worried and more happy with their occupations, which enhances well-being.

The business culture that resilient companies foster stresses even more the part of employee involvement as a mediator (Fekadu et al., 2024). Cultures that respect creativity, cooperation, and shared interests make workforces more inclined to cooperate effectively. Workers who collaborate and play to one another's strengths increase engagement, which results in increased resilience and more powerful influence processes (Biadgilign, 2023). Resilient companies often implement work-life balance policies since, along with others, they know that continuous involvement relies on staff well-being (Yusuf, 2019). Friendly work timings, health programs, and supporting administrative policies help workforces to better manage their professional and private lives (Mohammadi et al., 2023). This balance increases mental health and reduces burnout, hence increasing employee involvement even further. When employees feel that their well-being is a top priority, they are more likely to be involved and driven, which will increase productivity levels (Abbink, 2018). The increased productivity of labor force that results helps to create a positive cycle of well-being as happy employees are less likely to leave the company (Chaiya, 2025). When this stability fosters a sufficient corporate culture, which subsequently enhances resilience, a positive response loop of involvement and well-being is created.

Figure 1. Conceptual framework

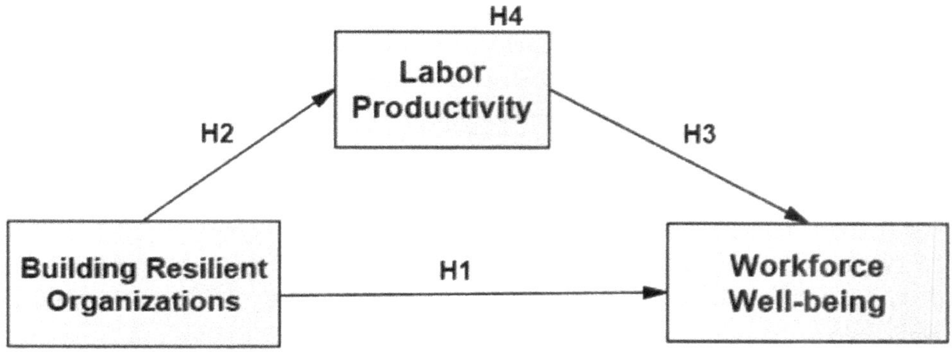

METHODOLOGY

This study looked at the effect of incubating resilient corporates on worker well-being, as mediated by labor force productivity, utilizing a quantitative research approach. A sample of 390 people from many companies inside the Gedeo Zone was selected to provide a varied representation of industries and organizational types. The major objective was to gather empirical data illuminating how company resilience, well-being of personnel, and labor force productivity interact. Using validated ratings and prior published material, a systematic questionnaire was developed to begin the research. The poll combined numerous elements evaluating the ideas of worker engagement, company resilience, alongside well-being of workforce. The questionnaire was pilot tested with a small sample of participants to ensure item clarity and consistency before distribution (Maru et al., 2020). Changes to the wording and structure of the questions in reaction to pilot study results increased the instrument's general quality (Ayele, 2019).

Chosen using a stratified random selection process to ensure representation across all company levels and functions, the 390 participants got the finished questionnaire once it was completed. Honest and accurate responses were promoted as participants were promised anonymity and confidentiality. Data was collected over a few weeks utilizing both offline and online methods in order to increase response rates. Exploratory factor analysis (EFA) was done utilizing the data gathering process to assess the dimensionality of the components (Sahle et al., 2019). The Kaiser-Meyer-Olkin (KMO) measure of sample adequacy was used to evaluate the data's suitability for factor analysis before EFA (Abebe & Kjørholt, 2009). When the KMO value exceeded 0.6, suggesting that the sample size was enough and the variables were sufficiently correlated for the analysis, the data were appropriate for EFA (Fetene & Kedir, 2024). The EFA was done with principal component analysis and varimax rotation to identify the underlying factor structure of the constructs (Riché et al., 2009). This method, which found distinct elements for labor force productivity, workforce well-being, and business resilience, included the theoretical framework created from the literature. Factor loadings demonstrated that the items grouped suitably, hence supporting the validity of the measurement scales.

Structural equation modeling (SEM) with AMOS software investigated the observable connections among the constructs after EFA (Fekadu et al., 2024). By allowing one to examine several interactions simultaneously, SEM gave studies a comprehensive knowledge of how company resilience affects employee involvement, which then affects employee well-being (Biadgilign, 2023). Model fit indices, which combined the root mean square error of approximation (RMSEA), comparative fit index (CFI), and chi-square statistic, were used to assess the overall fit of the model. A reasonable model fit was shown by CFI scores over 0.90 and RMSEA values under 0.08, suggesting the data backed the postulated links (Yusuf, 2019).

SCALE ADVANCEMENT

Ideas emphasizing flexibility, creativity, and taking the initiative to solve challenges helped to incubate the scale for gauging business resilience (Maru et al., 2020). Prominent dimensions were the following: "Adaptive Capacity," a measure of a company's adaptability; "Crisis Management," which stresses preparedness and reaction strategies; and "Learning Orientation," which summarizes the commitment to continuous development and experience-based learning, were among the most notable dimensions. Statements such as "Our company fast adapts to new challenges" and "We learn from past experiences to augment our processes" were created to assess these traits. The productivity of labor force scale came from the study already in place stressing the link between workforce resilience and engagement. "Efficiency," which measures output in relation to input, and "Quality of Work," which reflects the quality of the products or services produced, were notable features (Ayele, 2019). Items such as "The quality of my work meets corporate standards" and "I complete my tasks efficiently" were meant to increase capture these aspects. This approach ensured that the scale reflected both the qualitative and quantitative aspects of involvement. Combining data from studies on work-life balance, mental health, and job satisfaction, the scale for assessing worker well-being was created (Fetene & Kedir, 2024). "Job Gratification," which focuses on specific job satisfaction; "Work-Life Balance," which stresses the ability to balance personal and professional responsibilities; and "Emotional Well-being," which integrates mental health and emotional stability, were among the notable dimensions (Sahle et al., 2019). Items such as "I am satisfied with my job overall" and "I can balance my work and personal life influence" were included to ensure a thorough assessment of well-being. Expert consultations and pilot testing helped to ensure the validity and reliability of the scale by means of examination and augmentation of each item (Abebe & Kørholt, 2009). Comments from subject area specialists increased the relevance of the pieces and helped to clarify their phrasing. The pilot test's supply of early data steering changes increased the scale's psychometric characteristics (Chaiya, 2025).

DATA ANALYSIS

Table 1. KMO and Bartlett's Test

Kaiser-Meyer-Olkin Measure of Sampling Adequacy.		**.850**
Bartlett's Test of Sphericity	Approx. Chi-Square	1419.738
	Df	91
	Sig.	.000

Source: Authors

Table 1 shows the results of Sampling Adequacy. The KMO score of 0.850 showed a decent level of sample adequacy. According to KMO guidelines, values over 0.8 are considered commendable as they suggest the variables will generate distinct and reliable factors and that the information in question is appropriate for factor analysis. The results of the Bartlett Test of Sphericity of shows more proof that the data is appropriate for factor analysis with a repercussion level (Sig.) regarding.000, 91 levels of freedom (df), and an estimated chi-square value of 1419.738. The consequence level includes significant correlations among the variables, indicating that the causality matrix varies significantly from an identity matrix. This result is significant as it suggests that at least some of the variables are linked and suitable for outcome underlying factors, hence supporting the idea that factor analysis might be done on the dataset. Ultimately, the results of the KMO and Bartlett's Test indicate that the data is appropriate for factor analysis, providing a sufficient basis for further statistical research.

Table 2. Total Variance Explained

C	Initial Eigenvalues			Extraction Sums of Squared Loadings			Rotation Sums of Squared Loadings		
	Total	% of Variance	Cumulative %	Total	% of Variance	Cumulative %	Total	% of Variance	Cumulative %
1	4.1	32.35	32.35	4.1	32.35	32.35	3.0	21.51	21.51
2	1.7	9.87	42.22	1.7	9.87	42.22	2.6	15.91	36.42
3	1.7	9.87	51.10	1.7	9.87	51.10	1.8	14.68	51.10
4	.99	6.82	57.92						
5	.81	6.23	64.15						
6	.73	5.19	69.34						

Extraction: Principal Component Analysis.
Source: Authors

Table 2 shows the results of underlying fundamentals. The table shows the components taken from the study together with their original Eigenvalues, which utilization sums of lined up loadings, as well as rotation sums of squared loadings. The first component, with an initial eigenvalue of 4.51, contributed 32.235% of the overall variance. This is significant as it indicates that this component accounts for a significant proportion of the variation in the data. Following the first component, the overall share of variance increased to 32.235%, suggesting that this one factor sufficiently reflects a significant part of the dataset. With an eigenvalue of 1.37, the second component contributed 9.787% of the variance to the total cumulative variance of 42.022%. The third component, which boosted the cumulative percentage to 51.110% and accounted for 9.087% of the total variance, also had an eigenvalue of 1.27. Components four, five, and six added 6.782%, 6.223%, and 5.519% to the variance with corresponding relative eigenvalues of 0.949, 0.871, and 0.773. Though they contribute to the total variance, these components are less significant in characterizing the underlying structure of the data as they do not meet the conventional criterion of an eigenvalue greater than one.

Table 3. Covariances

Covariance			Approximation	S.E.	C.R.	P	Hy.
Productivity of labor force	**<-->**	**Building Resilient Corporates**	**.242**	**.021**	**8.231**	**0.00**	**H2**
Productivity of labor force	<-->	Well-being	.303	.032	7.692	0.00	**H3**
Building Resilient Corporates	<-->	Well-being	.254	.24	7.831	0.00	**H1**

Source: Authors

Table 3 shows the covariances among the well-being, creating resilient corporates, and labor force productivity constructs, highlighting the importance and strength of the relationships among these three variables. The covariances reveal the relationships between several structural model components. The relationship between Productivity of labor force alongside Building Resilient Corporates was found to be 0.242 with an estimated value (S.E.) of 0.021. With a probability value of 0.001 or lower (indicated by 0.00), the critical ratio (C.R.) of 8.231 for this connection supported Hypothesis 2 (H2) and was statistically significant. According to this strong positive correlation, labor force productivity tends to increase in lockstep with a company's resilience. Similarly, well-being and labor force productivity showed a correlation of 0.303 with an expected error of 0.032. The C.R. included Hypothesis 3 (H3) for this correlation, with a value of 7.692 and significant at p less than 0.001. By indicating that greater labor force level productivity is connected to better well-being of workforce, this result reveals the favorable advantages of involvement on workers' specific satisfaction and mental health. At last, a correlation of 0.254 with a standard error of 0.24 was discovered between Building Resilient Corporates and Well-Being. The the coefficient of variation about 7.831, which was significant at $p < 0.001$, supported Hypothesis 1. This good link shows the result of resilience in fostering a healthy workplace culture by suggesting that resilient companies foster a culture that promotes worker well-being.

Table 4. Validity Concern

	CR	AVE	MSV	MaxR(H)	BRO	LP	WB
BRO	0.732	0.643	0.219	0.852	**0.654**		
LP	0.758	0.657	0.132	0.763	0.171	**0.665**	
WB	0.793	0.603	0.274	0.797	0.383	0.193	**0.684**

Note: BRO= Building Resilient Corporates*; LP=* Productivity of labor force*; WB= Well-being*
Source: Authors

The validity problems are shown in Table 4. The Composite Reliability (CR) scores of each construct reveal reasonable degrees of internal consistency; BRO, LP, as well as WB all obtained CRs about 0.732, 0.758, as well as 0.793, respectively. Scores on the CR over 0.7 are especially considered acceptable as they suggest that all three constructions' dependability is adequate. The AVE values demonstrate the convergent validity of the constructs: WB has an AVE of 0.603, LP has an AVE of 0.657, along with BRO has an AVE of 0.643. The Maximum Mutual variation (MSV) values, which gauge how much one construct's variation is shared with another, were 0.219 for BRO, 0.132 for LP, alongside 0.274 for WB. These results include discriminant validity, which indicates that although there is some shared variation among the components, it is still below the AVE values for each construct. This suggests that every construct is different and that there is almost no correlation among the several ones. Reported Maximum Reliability (MaxR(H)) values for BRO, LP, alongside WB—0.852, 0.763, and 0.797—further showed the strength of the reliability metrics for these structures. Ultimately, the results in Table 4 reveal sufficient validity and dependability of the constructs of well-being, labor force productivity, and constructing resilient corporates.

Mediating role Examination

Figure 2. Structure equation Model

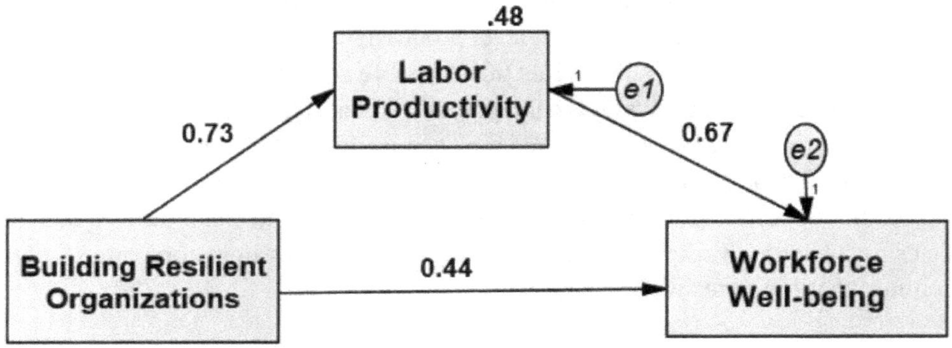

Model fit indices

Table 5. Model Fit

Sig.	Chi-Sq	RMR	Fitness Goodness	Fitness corporate	TLI	RMSEA
0.004	1.954	.036	0.930	0.912	0.911	.028

Source: Authors

Table 5 shows the model fit indices. Though low, the stated chi-square statistic (Chi-Sq) of 1.954 is especially predicted to be associated by a non-substantial the probability value for a well-fitting model. Computed at 0.036, the square root of the Mean Square as a way Residual (RMR) showed a decent match. Lower numbers indicate stronger residual resilience for the model. Recorded at 0.930, over the especially acknowledged threshold of 0.90, the accuracy of Fit coefficient (GFI) shows that the model fits the data well. It has been reported at 0.912 and designated as Fitness, the In contrast Fit Index (CFI) helped to bolster the model's suitability even further. Values above 0.90 show an excellent fit. The Tucker-Lewis Ranking (TLI), which was also acceptable at 0.911, further reinforced the conclusion that the model matches the observed data well. At last, the root average square error of Nearness (RMSEA) was 0.028, which is excellent since values below 0.08 show a decent match and values below 0.05 show a close fit. Ultimately, the fit indices shown in Table 5 reveal that the structural model fairly matches the data by including the model's specific dependability and validity and including the suggested connections among the components.

Table 6. Regression Examination

Relative			**Approx.**	**S.E.**	**C.R.**	**P**	**Ass.**
Productivity of labor force	**<---**	**Building Resilient Corporates**	**.864**	**.143**	**7.583**	**0.00**	**H2**
Well-being	<---	Productivity of labor force	.328	.131	2.853	0.00	H3
Well-being	<---	Building Resilient Corporates	.957	.192	7.958	0.00	H1

Source: Authors

Table 6 shows the results of the regression studies. The regression study revealed a strong positive correlation between labor force productivity and establishing resilient corporates with a normalized coefficient of 0.864. Statistically significant at $p < 0.001$ with a critical ratio (C.R.) of 7.583 and an estimated standard error of 0.143, this relationship is shown by 0.00. This result supports Hypothesis 2 (H2), which showed that increased business resilience significantly increases employee involvement. Along with, it was shown that, with a coefficient of 0.328, labor force productivity had a good impact on well-being. This regression is also significant at $p < 0.001$ with a C.R. of 2.853 and a standard error of 0.131. This result confirms Hypothesis 3 (H3), which states that more labor force productivity increases worker satisfaction. At last, the study revealed a sufficient correlation between well-being and building resilient corporates with a score of 0.957. Indicating consequence at $p < 0.001$ and including Hypothesis 1 (H1), this association's C.R. was 7.958 and its standard error was 0.192. This strong correlation suggests that resilient companies not only increase output but also significantly improve worker well-being.

Table 7. Mediating role Influence

	Influence	worth	Path Influence
Building Resilient Corporates → Well-being	Direct Influence	.44	Direct influence stated
Building Resilient Corporates → Productivity of labor force→ Well-being	Indirect Influence	.73*.67=.48	Indirect Influence Ensued
	Whole influence	.91	Partial mediation

Source: Authors

Table 7 shows the relationships. Building Resilient Corporates and well-being showed a significant positive correlation with a direct impact of 0.43. This direct impact underlines the effect of business resilience in improving well-being of workers as it suggests that resilient corporates directly affect increased labor satisfaction and mental health. As shown in Table 7, labor force productivity mediates the relationship between well-being and the development of resilient corporations. By showing the two types of impacts, the table provides a comprehensive knowledge of the interconnections among these components. Building Resilient Corporates showed a significant positive correlation with well-being, with a direct impact of 0.43. This direct impact underlines the effect of business resilience in improving the well-being of workers as it suggests that resilient corporates directly affect increased workforce satisfaction and mental health. The total effect of Building Resilient Corporates regarding well-being was found to be 0.91 by including both direct and indirect factors. Though resilience directly increases well-being, a large portion of that advantage is also obtained by increased labor force productivity, according to this overall effect, which indicates that the correlation is sufficient. According to the findings of the study, both direct and indirect routes are significant in the relationship between Building Resilient Corporates and Well-Being, which characterizes the function of Productivity of labor force as "Partial mediation." This study underlines how closely these aspects interact and how absolutely vital it is to create business resilience in order to get good workforce results.

DISCUSSION

The findings of this analysis provide significant new information on the linkages between the productivity of the work force, well-being, and the development of resilient corporations. They demonstrate a difficult interaction that highlights the need of resiliency in the workplace. First and foremost, the strong positive correlation (coefficient of 0.864) between the productivity of the labor force and the development of resilient corporations shows that resilient corporations are more likely to support high levels of staff involvement. These findings provide credence to the body of research that suggests resilience enables businesses to more effectively adapt to challenges and cultivate an environment in which employees are able to perform to the best of their abilities. In addition, the investigation revealed that the productivity of the workforce is a significant mediator between well-being and the development of resilient corporations. Even while resilience has a direct impact on the well-being of the workforce (with a coefficient of 0.43), a significant portion of that beneficial influence is reflected through increased engagement, as demonstrated by the adequate indirect influence (0.48). It is clear from this analysis that it is of the utmost importance to incorporate the productivity of the labor force as a significant component in both the satisfaction of the workforce and the mental health of the workforce, rather than only as a measuring tool for production. Additionally, the concept of partial mediation is further reinforced by the overall effect of Building Resilient Corporates on Well-Being, which is 0.91 and demonstrates that both direct and indirect channels are significant. This implies that both types of paths are relevant. It has been suggested that businesses that are interested in improving the well-being of their staff should give emphasis to resilience-building approaches. These tactics not only increase production but also foster a good culture in the workplace. When everything is taken into consideration, these results demonstrate how important business resilience is for fostering wellness and engagement, and they suggest that leaders and managers invest money in initiatives that are focused on resilience. By taking this action, they will be able to cultivate a workforce that is more engaged, healthier, and capable of thriving in the face of adversity, which will ultimately be to the advantage of both the workers and the corporation as a whole.

CONCLUSION

In a nutshell, this inquiry provides valuable information on the interconnected functions of well-being, productivity of work force, and the development of resilient corporations. The findings indicate that resilient corporations significantly increase the productivity of their staff, and that there is a significant relationship between the productivity of the workforce and the increased well-being of the workforce. There are strong direct and indirect relationships that exhibit the result of creating resilience within corporate cultures. According to the analysis, corporations that emphasize resilience are better equipped to support the engagement as well as particular pleasure of their workforces. This is the case because of the substantial direct along with indirect links. In addition, the results highlight how important it is for decision-makers and leaders to put strategies that develop resilience into practice in order to cultivate a healthy environment at work. By recognizing the numerous benefits of resilience, corporations have the ability to cultivate environments that not only improve employee engagement but also improve mental health. The implementation of an all-encompassing approach not only improves the resilience of the organization but also enhances the well-being of its employees, which ultimately leads to a workplace that is more affluent along with sustainable. For the purpose of shedding more light on the ways in which resilience promotes wellbeing as well as engagement, as well as providing more comprehensive understandings that may drive corporate processes and laws, future research should continue to investigate these correlations.

Managerial Implications

The findings of the study have significant management consequences for companies hoping to increase employee satisfaction and involvement. Executives should give top priority the development of robust business cultures able to adapt to unpredictability and hardship. Skill development initiatives that provide staff members stress-reduction strategies, adaptability, and problem-solving can assist to create a more competent workforce equipped to manage change. Managers should also work hard to foster a culture that appreciates open communication, cooperation, alongside mutual assistance. By fostering a supportive environment, companies may help employees feel valued along with involved, hence increasing their engagement. The results of the study underline even more how important it is to recognize and reward outstanding performance as doing so not only motivates employees but also supports a strong and efficient culture. Corporates can also implement wellness programs emphasizing mental and physical health, so ensuring that employees have access tools supporting their specific well-being. This combined focus on resilience and well-being will lead to higher staff satisfaction and retention rates. Ultimately, by investing in resilience-building initiatives that will result in more engaged, productive, and happy team, managers may place their companies for long-term progress in a corporate environment growing increasingly volatile.

Practical Implications

The results of this study have clear relevance for companies trying to improve employee happiness and operational effectiveness. First, companies wish to consider include resilience skill development program in their staff members' skill development program projects. This should include training sessions and lectures on stress management, adaptability, and problem-solving strategies, thereby equipping workers with the technical tools they need to meet obstacles. Companies also need clear routes for reaction and communication as this helps employees express concerns and share ideas. By means of open communication, corporates may create a cooperative and trustworthy culture that strengthens output and resilience. Regular team-building activities may help increase relationships among colleagues, hence supporting a good work culture. Apart from this, companies should regularly monitor and assess the well-being of their staff via feedback mechanisms and polls. Whether related to work-life balance, mental health support, or physical wellbeing, this knowledge may help to shape the invention of targeted wellness programs that meet the needs of certain workforces. Companies may ensure that these programs are complementing the different needs of their employees by tailoring them for the workforce. At last, to provide a good a case study for their employees, leaders should display tenacious attitudes and actions. Staff members are motivated to follow suit when supervisors show flexibility and a commitment to health. By using these practical strategies, companies can create a strong workforce that not only increases involvement but also supports a happier, healthier environment.

Theoretical Implications

By offering several significant theoretical implications, the results of this study contribute to the body of data now in existence on corporate conduct and well-being of workers. First, the connections that have appeared among labor force productivity, well-being, as well as constructing resilient corporates reinforce and extend current concepts of corporate resilience alongside engagement. By experimentally proving the mediation role of productivity of labor force, this study shows the intricate mechanisms by which resilience affects workforce results. It suggests that rather than being a corporate trait, resilience is a dynamic process that increases general resilience. The results also help to re-evaluate the conceptual models for worker well-being. Though this study indicates that corporate elements—especially resilience—greatly affect how well-being results, well-being has traditionally been seen in a vacuum. This insight motivates academics to look more closely at how corporate characteristics could act as precursors to well-being, hence generating the invention of more comprehensive models including several elements. Along with, the positive effects of resilience on well-being and involvement suggest that more research should show the underlying processes, considering moderating and mediating factors that might affect these relationships. Looking at the impact of outside settingal influences, business culture, and guidance ideologies might fit this description. The theoretical implications of the study, all things considered, not only enhance our understanding of how these ideas are connected to one another but also provide the basis for future empirical research initiatives that will show the complexity of corporate resilience as well as its effect on employee results.

Recommendations

Based on the results of the study, corporates wishing to increase well-being of staff, involvement, as well as resilience might make many recommendations. First and foremost, companies have to implement comprehensive resilience skill development initiatives stressing staff members' ability to be more flexible. These initiatives should include workshops on stress management, power communication, and problem-solving techniques so that employees have the technical expertise required to surmount challenges. Companies should also promote a culture of open responsiveness and communication so that employees feel free to share their thoughts and concerns. This may be accomplished by means of regular check-ins, response meetings, along with team-building activities fostering team member collaboration and confidence. By ensuring that employees feel recognized and valued, corporates may increase involvement and morale, which are essential for both engagement and well-being.

Apart from that, companies should support wellness initiatives catering to the different needs of their employees. This might include offering mental health services, welcoming work hours, and wellness programs encouraging exercise as well as work-life balance. Tailoring these programs to fit their particular needs can help to increase workforce satisfaction and attrition rates. Finally, important responsibilities for direction include setting alongside a case study of resilience as well as fostering a good culture. Leaders should behave flexibly and demonstrate commitment to the well-being of their staff members in order to provide a case study for their teams. Giving these ideas top attention helps companies to create a strong workforce that not only boosts output but also supports health and involvement at work.

Future Directions

Future research on company resilience, engagement, as well as staff well-being should emphasize a few key areas to help us better grasp these ideas. One interesting path is to investigate the exact mechanisms by which Building Resilient Corporates affects well-being and labor force output. This would include considering moderating variables to see how they affect the relationships discovered in this study, including team dynamics, guidance approaches, alongside corporate culture. Longitudinal research would also be useful in assessing the ways whereby resilience and its effects on well-being and involvement change with time. Such studies might highlight the long-term benefits of resilience-building initiatives, hence helping companies to grasp the long-term effects of these efforts on employee results. Future research should also emphasize how individual differences, including coping strategies or personality traits, could modify the relationship between company resilience and workforce outcomes. Knowledge of the relationships between business characteristics and human variables can guide tailored treatments increasing well-being and resilience. Along with, widening the inquiry to include a variety of sectors and organizational settings can show interesting contrasts stressing contextual foundations and best practices affecting the effectiveness of resilience strategies. Finally, adding qualitative approaches such as case studies or interviews might provide a better understanding of workers' real experiences and viewpoints on engagement, well-being, as well as company resilience. By exploring these prospective paths, investigators may promote a more extensive and nuanced understanding of the relationships among these vital structures.

REFERENCES

Abbink, J. (2018). Introduction: Promise and Peril in Africa—Advancement Narratives vs. Indigenous Problems. *The Crunch in Africa: Advancement Narratives vs. Indigenous Realities*, 1-28.

Abebe, A. (2024). A REVIEW OF WOMEN'S LAND TENURE SECURITY FOR SUSTAINABLE LAND ADMINISTRATION IN COFFEE PRODUCING REGIONS OF ETHIOPIA: IMPLICATIONS FOR FUTURE INVESTIGATION. *Investigation Report Series No. 39*, 1.

Abebe, T., & Kjørholt, A. T. (2009). Social actors and victims of exploitation: Working children in the cash economy of Ethiopia's South. *Childhood, 16*(2), 175–194.

Ayele, D. G. (2019). Challenges to rural livelihoods: A case investigation of Chichu, Gedeo, Southern Ethiopia. *Journal of Rural and Community Advancement, 14*(2).

Bedasa, Y., & Deksisa, K. (2024). Journal of Agriculture and Food Investigation. *Journal of Agriculture and Food Investigation, 15*, 100978.

Biadgilign, S. (2023). Coping strategies to mitigate food insecurity at household level: Evidence from urban setting in Addis Ababa, Ethiopia. *INQUIRY: The Journal of Health Care Corporate, Provision, and Financing, 60*, 00469580231206263. PMID: 37872773

Chaiya, C. (2025). Empowering climate resilience: A people-centered exploration of Thailand's greenhouse gas emissions trading and sustainable advancement through climate risk administration in community forests. *Heliyon, 11*(2). PMID: 39877598

Debisa, N. G. (2022). Building peace by peaceful approach: The role of Oromo Gadaa system in peace-building. *Cogent Social Sciences, 8*(1), 2023254.

Fekadu, E., Hebo, M., & Emirie, G. (2024). Influence of shift on the indigenous belief system of the Bayso Community of Gidicho Island, Southern Ethiopia. *International Journal of Anthropology and Ethnology, 8*(1), 16.

Fetene, M., & Kedir, A. (2024). Irrigation interventions and food security in the case of Ethiopia's East Hararghe lowlands. *Journal of Corporate and Fiscal Options, 7*(1), 39–47.

Kassaye, W. (2024). Traditional/Indigenous Education of Ethiopia. In *Education and Curriculum Advancement of Africa Vol. 1: Indigenization and Knowledge Production in Ethiopia* (pp. 89-160). Cham: Springer Nature Switzerland.

Maru, Y., Gebrekirstos, A., & Haile, G. (2020). Indigenous ways of setting protection in Gedeo community, Southern Ethiopia: A socio-ecological perspective. *Cogent Food & Agriculture*, *6*(1), 1766732.

Mohammadi, A., Omidi Najafabadi, M., & Poursaeed, A. (2023). A comprehensive sustainable advancement framework; community capitals and village-cooperative initiative. *Brazilian Journal of Biology*, *84*, e269509. PMID: 36856234

Riché, B., Hachileka, E., Awuor, C. B., & Hammill, A. (2009). Climate-related vulnerability and adaptive capacity in Ethiopia's Borana and Somali communities. *IISD report, 3*.

Sahle, M., Saito, O., Fürst, C., Demissew, S., & Yeshitela, K. (2019). Future land use administration influences on ecosystem services under different scenarios in the Wabe River catchment of Gurage Mountain chain landscape, Ethiopia. *Sustainability Science*, *14*, 175–190.

Yusuf, S. (2019). Drivers of ethnic conflict in contemporary Ethiopia. *Institute for security investigations monographs, 2019*(202), v-46.

KEY TERMS AND DEFINITIONS

Adaptability: The ability of a corporate or person to adapt to changing circumstances, challenges, or setting shifts. Adaptability is essential for resilient corporates to maintain resilience and well-being under changing and unpredictable circumstances.

Building Resilient Corporates (BRO): the process of incubating corporate cultures, practices, and structures that allow a corporate to flourish and adjust to shift, uncertainty, and obstacles. This entails cultivating adaptability, innovativeness, and a positive work culture.

Corporate: Culture: the common values, attitudes, and customs that influence how workers collaborate and interact within a corporate. The behavior, engagement, and particular resilience of workforces can all be adequately influenced by a great corporate culture.

Mediation: A term used in statistics to describe how a third variable, known as the mediator, explains the link among an independent and dependent variable. As per to this investigation, the association among well-being and incubating resilient corporates is mediated by productivity of labor force.

Productivity of labor force (LP): An indicator of how well labor produces products and services, usually expressed as output per hour of work. Augmented productivity of labor force, which frequently reflects the efficiency of corporate procedures and workforce involvement, shows that workers can accomplish more in a given amount of time.

Well-being (WB): a comprehensive idea that takes into account a person's mental, emotional, and physical well-being. Well-being in the workplace refers to the particular contentment and standard of living that workers feels, which is influenced by fundamentals including job gratification, work-life balance, and workplace assistance.

Workforce Engagement: how dedicated, enthusiastic, and passionate staff members are about their jobs and the corporate. Augmented retention rates, job gratification, and engagement are all correlated with high levels of workforce engagement.

Chapter 10
Ethical Leadership:
Effect on Workplace Well-Being Through Organizational Citizenship Behavior in Ethiopia Healthcare Sectors

Wako Jio
https://orcid.org/0009-0003-6192-459X
Bule Hora University, Ethiopia

Shashi Kant
https://orcid.org/0000-0003-4722-5736
Lovely Professional University, India

Hirut Assegid
https://orcid.org/0009-0003-6192-459X
Bule Hora University, Ethiopia

ABSTRACT

This chapter looks at how ethical leadership affects the well-being of employees at Ethiopian hospitals, focusing especially on how someone's willingness to help others at work (called organizational citizenship behavior) acts as a middleman among the leader's actions as well as the overall happiness of the staff. Using samples from 390 healthcare professionals, the study checked how good the data was for analysis through adequacy, as well as the outcomes showed that the data was good enough to look at the relationships among variables. CFA is done to check if the study's model is correct as well as to make sure the concepts like ethical leadership, OCB, as well as job life are accurately measuring what they aim to look at SEM is used through the AMOS to look at how the different constructs are connected through

DOI: 10.4018/979-8-3373-2292-6.ch010

each other. The outcomes show that ethical headship really helps publics do more for their organization, as well as this has a big positive effect on how happy as well as comfortable workers feel at work. This means that when headships act in an honest as well as fair

INTRODUCTION

Recently, more as well as more organizations around the world have noticed the importance of having ethical leaders (Su & Hahn, 2025; Fröhlich et al., 2025). Through growing issues such as a lack of resources as well as ethics in healthcare, the function of leadership is vital for ensuring employees are happy as well as productive (Sun et al., 2024). Ethical leadership begins through leadership studies, placing special attention on ethics as a guide for leaders' behavior as well as their effect on others (Kim et al., 2025). It has been manifested in empirical investigation that ethical leadership is related to more active engagement from staff, greater job satisfaction (Bayati et al., 2025), as well as boosts OCB in organizations (Güler & Ayık, 2025). The problem is that much of the investigation has been done in the West, making it unclear how these dynamics impact less developed countries, especially in healthcare (Lee, 2025).

In practice, several obstacles in Ethiopian healthcare organizations include frequent staff leaving (Chiboiwa et al., 2025), low spirit among workers, as well as weak leadership, which might hinder how the staff work as well as feel (Demissie et al., 2024). As such, a more careful study of ethical leadership is necessary to see how it fosters OCB as well as improves the well-being found in the workplace (Faisal et al., 2025). (Su & Hahn, 2025). This chapter studies the connections among ethical leadership, OCB, as well as workplace well-being in healthcare. To ensure the study's findings can help in policy as well as investigation, ample statistical methods such as SEM as well as CFA have been used. The main investigation question guiding this paper is: ***How can ethical leadership affect workplace well-being in the Ethiopian healthcare industry by means of OCB?*** This study not only fills theoretical gaps but also provides useful ideas for creating a better workplace in a vital area.

BACKGROUND OF THE STUDY

Understanding leadership effectiveness, particularly in healthcare settings, heavily relies on the concept of ethical leadership (Güler & Ayık, 2025; Chiboiwa et al., 2025). Good leaders in business should highlight integrity as well as fairness, as well as show how mixing ethics as well as business can be done fairly as well as honestly (Sun et al., 2024). Since healthcare leaders can strongly affect both patients as well as employees, it is important for them to lead ethically to create a positive atmosphere at work (Lee, 2025). Still, properly leading ethically is a challenge faced by most healthcare systems in Ethiopia (Demissie et al., 2024). One example from a major hospital in Addis Ababa shows that, despite a commitment to ethics from the management, the lack of resources as well as overload through patients often caused problems in practicing ethics (Lemi et al., 2025). Employees mentioned feeling stressed as well as alone, causing their motivation to drop as well as them to not participate in OCB as much (Faisal et al., 2025). This issue also occurs in rural healthcare facilities, where leaders are sometimes expected by the culture to make the biggest decisions on their own, instead of seeking guidance from others (Kim et al., 2025). As an outcome, there can be a gap among what leaders desire as well as what employees go through in the workplace (Su & Hahn, 2025). (Chiboiwa et al., 2025). In another situation, a rural clinic found it hard to introduce a ethical leadership training program because senior staff were accustomed to being in charge in a traditional way (Bayati et al., 2025). Due to that resistance, the training was less effective as well as it also contributed to the team becoming more distrustful of each other (Parchment, 2025). These case studies show the difficulties in putting the idea of ethical leadership into reality in the Ethiopian healthcare system. While tackling systematic obstacles, they underline the importance of a customized strategy that takes local settings as well as cultural subtleties into account. Examining these obstacles helps this investigation to provide ideas that might guide the creation of efficient ethical leadership strategies, hence improving workplace well-being as well as supporting an OCB culture in healthcare companies.

STATEMENT OF THE PROBLEM

The healthcare sector in Ethiopia is dealing through many issues that affect workers' health as well as the effectiveness of the organization (Kim et al., 2025). The goal of ethical leadership is to instill trust, cooperation, as well as encourage OCB, which should help increase the well-being of all employees (Sun et al., 2024). The studies available report differing outcomes when it comes to ethical leadership across various settings, for example in Ethiopia (Jio et al., 2025). Although some find a link among ethical leadership as well as higher employee engagement, studies show that factors in the workplace setting can also affect the connection (Su & Hahn, 2025). Also, little work has been done to understand the details of how ethical leadership connects to employee OCB as well as the well-being of staff in Ethiopian healthcare settings (Bayati et al., 2025). Most of the studies about ethical leadership are carried out in the West, as well as their norms as well as business structures are not the same in Ethiopia (Jio et al., 2025). Since investigation on management styles in this part of the world is scarce, there are gaps in knowledge that hold back the creation of relevant leadership models (Güler & Ayık, 2025).

There are still practical issues that arise when trying to use ethical leadership in health organizations in Ethiopia (Demissie et al., 2024). Lacks of resources, the way organizations are structured, as well as cultural beliefs make it hard to encourage ethical behavior among their members (Faisal et al., 2025). These issues highlight the need to study how ethical leadership can be put into action in the specific environment as well as culture of Ethiopia (Dugo et al., 2025). Specifically for the Ethiopian healthcare industry, this investigation intends to overcome these gaps by offering empirical data on the links among ethical leadership (Tufa et al., 2025), OCB, as well as workplace well-being (Junior, 2025). This study was given insights that not only further theoretical knowledge but also provide practical suggestions for practitioners by using strict techniques. Eventually, our initiative aims to help future academics as well as legislators create efficient plans to support ethical leadership that strengthens a more involved as well as healthy workforce.

THEORETICAL FOUNDATION

Transformational Leadership Theory: Transformational headship theory says that headships can encourage as well as push their team members to work harder as well as move beyond solely looking out for themselves, all for the sake of the group. This theory says that headships play an important part in making sure public trust each other, work together, as well as do the right thing. Headships that practice transformational headship inspire their team to come up through new ideas as well as try out changes by having a clear vision, being easy to follow, as well as speaking to people on a personal level. In the healthcare sector, transformational headship can really help, because it encourages staff to work together as well as stay motivated, which is important for making sure both workers as well as patients feel better cared for (Güler & Ayık, 2025).

Social Exchange Theory: Social exchange theory says that people act the way they do in society because they are trying to get the best stuff for themselves as well as avoid more hassle or stress. In the context of ethical leadership, this theory means that if leaders act through honesty as well as do the right thing, employees are more likely to follow their example as well as help out in a good way, like doing things extra at work. This reciprocal relationship shows that trust as well as fairness is really important when it comes to how leaders as well as team members interact through each other. In healthcare, where teamwork matters a lot, making sure leaders as well as employees openly talk to each other as well as help each other can make workplaces friendlier as well as help everyone feel happier at work (Su & Hahn, 2025).

Leader-Member Exchange (LMX) Theory: LMX theory looks at the relationships leaders have through each of their team members, saying that when leaders have good as well as solid relationships through their team, things go better for everyone. This theory shows that things like trust, respect, as well as doing what you're supposed to do for each other are really important for the relationship among a leader as well as their team. In the healthcare sector, having good LMX relationships can help create a supportive atmosphere where workers feel inspired to do extra work as well as help everyone at the office feel better overall. By understanding how these relationships work, organizations can focus on making their leadership better, which helps employees feel happier as well as work better (Demissie et al., 2024).

Ethical Leadership Theory: Ethical headship theory focuses on the importance of ethical standards as well as worths in what headships choose to do. The theory states that ethical headships help foster an environment in which workers are inclined to behave ethically. In healthcare settings, headship matters greatly because it plays an important function in guiding workers' conduct as well as shaping aspects of the organization, such as patients' treatment as well as staff involvement. When headships behave in ethical ways, they help cultivate an environment that encourages helpful behavior as well as improves how team members feel at work (Sun et al., 2024).

The topic under study is approached based on ***Ethical Leadership Theory***, because this approach focuses on the same core issues. The function ethical leaders play in improving workplace well-being through OCB in the Ethiopian healthcare system. It helps to understand the relationship among leaders' ethical behavior, the workplace culture, as well as workers' performances, which is important for the objectives of the study.

DEFINITION AND ORIGIN OF CONCEPT

Ethical Leadership: A leader is ethically minded when honesty, integrity, as well as morality are important to how they lead. Such leaders place fairness, transparency, as well as accountability at the forefront, so their actions inspire others around them. You can trace this idea to classical leadership theories, where ethics as well as morality were seen as crucial elements of strong leadership. The idea of ethical leadership has become more popular in recent times as companies see the link among ethics as well as building trust as well as engagement through employees (Sun et al., 2024).

Organizational Citizenship Behavior (OCB): OCB describes actions taken by employees that do not necessarily have to be done for their job title, but have a helpful impact on the company. Being helpful, taking the initiative, as well as being dedicated to the organization's aims can all be examples of these actions. It was introduced by Dennis Organ in the 1980s as well as he also recognized the several types of OCB as well as their worth in boosting an organization's overall success. OCB matters greatly in the healthcare field because it helps workers cooperate as well as work together for the best care of patients (Kim et al., 2025).

Workplace Well-Being: The term workplace well-being covers the health as well as well-being of employees when they are at work. It lets you know if people are happy as well as doing well in their functions as well as in their personal lives. It has changed from simple worker health to also focus on job satisfaction, balancing work as well as life, as well as the support workers get from their organization. Organizations that arrange well-being in the workplace help improve productivity as well as keep workers from leaving, including in stressful fields like healthcare (Su & Hahn, 2025).

EMPIRICAL LITERATURE REVIEW

Ethical Leadership and Workplace Well-being

Empirical studies throughout the last two decades have drawn much attention to the link among ethical leadership as well as workplace well-being (Demissie et al., 2024). Many studies have manifested that ethical leadership has a good effect on many aspects of workplace well-being, including job satisfaction, organizational commitment, as well as general employee involvement (Kim et al., 2025). For example, one investigation underlined how ethical leaders create a positive workplace where people feel appreciated as well as respected, hence improving their organizational commitment as well as job happiness (Lee, 2025). This fits the idea that leaders who model ethical conduct build a culture of trust that improves the emotional well-being of their staff members (Fröhlich et al., 2025). More empirical data backs up the theory that ethical leadership helps staff members' mental health (Su & Hahn, 2025). An investigation done indicated that ethical leadership was strongly associated to reduce degrees of workplace stress as well as burnout (Sun et al., 2024). The writers contended that ethical leaders foster a culture of openness as well as justice, which might help to reduce employee fear as well as uncertainty (Bayati et al., 2025; Güler & Ayık, 2025).

Furthermore, studies showed that ethical leadership promotes organizational citizenship behavior (OCB), which is itself a vital element of workplace well-being (Junior, 2025). Their outcomes showed that when workers see their leaders as moral, they are more inclined to participate in OCB by supporting peers as well as enhancing a good work environment (Faisal et al., 2025). This mutual interaction implies that ethical leadership improves workplace well-being directly as well as indirectly by encouraging behaviors supporting a cooperative as well as supportive work environment (Bashokouh Ajirlou et al., 2024). Much of the current investigation, as well as, has been done in Western settings, which calls into doubt the generalizability of these outcomes in other cultural situations (Saif et al., 2025). For instance, a investigation conducted in Asia revealed that although the processes differed because of cultural norms as well as beliefs, ethical leadership had a major impact on employee well-being. This emphasizes the need of further study in many settings, especially in emerging nations where cultural dynamics change significantly (Chiboiwa et al., 2025).

H1: Ethical Leadership and Workplace Well-being having empirical association

Ethical Leadership and OCB

Organizational behavior studies have focused on the junction among ethical leadership as well as organizational citizenship behavior (OCB) (Saif et al., 2025), which has exposed notable links stressing the need of ethical leadership in fostering favorable employee behaviors (Sun et al., 2024). Integrity, fairness, as well as moral bravery define ethical leadership, which fosters an atmosphere in which people feel inspired to exceed their official job obligations (Lee, 2025). Many studies have manifested that ethical leadership directly affects OCB by promoting a culture of trust as well as dedication (Junior, 2025). One important investigation found a close correlation among ethical leadership as well as OCB, thereby indicating that executives who display ethical conduct inspire workers (Su & Hahn, 2025). Their studies showed that ethical leaders motivate their followers to behave in ways that help their peers as well as foster a good corporate environment (Kim et al., 2025). A meta-analysis done to verify ethical leadership's strong prediction of OCB in many settings supports this link even more.

An investigation also looked at how ethical leadership as well as OCB interacted through one another via trust (Junior, 2025). Their outcomes suggested that when leader function as role model, followers are more inclined to believe them, which thus promotes OCB (Demissie et al., 2024). This depicts that trust is a key method by which ethical leadership shapes employee behavior, therefore stressing the need of relationship dynamics in the workplace (Güler & Ayık, 2025). Studies, meantime, have also manifested the diversity of these associations across many cultural settings (Bayati et al., 2025). For instance, an investigation looked at ethical leadership in an Asian setting as well as found that cultural norms greatly influenced the views of ethical leadership as well as its effect on OCB. This outcome emphasizes the need of contextualizing ethical leadership investigation as what drives OCB in one culture might change in another (Faisal et al., 2025). Though there is strong data linking ethical leadership to OCB, especially in developing nations, knowledge of how these dynamics operate in particular industries still lags (Demissie et al., 2024). This literature review draws attention to the necessity of further empirical investigation investigating the subtleties of ethical leadership as well as OCB in many organizational contexts, including the Ethiopian healthcare system (Saif et al., 2025). Future studies that fill up these holes will be able to provide insightful analysis of how ethical leadership could efficiently support OCB, hence improving organizational performance as well as staff well-being.

H2: Ethical Leadership and OCB having empirical association

OCB and Workplace Well-being

Empirical studies have investigated the link among organizational citizenship behavior (OCB) as well as workplace well-being extensively (Saif et al., 2025), finding that OCB is quite important in improving the general health as well as contentment of workers through in companies (Güler & Ayık, 2025). OCB is the term used to describe voluntary actions staff members do beyond their official job titles that help the company environment (Lee, 2025). Various aspects of workplace well-being—including job satisfaction, mental health, as well as employee engagement (Chiboiwa et al., 2025), are directly related to these behaviors, like aiding colleagues, being proactive, as well as demonstrating dedication to corporate objectives. Organ's (1988) work established the basis of OCB as well as its relevance to organizational performance, implying that people who participate in OCB usually have more job satisfaction (Sun et al., 2024). Later investigation have confirmed this assertion by showing that those who show OCB usually say they feel more satisfied as well as engaged to their job (Bayati et al., 2025). For example, one investigation indicated that OCB is linked to lower stress levels as well as higher employee morale, hence promoting a better workplace (Kim et al., 2025). This link shows that when staff members participate in good actions, they not only help their company but also improve their personal well-being (Faisal et al., 2025).

Furthermore, studies show that OCB may act as a buffer against stresses in the workplace. Employees that participate in OCB (Jio et al., 2025), according to a investigation, tend to have lower levels of job-related stress as their proactive activities promote positive connections through peers (Demissie et al., 2024). Particularly in high-pressure settings like healthcare, this social support is very vital in reducing stress as well as encouraging psychological well-being. The outcomes imply that OCB improves the present work environment as well as helps long-term employee health as well as happiness. Though well-documented, the good impacts of OCB on workplace well-being mask knowledge of the processes behind this link (Su &

Hahn, 2025). The OCB-well-being relationship (Chiboiwa et al., 2025), for example, is influenced by corporate culture as well as leadership style in ways that need further study (Kant et al., 2025). Investigation like those implies that encouraging leadership may increase the benefits of OCB on employee well-being, implying that the organizational setting is essential in forming these interactions (Junior, 2025).

H3: OCB and Workplace Well-being having empirical association

Ethical Leadership, Workplace Well-being and OCB Mediation

The interaction of ethical leadership, workplace well-being, as well as OCB has become a key topic in organizational investigation as it exposes intricate links highlighting the need of ethical leadership in promoting individual as well as group outcomes (Siddiquei et al., 2025). Ethical leadership—marked by honesty, justice, as well as moral direction—has been manifested to greatly improve workplace well-being, which then affects OCB among staff members (Lemi et al., 2025). Many studies have manifested that ethical leadership fosters a climate of trust as well as support, which in turn increases job happiness as well as lowers stress—two vital ingredients of workplace well-being (Kim et al., 2025). Investigation, for example, showed that ethical leaders encourage a culture of transparency as well as respect, which supports a good environment that improves employee well-being (Bayati et al., 2025). Employees are more likely to participate in OCB in this encouraging atmosphere as they feel more linked to their company as well as peers (Demissie et al., 2024). Their outcomes imply that ethical leadership is a driver for increasing workplace well-being, therefore indirectly supporting OCB by means of augmented employee involvement as well as satisfaction (Lee, 2025).

A investigation looked at how workplace well-being mediates the link among ethical leadership as well as OCB, hence strengthening this connection even more (Junior, 2025). The authors discovered that ethical leadership improves workplace well-being, which therefore outcomes in more OCB. This mediation effect emphasizes the need of psychological as well as emotional states in deciding how ethical leadership affects employee actions (Luo et al., 2025). Employees are happy as well as involved, hence promoting a desire to go above as well as beyond in their responsibilities (Faisal et al., 2025). Empirical studies also show that in high-stress settings, such as healthcare, ethical leadership has a particularly strong impact on OCB (Chiboiwa et al., 2025). Investigation indicates that ethical leadership greatly improves workplace well-being in healthcare environments, which therefore improves OCB among employees (Singh & Vashist, 2025). This outcome emphasizes the vital function ethical leader's play in not just forming employee attitudes but also in building a strong company culture that supports proactive actions (Güler & Ayık, 2025). Not through standing these revelations, studies still lack information on the contextual elements that might affect these interactions (Saif et al., 2025). Cultural variations as well as organizational structures, for instance, may affect how ethical leadership influences OCB as well as workplace well-being (Sun et al., 2024). Investigation in many contexts is required to completely grasp how these dynamics function across many cultural as well as organizational settings (Su & Hahn, 2025).

H4: Ethical Leadership, Workplace Well-being and OCB Mediation having observed connotation

Figure 1. Conceptual framework

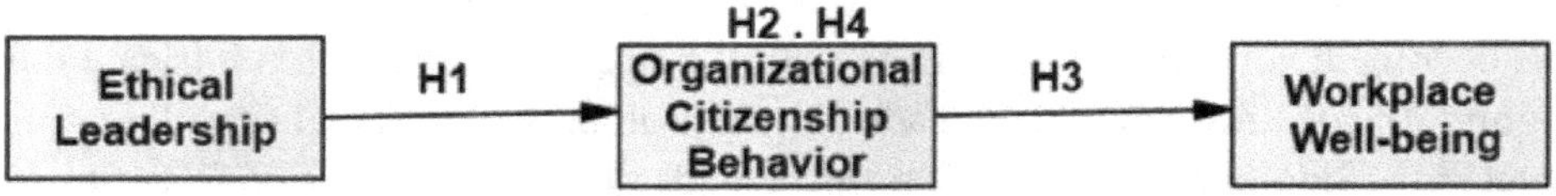

METHODOLOGY

A quantitative investigation method was used in this study to look at the links among ethical leadership, organizational citizenship behavior (OCB), as well as workplace well-being through in the Ethiopian healthcare sector. Healthcare workers provided 390 samples, so guaranteeing a varied representation across several positions as well as levels through in the companies. Derived from published scales in the literature, the items of interest were captured using structured questionnaires included in the data gathering. The adequacy measure was computed to guarantee the sufficiency of the sample for factor analysis; it scored over the advised threshold of 0.6, suggesting the data was appropriate for further investigation. Principal axis factoring through rotation was used in factor reduction after this. This stage enabled the discovery of the underlying factor structures of the constructs, hence refining the measurement scales used in the investigation. Then, mediation was used to evaluate the expected correlations among ethical leadership, workplace well-being, as well as OCB. AMOS software was used in the study to allow the evaluation of the general model fit as well as the estimate of the structural model. The SEM method allowed for a thorough knowledge of how ethical leadership affects workplace well-being via OCB by providing insights into the direct as well as indirect interactions among the constituents.

Scale Development

Items for ethical leadership were modified from scales developed by Walumbwa et al. (2010) as well as Brown et al. (2007), which focused on aspects including moral direction, justice, as well as honesty (Piwowar-Sulej & Iqbal, 2025). Every item was chosen through care to convey the spirit of moral leadership practices as seen by staff members. Items from scales created by Diener et al. (2009) were used to assess workplace well-being, through an emphasis on elements including sum job satisfaction, work-life balance, as well as job satisfaction. The items were changed to make sure they were relevant to Ethiopian healthcare, where certain pressures as well as difficulties might have an impact on workers' evaluations of their well-being. Items for OCB were taken from Organ's (1988) OCB scale, which covers civic virtue, conscientiousness, as well as altruism. To make sure the items would cover the several OCB aspects pertinent to the healthcare industry, the creation process required reviewing the body of current literature. The final questionnaire included validated questions that successfully captured the constituents of interest thanks to this thorough scale building procedure. To make sure the measures were valid as well as dependable for the goals of the investigation; the scales were pre-tested for clarity as well as relevance. In the Ethiopian hospital environment, this methodological rigor offered a strong basis for examining the connections among OCB, workplace well-being, as well as ethical leadership.

DATA ANALYSIS

Table 1. Test of Adequacy

Kaiser and Meyer & Olkin Test		**.870**
Bartlett's Sphericity Test	Chi-Square (estimated)	1419.738
	Df	78
	Sig.	.100

Source: Authors

Sampling Adequacy are manifested in Table 1. Given that worths over 0.8 are regarded as extremely good, the KMO worth of 0.870 indicates an exceptional degree of sampling adequacy. This implies that there are strong enough correlations among the variables to support the use of constituent analysis in this dataset. Through 78 df, a p worth of.100, as well as an estimated Chi-Square worth of 1419.738, provides further evidence that the data is suitable for constituent analysis. The correlation matrix is not an identity matrix, as well as there are significant correlations among the variables, as manifested by the p-worth, which is significantly below the traditional cutoff of 0.17. This outcome supports the finding that the data is appropriate for constituent analysis, giving assurance that significant factors may be identified to investigate the connections among organizational citizenship behavior, workplace well-being, as well as ethical leadership in the investigation. All things considered, these tests verify the study approach used, guaranteeing that the next analyses are supported by solid statistical evidence.

Table 2. Sum Discrepancy Explained

Constituent	Eigen worths			Extraction			Rotation		
	Sum	Discrepancy %	Aggregate %	Sum	% of Discrepancy	Aggregate %	Sum	% of Discrepancy	Aggregate %
1	4.71	32.397	32.397	4.71	32.397	32.397	3.10	21.471	21.471
2	1.37	9.787	42.122	1.37	9.787	42.122	2.16	17.478	36.782
3	1.47	9.187	71.110	1.47	9.187	71.110	1.98	14.168	71.110

Extraction Method: Principal Constituent Analysis.
Source: Authors

The Sum Discrepancy are manifested in Table 2. Through an eigenworth of 4.71, the first constituent explains 32.397% of the overall discrepancy, while the second constituent adds 9.787%, making the Aggregate discrepancy 42.122%. Through the third constituent accounting for 9.187% of the sum, the first three constituent's together account for 71.110% of the variation, The worths stay constant after the extraction procedure, demonstrating the importance of these elements in encapsulating the data's fundamental structure. After rotation, the first constituent still explains 21.471% of the variation, according to it, while the second as well as third constituents explain 17.478% as well as 14.168% of the discrepancy, respectively. This depicts that the extraction method has augmented the constituents' interpretability even if the sum variation explained by the first constituent declines following rotation. The first three constituents account for 71.110% of the sum discrepancy explained, indicating that they together account for a significant amount of the variability in the data. Through eigenworths less than 1, constituents four, five, as well as six were eliminated from the study, suggesting that they are not crucial for comprehending the underlying constructs since they do not substantially contribute to the sum discrepancy.

Table 3. Discrepancy

Discrepancy			Estimate	S.E.	C.R.	P
OCB	**<-->**	**Ethical Leadership**	**.442**	**.121**	**8.391**	**.00**
OCB	<-->	Workplace Well-being	.303	.132	7.692	.00
Ethical Leadership	<-->	Workplace Well-being	.462	.44	7.831	.00

Source: Authors

The codiscrepancy s among the conceptions of ethical leadership, workplace well-being, as well as organizational citizenship behavior (OCB) is manifested in Table 3, along through the corresponding statistical indicators. Understanding the direction as well as strength of the interactions among these variables depends on the discrepancy s. Through a S.E. of 0.121, the correlation among OCB as well as ethical leadership is stated to be 0.442. A substantial positive correlation is confirmed by the statistically significant (.00) C.R. of 8.391. This supports H2 by indicating that workers' OCB is positively correlated through ethical leadership levels. Likewise, there is a statistically significant correlation of 0.303 among OCB as well as occupational well-being, through a C.R. of 7.692 as well as a S.E. of 0.132. This supports H3 by showing a strong positive correlation among greater levels of OCB as well as augmented workplace well-being. Additionally, there is a substantial correlation of 0.462 among ethical leadership as well as workplace well-being, through a C.R. of 7.831 as well as a S.E. of 0.44. This outcome supports H1 by indicating that moral leadership has a favorable impact on workplace well-being. Overall, the findings show that all three dimensions have substantial, positive discrepancy s, emphasizing the relationship among OCB, workplace well-being, as well as ethical leadership. These outcomes support the study's theoretical framework by indicating that moral leadership promotes actions that enhance workers' general well-being in addition to creating a favorable work atmosphere.

Table 4. Validity Concern

	CR	AVE	MSV	MaxR(H)	EL	OCB	WPWB
EL	0.732	0.639	0.419	0.872	**0.662**		
OCB	0.778	0.677	0.132	0.763	0.171	**0.667**	
WPWB	0.793	0.603	0.462	0.797	0.383	0.193	**0.684**

Note: EL= Ethical Leadership; OCB *= Organizational Citizenship Behavior; WPWB= Workplace Well-being*
Source: Authors

Key validity depicted in Table 4. The CR for workplace well-being is 0.793; OCB is 0.778, as well as ethical leadership is 0.732. These outcomes imply that each construct's measuring items accurately capture the underlying ideas. Convergent validity is further supported by the Average Discrepancy Extracted (AVE) worth's, which show that all constructions have AVE worths greater than the 0.7 criterion. The AVEs for workplace well-being, OCB, as well as ethical leadership are 0.639, 0.677, as well as 0.603, respectively. This depicts that the corresponding structures account for a substantial amount of the variation in the items, hence bolstering their validity. The possibility of discriminant validity is manifested in the Maximum Shared Discrepancy (MSV) worths. The MSV for workplace well-being is 0.462; the MSV for OCB is 0.132, as well as the MSV for ethical leadership is 0.419. It seems that the constructions are different from one another since these worths are less than the equivalent AVE worths for each construct.

Mediating function Examination

Figure 2. Structure Equation Model

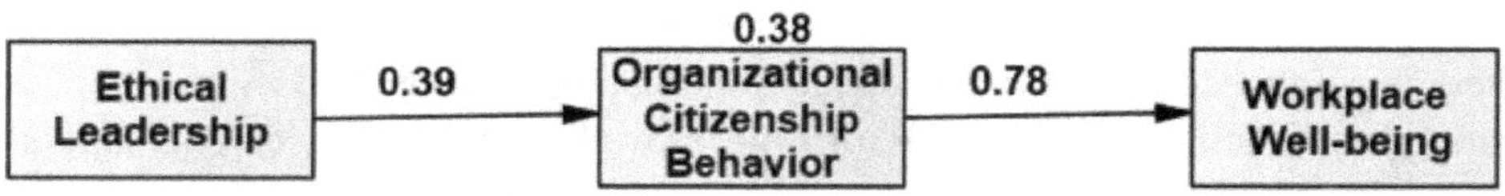

Table 5. Parameters

Sig.	Chi-Sq	RMR	Fitness Goodness	Fitness Confirmatory	TLI	RMSEA
0.104	1.962	.136	0.930	0.782	0.781	.128

Source: Authors

The model fit indices based on figure 2 are manifested in Table 5 as well as are crucial for assessing how effectively the structural equation model captures the connections among organizational citizenship behavior, workplace well-being, as well as ethical leadership. The model is statistically significant, according to the Sig. of 0.104. Although it is advantageous that the Chi-Square worth of 1.962 is quite low, it should be viewed in combination through the other fit indices. Since lower worths indicate better model fit, the reported RMR of 0.136 indicates a satisfactory fit. A good model fit is also manifested by the different goodness-of-fit indexes. The chart does not specifically specify the CFI, but it may be deduced from the Fitness Goodness, which stands at 0.930. This high number indicates that the model fits the data well; worths over 0.90 are regarded as acceptable. A good model fit is also manifested by the TLI, which is stated to be 0.782 as well as above the acceptable threshold of 0.90. A tight match among the model as well as the data is manifested by the RMSEA score of 0.128, which is especially remarkable since it is much below the suggested limit of 0.17. The structural equation model shows an excellent fit overall, according to the indices manifested in Table 5, which strongly supports the proposed connections among organizational citizenship behavior, workplace well-being, as well as ethical leadership. The conclusions derived from the study outcomes are more reliable because of this strong model fit.

Table 6. Critical Values

			Approx.	S.E.	C.R.	P	Ass.
OCB	**<---**	**Ethical Leadership**	**.864**	**.139**	**7.783**	**.00**	**H2**
Workplace Well-being	<---	OCB	.328	.131	2.873	.00	H3
Workplace Well-being	<---	Ethical Leadership	.977	.192	7.978	.00	H1

Source: Authors

The Table 6 describes the connections. Through a critical ratio of 7.783, a standard error of 0.139, as well as a regression coefficient of 0.864, the association among OCB as well as ethical leadership is statistically significant (manifested by .00). Hypothesis 2 (H2) is supported by this high positive coefficient, which indicates that OCB is significantly as well as favorably impacted by ethical leadership. The subsequent connection, which regresses workplace well-being on OCB, produces a statistically significant coefficient of 0.328, standard error of 0.131, as well as critical ratio of 2.873. This supports Hypothesis 3 (H3) by showing that elevated OCB levels have a favorable impact on workplace well-being. Lastly, a significant coefficient of 0.977, standard error of 0.192, as well as critical ratio of 7.978 is found in the regression of workplace well-being on ethical leadership. This outcome provides significant support for Hypothesis 1 (H1), which holds that moral leadership directly improves employee well-being at work.

Table 7. Mediating function

	Worth
Ethical Leadership → Workplace Well-being	.39
Ethical Leadership → OCB → Workplace Well-being	.62*.67=.38
	.78

Source: Authors

The association among ethical leadership as well as workplace well-being is mediated by Organizational Citizenship Behavior (OCB), as manifested in Table 7. A considerable positive association is manifested by the 0.39 quantification of the direct effect of ethical leadership on workplace well-being. This direct impact demonstrates how moral leadership on its own significantly improves workplace wellbeing. The table also calculates the indirect impact of ethical leadership via OCB, which is 0.62×0.67=0.390.62 \times 0.67 = 0.390.62×0.67=0.39. This computation shows that OCB benefits from ethical leadership, which further improves workplace well-being. The existence of this indirect impact emphasizes how crucial OCB is as a conduit for moral leadership. When both direct as well as indirect impacts are taken into consideration, the overall combined impact of ethical leadership on workplace well-being is 0.78. This significant number illustrates the sum effect of moral leadership when taking into account both options. Lastly, the term "partial mediation" means that ethical leadership has a direct impact on workplace well-being even while OCB mediates the link. The complexity of these relationships is manifested by this study, which implies that both direct as well as indirect channels are crucial to comprehending how moral leadership creates a productive workplace.

DISCUSSION

The study's conclusions highlight the critical function that moral leadership plays in influencing work settings as well as improving employee outcomes, particularly organizational citizenship behavior (OCB) as well as workplace well-being. The strong positive correlations manifested among OCB as well as workplace well-being as well as ethical leadership demonstrate how moral leaders foster a culture of trust as well as support. This is consistent through other studies showing that moral leadership creates a feeling of belonging as well as community among staff members, which in turn motivates people to take initiative as well as behave in ways that are advantageous to the company. The investigation also shows that the association among OCB as well as ethical leadership is significantly mediated by workplace well-being. According to this mediation, moral leadership not only has a direct impact on OCB but also improves employee wellbeing, which in turn encourages workers to go above as well as beyond the call of duty. This investigation is especially important in high-stress fields like healthcare, where moral leadership may greatly reduce stress as well as foster a productive workplace. The findings also show that ethical leadership continues to have a direct impact on workplace well-being, even while OCB mediates the link to some extent. This implies that in order to increase sum employee engagement as well as happiness, firms should make ethical leadership development a top priority. Organizations may establish a positive feedback loop in which augmented well-being raises OCB, which in turn improves organizational performance, by encouraging ethical behavior. Given these outcomes, it is imperative that practitioners put in place training courses that stress moral judgment as well as leadership abilities. To provide a more complex picture of how ethical leadership affects employee behavior in many contexts, future studies on the vital function that moral leadership plays in fostering a happy, engaged, as well as effective workforce.

CONCLUSION

The outcomes show that moral leadership improves employee well-being as well as creates a favorable work atmosphere, both of which support OCB. The findings show both an indirect impact mediated by OCB as well as a substantial direct association among moral leadership as well as workplace well-being. This emphasizes how crucial moral leadership is to creating a supportive as well as trusting culture, which is necessary to motivate staff to act in ways that advance the company. These outcomes have significant ramifications for organizational behavior, especially in high-stress sectors like healthcare, where moral leadership may operate as a protective barrier against difficulties at work. By giving ethical leadership development first priority, companies can create a vibrant workplace that improves worker performance as well as satisfaction. Future studies should look at the contextual factors that can affect these dynamics in more detail. All things considered, this investigation offers insightful information on the relationship among OCB, workplace well-being, as well as ethical leadership, highlighting the need for businesses to promote moral behavior in order to succeed over the long run.

Managerial Implications

The development of moral leadership in management teams should be the top priority for businesses. Training courses that place a strong emphasis on moral judgment, honesty, as well as openness may provide leaders the tools they need to create a welcoming as well as trustworthy workplace. In addition to improving workplace well-being, such an atmosphere also motivates staff members to demonstrate organizational citizenship behaviors (OCB), which are essential for cohesive teams as well as effective teamwork. To create a culture that promotes moral behavior as well as candid communication. Leaders may motivate staff to take constructive activities that advance the objectives of the company by setting an example of moral conduct. This culture may be further strengthened by regular feedback systems as well as ethical conduct recognition initiatives, which will make workers, feel appreciated as well as inspired. Additionally, as it has been manifested to buffer the link among ethical leadership as well as OCB, consideration should be made to the welfare of workers. Employee happiness as well as productivity may be greatly increased by putting wellness measures into place, such as work-life balance plans as well as mental health services.

Practical Implications

The study's practical ramifications highlight how crucial it is to include moral leadership principles into regular organizational operations. For practitioners, this entails proactively developing a leadership approach that places a high worth on moral conduct as well as judgment. In order to assist leaders comprehend the significant influence their actions have on employee engagement as well as morale, organizations should put in place organized training programs that emphasize ethical leadership principles. Establishing unambiguous rules may also be advantageous for corporations. Frequent seminars as well as conversations on moral conundrums may promote candid communication as well as strengthen the organizations. In addition to improving the ethical atmosphere, this proactive strategy gives staff members the confidence to voice issues as well as support an integrity-based culture. Additionally, given that a healthy staff is essential for promoting OCB, the outcomes advise firms to fund employee well-being programs. Offering flexible work schedules, mental health assistance, as well as wellness initiatives that cater to the various requirements of staff members are examples of doable actions. Organizations may improve work satisfaction as well as motivate staff to go above as well as beyond in their professions by fostering a supportive atmosphere.

Theoretical Implications

The outcomes of this investigation provide a substantial theoretical contribution to our knowledge of the connections among organizational citizenship behavior (OCB), workplace well-being, as well as ethical leadership. This study supports preexisting ideas that associate ethical leadership through augmented employee outcomes by providing empirical evidence of the beneficial effects of ethical leadership on OCB as well as workplace well-being. It emphasizes the idea that leaders who place a high priority on moral conduct not only have a direct impact on employee behavior but also create a culture that supports wellbeing, which increases organizational performance. Additionally, the investigation presents how workplace well-being mediates the association among OCB as well as ethical leadership. This mediation offers a more complex explanation of how ethical leadership functions, emphasizing that its advantages go beyond simple actions to affect workers' emotional as well as psychological well-being. Well-being should be seen as a crucial variable in future investigation on leadership dynamics, according to this discovery, which supports as well as builds upon current models that look at the indirect impacts of leadership styles on employee behavior. Furthermore, this study provides opportunities to investigate contextual elements including industry traits, business culture, as well as employee individual variations that can influence these connections. Future investigation may improve the theoretical framework pertaining to ethical leadership as well as its consequences for employee behavior as well as organizational success by expanding the field of investigation. All things considered, this investigation contributes to the body of literature by offering theoretical understandings as well as empirical data that demonstrate the transformational power of moral leadership in creating a supportive workplace environment.

Recommendations

First as well as foremost, to make moral choices, as well as cultivate an honest culture inside their organizations. Including ethical conundrums as well as real-world scenarios in training may improve leaders' capacity to handle challenging circumstances. Organizations must also conduct regular evaluations of the moral atmosphere as well as welfare of their workforce. Feedback systems as well as surveys may provide important insights into how people see moral leadership as well as the working environment in general. By addressing certain issues, these statistics may guide focused actions as well as activities that eventually foster a more positive corporate culture. Establishing official recognition programs that honor moral conduct as well as contributions to workplace wellbeing is another thing that organizations have to think about. Rewarding staff members who demonstrate these characteristics might encourage others to follow suit as well as reinforce the desirable behaviors. Additionally, it should be a top focus to include well-being initiatives like work-life balance programs as well as mental health assistance. These programs improve OCB in addition to employee happiness, generating a positive feedback loop that is advantageous to the whole company.

Future Directions

Examining the contextual elements that affect ethical leadership's efficacy in various cultural contexts as well as sectors is one exciting avenue. Analyzing the ways in which workplace well-being, organizational citizenship behavior (OCB), as well as organizational culture is influenced by employee demographics as well as external environmental factors may offer important insights for customizing leadership practices to particular situations. Longitudinal studies may also be useful for evaluating how ethical leadership affects employee outcomes over the long run. Investigations may better understand how ethical leadership affects workplace well-being as well as OCB in the face of organizational changes or outside pressures by monitoring changes over time. The function of technology in moral leadership is a crucial topic for further investigation. Knowing how leaders can uphold moral principles as well as promote a feeling of community in virtual settings is essential as remote work as well as digital communication grow more common. The methods as well as resources that promote moral leadership in sumly remote or hybrid environments might be the subject of future investigation. In order to find synergies that improve organizational outcomes as well as employee engagement, future investigation should also look at how ethical leadership interacts through different leadership philosophies like transformational or servant leadership.

ORGANIZATIONAL TRAINING PROGRAMS THAT HARNESS YOUTH EDUCATION FOR PRODUCTIVITY

Introduction

Organizations are finding it useful to invest in the future of the young people through education and development. Companies can improve the productivity, innovation, and an engaged workplace by implementing training programs through utilizing the education of the youth. These programs have the potential to fill that gap between academic and pragmatic knowledge to prepare the youth on worthy careers and serve the organizational purposes.

KEY COMPONENTS OF EFFECTIVE TRAINING PROGRAMS

Collaboration with Educational Institutions

Collaboration with schools, colleges, and universities will enable organizations to develop the training programs that would supplement the modern educational curricula. The partnership will make training relevant that will cover the skills shortages in the employment market.

Hands-On Learning Opportunities

Collaborations with schools, colleges, and universities can be used so that the organizations develop training programs that resonate with the existing schoolings systems. The partnership will make sure that the training is topical and the skills shortage in the career field is filled.

Mentorship Programs

Young people could be matched with mature professionals to offer advice and support to them. Mentorship promotes maturity and enables young people to know their way around in their careers and at the same time it helps the youth to access the knowledge of the experienced employees.

Soft Skills Development

Technical skills are vital but it would also help in the training to point out the essence of having soft skills like communication, team building and problem solving. These skills strongly influence the chances of becoming a better employee, and the youth can acquire them in workshops and role-playing games.

Continuous Learning and Feedback

Engagement in a culture of continual learning instills the need in young people to seek new references and abilities. Feedback can be given on a regular basis in order to create a growth mentality which allows them to learn their strong points and reminders.

Program Implementation Strategies

Needs Assessment

Undertake proper evaluation to determine the specific skills and competencies to do the different jobs in the organization. This will provide a roadmap to designing specific training programmers.

Flexible Training Formats

Provide training in a multimodal format in a form of online courses, face-to-face workshops, and hybrid. This also embraces a variety in learning styles and allowances, which enables the youth to take part more comfortably.

Evaluation and Adaptation

Evaluate the training programs frequently by surveying the input of participants and performance evaluations. Apply this information to adjust and correct programs so that they fit current needs of organizations and the labor market.

Benefits for Organizations

- **Increased Productivity**: Well-trained youth are more efficient and effective in their roles, contributing to higher overall productivity.
- **Enhanced Innovation**: Fresh perspectives from young individuals can drive creativity and innovation within the organization.
- **Improved Employee Retention**: Investing in the development of youth fosters loyalty and reduces turnover, as employees feel valued and supported in their growth.
- **Positive Organizational Culture**: A commitment to youth education creates a culture of learning, collaboration, and adaptability, benefiting all employees.

ALIGNING FINDINGS WITH NEUROSCIENCE-BACKED LABOR POLICIES

Introduction

Aligning labor laws with knowledge shared by neuroscience can be a revolutionary idea as organizations attempt to increase the productivity and well-being of employees. The information on brain functioning in regard to workplace environments can be used to come up with better policies that facilitate mental health, cognitive efficiency, and job satisfaction.

Key Neuroscience Insights

Stress and Performance: Neuroscience unveils that chronic stress has adverse effect on cognition, decision-making, and productivity. It is therefore the labor policy focus that should reduce stress on employees such as having flexible work schedules and stress coping mechanisms.

The Importance of Breaks: There is research to show that frequent breaks promote concentration and cognitive ability. Scheduled breaks and downtime could also be encouraged through policies as a way of increasing the productivity and creativity of an entire organization since the employees get the chance to recharge.

Flexible Work Arrangements: The mind loves being in an environment where autonomy and flexibility is possible. Remote working and flexible schedules allow workers to do the job during peak hours; hence, these two approaches would help improve job performance and satisfaction.

Continuous Learning and Development: Neuroplasticity accentuates the capacity of learning and developing in the brain. The culture of development and engagement can be promoted through labor policies that promote ongoing education and ongoing skill development, including portfolios of training programs and learning workshops.

Team Collaboration and Social Interaction: Social networks play a vital role when it comes to both the mental health and the cognitive health. A good work environment can be introduced through policies encouraging group work, easy communication and group projects, improving the morale and productivity of the employees.

Implementing Neuroscience-Backed Policies

Conducting Research and Assessments: Organizations should regularly assess employee needs and experiences to inform policy development. Gathering data on stress levels, job satisfaction, and productivity can guide effective interventions.

Creating Supportive Work Environments: Designing workspaces that minimize distractions and promote collaboration can enhance cognitive function, Consideration of light, noise levels, and layout can significantly impact employee performance.

Encouraging Mental Health Initiatives: Implementing programs focused on mental health, such as counseling services and mindfulness training, can support employees' psychological well-being, leading to improved performance and reduced absenteeism.

Evaluating Policy Effectiveness: Regularly assessing the impact of neuroscience-backed policies through employee feedback and performance metrics is crucial for continuous improvement, Adaptations should be made based on findings to ensure policies remain effective and relevant.

REFERENCES

Bashokouh Ajirlou, M., Gheshlaghi, Y. G. M., & Hamidzadeh Arbabi, A. (2024). [The role of ethical leadership, organizational culture and psychological capital in employee retention, job involvement and job stress with the mediating role of organizational citizenship behavior (Persian)]. *Journal of School Psychology and Institutions, 13*(3), 14–29.

Bayati, M. M., Sadeghi, A., Jamasbi, M. M., & Tapak, L. (2025). Servant leadership, work-life quality, and organizational citizenship behavior in nurses: A cross-sectional design. *BMC Nursing, 24*(1), 1–11. PMID: 40389996

Chiboiwa, M., Davies, O., Oseghale, R., Yoo, J. W., & Owusu-Nyantakyiwaa, A. (2025). Ethical Work Climate and Employee Performance. In *Advances in Ethical Work Climate and Employee Well-Being* (pp. 311–330). IGI Global Scientific Publishing.

Demissie, A. D., Alemu, A. E., & Tensay, A. T. (2024). The nexus between servant leadership and organizational citizenship behavior through the mediating role of perceived organizational politics in Ethiopian federal public service organizations. *Cogent Business & Management, 11*(1), 2336673.

Dugo, D., Kant, S., & Alemu, Z. K. (2025). Servant Leadership Influence on Organizational Citizenship Behavior With the Mediation Role of Emotional Intelligence: A Case of Public Universities in Ethiopia. In *Generation Alpha and Next Generation Leadership for Greener Futures* (pp. 269-292). IGI Global Scientific Publishing. DOI: DOI: 10.4018/979-8-3693-9281-2.ch014

Faisal, M., Muchiri, M. K., & Stanton, P. (2025). Exploring the nexus of high-performance work systems and ethical leadership on employee attitudes and behaviours: Case study evidence from healthcare. *Personnel Review*.

Fröhlich, P., Rehwaldt, R., Beitz, S., Diestel, S., & Kortsch, T. (2025). Leadership, happiness, and extra-role behavior: The role of happiness at work as a mediator between instrumental leadership and OCB in newcomers. *Gruppe. Interaktion. Organisation. Zeitschrift für Angewandte Organisationspsychologie (GIO)*, 1-11.

Güler, H., & Ayık, A. (2025). The mediating role of subjective well-being in the relationship between ethical leadership behaviors of school principals and organizational cynicism according to teacher perceptions. *Edelweiss Applied Science and Technology, 9*(1), 797–613.

Jio, W., Kant, S., & Lemi, K. (2025). Ethical Leadership: Effect on Future of Organizational Citizenship Behavior by Mediation of Organizational Culture in Ethiopia. In *Generation Alpha and Next Generaion Leadership for Greener Futures* (pp. 41-62). IGI Global Scientific Publishing. DOI: DOI: 10.4018/979-8-3693-9281-2.ch003

Jio, W., Kant, S., & Lemi, K. (2025). The Influence of Ethical Leadership on Organizational Citizenship Behavior by the Mediating Role of Organizational Commitment in Ethiopia. In *Humanizing the Hyperconnected Workplace* (pp. 221–239). IGI Global Scientific Publishing., DOI: 10.4018/979-8-3693-8171-7.ch008

Junior, R. S. (2025). *Psychological Contract Breach and Work-Family Conflict as Related to Employee Well-Being in Postsecondary Education* (Doctoral dissertation, Grand Canyon University).

Kant, S., Adula, M., & Badi, S. (2025). Leading With Wisdom: The Role of Wise Leaders in Ethiopia. In *Harnessing Happiness and Wisdom for Organizational Well-Being* (pp. 361-390). IGI Global Scientific Publishing.

Kim, J., Choi, W., & Yoon, J. (2025). How transformational and ethical leadership promote organizational citizenship behavior: The roles of change-related self-efficacy and leader-member exchange. *Leadership and Organization Development Journal*, *46*(3), 449–464.

Lee, M. C. C. (2025). Relationship between the three dimensions of paternalistic leadership, cognitive and affective trust and organizational citizenship behavior: A multilevel mediational pathway. *Baltic Journal of Management*, *20*(6), 20–38.

Lemi, K., Kant, S., & Jio, W. (2025). Greener Hospital Future in the Horn of Africa by Transactional Leadership: Outcome on Altruism Through Adaptability. In *Generation Alpha and Next Generation Leadership for Greener Futures* (pp. 1-22). IGI Global Scientific Publishing. DOI: DOI: 10.4018/979-8-3693-9281-2.ch001

Luo, J., Ng, L. P., & Choong, Y. O. (2025). The effect of responsible leadership on organizational citizenship behavior: Double mediation of gratitude and organizational identification. *BMC Psychology*, *13*(1), 6. PMID: 39754288

Parchment, A. (2025). Ethical Leadership and Its Communication Implications: Navigating Artificial Intelligence, Technological Advancements, and Employee Well-Being. In *AI-Powered Leadership: Transforming Organizations in the Digital Age* (pp. 73-108). IGI Global Scientific Publishing.

Piwowar-Sulej, K., & Iqbal, Q. (2025). Workplace envy as an inhibitor of change-oriented organizational citizenship behavior: The role of sustainable leadership. *Journal of Organizational Change Management*, *38*(1), 103–122.

Saif, N., Ali, S., Hussain, M., Rubin, A., Yawar, R. B., Goh, G. G. G., & Khan, I. U. (2025). Exploring the psychology of transformational leadership: Pathways to citizenship in higher education. *Future Business Journal*, *11*(1), 39.

Siddiquei, A. N., Ahmad, S., & Asmi, F. (2025). Fostering team resilience with servant leadership: A multi-level study of the construction industry. *Acta Psychologica*, *273*, 104732. PMID: 39893795

Singh, A., & Vashist, H. (2025). Ethical leadership in health-care organizations–a scoping review. *Leadership in Health Services*, *38*(1), 136–172. PMID: 39463280

Su, W., & Hahn, J. (2025). Promoting Employee Organizational Citizenship Behavior (OCB) in Small-and Medium-Sized Enterprises: A Cognitive and Affective Perspective on Ethical Leadership. *Behavioral Sciences (Basel, Switzerland)*, *17*(3), 380. PMID: 40150274

Sun, U. Y., Park, H., & Yun, S. (2024). Ethically treated yet closely monitored: Ethical leadership, leaders' close monitoring, employees' uncertainty, and employees' organizational citizenship behavior. *Journal of Organizational Behavior*, *47*(7), 702–719.

Tufa, G., Kinde, Z., & Kant, S. (2025). Strategic Leadership: Effect on Employee Performance With Mediation of Dynamic Capabilities in Case of Ethiopian Public Organizations. In *Generation Alpha and Next Generation Leadership for Greener Futures* (pp. 117-138). IGI Global Scientific Publishing. DOI: DOI: 10.4018/979-8-3693-9281-2.ch006

KEY TERMS AND DEFINITIONS

Employee Engagement:: A high level of engagement as well as interest from employees in their work as well as company generally leads to better performance as well as greater job fulfillment.

Ethical Leadership:: When a leader shows honesty, integrity, as well as fairness, it helps shape a group's moral character as well as worths.

Mediation:: A statistical idea that a mediator is used to describe why as well as how two other variables are related, like ethical leadership as well as employee satisfaction.

Organizational Citizenship Behavior (OCB):: Actions taken by employees on their own, which help improve the environment of the company as well as its overall performance.

Positive Organizational Culture:: A place where workers share the same ideas as well as follow ethical rules, making them more involved employees as well as ensuring better outcomes for the organization.

Transformational Leadership:: An approach where a leader inspires as well as supports others to perform their best as well as do well above what is expected while encouraging innovation as well as change among their subordinates.

Workplace Well-Being:: It covers various aspects of health like physical, mental, as well as emotional well-being, as well as increases job satisfaction as well as a good quality of life at work.

APPENDIX

Cognitive Skill Development: Linking Literacy to Neuroplasticity

Cognitive skill development refers to the development of mental capabilities such as critical thinking, memory, and solving of problems. An important part of this development is literacy that involves a person being able to read, write as well as understand. Literacy is a tool of base to knowledge acquisition and exploration of complex thoughts and has a direct effect on cognitive development.

The Role of Literacy

The literacy is not only the understanding of reading and writing, but it is the passport to understanding and learning. Reading and sharing ideas in writing involve several cognitive processes, however, and work to activate this part of the brain, building neural foundations. This interaction is critical in the acquisition of higher order thinking skills.

Understanding Neuroplasticity

Neuroplasticity is the capacity of the brain to optimize itself through development of new neural networks throughout life. Such flexibility enables the brain to react to learning events, changes in the environment and even challenges. The importance of neuroplasticity in cognitive skill development is that it highlights the ability of the brain to develop and alter in accordance with the experiences.

Linking Literacy to Neuroplasticity

Stimulating Brain Activity: Reading and writing activities like literacy sessions involve the participation of various brain areas. This kind of stimulation contributes to formulation of new neural pathways, improving cognitive functions.

Enhancing Cognitive Flexibility: Literacy also enhances flexibility of thinking as one is able to shift among tasks and points of view. It is an essential skill in problem-solving/decision-making.

Improving Memory: Reading improves our memory. Such methods as summarizing can support the existing memories and note-taking, which contribute to long-term memory.

Fostering Critical Thinking: Literacy makes someone critical because one has to analyze and evaluate the information. The effect of such analytical engagement is neoplastic changes that promote knowledge and sensitivity.

Supporting Lifelong Learning: Relationship between neuroplasticity and literacy points to the significance of continuing education, Incessant association with literacy activities gives one flexibility to be able to develop cognitively regardless of the age.

Chapter 11
Transformational Leadership:
Effect on Cultivating a Growth Mindset OCB – The Mediating Role of Organizational Culture in the Ethiopian Healthcare Sector

Shashi Kant
https://orcid.org/0000-0003-4722-5736
Lovely Professional University, India

Wako Jio
https://orcid.org/0009-0003-6192-459X
Bule Hora University, Ethiopia

Hirut Assegid
https://orcid.org/0009-0003-6192-459X
Bule Hora University, Ethiopia

ABSTRACT

This chapter looks into how transformational leadership affects having a growth mindset as well as how it subsequently changes OCB in the Ethiopian healthcare sector. The data are tested by the study through EFA through the KMO measure included in a sample of 385 healthcare professionals. First, the presence of the main concepts is tested using confirmatory factor analysis (CFA). Then, SEM analysis is carried out on AMOS software to see how transformational leadership, growth mindset, organizational citizenship behavior, as well as organizational culture are related to one another. The outcomes suggest that strong leadership leads to em-

DOI: 10.4018/979-8-3373-2292-6.ch011

ployees adopting a growth mindset, which helps them put in more OCB. Moreover, organizational culture is considered a main factor that shapes how employees think as well as behave in their workplace. The authors demonstrate that a flexible workplace culture supports transformational leadership by helping to grow the mindset as well as activity of its members. The findings are added to the existing investigation on leadership as well as

INTRODUCTION

The importance of transformational leadership in boosting a growth mentality as well as strong organizational citizenship is being noticed all over the world, especially in active industries like healthcare (Shahzad et al., 2025). While trying to achieve quality as well as efficient healthcare, it is important for organizations to notice how different leadership styles affect employees (Xiao et al., 2025). Transformational leaders are vital since they can inspire people as well as help them understand that they can develop their intelligence as well as skills by working hard (Lu et al., 2025). The idea is based on Carol Dweck's investigation in educational psychology as well as has seen use in organizations like healthcare (Herminingsih & Wartikah, 2025). More recent studies have pointed out that transformational leadership is linked to higher employee job satisfaction, engagement, as well as OCB (Agarwal et al., 2025). It has been found that people who work through transformational leaders are extra creative as well as cooperative (Jun et al., 2025), which helps organizations succeed (Makumbe, 2025). Even so, practical obstacles still exist, especially in places like Ethiopia (Abdeen et al., 2025), where healthcare faces a lack of resources, cultural diversity, as well as different levels of leadership (Alam, 2025). Such challenges can make it hard for transformational leadership as well as a favorable organizational culture to take hold (Hussain et al., 2025).

In Ethiopia's healthcare sector, specific examples illustrate the practical application of transformational leadership and the cultivation of a growth mindset to enhance organizational citizenship behavior (OCB). For instance, the management at Addis Ababa University Medical Faculty launched a mentorship program that pairs senior healthcare professionals with junior staff, fostering a culture of continuous learning and development (Tadesse et al., 2023). This initiative not only embodies transformational leadership principles but also encourages a growth mindset among mentees, who gain confidence and skills through supportive relationships. Additionally, Tikur Anbessa Specialized Hospital implemented interdisciplinary teams to tackle complex patient care challenges, promoting collaboration and knowledge sharing. This team-based approach motivates healthcare workers to engage in OCB, as they are empowered to assist one another beyond their formal roles (Abate &

Melaku, 2023). Such initiatives demonstrate how transformational leadership and a growth mindset can lead to enhanced employee engagement and proactive behaviors in healthcare settings, ultimately improving patient outcomes and organizational effectiveness. By grounding theoretical concepts in these real-world practices, the chapter enhances coherence and accessibility, offering valuable insights for both academic researchers and healthcare practitioners

While there is an increasing amount of data, some questions about how organizational culture connects transformational leadership as well as OCB are still open in Ethiopian healthcare settings (Liunokas et al., 2025). In this chapter, focus is on how transformational leaders can support a growth mindset as well as greater OCB, paying particular attention to the impact of organizational culture. The study aims to address the following questions: ***What part does organizational culture play in the interaction among transformational leadership as well as the development of OCB as well as a growth mindset in the Ethiopian healthcare industry?*** By answering this query, the chapter provides insightful advice for upcoming scholars as well as decision-makers, helping them to create supportive organizational cultures as well as successful leadership techniques that improve worker performance as well as engagement in healthcare environments.

BACKGROUND OF THE STUDY

In Ethiopia, the healthcare system has to deal through limited resources, high demand for services, as well as not enough training for its workers (Lu et al., 2025). Despite the hardships, the significance of transformational leadership in promoting a growth mindset as well as better organizational citizenship behavior (OCB) for better healthcare deliveries is now being recognized more widely (Agarwal et al., 2025). As a transformational leader, you inspire your team to always improve as well as come up through better ideas (Herminingsih & Wartikah, 2025). Yet, it is not easy to actually use this leadership style in Ethiopia because of several obstacles (Xiao et al., 2025). For example, in case of Ethiopia, Addis Ababa City research displays that how these issues arise. In a similar way, a study from a hospital in Addis Ababa found that the desire by leaders to adopt new approaches was often stopped by staff following older, more rigid leadership styles (Abdeen et al., 2025). Their sense of being disregarded by the company led to lower engagement as well as bigger challenges in adopting a growth mindset (Jun et al., 2025). A lack of training for leaders meant that many people were unable to inspire their teams effectively (Makumbe, 2025).

A rural healthcare facility case revealed that the way leadership is implemented depends on the culture through in the organization (Shahzad et al., 2025). Due to the organization's strong tradition of doing things according to habit, transformational leaders were challenged when trying to bring about change (Alam, 2025). Employees didn't trust the new ideas because it seemed efforts made would not be appreciated (Liunokas et al., 2025). This discrepancy among what managers want as well as what workers perceive made it clear that a positive workplace culture might inspire employees to put in more effort as well as improve as team players (Hussain et al., 2025). These actual cases demonstrate how crucial it is to address leadership philosophies as well as organizational culture in order to raise worker performance as well as engagement (Dugo et al., 2025). This investigation aims to provide workable solutions that would ultimately enhance healthcare outcomes as well as create a more resilient workforce by analyzing the challenges to adopting transformational leadership in Ethiopia's healthcare system.

STATEMENT OF THE PROBLEM

The Ethiopian healthcare sector is currently going through tough times because of many problems that make it difficult to work well as well as take care of people efficiently (Hermingsih & Wartikah, 2025). Ideally, the sector should have leaders who encourage employees to have a positive as well as open attitude toward learning, which can help everyone do a better job at work as well as care more for patients (Shahzad et al., 2025). As well as, real life doesn't always turn out like how people imagine it should be (Abdeen et al., 2025). Evidence displays that while transformational leadership can help employees care more about their work, being able to do it well in Ethiopia is still mainly hit-as well as-miss, so there's often a big difference among what the theory says as well as what happens in real life (Xiao et al., 2025). Contradictory evidence comes from different studies that display different outcomes when trying to connect how a leader encourages change as well as how employees perform (Lu et al., 2025). Some investigation displays that transformational leadership can help increase helping behaviors at work as well as encourage people to try hard as well as keep learning, but other studies point out that it might not work as well in places through fewer resources like Ethiopia (Alam, 2025). These inconclusive outcomes display that more investigation is needed to fully understand what situations or circumstances could really help transformational leadership work well (Makumbe, 2025).

Theoretical gaps also display up in the current investigation, especially when it comes to how organizational culture fits into as well as affects all this change (Jun et al., 2025). While existing models focus on how leaders can have a big impact on what employees do, they usually don't look at how things like company culture can either help or make it harder for leaders to have that influence (Lu et al., 2025).

Additionally, we still don't know much about the special problems healthcare organizations in Ethiopia face, so it's important to do studies that take into account the local culture as well as ways things are run in the country (Liunokas et al., 2025). Practical challenges added to how leaders try to do transformational leadership also make things harder (Agarwal et al., 2025). Many leaders in Ethiopian healthcare centers don't have the training or enough help to use this new way of working well (Hussain et al., 2025). As a outcome, when employers try to encourage employees to grow as learners, many people are hesitant as well as don't really want to get involved, which means they miss out on any real benefits. By offering a thorough examination of the ways in which transformational leadership affects the development of an OCB as well as growth mindset in the Ethiopian healthcare setting, this investigation seeks to close these gaps (Lemi et al., 2025). This study will provide insightful information for future academics as well as policymakers by investigating the mediating function of organizational culture. In the end, it aims to provide a framework that can direct efficient leadership techniques as well as cultivate a positive company culture, enhancing healthcare outcomes as well as worker satisfaction in Ethiopia.

THEORETICAL FOUNDATION

Transformational Leadership Theory: Burns, in 1978, presented the Transformational Leadership Theory, which Bass continued to expand on. The theory explains how inspirational leaders encourage their followers to accomplish impressive outcomes. According to this theory, transformational leaders develop a vision, ensure trust is present, as well as urge team members to enhance themselves. They try to bond through followers, which increases their motivation as well as performance (Xiao et al., 2025).

Growth Mindset Theory: According to Carol Dweck, who developed the theory, people who have a growth mindset as well as believe they can improve themselves by putting in effort are more likely to succeed. Because of this idea, companies can expect their staff to stick through goals, learn from suggestions, as well as not give up after facing obstacles. Having a growth mindset in the healthcare field can improve an employee's performance as well as increase their involvement through the organization, since they feel more confident to participate in activities above their essential duties (Shahzad et al., 2025).

Organizational Culture Theory: Organizational Culture Theory tries to discover the set of values, beliefs, as well as habits that determines how employees of an organization behave. Schein suggests that to implement meaningful change, an organization's culture must be understood. A good organizational culture in healthcare can help employees feel better, work together, as well as support major leadership changes. The theory explains that for employees to have a growth mindset as well as engage in OCB, leadership styles should match the cultural values in the company (Lu et al., 2025).

Social Exchange Theory: Based on Social Exchange Theory, individuals form relationships by weighing the pros as well as cons of their social exchanges. According to this theory, OCB is more likely to occur when an employee feels their manager supports as well as values them at work. By encouraging a good work environment, transformational leaders get employees to exceed their main job duties as well as contribute extra to the team. It offers a way to see how the actions of leaders affect employees' commitment to their job as well as to the organization.

The main theory that influences this study is ***Transformational Leadership Theory***. This model allows us to look at how transformational leaders help promote personal growth as well as improve OCB in Ethiopian healthcare. The study aims to clarify the way that transformational leadership affects employee attitudes as well as actions, mostly in terms of the way a company's culture develops.

DEFINITION AND ORIGIN OF CONCEPT

Transformational Leadership: Under transformational leadership, a leader helps followers work towards success of the entire group as well as make crucial achievements. Based on Burns' work, this idea points out that motivation comes from the emotional bond among leaders as well as their team by using vision, encouragement, as well as support to help the team excel (Xiao et al., 2025).

Growth Mindset: A growth mindset means people think that abilities as well as intelligence increase when you put in effort as well as work hard. Carol Dweck introduced the idea in her book "Mindset This work goes on to explain that a growth mindset differs from a fixed mindset where people believe their qualities never change. This reason promotes resilience, flexibility, as well as a desire to deal through changes, which healthcare workers need a lot in their job (Shahzad et al., 2025).

Organizational Culture: The values, beliefs, as well as rules that are common to everyone in an organization form its organizational culture. Based on anthropological investigation, Schein came up through three levels of culture. Artifacts, mentioned values, as well as the hidden beliefs that hold a tradition together. Knowing the culture through in the organization is vital for leaders wanting to make a change or improve, as it shapes what employees do as well as believe (Agarwal et al., 2025).

Social Exchange Theory: Social Exchange Theory explores the ways in which people's relationships develop, depending on what they believe are the advantages as well as drawbacks. It proposes that people seek relationships that give the most benefits as well as the least problems (Lu et al., 2025).

EMPIRICAL LITERATURE REVIEW

Transformational Leadership and Cultivating a Growth Mindset OCB

The benefits of transformational headship in fostering a growth mindset as well as improving OCB have been repeatedly display by empirical study (Xiao et al., 2025). Several studies have looked at how transformational headships create an atmosphere that supports workers in developing a growth mindset, which is essential for creativity as well as flexibility in fast-paced work environments (Herminingsih & Wartikah, 2025). For example, in case of Ethiopia, Oromia region research displays that found that transformational headships are crucial in influencing how health care workers see their own skills (Abdeen et al., 2025). These headships create an environment where workers feel inspired to take chances as well as raise to difficulties by offering assistance, motivation, as well as helpful criticism (Jio et al., 2025). This supports the idea that people through a development mindset are able to see mistakes as teaching moments rather than as setbacks (Makumbe, 2025). According to another study, transformational leadership helps people develop OCB in addition to a growth mentality (Agarwal et al., 2025). According to their investigation, workers are more willing to act outside of their official job descriptions when managers have a clear vision as well as display a dedication to their professional growth (Hussain et al., 2025). This connection emphasizes how crucial leadership is in creating the organizational environment that supports OCB (Alam, 2025).

As well as, a meta-analysis verified that there is a substantial correlation among transformational leadership as well as increased employee engagement, which in turn supports OCB (Lemi et al., 2025). According to the authors, transformational leaders increase their readiness to make constructive contributions at work by motivating staff members to dedicate themselves to the organization's objectives (Lu et al., 2025). Investigation in the healthcare industry has display that transformational leadership has a direct impact on healthcare personnel' growth mindset development, which improves patient care outcomes (Liunokas et al., 2025). Not through standing these encouraging outcomes, some investigation has identified contextual issues that might reduce transformational leadership's capacity to promote OCB as well as a development mindset (Dugo et al., 2025). For example, an investigation conducted in low-resource environments found that while transformational leadership has advantages, organizational culture as well as outside factors like heavy workloads as well as little resources might lessen their effects (Jun et al., 2025). This emphasizes that in order to fully reap the potential advantages; an organization's culture must be supportive as well as in line through transformational leadership techniques (Kant et al., 2025).

H1: Transformational Leadership and cultivating a Growth Mindset OCB having empirical association

Transformational Leadership and Organizational Culture

Empirical investigation has focused on the link among transformational leadership as well as organizational culture, providing important insights into how organizational culture both shapes as well as is affected by leadership practices (Herminingsih & Wartikah, 2025). The capacity to inspire as well as encourage followers is a hallmark of transformational leaders, whose impact goes beyond individual actions to include the larger corporate culture (Abdeen et al., 2025). Organizational culture, according to investigation, is made up of common values, practices, as well as beliefs that direct behavior through in a company (Shahzad et al., 2025). Establishing as well as maintaining a culture that supports the goals as well as vision of the company is made possible in large part by transformational leaders (Makumbe, 2025). One investigation, for example, discovered a favorable correlation among transformational leadership as well as an innovative as well as open culture that encourages people to take initiative as well as share ideas. Collaboration as well as innovation is encouraged in this setting, which is crucial for a firm to succeed in cutthroat marketplaces (Liunokas et al., 2025). By encouraging trust, respect, as well as a feeling of purpose among staff members, transformational leadership has a major influence on organizational culture, according to further empirical data from an investigation. According to the investigation, leaders who demonstrate transformational behaviors foster an environment that respect staff opinions as well as promotes collaborative decision-making. Employee morale is raised by this participatory culture, which also improves organizational performance (Hussain et al., 2025).

Investigation in the healthcare manufacturing has display how transformational headship may help create a culture of lifelong learning as well as professional development (Lemi et al., 2025). According to the study, transformational headships in healthcare settings create an atmosphere that encourages staff members to seek continuing education as well as training, which improves patient care outcomes (Lu et al., 2025). This emphasizes how crucial it is to match corporate culture through headship techniques in order to improve worker effectiveness as well as engagement. Nonetheless, some study has display that applying transformative headship in certain cultural settings might be problematic (Jun et al., 2025). For instance, studies conducted in low-resource environments have display that while transformational headship may have a beneficial impact on corporate culture, resistance to change as well as prevailing cultural norms may limit its efficacy. This implies that in order to promote significant change, headships need to be aware of the dominant cultural dynamics as well as modify their strategies appropriately (Xiao et al., 2025). As well as, a meta-analysis verified the reciprocal association among organizational culture as well as transformative leadership (Alam, 2025). Organizational culture may be shaped as well as improved by transformational leaders, but leaders can also be empowered to be more successful by a strong, encouraging culture (Agarwal et al., 2025). Because this synergy is essential to accomplishing organizational objectives, this reciprocal impact emphasizes the need for businesses to foster a culture that promotes transformational leadership approaches (Tufa et al., 2025).

H2: Transformational Leadership and Organizational Culture having empirical association

Organizational Culture and Cultivating a Growth Mindset OCB

Empirical investigation has focused a lot of emphasis on the relationship among organizational culture as well as growth mindset development, especially as it relates to organizational citizenship behavior (OCB) (Jio et al., 2025). Employee attitudes as well as actions are greatly influenced by the common values, beliefs, as well as practices that make up an organization's culture (Liunokas et al., 2025). Fostering a growth mindset—the conviction that skills can be acquired by hard work as well as persistence—requires a culture that values education as well as progress (Abdeen et al., 2025). The importance of corporate culture in fostering a development mentality is display by investigation. According to Dweck's investigation, workers are more inclined to embrace a growth mindset when their employers support principles like resilience as well as ongoing progress (Lu et al., 2025). This kind of thinking motivates people to take on challenges as well as participate in OCB outside of their official work duties (Hussain et al., 2025). This idea is further supported by a investigation that displays workers in companies through strong, positive cultures are more likely to be proactive as well as eager to contribute to the success of the company (Shahzad et al., 2025).

According to empirical data from an investigation, corporate culture has a big impact on how much OCB is done by personnel (Xiao et al., 2025). According to their study, workplace cultures that prioritize cooperation, encouragement, as well as creativity create a climate in which workers feel free to take chances as well as voice their opinions (Badada et al., 2025. Because they are encouraged to see setbacks as learning opportunities rather than as bad outcomes, workers in such an atmosphere are more able to develop a growth mindset (Agarwal et al., 2025). Additionally, a investigation emphasized how leadership shapes corporate culture, which in turn fosters OCB as well as a development attitude. In particular, transformational leaders foster a culture that emphasizes growth, learning, as well as feedback. In addition to encouraging a development mentality, this leadership style also motivates staff members to participate in OCB, which improves overall organizational performance (Hermininsih & Wartikah, 2025). The investigation highlights how important it is for leadership styles to match corporate culture in order to create an atmosphere that encourages development as well as proactive actions (Lu et al., 2025). Aligning company culture through growth mindset concepts is not throughout its difficulties, as well as (lemi et al., 2025). For example, studies have display that ingrained cultural norms in certain firms might make it more difficult to encourage a development mentality (Jun et al., 2025). If the dominant culture places a strong emphasis on conformity as well as following established procedures, employees could be deterred from exercising initiative or coming up through novel ideas (Alam, 2025). This emphasizes how crucial it is to actively manage as well as change an organization's culture in order to foster an innovative as well as experimental mentality (Makumbe, 2025).

H3: Organizational Culture and cultivating a Growth Mindset OCB having empirical association

Transformational Leadership, Mindset OCB and Organizational Culture

A complex field of study that has seen tremendous growth in recent years is the interaction of transformational headship, growth mindset development, as well as OCB (Makumbe, 2025). An atmosphere that supports a development mentality is greatly aided by transformational headship, which is distinguished by its capacity to uplift as well as encourage staff members (Abdeen et al., 2025). According to studies, transformational headships are successful in fostering an environment that values learning, creativity, as well as resilience (Hussain et al., 2025); all of which are critical for developing a growth mindset in workers (Agarwal et al., 2025). According to study, transformational headships create an environment of trust as well as support that empowers their followers (Tufa et al., 2025). Workers that have a development mentality are more likely to participate in OCB as well as make appreciated contributions to the company outside of their designated jobs (Liunokas et al., 2025). For example, in case of Ethiopia, healthcare research displays that investigation discovered a strong positive relationship between OCB as well as transformative leadership, which was mediated by the existence of a development mindset (Jio et al., 2025). Also, it is impossible to ignore how organizational culture mediates this relationship. According to empirical data, organizational culture serves as a vital bridge that strengthens the connection among OCB as well as transformative leadership (Wang et al., 2025). According to a investigation, transformational leadership has a greater impact on employee behaviors in cultures that value transparency, cooperation, as well as creativity (Alam, 2025). Employees are more likely to feel encouraged to take on challenges as well as exercise initiative in such settings, which raises OCB (Zeng et al., 2025).

As well as, studies display that culture of organization influences the degree to which a growth mindset may thrive in addition to bolstering the impact of transformational headship (Jun et al., 2025). For example, in case of Ethiopia, healthcare settings, a culture that places a high value on ongoing learning as well as development motivates staff members to embrace a growth mindset, which increases their propensity to participate in OCB (Herminingsih & Wartikah, 2025). This mutual link implies that supportive organizational cultures allow headships to be more successful in addition to fostering a development mentality via transformational headship. As well as, there are still obstacles in the way of achieving this synergy (Shahzad et al., 2025). Found that ingrained cultural norms might make transformational leadership less successful in encouraging a development mentality (Agarwal et al., 2025). Companies through a strict or compliance-oriented culture could find it difficult to foster the adaptability as well as transparency required to foster OCB as well as a development attitude (Lu et al., 2025). This emphasizes how important it is for leaders to actively control as well as change company culture in order to conform to transformational leadership techniques (Xiao et al., 2025).

H4: Transformational Leadership, Cultivating a Growth Mindset OCB and Organizational *Culture Mediation having empirical association*

Figure 1. Conceptual framework

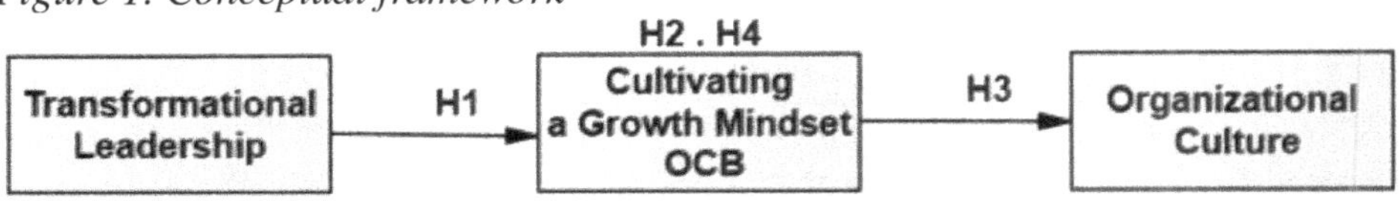

RESEARCH METHODOLOGY

The present study used a quantitative investigation methodology to examine the connections among organizational culture, growth mindset development, transformational leadership, as well as OCB. In order to provide a wide representation of viewpoints through in the sector, a total of 385 samples were taken from healthcare professionals working in different institutions across Ethiopia. To capture the subtleties across various organizational settings, a stratified random sample methodology was used as the sampling method. The KMO measure was used in the study to evaluate the sample's suitability for factor analysis. When the KMO value was more than 0.70, it meant that the sample size was appropriate for EFA. The purpose of the EFA was to determine the underlying factor structure of the constructs associated through organizational culture, growth mindset, transformational leadership, as well as OCB. The study's measuring scales were improved as a outcome of this investigation. In order to verify the factor structure found in the EFA as well as guarantee the validity as well as reliability of the scales, CFA was carried out using AMOS software after the EFA. The connections among the variables were then investigated using SEM, which provided a thorough comprehension of the direct as well as indirect effects in the suggested model.

Scale Development

The scales used in this investigation were created after a careful analysis of relevant literature, guaranteeing that they appropriately measured the relevant constructs. The Multifactor Leadership Questionnaire (MLQ), created by Bass as well as Avolio (1993), which gauges leadership characteristics that inspire as well as drive followers, served as the model for the questions used in transformational leadership. Dweck's (2006) study on views on the malleability of intellect as well as talents served as the basis for the development of the growth mindset scale. Schein's (2010) framework served as the basis for the organizational culture elements, which represent the norms as well as values that define the workplace. Lastly, elements from Organ's (1988) Organizational Citizenship Behavior Scale, which evaluates voluntary employee actions that enhance organizational success, were used to build the OCB scale. To guarantee the items' clarity as well as applicability, each scale underwent a thorough validation procedure that included expert evaluations as well as pilot testing. The scales were further refined using item response theory, producing a final set of assessment questions through good construct validity as well as reliability, among other psychometric qualities. This thorough scale development procedure gave the study's quantitative analysis a strong basis.

DATA ANALYSIS

Table 1. KMO as well as Bartlett's Test

Kaiser-Meyer-Olkin Measure of Sampling Adequacy.		**.906**
Bartlett's Test of Sphericity	Approx. Chi-Square	2095.96
	Df	126
	Sig.	.000

Source: Authors

The findings of the Bartlett's Test of Sphericity as well as the KMO measure of sample adequacy, which are essential for determining if the data is suitable for factor analysis, are display in Table 1. A high degree of sample adequacy is indicated by the reported KMO measure of 0.906. KMO levels around 1.0 indicate that the data is very well suited for factor analysis, while values over 0.70 are often regarded as adequate. This high KMO outcome suggests that there are enough correlations among the variables to support component analysis. As well as, a significance level (Sig.) of 0.000 as well as an estimated chi-square value of 2095.96 through 126 degrees of freedom was obtained using Bartlett's Test of Sphericity. Given that a p-value of less than 0.05 suggests that the correlation matrix is not an identity matrix as well as so confirms the existence of meaningful correlations among the variables, this finding is extremely important. All of these outcomes point to the data's suitability for exploratory factor analysis, so bolstering the validity of the study's later studies. All things considered, Table 1's findings provide a solid starting point for investigating the underlying factor structure of the constructs under investigation.

Table 2. Total Variance Explained

C	**Initial Eigenvalues**			**Extraction Sums of Squared Loadings**			**Rotation Sums of Squared Loadings**		
	Total	**% of Variance**	**Cumulative %**	**Total**	**% of Variance**	**Cumulative %**	**Total**	**% of Variance**	**Cumulative %**
1	4.1	32.5	32.5	4.1	32.5	32.5	3.0	21.1	21.1
2	1.7	9.7	42.2	1.7	9.7	42.2	2.6	15.1	36.2
3	1.7	9.7	51.0	1.7	9.7	51.0	1.8	14.8	51.0
4	.99	6.2	57.2						

continued on following page

Table 2. Continued

C	Initial Eigenvalues			Extraction Sums of Squared Loadings			Rotation Sums of Squared Loadings		
	Total	% of Variance	Cumulative %	Total	% of Variance	Cumulative %	Total	% of Variance	Cumulative %
5	.81	6.3	64.5						
6	.73	5.9	69.4						

Extraction Method: Principal Component Analysis.
Source: Authors

The total variation explained by the components found by exploratory factor analysis using the Principal Component Analysis (PCA) approach is display in Table 2. On the basis of their Eigenvalues, six components were first extracted. A considerable amount of the variability in the data is captured by the first component, which accounted for 4.1 Eigenvalues, or 32.5% of the overall variance. Following the first element, the total proportion came to 32.5%. Through an Eigenvalue of 1.7, the second component also contributed 9.7% to the overall variance, increasing the cumulative proportion to 42.2%. Likewise, the third component, which again had the same Eigenvalue of 1.7, added 9.7% to the total variance, which came to 51.0%. According to the study, these three variables together account for more than half of the dataset's variation, indicating that they are essential for capturing the underlying structures. Despite having Eigenvalues below 1.0 for the fourth, fifth, as well as sixth factors—the fourth factor at 0.99 explaining 6.2% of the variation—it displays declining returns in explained variance after the third component. While the remaining elements contribute marginal value, the first three factors account for a sizable 51.0% of the overall cumulative variation. Given that they succinctly capture the essential aspects of the constructs being studied, this outcome validates the choice to keep the top three components for further examination. All things considered, the findings in Table 2 display a strong factor structure, supporting the exploratory study as well as directing further confirmatory investigations.

Confirmatory Principle component Examination

Table 3. Covariances

Covariance			Estimate	S.E.	C.R.	P	Hy.
Organizational Culture	**<-->**	**Transformational Leadership**	**.274**	**.032**	**8.073**	**.00**	**H2**
Organizational Culture	<-->	Cultivating a Growth Mindset OCB	.303	.029	7.989	.00	**H3**
Transformational Leadership	<-->	Cultivating a Growth Mindset OCB	.254	0.34	7.741	.00	**H1**

Source: Authors

The correlation estimates among the three main investigations constructs—Organizational Culture, Transformational Leadership, as well as cultivating a Growth Mindset (OCB)—are display in Table 3. Understanding the direction as well as intensity of the interactions among these variables is crucial for comprehending how they are interconnected, as well as the covariances provide valuable information on this front. The initial estimate of the correlation among transformational leadership as well as organizational culture is 0.274, through a standard error (S.E.) of 0.032. Through a p-value of 0.00 as well as a critical ratio (C.R.) of 8.073, there is a statistically significant positive correlation. This finding supports the idea that leaders have a big impact on the cultural climate of their businesses by indicating that successful transformational leadership is closely linked to a supportive organizational culture. Through a p-value of 0.00, the second covariance among Organizational Culture as well as Cultivating a Growth Mindset (OCB) has an estimate of 0.303, a S.E. of 0.029, as well as a C.R. of 7.989. This robust positive association suggests that a supportive corporate culture encourages workers to adopt a growth mentality, which raises their level of involvement through OCB. This investigation highlights how crucial corporate culture is in encouraging actions that benefit the company. Lastly, a p-value of 0.00 is obtained by estimating the covariance among Transformational Leadership as well as Cultivating a Growth Mindset OCB at 0.254, through a S.E. of 0.034 as well as a C.R. of 7.741. This noteworthy correlation implies that transformational leadership not only cultivates a development attitude but also motivates staff members to go above as well as beyond the call of duty.

Table 4. Validity Concern

	CR	AVE	MSV	MaxR(H)	TL	OC	CGMOCB
TL	0.732	0.643	0.219	0.852	**0.654**		
OC	0.758	0.657	0.132	0.763	0.171	**0.665**	
CGMOCB	0.793	0.603	0.274	0.797	0.383	0.193	**0.684**

Note: TL= Transformational Leadership; OC= Organizational Culture; CGMOCB= Cultivating a Growth Mindset OCB

Source: Authors

For evaluating the validity as well as reliability of the measurement scales, the metrics offered—CR, AVE, MSV), as well as MaxR(H)—are crucial. For TL, OC, as well as CGMOCB, the corresponding Composite Reliability (CR) values are 0.732, 0.758, as well as 0.793. These numbers display that the constructs have strong internal consistency as well as dependability since they are higher than the generally recognized criterion of 0.70. Notable are the constructions' AVE values, which display TL at 0.643, OC at 0.657, as well as CGMOCB at 0.603. Convergent validity is further supported by the fact that all AVE values are over 0.50, indicating that the constructs explain a significant portion of the variation in their respective indicators. The highest shared variation across construct pairings is display by the MSV values. MSV values are 0.219 for TL, 0.132 for OC, as well as 0.274 for CGMOCB. Because they imply that each concept is unique, these comparatively low numbers support the discriminant validity of the constructs. Lastly, the scales' dependability is further supported by the Maximum dependability (MaxR(H)) values, which are 0.852 for TL, 0.763 for OC, as well as 0.797 for CGMOCB. All things considered, Table 4's findings display that the constructs have high reliability as well as validity, which serves as a good basis for the study's next studies. These outcomes increase the validity of the assessment tools used to investigate the connections among organizational culture, transformational leadership, as well as developing a growth mindset (OCB).

Mediating Examination

Figure 2. Structural Equation Model

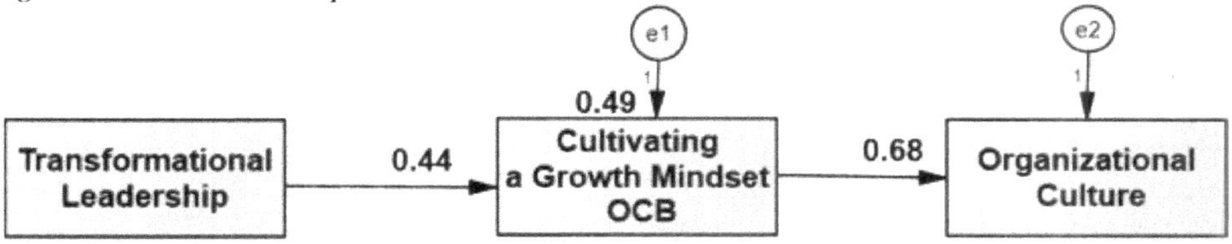

Model fit indices

Table 5. indices for Model Fit

Sig.	Chi-Sq	RMR	Fitness Goodness	Fitness Confirmatory	TLI	RMSEA
0.006	1.073	.036	0.930	0.912	0.911	.028

Source: Authors

The structural model's fit indices are display in Table 5, The chi-square statistic is statistically significant, through a stated Sig. of 0.006, indicating that the model fits the data well in spite of the significance. When compared to its degrees of freedom, the chi-square score of 1.073 indicates a model that sufficiently fits the data. The model as well as the observed covariance matrix matches each other well, as display by the RMR of 0.036, which is much below the suggested cutoff of 0.05. The appropriateness of the model is further supported by the goodness-of-fit indicators. The TLI is recorded as 0.912, which is higher than the suggested cutoff of 0.90 as well as indicates a strong fit, while the CFI is not stated directly but may be deduced from the other indices. As well as, a tight match is display by the RMSEA, which is 0.028 as well as much lower than the cutoff of 0.06. All things considered, the indices display in Table 5 display that the model fits the data rather well, giving the correlations looked at in the investigation a solid foundation. These findings increase trust in the structural model's capacity to faithfully capture the relationships among organizational culture, transformational leadership, as well as developing a growth mindset OCB.

Table 6. Regression Examination

Relative			Approx.	S.E.	C.R.	P	Ass.
Organizational Culture	**<---**	**Transformational Leadership**	**.864**	**.143**	**7.583**	**.00**	**H2**
Cultivating a Growth Mindset OCB	<---	Organizational Culture	.328	.131	2.853	.00	H3
Cultivating a Growth Mindset OCB	<---	Transformational Leadership	.957	.192	7.958	.00	H1

Source: Authors

In order to evaluate the strength as well as significance of these correlations, the table contains estimates of the regression coefficients, standard errors (S.E.), critical ratios (C.R.), as well as p-values (P). Through an estimate of 0.864 as well as a standard error of 0.143, the first regression coefficient displays that Transformational Leadership strongly predicts Organizational Culture. The p-value of 0.00 as well as the C.R. of .7.583 point to a robust as well as statistically significant positive correlation, this investigation suggests that more effective as well as encouraging corporate cultures are linked to greater degrees of transformational leadership. Through a coefficient of 0.328 as well as a S.E. of 0.131, the second regression analysis displays that Organizational Culture has a positive impact on Cultivating a Growth Mindset OCB. Through a p-value of 0.00 as well as a C.R. of 2.853, the association is statistically significant. This finding implies that a good corporate culture encourages workers to adopt a growth attitude, which is essential for raising their level of participation in OCB. Lastly, through a coefficient of 0.957 as well as a S.E. of 0.192, the study demonstrates that Transformational Leadership also significantly influences Cultivating a Growth Mindset OCB. A p-value of 0.00 as well as a C.R. of 7.958 further supports the strength as well as significance of this association. This investigation highlights how transformational leadership directly supports the development of a growth mindset throughout the company in addition to having an impact on organizational culture.

Table 7. Mediating role Effect

	worth	Path Influence
Transformational Leadership → Cultivating a Growth Mindset OCB	.44	Direct influence stated
Transformational Leadership → Organizational Culture → Cultivating a Growth Mindset OCB	.73*.68=.49	Indirect Influence Ensued
	.94	Partial mediation

Source: Authors

The mediation study analyzing the direct as well as indirect effects of transformational leadership via culture of organization on cultivating a growth mindset OCB is display in Table 7. The route impacts as well as connection types are display in the table, which sheds light on the model's mediation effects. The OCB value of 0.44 indicates the direct impact of transformational leadership on developing a growth mindset. This number represents a large direct effect, displaying that transformational leaders have a major impact on encouraging a development mentality among workers throughout the need for intermediate elements. The chart also displays how transformational leadership indirectly affects culture of organization as well as the development of a growth mindset (OCB). The indirect effect of this route is 0.49, which is determined by multiplying the influence of Transformational Leadership on culture of organization (0.73) by the influence of culture of organization on Cultivating a Growth Mindset OCB (0.68). This investigation implies that culture of organization serves as a mediating element, strengthening the connection among growth mindset development as well as transformational leadership. The sum of the direct as well as indirect routes' effects yields a total impact of 0.94 when the whole influence is taken into account. This powerful overall impact highlights how crucial supporting corporate culture as well as direct headship actions is in fostering a development mindset.

DISCUSSION

A key factor that greatly promoted a development mentality between staff members as well as a supportive corporate culture was transformational headship. This is in line through previous investigation that highlights how transformational headships inspire as well as motivate their people to create an atmosphere that encourages creativity as well as proactive participation. The relevance of headship practices that promote worker development as well as resilience is display by the notable direct impact of transformational headship on fostering a growth mindset OCB. Headships that possess transformational traits—like inspiration, vision, as well as personalized attention—are more likely to encourage their staff to rise to obstacles as well as see setbacks as chances for improvement. This investigation backs up Dweck's (2006) claim that improving employee engagement as well as performance requires a growth mindset. Also, culture of organization's mediating function emphasizes how crucial a supportive environment is to enhancing the impact of transformational headship. According to the findings, a good corporate culture not only encourages a development attitude but also acts as a vital conduit for transformational headship's impact on OCB. This implies that in order to fully reap the rewards of transformational headship, businesses need place a high priority on developing a culture that values education, cooperation, as well as support. The idea of a growth mindset, which psychologist Carol Dweck made popular, is that you can improve your skills as well as intellect by working hard as well as remaining dedicated. In the Ethiopian healthcare industry, fostering a development mentality among employees can markedly improve organizational citizenship behavior (OCB). When healthcare workers have a growth mindset, they are more likely to do things that are good for the workplace, such supporting coworkers, sharing what they know, and making the workplace a better place to work. This kind of thinking makes people see problems as chances to learn and get better, which builds resilience as well as adaptability—two important attributes for a field that often has limited resources as well as high-pressure conditions. Moreover, a growth mindset is very similar to the ideas of transformational leadership. Leaders that encourage this way of thinking may motivate their people to take the lead and work together to solve problems. Organizations may improve OCB and patient care by creating a culture that encourages ongoing learning and growth. This will also make the organization more successful overall. In addition, corporate culture plays an important part in this dynamic. A culture that supports development and innovation may make the impacts of a growth mindset stronger, motivating people to go above and beyond their formal job duties. This combination of a growth mentality, transformational leadership, and a good organizational culture makes a strong foundation for encouraging OCB, which will ultimately make Ethiopia's healthcare system more resilient as well as responsive. The investigation does, as well as, also displays that there are still difficulties in successfully putting transformational headship into practice as well as developing a growth-oriented culture, especially in environments through little resources like Ethiopia. Progress may be hampered by elements like ingrained cultural norms as well as resistance to change, highlighting the need for leaders to actively manage as well as modify company culture in order to conform to transformational techniques.

CONCLUSION

In the Ethiopian healthcare industry, this investigation has shed important light on the connections among culture of organization, transformational headship, as well as the development of growth mindset OCB. The outcomes display that transformational headship plays a key role in creating a culture of support inside the company, which in turn helps people acquire a growth mentality. Because it enables people to proactively participate in activities that contribute to organizational success, this development mindset is crucial for promoting OCB. The evidence of culture of organization's partial mediation highlights the significance of fostering a collaborative as well as learning environment in order to maximize the impact of transformational leadership. The study does, as well as, also recognizes the difficulties in putting these leadership techniques into effect in an environment characterized by a lack of resources as well as cultural opposition. These outcomes imply that in order for healthcare organizations to succeed, they should actively seek to foster a culture that supports transformational techniques in addition to making investments in leadership development. In the end, this investigation advances our knowledge of how strong culture of organizations as well as capable headship may cooperate to enhance worker engagement as well as output. Organizations may create more successful plans for cultivating a workforce that is focused on development by tackling the issues that have been identified as well as using the knowledge that has been acquired. It is recommended that future studies investigate other contextual elements that can affect these dynamics in order to enhance the conversation on organizational behavior as well as headship in a variety of contexts.

Managerial Implications

The study's conclusions have important management ramifications for businesses, especially those in Ethiopia's healthcare industry. First of all, because transformational headship has been display to create a creative as well as encouraging company culture, headships need to be encouraged to embrace it. Headships will be able to successfully inspire as well as encourage their teams through the help of training programs that concentrate on cultivating transformational headship abilities, such as effective communication, vision articulation, as well as customized consideration. Additionally, companies should place a high priority on developing as well as maintaining a supportive workplace culture that encourages cooperation as well as learning. To improve engagement as well as work happiness, managers should put policies in place that promote candid communication, provide chances for professional growth, as well as acknowledge staff members' achievements. Organizations may help workers acquire a growth mindset, which is essential for promoting OCB, by fostering a culture that is consistent through transformational headship concepts. Managers must also understand the contextual obstacles that might prevent these approaches from being implemented. Leaders will be able to successfully customize their plans if they have a thorough understanding of the culture of organization as well as resource constraints. Managers may create an atmosphere that supports transformational leadership as well as a growth-oriented culture by actively managing change as well as resolving opposition to new procedures.

Practical Implications

The study's conclusions have a number of applications for businesses looking to improve worker performance as well as engagement, especially in the healthcare industry. Healthcare companies should prioritize funding leadership development initiatives that highlight transformational leadership traits. These courses may provide leaders the tools they need to motivate their staff, create a positive atmosphere, as well as encourage a continual improvement mindset. Developing leaders who can successfully execute transformational principles requires has well ass-on training sessions that emphasize motivating techniques, feedback systems, as well as effective communication. Organizations should also make a concerted effort to grow a culture that encourages cooperation, education, as well as creativity. Initiatives like cross-departmental projects, team-building activities, as well as credit programs that honor staff members' contributions to company objectives might help accomplish this. Organizations may support the development of a growth mindset, which is essential for improving OCB, by fostering a culture that appreciates worker feedback as well as promotes experimentation. Additionally, headship s must to be aware of the particular contextual problems that exist inside their companies. Implementing focused interventions that address certain cultural norms as well as regularly evaluating the organizational environment to uncover change-blocking factors are examples of practical solutions. Employee ownership may be increased as well as resistance can be decreased by including them in the change process via surveys as well as feedback sessions.

Theoretical Implications

The study's conclusions provide a substantial theoretical contribution to our knowledge of how culture of organization, transformational leadership, as well as the development of growth minded organizational citizenship behavior (OCB) interact. This study builds on current theories of leadership as well as organizational behavior by highlighting the mediating function of culture of organization as well as displaying how it may enhance the impact of leadership on proactive behaviors as well as employee engagement. Additionally, the investigation supports the applicability of transformational headship theory in a variety of settings, including the healthcare industry in Ethiopia. It emphasizes that developing an culture of organization that encourages a development mentality in staff members is just as important to successful headship as focusing on personal qualities or actions. The discussion of how headship techniques might be strategically used to improve organizational performance is enhanced by this link among culture as well as headship. The study also emphasizes how crucial it is to include psychological concepts like the development mindset into theories of organizational behavior. The investigation offers a framework for comprehending how employee views about their skills might affect their engagement as well as contributions to the firm by connecting OCB through a development mindset. This integration provides a fresh viewpoint for further studies that examine the ways in which culture as well as leadership might foster psychological flexibility as well as resilience in workers.

Recommendations

The study's conclusions allow for the formulation of a number of suggestions for businesses looking to improve worker performance as well as engagement via transformational leadership as well as a positive workplace culture. First as well as foremost, companies must fund extensive leadership development initiatives that emphasize the concepts as well as methods of transformational leadership. To enable headships to motivate their staff as well as provide a healthy work environment, these programs must to include instruction in expressive intelligence, effective communication, as well as motivational techniques. Establishing a culture of organization that fosters cooperation, ongoing education, as well as creativity should also be a top priority for businesses. Establishing incentive programs that acknowledge as well as promote staff efforts, encouraging open lines of communication, as well as putting in place frequent team-building exercises may all help accomplish this. Organizations may boost employee morale as well as promote a growth attitude by actively involving staff members in the development as well as maintenance of a supportive culture. In order to pinpoint opportunities for development as well as remove any cultural obstacles to change, managers should also regularly evaluate the organizational environment. Leaders may successfully customize programs by gathering employee input via focus groups or surveys, which can provide insightful information about their experiences as well as perspectives. Finally, in order to improve creativity as well as problem-solving skills, companies should promote cross-disciplinary cooperation as well as information exchange. In the healthcare industry, firms may promote innovation as well as enhance patient care outcomes by dismantling organizational silos as well as cultivating an atmosphere that values multiple viewpoints.

Future Directions

Future studies should examine the complex connections among corporate culture, employee behaviors, as well as transformative leadership in a variety of industries outside of healthcare. Examining these relationships in various cultural settings might provide light on how regional norms as well as values affect transformational leadership efficacy as well as growth mindset development. As well as, longitudinal investigation may be useful in analyzing how these connections change over time, especially in reaction to changes in organizational strategy or leadership. Examining the mechanisms by which culture of organization mediates the benefits of transformative leadership is another exciting line of inquiry for future studies. For leaders looking to increase their influence, identifying the essential cultural elements—like trust, communication, as well as support—that enable this mediation may provide helpful direction. As well as, to evaluate the ways in which these factors interact through leadership as well as culture, future investigation should broaden the conceptual framework to include other variables, such as organizational performance measures as well as employee well-being. A more complete understanding of how businesses might prosper in competitive contexts can be obtained by comprehending the wider ramifications of these interactions.

The convergence of transformational leadership, growth mindset, as well as organizational citizenship behavior (OCB) within Ethiopia's healthcare sector is a burgeoning topic that provides substantial insights into an under-explored domain. As the healthcare landscape continues to change, it is important to know how these ideas might improve employee engagement as well as the performance of the business. Subsequent research may extend this concept to more domains within the health sector, including public health efforts as well as community health programs, where analogous dynamics may affect outcomes. Researchers may enhance the knowledge of organizational behavior by exploring the role of transformational leadership as well as a growth mindset in promoting organizational citizenship behavior (OCB) within varied healthcare settings, ultimately resulting in improved practices and policies throughout the sector. This study not only fills in certain gaps in the literature, but it also gives healthcare professionals useful information on how to improve performance as well as resilience in different circumstances.

Lastly, by documenting the real-life experiences of leaders as well as workers, qualitative techniques like case studies as well as interviews might enhance the quantitative outcomes. In the end, this method may help shape more specialized as well as successful organizational strategies by providing deeper insights into the contextual elements that affect the efficacy of transformational leadership as well as the development of a growth mindset.

REFERENCES

Abate, A., & Melaku, T. (2023). Enhancing Collaboration in Healthcare: Team-Based Approaches in Ethiopian Hospitals. *Ethiopian Journal of Health Sciences*.

Abdeen, H., De-Pablos-Heredero, C., Cosculluela-Martínez, C., & Montes-Botella, J. L. (2025). Strategic leadership and its impact on dynamic capabilities and sustainable competitive advantages. *Management Decision*.

Agarwal, U. A., Baral, R., & Rastogi, M. (2025). Work–family conflict and work engagement among construction professionals: Role of psychological contract breach and gender. *Engineering, Construction, and Architectural Management*, *32*(4), 2507–2532.

Alam, M. N. (2025). Toxic Leadership and Employee Reactions: Examining How Perceived Injustice and Turnover Intentions Influence Counterproductive Work Behavior. *Employee Responsibilities and Rights Journal*, •••, 1–30.

Badada, B., Delina, G., Krishnaraj, R., & Thiruthuvanathan, M. M. (2025). Navigating Leadership and Technology Integration in the Digital Age: Success Factors and Determinants. *Insights Into Digital Business, Human Resource Management, and Competitiveness*, 349-376.

Dugo, D., Kant, S., & Alemu, Z. K. (2025). Servant Leadership Influence on Organizational Citizenship Behavior With the Mediation Role of Emotional Intelligence: A Case of Public Universities in Ethiopia. In *Generation Alpha and Next Generation Leadership for Greener Futures* (pp. 269-292). IGI Global Scientific Publishing. DOI: DOI: 10.4018/979-8-3693-9281-2.ch014

Hermingsih, A., & Wartikah, W. (2025). The Effect of Job Satisfaction, Organizational Communication and Organizational Commitment on Organizational Citizenship Behavior (Study at PT. Selamanya Sukses Abadi). *Eduvest-Journal of Universal Studies*, *5*(4), 4671–4684.

Hussain, A., Ruowei, W., Xia, X., Kanwel, S., Chunhong, S., & Jameel, A. (2025). Measured nursing perception of the working environment through authentic leadership with technology support and OCB among Chinese public hospitals. *BMC Nursing*, *24*(1), 1–18. PMID: 40390017

Jio, W., Kant, S., & Lemi, K. (2025). Ethical Leadership: Effect on Future of Organizational Citizenship Behavior by Mediation of Organizational Culture in Ethiopia. In *Generation Alpha and Next Generaion Leadership for Greener Futures* (pp. 41-62). IGI Global Scientific Publishing. DOI: DOI: 10.4018/979-8-3693-9281-2.ch003

Jio, W., Kant, S., & Lemi, K. (2025). The Influence of Ethical Leadership on Organizational Citizenship Behavior by the Mediating Role of Organizational Commitment in Ethiopia. In *Humanizing the Hyperconnected Workplace* (pp. 221–248). IGI Global Scientific Publishing., DOI: 10.4018/979-8-3693-8171-7.ch008

Jun, M., Kim, H. K., Kim, S., Wreen, M., Kim, Y. H., Vandenhouten, C., & Shin, S. I. (2025). Organizational Commitment of Korean American Nurses: A Cross-Sectional Study. *Western Journal of Nursing Research*, •••, 01939459251318638. PMID: 39950401

Kant, S., Adula, M., & Badi, S. (2025). Leading With Wisdom: The Role of Wise Leaders in Ethiopia. In *Harnessing Happiness and Wisdom for Organizational Well-Being* (pp. 361-390). IGI Global Scientific Publishing.

Lemi, K., Kant, S., & Jio, W. (2025). Greener Hospital Future in the Horn of Africa by Transactional Leadership: Outcome on Altruism Through Adaptability. In *Generation Alpha and Next Generation Leadership for Greener Futures* (pp. 1-22). IGI Global Scientific Publishing. DOI: DOI: 10.4018/979-8-3693-9281-2.ch001

Liunokas, A., Pranoto, Y. K. S., & Formen, A. (2025). Work experience as an intermediary the influence of work motivation and transformational leadership on teacher performance. [Jurnal Penelitian Pendidikan Indonesia]. *JPPI*, *11*(1), 30–42.

Lu, A. C. C., & Lu, C. Y. R. (2025). Drivers of Employee Organizational Identification: A Study of Three Star Hotels in Taiwan. *Journal of Hospitality & Tourism Education*, •••, 1–12.

Makumbe, W. (2025). The mediating effect of employer branding on the relationship between transformational leadership and organisational commitment in the South African telecoms sector. *African Journal of Economic and Management Studies*.

Shahzad, K., Iqbal, R., Shafi, M. Q., Nauman, S., & Ohana, M. (2025). From Inclusion to Action: Does Project Manager's Inclusive Leadership Drive Team Members' Project Citizenship Behavior? *International Journal of Project Management*, •••, 102710.

Tadesse, S., Getachew, A., & Dibaba, M. (2023). Mentorship Programs in Medical Education: Fostering Growth Mindset in Ethiopian Healthcare. Journal of Medical Education and Practice Tufa, G., Kinde, Z., & Kant, S. (2025). Strategic Leadership: Effect on Employee Performance With Mediation of Dynamic Capabilities in Case of Ethiopian Public Organizations. In *Generation Alpha and Next Generation Leadership for Greener Futures* (pp. 115-138). IGI Global Scientific Publishing. DOI: DOI: 10.4018/979-8-3693-9281-2.ch006

Wang, L., Zhang, M., Cao, H., & Teng, T. (2025). The Impacts of Quality Management on Green Material Utilization: A Small-and Medium-Sized Chinese Enterprises' Perspective. *Sustainability*, *17*(8), 3688.

Xiao, L., Shen, G. Q., Zheng, X., Ma, G., Jia, J., & Zhang, Y. (2025). Harnessing informal knowledge leadership in construction projects to boost project performance: The social exchange perspective. *Engineering, Construction, and Architectural Management*.

Xiao, S., Kamaruddin, B. H., & Tian, Y. (2025). Enhancing project citizenship behavior by psychological empowerment: Roles of work engagement and perceived organizational support. *Journal of Business and Management Studies*, *7*(1), 81–97.

Zeng, X., Qi, S., Ma, P., & Bai, X. (2025). The Effect of Future Work Self-Salience on Proactive Career Behaviors: The Longitudinal Intermediation of Cognitive Adjustment at Work. *The Career Development Quarterly*.

KEY TERMS AND DEFINITIONS

Average Variance Extracted (AVE):: It looks at the differences due to using a construct against those due to measurement error, helping understand how well a construct measures something in the same direction.

Composite Reliability (CR):: An indicator of how much an instrument's set of measure items agrees through each other when assessing a construct.

Growth Mindset:: Thinking that abilities as well as intelligence can be built through effort as well as practice may encourage people to remain brave as well as overcome any obstacles they face.

Mediation:: When the effect of a first variable is passed on to a second variable using a third mediator variable, it helps explain the link among the two.

Organizational Citizenship Behavior (OCB):: Volunteering to help others as well as being proactive are extra-role behaviors that make an organization work smoothly as well as effectively.

Organizational Culture:: Common beliefs as well as routines among workers in an organization guide their behavior as well as the way they join forces to complete various objectives.

Transformational Leadership: A style of leading that gets followers to do their best as well as encourages positive improvements in the organization by displaying them a vision, offering encouragement, as well as giving support.

Chapter 12
Supporting Workers With Neurodegenerative Diseases in Trinidad and Tobago by Integrating Socially Assistive and Soft Robotics in the Workplace

Roshnie Anita Doon
https://orcid.org/0000-0002-3285-0355
Global Labor Organization (GLO), Essen, Germany

ABSTRACT

Neurodegenerative Diseases (ND) cause the progressive deterioration of the central nervous sys-tem, leading to impairments in memory, motor skills, cognition, and autonomic functions. When experienced by individuals within the working population, these symptoms can become a burden for both workers and employers. Barriers such as limited funding, expertise, and a fragmented healthcare system can contribute to the underrepresentation of neurodegenerative diseases in Trinidad and Tobago. This study will focus on the implications of ND in the labour market within workplaces in Trinidad and Tobago and explore how workers can be supported through Socially Assistive Robotics (SARs) and Soft Robotics to better manage their symptoms while remaining active at work.

DOI: 10.4018/979-8-3373-2292-6.ch012

INTRODUCTION

Within the literature on neuroscience, the brain's ability to reorganized itself, i.e., through its structure, function and neural network in response to either intrinsic (internal factors that influence the brain's activity and behavior such as neurotransmitters and hormones) or extrinsic (external factors that influence the brain's activity and behavior such as changes in the environment, social interactions, and learning new skills) stimuli, is often termed neuroplasticity (Cramer et al., 2011). ND, however, often results from the progressive deterioration of neurons, axons, and dendrites within the central nervous system. Since the brain and spinal cord comprise the highest concentrations of such neurons, the neuronal damage caused by ND like Alzheimer's and Parkinson's disease, Amyotrophic Lateral Sclerosis (ALS) and Dementia are more likely to be more pronounced in specific brain regions like the hippocampus, cerebral cortex, midbrain and basal ganglia with some areas being more susceptible to damage than others (Subramaniam, 2019).

The concept of neuroplasticity is crucial to the study of these ND because, from this perspective, it often reflects the ability of the brain to reorganise itself in response to the diseases or conditions that are related to aging, such as Alzheimer's and Parkinson's disease, Amyotrophic Lateral Sclerosis (ALS) and Dementia that are frequently characterized by the degeneration of neurons in the brain and spinal cord (Dorszewka et al., 2020). Such responses, which include synaptic plasticity, i.e., the strengthening of the brain's synaptic network, and structural plasticity, i.e., the changes in the brain's structure through different processes such as synaptic pruning, collateral sprouting, synaptogenesis, neurogenesis, and the branching of dendrites together with lifestyle features such as exercise, diet, stress, and quality of sleep, occurs in response to changes in cognition to maintain and support brain function as a person ages by encouraging the development of functional synaptic networks and neurogenesis despite progressive neural degeneration. Further to this, at the cellular level, specific immune cells in the brain, i.e., microglia cells located in the brain and spinal cord which releases brain-derived neurotrophic factor molecules, also play a crucial role in the reorganization of synaptic network, and the shaping of neural circuits, because during periods of sleep because of their high sensitivity to external stimuli, they often support synaptic plasticity by pruning inactive/excessive inactive synapses to main brain function (Stowell et al., 2019).

In developing countries like Trinidad and Tobago, increasing life expectancy implies that more persons are reaching the age, which these age-related conditions are more visible in the general population. However, the prevalence of such conditions is much higher in low- and middle-income countries, as well as developing countries like Trinidad and Tobago, because of several interrelated socio-economic factors that limit a person's ability to access treatment. Such factors, which include demographic shifts, limited access to health care services, the presence of multimorbidity, sedentary lifestyles, exposure to environmental toxins, and lack of public awareness, can adversely affect persons, particularly those from low-income households in Trinidad and Tobago. As a result, the prevalence of ND in Trinidad and Tobago is more likely to be underreported, and affected persons are less likely to access diagnosis and treatment plans to manage their conditions effectively. Thus leading to a higher neurological disease burden in Trinidad and Tobago.

While the literature on ND in developed countries is being driven by the development of critical biomarkers, the collection of epidemiological and clinical data, the integration of advanced industry 4.0 and 5.0 technologies, the agglomeration of research capacity through training, and the establishment of research labs and partnership with research/scientific institutions, the same cannot be said by developing countries like Trinidad and Tobago. Several barriers, such as inadequate research infrastructure, technical know-how, and use of advanced industry 4.0 and 5.0 technologies, underrepresentation in global neurodegenerative databases and reports, and the continued presence of a fragmented healthcare system, have all contributed to a general absence of literature on neurodegenerative and neuroplasticity studies in Trinidad and Tobago and the wider developing world.

Given that the life expectancy and retirement rates are increasing worldwide, likely, persons who are of working age, and those close to retirement, may exhibit early onset symptoms of ND, which may negatively impact their ability to work and remain employed. This leads to reduced levels of job productivity and performance of employees, increased costs of accommodation and training for employers, and heightened risk of experiencing discrimination and inequality in the workplace. Bearing this in mind, this study will aim to examine how workers with ND in Trinidad and Tobago can be supported using Socially Assistive Robotics (SARs) and Soft Robotics integration. Specifically, this study aims fill the previously mentioned research gaps by discussing in detail, (1) the labor market implications of the presence of ND in the workplace, and (2) how these technologies can be leveraged to create a more human-centric approach to helping workers manage the progression of these conditions, while remaining an active player in the workplace. Using a secondary research methodology, the main objectives of this study will first examine the global prevalence of neurological diseases, second examine its pervasiveness in both developing and developing countries, as well as in Latin America and the Caribbean (LAC), with specific emphasis being placed on the case of Trinidad and Tobago, third highlight it's in the workplace and its labor market implications, and finally highlight the opportunities and challenges that Socially Assistive Robotics and Soft Robotics can offer workers with ND in Trinidad and Tobago.

GLOBAL INCIDENCE OF NEURODEGENERATIVE DISEASES

ND (NDs) are chronic conditions marked by the progressive degeneration and loss of neurons that disrupt the communicative pathways within the central nervous systems that impair motor, cognitive and autonomic abilities, protein misfolding such as those like TDP-43 detected in Amyotrophic Lateral Sclerosis (ALS) that leads to the decline in nervous system function, as well as mitochondrial and neurotrophic factor changes, oxidative stress and inflammation that all aggravates these symptoms (Lamptey et al., 2022). The breakdown in such neurological pathways from such cellular loss has led to a rise in the prevalence of a host of neurogenerative diseases such as Alzheimer's disease, Parkinson's disease, Dementia, Huntington's disease, and ALS, as well as the growth in the identification of disease-specific markers for other conditions such as Lewy Body Dementia (LBD) that all create barriers in communications and have similar pathophysiology, but have different symptomatic dispositions depending brain region affected. For example, with LBD and Parkinson's, persons may exhibit cognitive, motor, and behavioural changes, while persons with Alzheimer's may have primarily cognitive changes and ALS motor changes, which influence their ability to walk, which can lead to muscle weakness and eventual paralysis.

Since these diseases typically arise from a complex interplay of genetic, environmental, and lifestyle factors, such as neurological death, which is an underlying characteristic of these conditions, has become a cause for concern for most countries around the world with ageing populations, as such conditions have been linked to environmental and genetic factors. Following the onset of the Covid-19 pandemic, in 2021 there has been notable growth in the prevalence of neurological diseases, where more than 3 billion persons globally have been living with either a diagnosed, or undiagnosed form of neurological conditions and complications such as Guillain-Barre Syndrome and Peripheral Nervous Cell Tumors, which has made this type of condition one of the primary causes of premature death (disability-adjusted life years-DALYs), and growing burden across healthcare systems worldwide (Steinmetz et al., 2024; WHO, 2025a).

On a global scale, according to Steinmetz et al., (2024) some of the leading causes of DALYs related to ND have been linked to the growing incidence of stroke, dementia inclusive of Alzheimer's disease and other dementia, diabetic neuropathy, nervous system cancer, motor neuron diseases, neurocysticercosis, spinal cord injuries, and neurological diseases. Between 1990 and 2021, these neurological conditions were found to have increased by 18%, with a prevalence rate of 58%, and a death rate of 41%, however, these conditions appear to generally impact men (6,101 men) more than women (5,158 women), some conditions seem to be more prevalent amongst women (43,458 women), as demonstrated by the higher number of women presenting and dying from specific conditions such as Alzheimer's disease and other dementias, (prevalence-769 women, death-27 women), migraines (prevalence-17,092 women), multiple sclerosis (prevalence-29 women), and neurocysticercosis (prevalence-55 women) (Steinmetz et al. 2024). Notwithstanding this, the growth in overall neurological conditions (25%) was found to be a main contributing factor to the increase in the risk of neurological conditions like Alzheimer's disease, Parkinson's disease, Dementia, Multiple Sclerosis, with the major risk factors begin linked to high systolic blood pressure, high fasting glucose, high smoking and alcohol use, as well as environmental contaminates such as high exposure to ambient particulate matter (PM) pollution resulting from traffic, and industrial activities.

PREVALENCE OF NEURODEGENERATIVE DISEASES IN DEVELOPED AND DEVELOPING COUNTRIES

Developed Countries

When looking at developed countries in the Western Hemisphere, such as Europe and North America, the prevalence of ND has increased. In developed countries, the principal conditions with neurological health losses have been mainly due to the high incidence of stroke, preterm birth, nervous system cancer, migraine, dementia, and neonatal encephalopathy in Central and Eastern Europe, with similar findings being made in North America and Western Europe with the inclusion of autism spectrum diseases (Steinmetz et al., 2024). Such growth in the prevalence of ND in developed countries has been thought to be driven by a multitude of interconnected risk factors, such as a progressively ageing population, and higher life expectancy which were both linked to age-dependent conditions like Alzheimer's and Parkinson's that share high levels of oxidative and nitrosative stress, Sedentary lifestyle and changes in diet that enhances the occurrence of chronic metabolic diseases such as diabetes (Type I), Obesity, and Metabolic Syndrome.

Although Deuschl et al., (2020) explain that the prevalence of neurological diseases in 2017 particularly in Europe was greater amongst senior men than women (ages 80-84), the total number of DALYs was higher amongst women for conditions like dementia, and migraine and multiple sclerosis, and for men for conditions like stroke and Parkinson's disease. The critical nature of such conditions and such high prevalence rates, may also be a reflection in the shifts in clinical settings of developed countries where greater emphasis is being placed on early detection and advanced diagnosis of conditions through biomarker identification, and genetic testing; targeted treatment plans and preventative strategies, through better access to healthcare services that reduce the comorbidities (cardiovascular diseases, metabolic diseases, sleep diseases, and mental health conditions) that often have shared pathways with ND, and greater investment in ND research and development at universities and research institutions such as the EU Joint Programme-Neurodegenerative Disease Research (JPND) that focuses on understanding the diseases mechanisms, developing treatments therapies and diagnostic tools for specific conditions, as well as enhancing patient care (JPND Research, 2023).

Developing Countries

Unlike developed countries, it is estimated that more than 90% of the neurological disease burden, and 84% of deaths and losses from ND have occurred in low and middle-income countries around the world (WHO, 2025a). Even though the main ND that impact the developing countries of North Africa, the Middle East, Southeast Asia, East Asia, Oceania and Sub-Saharan Africa, are stroke, neonatal encephalopathy, dementia (inclusive of Alzheimer's), migraines, and diabetic neuropathy, with the inclusion of meningitis and epilepsy, are strikingly similar to that exhibited by developed countries, the burden of these conditions remain largely obscured in developing countries.

Such a stark similarity in the range of conditions exhibited, but divergence in the prevalence and deaths from ND, between both developed and developing countries, implies that there is a vast disparity in access to healthcare services and resources needed for early intervention and treatment. Moreover, such variations may also be indicative of the divergence in risk factors, as persons from developing countries may have significantly more medical conditions, different family histories of ND, exposure to environmental pollutants, and often lower levels of education and household income that influences not only their rate of diagnosis and treatment but also adaptation of preventative healthcare as their diet may have higher levels of processed foods, type of lifestyle which may be more sedentary with limited social engagements, (Espinoza Jeraldo et al., 2024).

Given that developing countries tend to have limited resources to invest in the field of neurology, ND may be more likely to be underdiagnosed, underreported, and underrepresented in world reports, which suggests that their prevalence rates may be lower than those of developed countries (Acosta et al., 2024). However, such a conclusion may be misleading, as developing countries may experience several challenges such as limited access not only broad systemic patient registries and disease registries, reliable diagnostic equipment, and tools, (brain scans, biofluid markers, digital speech biomarkers, phenotyping), and models (pharmaco-epidemiological and Bayesian Multiparameter Evidence Synthesis models), and continued heavy reliance on clinical diagnosis, but also a lack of robust prevalence data, epidemiological data and country-specific data/estimates that may support the identification and diagnosis of ND, specialized healthcare professionals (neurologists and specialists trained in ND) and advanced laboratory infrastructure, as well as rehabilitation services to support the long-term management of such conditions (Garcia & Johann, 2025). In addition, weak public health care measures and policies that target the populations knowledge on prevention strategies, early detection, diagnosis and treatment of such conditions, lack of large scale and affordable screening tools that are low cost, and non-invasive which are specifically designed for resource-constrained countries, and limited access to medication to treat conditions also become barriers to treatment and management.

Considering these many trials, the high disease burden that ND presents to developing countries worldwide may also worsen global health inequality, as persons in low-middle-income developing countries may not be able to access modern screening and monitoring tools used in the diagnosis of ND, but also clinical trials that include advanced medical intervention strategies. As a result, not only are dementia cases projected to triple in low, and middle-income developing countries by 2050 compared to those of high-income countries, but the exclusion of persons from developing countries, from clinical trials by universities and research institutions that tend to exclusively invest in patient populations from high-income countries, is likely to make certain that these persons remain underserved and less knowledgeable about their conditions as these countries receive less global funding/investment for research and treatment (Garcia & Johann, 2025). This then makes not only underprivileged persons from developing countries less prepared to manage these deteriorating conditions as they may have less access to lifesaving and symptom-managing drugs, effective treatment and management plans, but medical practitioners from developing countries may face significant challenges in managing ND such as Alzheimer's disease and Parkinson's disease (Wang et al., 2024).

These problems may stem from a lack of access to advanced training programs that cover the latest diagnostic techniques, which can hinder early detection and accurate diagnosis, to not being familiar with the most current treatment protocols and evidence-based practices, as well as knowledge gaps in the academic literature which may not be representative of population-based studies in developing countries, which are essential for providing effective care through advancing research and development (Lekoubou et al., 2014). As a result, patients in these resource-constrained countries may not receive the comprehensive care necessary for managing their conditions effectively, which worsens their risks of human rights violations, as they may easily fall victim to discrimination and stigmatization surrounding conditions like dementia. This is a particularly important issue, as is the cultural background of developing countries, which includes (1) the perceptions of illness and ageing as being a normal aspect of ageing rather than a medical condition, (2) the reliance on traditional forms of medicine and community support rather than modern medical institutional care, that may reduce the rate of detection and treatment, and finally (3) the lack of culturally sensitive screening tools that take into account the language, ethnicity, education, history, and religious aspects of different cultures, may all contribute to not only the inaccurate diagnosis of conditions but also the recognition and management of these conditions, which widens healthcare disparities (Vila-Castelar et al., 2022).

NEURODEGENERATIVE DISEASES IN LATIN AMERICA AND THE CARIBBEAN (LAC): THE CASE OF TRINIDAD & TOBAGO

Latin America and the Caribbean (LAC)

Given the challenges posed by the high incidence of ND in developing countries, it is crucial to highlight the healthcare outcomes observed in Latin America and the Caribbean (LAC) and particularly Trinidad & Tobago, upon which this study is based. However, before examining the case of Trinidad and Tobago, it is important to look at the prevalence of conditions in LAC. According to Steinmetz et al. (2024), in 2021, the leading ND in LAC were found to be linked to strokes, neonatal encephalopathy, migraines, dementia (inclusive of Alzheimer's disease), diabetic neuropathy, and, to a lesser extent, meningitis. However, in the specific case of the Caribbean, several other ND, such as epilepsy, Parkinson's disease, and cancers of the nervous system, seem to rank higher than in other countries in Latin America (Steinmetz et al., 2024).

The increasing prevalence of ND, particularly Alzheimer's and Parkinson's disease, highlights the rapid ageing of the population in LAC. This trend is reflective of the global increase in the elderly population, with the proportion of people aged 65 and older becoming increasingly at risk of disability due to these conditions (WHO, 2025b). The pervasiveness of these conditions and their associated neuropsychiatric symptoms in Latin America is the leading cause of disabilities in its elderly demographic, with a high prevalence of Parkinson's disease (34.3%), Dementia (56,1%), and Parkinsonism-dementia (61.2%) (PAHO, 2023). There is, however, an expectation that the burden of these diseases is to rise considerably within the next 10 years for countries such as Peru, Cuba, Venezuela, Dominica Republic (estimated 74% increase by 2050), Mexico, Puerto Rico, Uruguay (estimated 80% increase by 2050), and more than 300% by 2050 in Nicaragua, Guatemala, and Belize (Rodriguez-Salgado et al., 2023). Individual studies such as Acosta et al. (2021), and Ashby-Mitchell et al., (2020), further find that in the case of the Caribbean dementia is most prevalent in countries such as the Dominican Republic (11.7%), Puerto Rico (11.6%), Cuba (10.8%), and Jamaica (34.46%), while Lefebvre (2019) finds that there is a high incidence of Parkinsonism in Martinique that were related to the exposure to environmental toxins.

A similar outcome is anticipated in the Caribbean (estimated increase of 152% by 2050), particularly where the presence of age-related diseases such as dementia and Late-Onset Alzheimer's Disease (LOAD) is concerned (PAHO, 2023). Interestingly, in the case of LAC and individual Caribbean islands like Barbados, the major risk factors associated with such a widening of cases were persons' gender, education, geographic location, public awareness, and lifestyle (Ashby-Mitchell, 2018). More women (8.97%) than men (7.26%) from rural areas with lower levels of education appear to be negatively impacted by dementia when compared to urban areas with higher levels of education (Xiang et al., 2021; Ribero et al., 2022). Custodio et al. (2024) further add that the prevalence of such conditions may also be influenced by the socioeconomic disparities among minority groups, which influence their level of awareness and stigma surrounding ageing. Such conditions, while having mild-to-clinically insignificant symptoms, appear to have their most pronounced manifestations amongst patients in LAC with an increased risk of becoming institutionalized, with indicators of sleep disturbances, and mental health diseases such as anxiety and depression.

These risk factors can be further worsened by the lack of effective policies and plans, as only a few countries in LAC, such as Chile, Mexico, Costa Rica, and Cuba, have national dementia plans to deal with the growth of these and other conditions (PAHO, 2023). The reason why so many countries, inclusive of LAC, lack comprehensive national dementia plans may be reflective of a much more global challenge. Most of the national policies and strategies are concentrated in Europe, with 20% of the WHO member states having plans. This implies that global progress on addressing these problems has been slow (WHO, 2025b). Such a deterioration in the implementation of these plans in LAC is likely to be due to several barriers, such as (1) the lack of urgency, addressing the frequency of ND may not be a top priority for governments, leading many plans to become either expired or, underutilized (2) the lack of resources to dedicate to the implement and sustain these plans, as reflected by the use of dedicated budgets (3) the lack of effective implementation and evaluation of the plans strategic targets, due to a lack of legislative support to sustain efforts (WHO, 2017a).

One of the most crucial barriers to the implementation of these plans is the lack of epidemiological data in LAC. The LAC has fallen short of collecting core data needed to strengthen not only the response and treatment of these conditions but also the research literature and published data on brain ageing and cognitive decline. According to Nakamura et al. (2015), many persons in low and middle-income households will likely remain undiagnosed, as highlighted by the case of Brazil, where 77% of elderly persons, and 3 million persons in LAC with dementia were not diagnosed, because of the lack of data and research in the field. The lack of historical explanations by Xiang et al. (2021) limits the use and design of evidence-based models, replication studies, and meta-analysis studies needed to estimate the burden of ND, like dementia, in LAC. Similar concerns have been presented by Lopera et al. (2023), who, apart from calling attention to the dearth of high-quality epidemiological data in LAC needed to both inform the implementation of healthcare policies and the standardisation/harmonization of clinical practices, also emphasized several other structural barriers to diagnosis and treatment.

Similar to other developing countries around the world, the diagnosis and treatment of ND in LAC have been constrained by many problems including, such as the presence of fragmented healthcare systems that lead to inconsistent treatment and misdiagnosis, limited research funding to support research on underrepresented groups in LAC to understand better the genetic heterogeneity that may influence their diagnosis, and insufficient infrastructure to support the use of modern diagnostic tools and procedures such as Cerebrospinal fluid (CSF) biomarkers, Magnetic Resonance Imaging (MRI), and Positron Emission Tomography (PET) imaging (Lopera et al., 2023). Moreover, the lack of standardized diagnostic protocols that have been adapted to reflect the demographic characteristics, genetic predispositions, lifestyle factors and overall diversity of persons living in LAC, and the lack of insurance coverage for ND, and the high cost of medications makes it difficult for persons living in vulnerable situations to access healthcare (Custodio et al., 2024). Thus, worsening healthcare disparities in cases where public healthcare systems provide support, coverage is often minimal and inconsistent, which can lead persons to bear direct and indirect costs of accessing healthcare.

Apart from the technical issues ingrained within fragile healthcare systems in both the LAC and the wider world, the social determinants of health, such as socioeconomic status, employment, housing and education, from the perspective of persons with ND, also play a vital role in their access to care and resources. Persons who hold a lower socio-economic status, as evident from their minimal levels of education, engagement in precarious forms of work, and may lack access to a safe form of housing and basic amenities, are more at risk of developing NDs like Alzheimer's and Dementia (Besser et al., 2024). Further to this, in situations where persons exhibit signs of NDs before and even after their formal diagnosis is made, their economic instability and insufficient access to education may prevent them from maintaining a healthy lifestyle by accessing community support through green spaces that enable them to build and maintain a strong social network, as well as healthcare services for early ND and chronic disease management.

Adkins-Jackson et al. (2023) state that these social determinants of health are core determinants of the disparities and inequalities experiences by persons with ND such as Alzheimer's and Dementia, as these determinants tend to accrue over time, and may adversely impact the person over their lifetime, which deepens structural issues such as classism, racism, discrimination and social exclusion. In instances where persons with ND experience discrimination and social exclusion in a home, community and healthcare setting, it may result in negative healthcare outcomes, as early diagnosis and treatment may become delayed.

Trinidad and Tobago

ND in Trinidad and Tobago have become a growing public health concern, impacting the quality of life for many persons and placing a significant burden on families and the already fractured healthcare system. As the population of people of an advanced age widens, it has been found by the WHO (2012) that compared to other countries around the world, and the LAC region, the prevalence of Dementia in the Caribbean region, is the highest in the world (8.1%), and is projected to by 215% between 2010-2050. Such an alarming statistic, together with other conditions such as Alzheimer's and Parkinson's disease, which also have a higher prevalence in Trinidad and Tobago, implies that there may be many factors at play that enhance a person's risk of developing these ND late in life.

A brief examination of the literature on ND for the case of Trinidad and Tobago reveals that much emphasis has been placed on research related to Dementia and, to a lesser extent, Alzheimer's disease. Focusing on the former, persons with dementia in Trinidad and Tobago were an average of 79 years old, had acquired less than 10 years of schooling (mainly had a primary school education), and were previously employed as skilled laborers in labor-intensive industries such as agriculture, and construction, as well as occupations such as maintenance technicians, woodworking, and landscaping (Davis et al., 2021).

Further to this, dementia patients in Trinidad and Tobago were also observed to have sensory (visual, hearing), as well as musculoskeletal (arthritis, muscular dystrophy), and other neurological impairments (stroke, epilepsy) that led to a significant reduction in their physical exercise (Davis et al., 2021). This group were more likely to have dementia if they had a stroke, but they had notable protective factors, such as a high level of education and specialized training, being employed in a professional capacity, and having no prior comorbidities, and were less likely to develop dementia (Davis et al., 2019). When considering the ethnic background and marital status of patients in the oldest-old age range of 85-102 in Trinidad and Tobago, the prevalence of dementia was higher amongst those of East Indian heritage than those of African heritage, and who were widowed (Davis et al., 2016).

Regarding Alzheimer's disease even though there is a dearth of knowledge on the condition in Trinidad and Tobago, recent studies by Rosano et al. (2024), show that women with an average of 60+ years and African ancestry in Tobago have a higher occurrence of Alzheimer's due to the presence of metabolic risk factors and a sedentary lifestyle, when compared to men with a similar background but with hypertension. However, the findings of the study were limited in that the biomarkers used were of a weaker dilution which impacts its measurement, and is prone to alterations due to changes in the patient's blood and cardiovascular system, as well as the lack of data on the apolipoprotein E (APOE) ε4 genotype and cognitive information limits the ability of medical professionals to link the patients genotype with their cognitive performance (Rosano et al., 2024). In earlier studies done by McRae et al. (2006), found that a vast majority of patients with an average of 75 years in Trinidad and Tobago presenting with memory impairments (short/long-term memory loss) had either Alzheimer's disease or Alzheimer's disease with Vascular Dementia, also known as mixed dementia, while most patients were of had an African heritage.

The major risk factors for dementia, and Alzheimer's disease in Trinidad and Tobago, are thought to be linked to not only the advanced age of persons which is largely due to age-related changes, socioeconomic changes, and psychological and behavioural changes, but most importantly to the elevated pervasiveness of cardiovascular diseases (ischemic heart disease, atherosclerosis, heart attack, stroke) together with other metabolic (hypertension, hyperlipidemia, obesity, and insulin resistance), and vascular (smoking, family history of coronary heart disease, diabetes mellitus, and kidney disease) risk factors (Davis et al., 2018). In the case of persons 70 years and older in Trinidad and Tobago, due to the presence of these risk factors, this group of persons are more likely to present with systemic multimorbidity, i.e., the presence of multiple long-term chronic health conditions, which may lead to a faster deterioration in the health and well-being, and an increased risk of ND, particularly dementia (La Foucade et al., 2020).

The presence of multimorbidity creates a serious challenge as earlier studies by Davis et al., (2018), revealed that there is indeed a growing incidence of dementia in Trinidad and Tobago that is related to a high frequency of cardiovascular risk factors, especially for persons in the early-old age group of 70-74 years, but is lower in patients with hyperlipidemia and hypertension. Davis et al. (2021) go on to explain that the risk factor for developing dementia is likely to increase in the middle-aged age group of 75-84 years if the person presents with more than four of the previously mentioned types of multimorbidity conditions. The presence of several of these conditions at once, according to Zhou et al. (2021), makes treating underlying multimorbid conditions, as well as ND like dementia, even more complicated, as multiple risk factors need to be managed simultaneously. In that the management of multiple conditions can not only lead to drug interactions and heightened cognitive decline, but can also become overwhelming for patients, leading to difficulties in maintaining individualised care plans, especially if residing in a low-income household.

The barriers experienced by persons with ND in Trinidad and Tobago were observed to be parallel to those of the LAC, in that there is a similar lack of reliable and updated data/statistics on the prevalence of conditions like dementia, and the monitoring of patients, leading to its misinterpretation and undervaluation (WHO, 2017b). Further to this, while some progress has been made in the development of a potential serum screening biomarker and dementia diagnostic tools, the overall absence of established, emerging as well as updated biomarkers/diagnostic tools for Alzheimer's and Dementia, together with the lack of investigational pharmacological trials, poses a significant challenge to the detection of cognitive decline in Trinidad and Tobago (Davis et al., 2007; WHO, 2017b).

Additionally, these issues are further amplified by the dearth of policy frameworks/guidelines to support the standards needed for dementia care, the lack of a dedicated national dementia plan for Trinidad and Tobago to improve the overall health and well-being of patients, and finally the absence of mechanisms across sectors to support the care of patients, lack of social protection programmes designed specifically for persons with ND to provide financial support, caregiving assistance, and accessibility to services (WHO, 2017b). As a result, not only would there be delays in the diagnosis of conditions, but there would also be an erosion in the quality of care, limited research into new and preventative treatments that are specific to patients in Trinidad and Tobago, a reduction in advocacy and widening of socio-economic inequalities amongst patients (WHO, 2012, 2017a).

THE PRESENCE OF NEURODEGENERATIVE DISEASES IN WORKPLACES AROUND THE WORLD

As demonstrated through the previous extensive discussion on the prevalence of ND, this class of conditions has undoubtedly become a major cost to healthcare systems around the world. Although these diseases typically affect older age groups, i.e., persons ages 65 and older, depending on the specific disease, it has been known to impact younger persons, through its early onset by exhibiting signs from infancy to adulthood. With the growing prevalence of ND amongst the aging population, as well as their early onset manifestations in persons of working age, their occurrence within the workplace has become a significant concern for employers. In that, it not only impacts the career and employment trajectories of workers, but also the workplace dynamics as employers may need to make special accommodations for such workers, and overall productivity and competitiveness of businesses, as workers may seek early retirement even as the individual cost of managing their condition intensifies. Bearing this in mind, this section will explore the existing literature to provide greater insight into the workplace dynamics and overall productivity challenges experienced by workers with Alzheimer's and Parkinson's disease, as well as Dementia.

Alzheimer's in the Workplace

The ability of individuals with Alzheimer's disease to remain in the workforce largely depends on the disease's progression, the specific symptoms they may exhibit, and the nature of their occupation. Despite facing the many challenges posed by the condition, many workers find ways to continue their professional lives, adapting to their evolving capabilities and seeking support where needed from either their employer, the public healthcare system, or caregivers. Focusing on the complexity of the occupation of the worker, Andel et al. (2005) finds that in the case of Swedish workers, the prevalence of Alzheimer's disease was less prevalent and had lower levels of risk amongst workers whose occupations required them to do more complex tasks with colleagues, where as if the tasks were not as complex and had less interaction the risk of Alzheimer's disease was higher. Such findings were also like previous studies such as Smyth et al. (2004), who in comparing the functions of workers in the United States (US), with that of their occupational traits across different age groups finds that Alzheimer's disease is likely to influence the workers ability to undertake mentally demanding occupations at an early age, i.e., workers in their 20s and 30s, however workers in older age groups who engage in occupations with complex mental requirements presented higher levels of neuronal activation which preserved neural pathways in the presence of Alzheimer's disease.

Environmental factors, such as exposure to Electromagnetic Fields (EMF), either residential or occupational, are a contributing risk factor to the prevalence of Alzheimer's disease amongst workers and residents. As Bonvicini et al. (2009) find that in the region of Reggio Emilia, Italy, the risk of Alzheimer's was not only higher when residents were exposed to EMF exposure from high-voltage power lines but also increased as persons aged. Similar findings were also made by Qiu et al. (2004), who found that Swedish men were at a higher risk of developing Alzheimer's when exposed to elevated levels of Extremely-Low-Frequency MF (ELF-MF) during their long-term occupations in the long-term, as they age.

Hansson Mild et al. (2023), explains that although the manifestations of Alzheimer's disease tends to vary based on both the intensity of EMF exposure, and the origin of such exposure (whole or partial body) such as in either an occupational, residential, or laboratory setting, it is worth mentioning that while workers such as welders and electric train drivers, may be exposed to EMF, such contact to high-EMF is often sporadic and isolated events followed by periods of minimal exposure to

ELF-MF. Furthermore, the different types of EMFs such as x-rays used for imaging bones/fractures, and radiofrequency EMF's used in Magnetic Resonance Imaging (MRI) scanners that are used in medical occupations, even though not-directly related to Alzheimer's disease, have according to Ghadimi-Moghadam et al. (2018), been linked to cognitive effects such as disruptions in sleep, concentration and attention due to prolong exposure.

Apart from Alzheimer's disease itself, workers in the US have also exhibited symptoms of Alzheimer related disorders, such as Dementia. In several instances, different forms of dementia, i.e., Mixed Dementia, Vascular Dementia, Lewy Body Dementia, Frontotemporal Dementia, and Huntington's Disease Dementia, have been found to occur alongside Alzheimer's. As a result of this, in most cases, workers showed signs of cognitive loss, having encountered problems when completing tasks due to either a lack of attentiveness or absent-mindedness, leading to a notable decline in their performance and motivation. While 86% of employers in the US have not let an employee go due to their symptomatic behaviour, there were instances where older employees were laid off due to cognitive decline (Cox & Pardasani, 2013). Further to this, full-time workers who reported positive experiences with their employers following the disclosure of their condition were more likely to have done so only if they felt supported, i.e., by having special accommodations such as supportive training programmes, flexible work tasks and schedules made so that they can continue working, retire early, or shift into an active social life following such retirement (Gibson et al., 2024). In contrast, the opposite trend was observed for workers with less support and significant financial distress

Such changes in employment patterns were revealed by younger workers in Denmark, who when demonstrating signs of Young-Onset Alzheimer's (YOAD) at moderate or severe stages, who requested not only frequent long-term sick leaves, but were more likely to become unemployed either through being laid-off or quitting, in the year before their formal diagnosis (Damsgaard et al., 2024). These trends appeared to occur when the cognitive demands of the workplace exceeded the cognitive resources available to young workers. When such an imbalance occurs, workers with YOAD-related symptoms are more likely to experience signs of chronic stress and burnout in the workplace, such as an increase in emotional and physical exhaustion, as well as absenteeism, reduced productivity, and motivation, when they cannot perform work tasks efficiently. This can adversely impact their working memory due to a loss of brain tissue resulting from reduced activity in the parietal lobes, and an increased risk of mild cognitive decline that is more than what is expected at their current age (Oliveri et al., 2020).

Dementia in the Workplace

In a similar manner to Alzheimer's, dementia has also been noted to have a significant impact on the occupational competence, participation, and identity of workers who show signs of Young Onset Dementia (YOD), or Mild Cognitive Impairment (MCI) during either young or middle adulthood while actively engaged in employment. In countries such as Japan, Edahiro et al. (2023), finds that 57.2% of workers with YOD who were employed, showed signs of memory loss, difficulties in word generation, heightened levels of irritability, and increased mistakes while on the job., while Edgall (2018), finds that in Scotland and the United Kingdom (UK), there were 3,200 persons and 40,000 persons respectively with dementia under the age of 65. Comparable findings were also made by Evans (2016), who found that the challenges that workers with YOD in Australia experience, such as forgetfulness, disorganization, deteriorating job performance, and the inability to manage work relationships, often go unnoticed by employers due to the slow progression of the disease. Such elevated levels of dementia within the working population imply that employers need to implement strategies and rehabilitation frameworks such as the Australian Side-by-Side workplace engagement program to support the needs of affected young workers, as its impact on workplace productivity is likely to be substantial (Robertson et al., 2013).

Within most workplaces, however, while there may be legislative guidelines to support workers with physical and mental health issues, they may have very little applicability for persons experiencing YOD and MCI. For this reason, employers in countries such as Canada, Finland and Sweden have begun to provide additional support for affected workers, by providing person-centred therapy and advice, raising awareness, offering flexible work arrangements in terms of hours worked and job tasks, individually tailored occupational health services, and creating a more positive culture in the workplace in which those with dementia feel supported and valued (Bak et al., 2023). Several of these strategies and interventions have been guided by policy and regulatory frameworks put in place in countries like Canada, which include the National Strategy for Alzheimer's Disease and Other Dementia, the Canadian Charter of Rights for People with Dementia, the Canadian Charter of Rights for People with Dementia, as well as the Dementia Strategy for Canada (MOJ 2025; PHAC, 2019; ASC, 2025). In the case of Europe, which includes countries like Finland, Sweden, Italy, Norway and the Netherlands, the European Union (EU) Joint Programme-neurodegenerative disease (EU-JPND_ has been instrumental in developing an ongoing project to support national neurodegenerative disease plans, research programmes such as Deltaplan for Dementia in the Netherlands (EU-JPND, 2023).

Apart from these country/region specific strategies and frameworks, labor protection for persons with neurodegenerative diseases is also guided by international standards put forward by the United Nations Convention on the Rights of Persons with Disabilities (CRRP-Article 27), the ILO's strategy on the Right to Decent Work of Persons with Disabilities, as well as the United Nations Human Rights Office of the High Commissioner (OHCHR), Convention on the Rights of Persons with Disabilities (UN-DESA, n.d.; O'Reilly, 2007; UN-OHCHR, 2006). With these strategies and frameworks in place, even though dementia, and other neurodegenerative diseases highlighted in the article is likely to present itself early in the working life of workers, they can still be able to work at their full potential, especially if the human rights tools embedded within the United National Convention on the Rights of Persons with Disabilities (UNCRPD) concerning accessibility, and assistive technology are used to support their rights.

Such an approach, however, may not be feasible for all employers and countries due to a lack of financial/technological/human capacity, as well as workplace and country-level policy guidelines to mitigate against the impact of dementia on the working environment. As a result, even in countries such as the UK and Scotland, where workers are entitled to have reasonable work adjustments made to accommodate their condition under the 2010 Equality Act, their career development needs are frequently left unaddressed, having to make their workplace adaptations (Egdell et al., 2018). Without such provisions, it is possible that the basic human and employment rights of those with YOD and MCI may not be consistently supported, resulting in reduced opportunities and elevated levels of discrimination and social exclusion due to stigma and a lack of understanding by employers (Ritchie et al., 2022). This then worsens the workers' ability to address the adverse effects of their condition in the workplace, such as their job performance and satisfaction, as well as eroding their work-life balance, as they become inundated by additional out-of-pocket expenses associated with their diagnosis.

Parkinson's in the Workplace

Achler et al. (2022) explain that 38% of workers in the US who exhibit symptoms of Parkinson's disease have experienced financial hardship, which negatively impacts their perceptions of job performance and satisfaction, as well as their access to opportunities for career advancement. Such economic distress, when paired with the degenerative symptoms of Parkinson's disease that often appear slowly and subtly, was found to have contributed not only to the stigma and discrimination against such groups of workers but also worsened their mental health outcomes (higher levels of psychiatric distress) and overall quality of life.

In such instances, workers with longer durations of Parkinson's disease and who were closer to retirement age were more likely to retire earlier than their colleagues, due to the adverse effect that the condition has on their working capacity and the lack of support provided to them by their workplaces. This as, many older patients may not be able to access supportive/integrative care such as physiotherapy and psychological therapy that is often tailored to the needs of the worker, which had it been provided, may have improved their healthcare and employment outcomes, as motor and non-motor symptoms together with psychological features can be managed simultaneously to reduce the burden of the disease. Thus, it allows workers to actively participate in the labor market.

Comparable to the case of workers with Parkinson's disease in the US, Collett et al. (2022), finds that like that of workers in the US as the condition progressed, those in Europe who were laid-off displayed adverse symptoms such as increased fatigue, and sleep disturbances, as well as an erosion in their employment potential due to a decline in self-efficacy and social support. As workers' careers and employment prospects diminished, they often reported higher levels of psychiatric distress within the 5 years before being laid off, particularly if younger at diagnosis, mainly because, despite their condition, they preferred to be employed or involved in some form of work. The reasoning behind the desire for workers with Parkinson's disease stems primarily from the direct and indirect costs of healthcare, which have led to an increase of 12-14% in healthcare claims during the first year of diagnosis, and the loss in earnings of $174 to $1104 by the third year of diagnosis, which accounted for between 20-30% of workers annual income (Jerry et al., 2023). For such workers to be retained in the workplace, much the same adjustments are needed for other ND, such as workplace modifications, as well as social support from employers and colleagues.

In cases where younger persons are affected by Parkinson's Disease, they are said to have either Young-Onset Parkinson's Disease (YOPD), which usually occurs between the ages of 21-40, or Juvenile Parkinson's disease, which occurs below the age of 21 (Klepac et al., 2013). Unlike older workers with Parkinson's disease, younger workers with YOPD indicated that the condition made it difficult for them to remain in paid employment, due to high levels of stress, inadequate support by employers, restricted opportunities for early retirement, stigma and cultural biases experienced within the workplace, which may have worsened their feelings of shame, embarrassment and social isolation (Crooks et al., 2025).

In the case of Parkinson disease amongst workers in Ireland, while the percentage of persons working declined considerably between 5-10 years of diagnosis, the condition according to Murphy et al. (2013), was observed to place YOPD workers at an elevated risk to experience considerable economic hardship such as poverty, unemployment, limited access to health care and social inequality. YOPD workers may be more likely to exhibit higher levels of perceived stigma, which in turn may limit their ability to seek help due to fear, and discrimination, increase their isolation and withdrawal within the workplace, and enlarge their feelings of grief, shame, negative criticism, self-talk, and considerable concealment of their diagnosis from peers and family members.

NEUROGENERATIVE DISEASES AND THEIR IMPLICATIONS FOR WORKPLACES IN TRINIDAD AND TOBAGO

Labor Market Implications of ND in Trinidad and Tobago for Employers

The prevalence of ND among the working population in countries from the Global North, as discussed, has undoubtedly had a significant impact on the working population. However, while most of the literature on the labor market implications of ND has focused on the experiences of workers from the Global North, there is a dearth of knowledge regarding its presence among the workforce in developing countries like Trinidad and Tobago. Even though ND is likely to have a severe impact on the economically active population of Trinidad and Tobago, the lack of labor market research and data collection often limits greater insights into its presence in the workplace. As a result, there is a notable gap that exists in the literature that reflects its possible labor market implications.

The literature on ND in Trinidad and Tobago has traditionally focused on the pervasiveness of Dementia, its epidemiology and risk factors, rather than the challenges that it may pose in the context of the workplace. Furthermore, such a research focus is also likely to have been influenced by the financial and human resources available to identify, track, and capture data on cognitive decline and neuropsychiatric symptoms exhibited by workers in Trinidad and Tobago. Given the complexity with which ND evolves and progresses, and the lack of labor market-related studies, this may be reflective of the funding priorities amongst scientific research institutions and the government funding policies in Trinidad and Tobago. Firstly, funding may be diverted by scientific institutions to better understand how ND mechanisms present themselves in Trinidad and Tobago's population, rather than examining the challenges that it may pose in the context of the workplace. Secondly, the complex rules, procedures, and delays associated with Trinidad and Tobago's bureaucratic system to apply for state funding, and fluctuations in state funding itself, can disrupt long-term research labor market research on ND.

Considering the absence of data and research studies highlighting the prevalence of ND in Trinidad and Tobago's labor market, the presence of such non-disclosed conditions among the working population is likely to have significant implications for employers, as workers may become diagnosed while still engaged in active employment. Thus, for purposes of this study, emphasis will be placed on briefly examining how these ND, i.e., Alzheimer's, Dementia, and Parkinson's, may become a challenge for employers in Trinidad and Tobago to manage, in the areas of (1) Productivity and Efficiency, (2) Employee Turnover and Absenteeism, (3) Workplace Management Issues, (4) Competitiveness and Innovation, and (5) Workplace Transformation.

Productivity & Efficiency

As the population of Trinidad and Tobago continues to age, an increasing number of workers may experience symptoms of cognitive decline while at work. As ND conditions continue to impair the ability of workers to perform their jobs, leading to reduced productivity, many may be faced with either taking early retirement packages or accessing frequent long-term sick leaves in the case of YOPD (Damsgaard et al., 2024). This then creates a burden for employers because the absence of workers, particularly those at a younger age with early onset of ND, may adversely impact the production of goods and services through reduced levels of productivity and efficiency within the workplace at production sites in Trinidad and Tobago (Ritchie et al., 2017). Depending on the severity of the diseases progression, in situations where symptomatic workers can engage in job tasks, this also poses a challenge to employers as workers with impaired memory, concentrating, decision making skills, and motor symptoms of Parkinson's such as tremors can make serious errors in judgement that can compromise the quality of the work being done, and the safety of both them and co-workers where machinery or harmful materials are being used daily (Mantri et al., 2024).

Employee Turnover & Absenteeism

The presence of workers with ND is also expected to have an impact on the overall performance of the workplace, which can be observed through changes in employee turnover, absenteeism, and the incidence of workplace management issues. As workers with ND in Trinidad and Tobago seek treatment and support for their condition, this may increase the rate of employee turnover and absenteeism in the workplace (Oliveri et al., 2020). With such frequent absence, labor costs may increase as co-workers may need to take on the additional responsibilities to maintain project deadlines, while employers either hire and train replacement employees or facilitate cross-departmental training to ensure that the productive efficiency of departments is preserved (Pauly et al., 2002). With such notable changes, the dynamics of teams and departments may be negatively impacted, as the additional strain on both employers and co-workers not only disrupts the workflow in these areas, leading to further declines in productivity and project delays, but also reduces collaboration amongst workers due to burnout and chronic stress.

Workplace Management Issues

The recurrent absenteeism of workers with ND who play a crucial role in the workplace can diminish the morale of co-workers and increase workplace management issues within businesses in Trinidad and Tobago. Problems such as a breakdown in communication, feelings of demotivation, frustration, and resentment amongst workers, together with a poor work-life balance, can bring about not just conflicts between workers, bullying, and harassment, but also produce a dysfunctional workplace culture with heightened levels of workplace deviance (Salin, 2003). The perpetuation of such negative traits, when linked to workers with ND, can damage the reputation of both organizations operating within industries and industries themselves, as such issues may make it difficult to attract and retain workers. According to Kumai et al. (2021), workers with conditions such as dementia have engaged more in short-term employment, having been forced to change their job frequently due to instances of bullying in the workplace by colleagues and employers who had no training on how to deal with workers with ND.

One of the direct consequences of such workplace bullying is that it stifles the creativity and innovativeness of existing workers with ND who may still be employed, as well as current and even new workers who may be observers, i.e., passive witnesses, to such incidents. During such intense levels of stress, workers with ND may become reluctant to collaborate and engage with colleagues because of a lack of trust and respect, but exposure to such a toxic environment may worsen their symptoms. This stems from the fact that the heightened release of stress hormones such as cortisol can disrupt normal brain function, as well as change the structure and volume of the brain in areas such as the ventrolateral prefrontal cortex and the inferior temporal gyrus (Du Plessis et al., 2019). Such changes can worsen the cognitive deficits of workers with ND, as persistent bullying and victimization can further worsen workers' ability to focus, remember, and manage complex tasks. This then makes them even more vulnerable to becoming victims of bullying within the workplace, as well as developing mental health disorders.

Competitiveness & Innovation

While workplace bullying of workers with ND in Trinidad and Tobago, together with issues of absenteeism and employee turnover, can have a distressing effect on the victims and coworkers, it can have long-term, often irreversible effects on a business's competitiveness and innovation. In that it can damage the reputation of businesses and have a severe financial impact on employers' ability to attract and retain key stakeholders such as suppliers, shareholders, major investors and regulatory bodies, and government entities. This adversely affects employers' ability to innovate and compete with regional or global rivals, as stakeholders and investors may perceive these businesses as having less value, particularly if affected workers with ND use digital mass media such as social media as an outlet to express their grievances in the public domain (Walsh et al., 2019). When workers with ND use social media to share their experiences, such as the lack of employment accommodations, bullying, biasness, and discrimination, as the situation escalates, the employer is likely to encounter even more challenges as legal and journalistic investigations begin, as workers seek legal recourse.

The impact of such negative publicity may cause businesses in Trinidad and Tobago to lose not only investment opportunities but also incur additional losses through a reduction in sales and revenue, as well as legal fees, as their credibility in the regional/global market weakens. This is of specific importance because, as Choi et al. (2018) explain, stakeholders who focus on the responsiveness of businesses/employers to philanthropic, ethical, and social responsibilities tend to devalue businesses that exhibit such negative traits towards workers with ND. With reduced financial resources and eroding organisational health, the capacity of businesses in Trinidad and Tobago to engage in innovative projects to boost their growth and efficiency may become limited as their declining market value may signal to stakeholders that investment may be riskier due to higher levels of liability.

Workplace Transformation

Another important labor market implication that employers in Trinidad and Tobago are likely to incur because of having workers with ND as a part of their workforce concerns the transformation that the workplace and its personnel must undergo to effectively integrate such a segment of vulnerable workers. One such adjustment is by providing support for workers with ND, co-workers, and employers through the provision of education, training, and awareness campaigns on such conditions. In doing so, workers with ND, co-workers, and employers can better understand and recognize the symptoms of such conditions, which may enable workers to feel supported and comfortable seeking early treatment. Early presentation and identification in the workplace can allow employers sufficient time to reallocate tasks and responsibilities of the impacted worker to coworkers, to minimize disruptions in workflow (Thomson et al., 2018).

Based on the early identification of workers symptoms, reasonable workplace accommodations such as implementing flexible working hours, remote and part-time work options, adjustments in job tasks and schedules, using assistive technologies and memory aids, as well as physical changes to the working environment such as the personalization of workspaces, having designated quiet spaces, ergonomic adjustments to facilitate workers with physical limitation can also be integrated into the working environment (Egdell et al., 2021). Even though the initial cost of making such accommodations, and training can be significant, in the long-term it can be quite beneficial to businesses in Trinidad and Tobago, as it not only reflects positively on the corporate culture and value of the business which may be viewed in a more positive light by potential investors, but it allows employers to retain highly skilled and educated workers to reduce costs associated with high employee turnover and absenteeism, and continue to support workers who are often the sole breadwinner in their households (Bleecker & Barnes, 2015).

In addition, another important adjustment within the workplace includes designing workplace policies and procedures that are tailored to the needs of workers with ND in Trinidad and Tobago. This can include codes of conduct that clearly explain the expectations of workers with ND, and include policies that address issues such as discrimination, bias, stigma, and bullying of this specific group of workers (Cichy et al., 2015). Recruitment policies (hiring, training, leave, termination, discipline, performance, and compensation) that consider the condition of workers with ND, so that as the condition progresses, changes in worker performance are not viewed as poor work ethic, but rather effects of the ND (Thomson et al., 2018). This ensures that workers with ND in Trinidad and Tobago are not only treated with fairness during the performance and evaluation process, and when accessing time-off, professional development programmes, career guidance programmes, and workers' compensation during early retirement, but can also access proactive diagnostic and psychosocial support within their work environment (Bolger et al., 2023).

Finally, employers must also consider the legal and ethical considerations when making such accommodations so that workers with ND are not only aware of their rights as employees, but also their right to trade union representation to protect against unfair dismissal (Cichy et al., 2015). In the case of Trinidad and Tobago, while persons with ND may be considered disabled based on the loss of their physical and mental faculties, and be protected against discrimination, and entitled to reasonable accommodation in the workplace under the Equal Opportunity Act, Chapter 22:03, Act 19 of 2000, and the Equal Opportunities Commission (EOC) guidelines on how to treat workers with disabilities, there are no specific directive to address the impact and accommodations needed by workers with ND in Trinidad and Tobago (EOCTT, 2018; MAGLA, 2001).

THE TECHNOLOGICAL NEUROREHABILITATION OF WORKERS WITH NEURODEGENERATIVE DISEASES IN TRINIDAD & TOBAGO

The presence of neurodegenerative diseases within Trinidad and Tobago's workforce presents employers with a unique group of challenges. As the symptoms of such conditions progress, the ability of workers to undertake tasks within the workplace may deteriorate to the point that it hampers their performance. However, such a decline can be reduced and managed with special workplace accommodations that not only support the prolonged employment of workers but also the management of their condition. Apart from the traditional workplace accommodations discussed earlier, the adoption of Industry 4.0 and 5.0-based technologies, such as Gamification (Mobile Video Games), Virtual Reality Glasses, Robotics, and Artificial Intelligence, can play a significant role in the rehabilitation of workers with neurodegenerative diseases. The collaborative use of these technologies by all stakeholders, i.e., workers, therapists, caregivers, and employers, can allow these complex interventions to be brought into the workplace (Saha, 2024). However, for purposes of this article, emphasis will be placed on the possible applications of the neurorehabilitation of workers with neurodegenerative diseases in Trinidad and Tobago using Socially Assistive Robotics and Soft Robotics. In many instances, as discussed below, these robots have been primarily used for rehabilitation in therapy settings. However, this type of technology can also be used in workplace settings to encourage workers with NDs to collaborate with other workers, engage in light physical activity to reduce sedentary behaviours in the workplace, while providing personalised health interventions in the workplace, and support social interactions amongst workers with NDs.

Socially Assistive Robots (SARs)

Within the literature on neurology, the use of robots and robotic systems has long played a key role in the diagnosis and management of several neurodegenerative diseases such as Alzheimer's, Parkinson's and Dementia. When integrated with artificial emotional intelligence (AEI), robotic systems such as Socially Assistive Robotics (SAR) can provide workers with dementia and Parkinson's in Trinidad and Tobago with the necessary social and emotional stimulation to reduce feelings of loneliness and isolation within the workplace (Abdollahi et al., 2023). To facilitate this, SARs can be personalised to analyse the vocal tones, facial expressions, and body language of workers, allowing for learning from their interactions and simulating suitable conversations in a way that is sympathetic and sensitive to the workers' emotional and physical states (Ahmad et al., 2017).

In this regard, SARs are designed by developers in a way to encourage patients with cognitive decline to have positive perceptions of such robots. Thus, these SARs often have either a pet-like appearance, mirroring the movement and behaviour of animals such as cats (HASBRO), sheep, and seals (PARO) through pet therapy, or a humanoid appearance, such as MATY and PEPPER, which adopts a multisensorial approach that enables Alzheimer's and Dementia patients through consistent communication and routine development to become more autonomous, and reduce instances of social isolation (Simao & Guerreiro, 2019; Dosso et al., 2022; Varrasi et al., 2019; Feng et al., 2019; Hung et al., 2019).

Although most SARs research has been concentrated on its use in healthcare systems, rehabilitation centres, and assisted living facilities, if incorporated within the working environment based on the progression of the specific type of neurodegenerative disease, it can present workers with neurodegenerative diseases in Trinidad and Tobago with several benefits. In that, through greater instances of positive social interactions with SARs, such humanoid robots like Nadine, when used as assistants, are better able to understand and adapt to workers' responses, while also supporting the treatment and management of neurodegenerative conditions (Vishwanath et al., 2019). To do this, robots can set reminders for workers to take medication, short breaks (break management), practice relaxation techniques, assist with routine tasks, as well as outline emergency protocols for use within the workplace (Van Assche et al., 2024). Where workers with Parkinson's show signs of motor symptoms, the use of SARs can monitor and track changes in the workers' motor function, posture, balance, and gait throughout the workday, to create individualized rehabilitation plans such as dance-based therapy, reminiscence therapy, and exergaming without disrupting their workflow (Bevilacqua et al., 2025).

Such information can then be used to support the adoption of these forms of therapy, and automated robotic devices at rehabilitation centres in Trinidad and Tobago, for gait training exercises to reduce the Freezing of gait, i.e., a Parkinsonian symptom that is characterised by the sudden involuntary inability to walk. This is of paramount importance as even with dance therapy, when facilitated with the aid of SARs that are enriched by AI-based exergames, apart from being an enjoyable activity that workers with Parkinson's and Dementia can engage in, it can significantly improve their motor, emotional, and cognitive skills (Li et al., 2022). According to Carda et al. (2012), such therapies may improve workers' motor function over time by stimulating both the cerebral cortex and cerebello-premotor pathways. This may lead to an increase in neuroplasticity, as the use of SARs in the rehabilitation process can enhance mobility by facilitating cortical reorganisation (Keci et al., 2019).

Further to this, it is possible that workers with neurodegenerative conditions like Dementia and Parkinson's with mild to moderate cognitive impairment would be more likely to demonstrate higher levels of well-being and independence by engaging more with co-workers to build stronger social connections and developing healthier coping mechanisms in response to stress in the working environment. For workers with major neurocognitive disorders, the short-term and customizable use of SARs such as PARO and mental health coaches such as Misty in the workplace may bring about more positive emotions and reduced potential for developing mental health problems (Spitale et al., 2023). Later studies done by Hung et al. (2019) confirm these views, as the behavioural and psychological symptoms of patients with dementia were observed to be less severe and more controlled when using PARO, resulting in less negative emotions and agitation. Such positive outcomes are, however, strongly dependent on the stage of the disease's progression, as even with the use of SARs, workers who are in an advanced stage of dementia may experience a decline in their quality of life and worsening cognitive changes (Soler et al., 2015).

Despite such converging findings, the benefits of using assistive technologies like SARs in the workplace may also reduce the burden on both caregivers and the healthcare system in Trinidad and Tobago. The use of assistive technologies in the rehabilitation of workers with Dementia and mild forms of cognitive impairment may also improve the quality of life and overall healthcare outcomes for caregivers, as it reduces burnout and chronic stress associated with financial strain due to increased medical expenses (Lee-Cheong et al., 2022). However, medical care professionals have raised the concern that the use of SARs when embedded with AI-based technologies may not only make their jobs obsolete, eroding the doctor-patient relationship, but their continued use may depersonalise and dehumanise healthcare (Koh et al., 2023). This is mainly because the focus on data-driven decision-making and the lack of transparency during this process can weaken dementia patients' trust, as there is less shared decision-making and individualised care.

Soft Robotics

In addition to SARs, soft robotics and soft robotic applications of SARs are also being explored in the rehabilitation and management of neurogenerative diseases. Based on polyfunctional designs, these types of robotics, which are typically composed of components like hydrogels, elastic polymers, self-healing polymers, and elastomeric materials that are delicate and bendable, allow for the easy movement and adaptability of robots in complex environments, where traditional "hard robotics" cannot be used (Whitesides, 2018). Due to the nature of the components themselves, soft robotics is often considered to be more adaptable, reliable, and affordable than traditional robots, because they can be easily made from low-cost materials using less complex manufacturing equipment. As a result of this, the manufacturing costs associated with soft robotics are lower, as production can be easily scaled up or down based on demand, while minimising waste from defects and damages, as the materials used are easily biodegradable (AboZaid et al., 2024).

Bearing in mind the sustainability and affordability of soft robotics, it can be applied in a wide range of areas from healthcare to agriculture. Concentrating on the area of health care, soft robotics can be used in the rehabilitation of persons with neurodegenerative diseases in Trinidad and Tobago because the features of soft robotics provide a greater level of comfort and quality of care. Although the literature on the use of soft robotics in neurodegenerative diseases is limited, its unique features have been found to improve the quality of sleep and well-being of workers with mild cognitive impairments from Dementia, Alzheimer's, and Parkinson's (Biswas et al., 2024). Mazumdar et al. (2025) explain that in specific cases where soft robotics has been used together with AI-based technologies such as the Internet of Things (IoT), Machine Learning (ML) algorithms, and Augmented Reality (AR), for patients with Parkinson's disease, functional materials like Shape-Memory Polymers (SMP) Hydrogel, Conductive Hydrogel, Neurotrophic Factor-Loaded Hydrogels as well as 2D, 3D, and 4D material-integrated hydrogel. This is undertaken by using the materials to deliver therapeutic proteins to encourage the regeneration of neural cells and pathways by mirroring the extracellular network that provides the support needed for cellular regeneration, in addition to sensors and actuators.

Further to this, the process can been facilitated using wearable technology encapsulated within watches, spectacles, shoes, and robotic wheelchairs, as well as innovative textiles such as gloves and soft exosuits to provide support to workers in Trinidad and Tobago who suffer from neurogenerative disorders, in a way that provides comfort, is unobtrusive, and conforms to the natural movement and structure of their body (Havard Biodesign Lab, n.d; Mazumdar et al., 2025). Together with sensors and actuators installed in these wearables, it not only helps with hip flexion and improving gait quality, but also records information on the progression of Parkinson's disease through changes in gait (episodes of tremors and freezing of gait).

This data can then be used to develop predictive models to anticipate changes in symptoms, allowing for the provision of both proactive and adaptive support in real-time. Similarly, the use of smart gloves by workers with dementia can be used to encourage hand function and improve their range of motion, while smart watches can be used to monitor movement, blood glucose, pressure, and sleep quality (Biswas et al., 2023). The application of soft robotics can also be applied to workers who exhibit symptoms of Amyotrophic Lateral Sclerosis (ALS) and Alzheimer's disease. According to Proietti et al. (2023), these soft robotics are designed using lightweight textiles that, when worn by workers, target the functionality of the shoulders, arms, forearms, and hands. These upper limb soft robotics aims to restore arm function by improving workers' fine motor skills, like writing, as well as grasping objects and extending arm movement.

These improvements that workers experience are also an indication that their brain plasticity has improved. In that the consistent use of soft robotics, enables new forms of non-verbal communication that not only engages the sensorimotor pathways, but also causes variations in the brain's ability to adapt to new experiences, and situations in response to both the workers environment and their internal state that encompasses their level of cognition, emotional health, and behavioural patterns (Harrison et al., 2022). Since soft robotics is often designed to mimic the natural structure and movement of the worker's body, such personalisation can facilitate both cognitive and neural plasticity through bioenergic changes (Roto Cataldo et al., 2023). This results in a growth of neurons and synapses (neurogenesis), the formation and reorganisation of neural circuits, and changes in brain structure and function. Thus, allowing workers to regain their motor skills through enhanced neuroplasticity which may positively impact their job performance and productivity.

Barriers & Ethical Considerations

Regardless of the many benefits that robotics promises for workers with neurodegenerative diseases such as Dementia, the fusion of industry 4.0 and 5.0-based technologies with healthcare is not without challenges, as the use of such technologies can encourage the stigmatization and discrimination of workers in Trinidad and Tobago. In this regard, as co-workers may become aware of colleagues' health conditions, they may perceive that the use of such robots is limited to workers with cognitive problems (Gerlowska et al., 2020). Such perceptions can lead to the exclusion of workers from work-related discussions, social gatherings, and opportunities for professional networking, which worsens their feelings of isolation, loneliness, and inadequacy. Considering this problem, workers with neurodegenerative diseases in Trinidad and Tobago may become reluctant to accept the support provided using SARs and Soft Robotics, especially if there is a mismatch between the SARs' function and their needs.

The use of robotics together with Artificial Intelligence (AI) based technologies may also create and exacerbate the presence of dependency syndrome amongst workers with neurodegenerative diseases in Trinidad and Tobago. Given that such robots are deeply integrated into the daily lives of workers with Alzheimer's and Dementia, as they rely heavily on them to perform both basic and complex tasks, it is possible that workers can develop an emotional attachment and emotional dependency on the robots, even as the level of loneliness and depressive symptoms declines (Koutentakis et al., 2020). As a result, workers' ability to independently manage tasks and challenges in the workplace may diminish, leading to a decline in job satisfaction and performance over time.

In addition to these issues, the deployment of robotics for the rehabilitation of workers with neurodegenerative diseases in Trinidad and Tobago also raises ethical concerns. In that, it is possible that workers with cognitive decline may not fully understand how to use the robots, and how the information collected from the monitoring of their daily activities by SARs is used to give proper informed consent (Koh et al., 2021). Thus, when SARs are integrated into the workplace, it is possible that due to their cognitive decline, workers may feel pressured into accepting robotic assistance out of fear of being reprimanded by employers. This deception of care, together with unconditional and non-consensual algorithmic surveillance, reduces not only their privacy but can also lead to negligence of workers with ND in the workplace, which poses a legal and ethical issue.

Apart from this, the use of SARs in the workplace can also contribute to the expansion of distributional injustice. Given the high cost associated with implementing SAR systems, only high-revenue-earning companies or individuals with a high household income may be able to access these technologies (Boada et al., 2021). This may exacerbate existing inequalities and social exclusion, as the prevalence of ND is higher amongst persons in low-income households and disadvantaged groups who may have limited access to healthcare and social support services. In addition to this, based on the type and quality of the data used to train SARs, it is possible that robots that use algorithms can exhibit bias and discriminatory behaviours (Hurtado et al., 2021). This can also lead to the exclusion of vulnerable groups, reducing their trust and acceptance of these technologies. Thus, ethical frameworks for the use of SARs in the workplace must emphasise that the use of these technologies is in a supportive capacity to help persons with ND as a form of ongoing access to health care while they remain active contributors to the labor force. The ethics of technology design for SARs for persons with ND in the workplace, such as being designed in a way that maintains the autonomy, privacy, and safety of the worker, while ensuring accountability and minimizing deception, are all critical to the successful use of such technologies in the workplace (Boada et al., 2021).

In such instances, the implementation of SARs in workplaces in Trinidad and Tobago must follow strict guidelines to not only maintain the safety and privacy of such a vulnerable group of workers with conditions such as dementia, but also their human dignity (Rouaix et al., 2017). One way to ensure that the privacy of workers is maintained would be to use robust security measures that use privacy-enhancing technologies such as Federated Learning, Homomorphic encryption and Secure Multi-Party Computation systems to (1) protect the workers data by making it unreadable during its transfer, (2) remove personal identifying information to reduce tracking, and (3) to restrict access to only authorized users to prevent modification (Sylvester et al., 2023).

To do so, the use of robotics for the rehabilitation of workers with neurodegenerative diseases by employers in Trinidad and Tobago must adhere to national and international legal frameworks, meet essential occupational health and safety requirements, and address issues of strict liability (De Micco et al., 2024). In the case of the latter, strict liability is essential as any disruption in the programming of robots, and data storage and processing errors of the information collected, can have severe consequences for the employer and manufacturer of such robots, who can be held liable for any harm to the worker resulting from the risk of using the robots. As a result, employers may be less willing to invest in the provision of SARs and Soft Robotics to support the needs of workers with neurodegenerative diseases in Trinidad and Tobago (Koh et al., 2021). This is of great importance, as the integrated use of such robotics, whose tasks may be heavily automated, may adversely impact the occupational health of workers with neurodegenerative conditions by creating feelings of inadequacy and displacement, workplace anxiety, which can worsen cognitive symptoms, and further erode cognitive function (Riso and Adăscăli ei, 2024).

CONCLUSION

Neurodegenerative diseases such as Alzheimer's, Parkinson's and Dementia are quickly becoming a health challenge, because as Trinidad and Tobago's population continues to age at a rapid pace, so too does its prevalence. This then places a significant burden on both the patient's carers and the domestic healthcare system to treat and manage the symptoms of these conditions as their cognitive function declines. Regardless of when patients become symptomatic, the presence of such diseases implies that their brain function has been compromised due to damage to the neural networks, leading to synaptic dysfunction. Such gradual and irreversible damage, together with structural changes to the brain, reduces the neuroplasticity of the patient's brain and their subsequent ability to adapt and change to new situations, experiences and stimuli.

As the disease progresses, the inability of the patient's brain to maintain neuroplastic responses often diminishes, and depending on when the symptoms begin to manifest, it can have a marked impact on their performance in the workplace. However, in developing countries like Trinidad and Tobago, there is a dearth of knowledge on not only the prevalence of neurodegenerative disease within the workplace but also strategies to effectively address this growing issue. Such an absence of information reflects bigger problems afflicting the field of neurology in the Caribbean, as research institutions may suffer from a lack of research infrastructure, limited expertise, insufficient funding and a perpetual underrepresentation in global datasets on neurodegenerative conditions.

Considering this research gap, this study examines the labor market implications of neurodegenerative diseases and the strategies that can be implemented to support such workers. In the case of Trinidad and Tobago, there are several interconnected labor market implications that were found to impact the workers' performance. These repercussions included adverse impacts on the productivity and efficiency of both the worker and the overall business due to early retirement and frequent sick leaves. This may have led to higher levels of employee turnover and absenteeism, which may have directly contributed to higher business operations costs, as employers may now have to hire and train new workers, as well as pay higher overtime costs to existing workers. The additional responsibilities assigned to existing workers may also breed resentment against workers with neurodegenerative diseases in Trinidad and Tobago. As the prevalence of workplace deviance grows, it may reduce the competitiveness and innovativeness of both the workers and the business.

While the transformation of workplaces in Trinidad and Tobago may come at a cost to employers to accommodate the needs of workers with neurodegenerative diseases, the integration of such changes may reduce the intensity of the previously mentioned implications. Physical modifications to the workplace, amendments to workplace guidelines and legislation, together with the integration of industry 4.0 and 5.0 technologies such as Socially Assistive Robotics and Soft Robotics, have the potential to revolutionise workplaces in Trinidad and Tobago, not only for workers with neurodegenerative diseases but the entire business environment. The use of SARs and Soft Robotics by workers with neurodegenerative diseases in Trinidad and Tobago helps workers to remain an active part of their working environment, as it not only provides them with physical and cognitive support but also helps them to maintain their autonomy and improve their mental health.

The provision of such robotics within the workplace, while reducing the burden on both caregivers and the healthcare system, is also likely to have a positive effect on the workplace dynamics and the society at large in Trinidad and Tobago, as the use of such robotics counters the negative perceptions associated with cognitive decline. As workers increase their communication and engagement using such robotics, there will also likely be a decline in the stigma and discrimination against this vulnerable group of workers. This shift is likely to be facilitated by a change in perception, in which co-workers become more empathetic, focusing on the positive contributions of colleagues. This then demonstrates that the working environment in Trinidad and Tobago, aided using robotics, is becoming more inclusive towards workers with neurodegenerative diseases.

Competing Interests:

The author of this publication declares there are no competing interests.

Funding:

This research received no specific grant from any funding agency in the public, commercial, or not-for-profit sectors. Funding for this research was covered by the author of the article.

REFERENCES

Abdollahi, H., Mahoor, M. H., Zandie, R., Siewierski, J., & Qualls, S. H. (2023). Artificial Emotional Intelligence in Socially Assistive Robots for Older Adults: A Pilot Study. *IEEE Transactions on Affective Computing*, *14*(3), 2020–2032. DOI: 10.1109/taffc.2022.3143803 PMID: 37840968

AboZaid, Y. A., Aboelrayat, M. T., Fahim, I. S., & Radwan, A. G. (2024). Soft robotic grippers: A review on technologies, materials, and applications. *Sensors and Actuators. A, Physical*, *372*, 115380. DOI: 10.1016/j.sna.2024.115380

Achler, S., Kocherginsky, M., Bega, D., Heinemann, A., Johnson, K., & Rafferty, M. (2022). Employment Challenges in Parkinson's Disease are Associated with Financial Hardship and Reduced Quality of Life. *Archives of Physical Medicine and Rehabilitation*, *103*(12), e184. DOI: 10.1016/j.apmr.2022.08.934

Acosta, D., & Llibre-Guerra, J. J., & Jimenez-Velazquez. (2021). Dementia Research in the Caribbean Hispanic Islands: Present Findings and Future Trends. *Frontiers in Public Health*, *8*, 1–7. DOI: 10.3389/fpubh.2020.611998 PMID: 33537283

. Acosta, I., Salgado, A. M. R., Kim, D. J., Zitser, J., Sosa, A. L., Acosta, D., ... & 1066 Dementia Research Group. (2024). 3-Prevalence and impact of neuropsychiatric symptoms in normal ageing and neurodegenerative syndromes: A population-based study from 6 Latin America centres. *International Psychogeriatrics*, 36(S1), 19-20. https://doi.org/DOI: 10.1017/S1041610224001091

Adkins-Jackson, P. B., George, K. M., Besser, L. M., Hyun, J., Lamar, M., Hill-Jarrett, T. G., Bubu, O. M., Flatt, J. D., Heyn, P. C., Cicero, E. C., Zarina Kraal, A., Pushpalata Zanwar, P., Peterson, R., Kim, B., Turner, R. W., Viswanathan, J., Kulick, E. R., Zuelsdorff, M., Stites, S. D., & Babulal, G. (2023). The structural and social determinants of Alzheimer's disease related dementias. *Alzheimer's & Dementia*, *19*(7), 3171–3185. DOI: 10.1002/alz.13027 PMID: 37074203

Ahmad, M. I., Mubin, O., & Orlando, J. (2017). A Systematic Review of Adaptivity in Human-Robot Interaction. *Multimodal Technologies and Interaction*, *1*(3), 1–25.

Alzheimer's Society of Canada (ASC). (2025). The Canadian Charter of Rights for People with Dementia. https://alzheimer.ca/en/get-involved/change-minds/canadian-charter-rights-people-dementia

Andel, R., Crowe, M., Pedersen, N. L., Mortimer, J., Crimmins, E., Johansson, B., & Gatz, M. (2005). Complexity of Work and Risk of Alzheimer's Disease: A Population-Based Study of Swedish Twins. *The Journals of Gerontology. Series B, Psychological Sciences and Social Sciences*, *60B*(5), 251–P258. PMID: 16131619

Ashby-Mitchell, K., Burns, R., & Anstey, K. (2018). The proportion of dementia attributable to common modifiable lifestyle factors in Barbados. *Revista Panamericana de Salud Pública = Pan American Journal of Public Health*, *42*(e17), 1–7. DOI: 10.26633/RPSP.2018.17 PMID: 31093046

Ashby-Mitchell, K., Willie-Tyndale, D., & Eldemire-Shearer, D. (2020). Proportion of Dementia Explained by Five Key Factors in Jamaica. *Journal of Alzheimer's Disease*, *78*(2), 603–609. DOI: 10.3233/JAD-200601 PMID: 33016910

Bak, K., Kokorelias, K. M., Nygard, L., Issakainen, M., Maki-Petaja-Leinonon, A., Nedlund, A.-C., Ryd, C., & Astell, A. J. (2023). Workplace accommodations for individuals with dementia: Experiences from three countries. *Alzheimer's & Dementia*, *19*(19), e080696. DOI: 10.1002/alz.080696

Besser, L. M., Edwards, K., Lobban, N.-S., Tolea, M. I., & Galvin, J. E. (2024). Social Determinants of Health, Risk and Resilience against Alzheimer's Disease and Related Dementias: The Healthy Brain Initiative. *Journal of Alzheimer's Disease Reports*, *8*(1), 637–646. DOI: 10.3233/adr-230155 PMID: 38746641

Bevilacqua, R., Maranesi, E., Benadduci, M., Cortellessa, G., Umbrico, A., Fracasso, F., Melone, G., Margaritini, A., La Forgia, A., Di Bitonto, P., Potenza, A., Fiorini, L., La Viola, C., Cavallo, F., Leone, A., Caroppo, A., Rescio, G., Marzorati, M., Cesta, A., & Rossi, L. (2025). Exploring Dance as a Therapeutic Approach for Parkinson Disease Through the Social Robotics for Active and Healthy Ageing (SI-Robotics): Results from a Technical Feasibility Study. *JMIR Aging*, *8*(e62930), 1–13. DOI: 10.2196/62930 PMID: 39810317

Biswas, B., Coulhard, E., & Roudaut, A. (2024). *Soft Robotics and Neurodegenerative Disease: Enhancing Sleep and Mental Wellbeing in Dementia Patients through Innovative Care Solutions.* In Companion of the 2024 Computer-Supported Cooperative Work and Social Computing (CSCW Companion '24), San Jose, Costa Rica, pp. 162-167. https://doi.org/DOI: 10.1145/3678884.3685937

Biswas, N., Chakrabarti, S., Jones, L. D., & Ashili, S. (2023). Smart wearables addressing gait disorders: A review. *Materials Today. Communications*, *35*(6), 106250. DOI: 10.1016/j.mtcomm.2023.106250

Bleecker, M. L., & Barnes, S. K. (2015). Chapter 28-Accommodation of workers with chronic neurologic disorders. *Handbook of Clinical Neurology*, *131*, 505–516. DOI: 10.1016/B978-0-444-62627-1.00028-7 PMID: 26563806

Boada, J. P., Maestre, B. R., & Genis, C. T. (2021). The ethical issues of social assistive robotics: A critical literature review. *Technology in Society*, *67*(101726), 1–13. DOI: 10.1016/j.techsoc.2021.101726

Bolger, E., Egdell, V., & Ritchie, L. (2023). Dementia in the workplace: Implications for career development practice. *British Journal of Guidance & Counselling*, *51*(1), 74–83. DOI: 10.1080/03069885.2021.1902471

Bonvicini, F., Vinceti, M., Fabbi, S., Teggi, S., Poli, M., & Zanichelli, P. (2009). Residential Exposure to Electromagnetic Fields and Risk of Amyotrophic Lateral Sclerosis in region Emilia, Italy. *Epidemiology (Cambridge, Mass.)*, *20*(6), S196. DOI: 10.1097/01.ede.0000362658.53867.c9

Carda, S., Invernizzi, M., Baricich, A., Comi, C., Croquelois, A., & Cisari, C. (2012). Robotic Gait Training is not Superior to Conventional Treadmill Training in Parkinson Disease: A Single-Blind Randomized Controlled Trial. *Neurorehabilitation and Neural Repair*, *26*(9), 1027–1034. PMID: 22623206

Choi, Y. H., Park, H. J., & Choi, S.-J. (2018). Impact of Emotional Harassment on Firm Value. *Frontiers in Psychology*, *9*, 1–15. DOI: 10.3389/fpsyg.2018.02333 PMID: 30564166

Cichy, K. E., Li, J., McMahon, B. T., & Rumrill, P. D. (2015). The workplace discrimination experiences of older workers with disabilities: Results from the national EEOC ADA research project. *Journal of Vocational Rehabilitation*, *43*, 137–148. DOI: 10.3233/JVR-150763

Collett, J., Brusco, N., Cordell, N., Cockroft, A., Lawrie, S., Coe, S., Reed, A., & Dawes, H. (2022). Lost employment potential and supporting people with Parkinson's to stay in work: Insights from a Pan European Cross-Sectional Survey. *Disability and Rehabilitation*, *45*(5), 832–839. PMID: 35249423

Cox, C. B., & Pardasani, M. (2013). Alzheimer's in the Workplace: A Challenge for Social Work. *Journal of Gerontological Social Work*, •••, 1–14. DOI: 10.1080/01634372.2013.821693 PMID: 24007202

Cramer, S. C., Sur, M., Dobkin, B. H., Sanger, T. D., Trojanowski, J. Q., Rumsey, J. M., Hicks, R., Cameron, J., Chen, D., Chen, W. G., Cohen, L. G., DeCharms, C., Duffy, C. J., Eden, G. F., Fetz, E. E., Filart, R., Freund, M., Grant, S. J., Haber, S., & Vinogradov, S. (2011). Harnessing neuroplasticity for clinical applications. *Brain*, *134*(6), 1591–1609. DOI: 10.1093/brain/awr039 PMID: 21482550

Crooks, S., Mitchell, G., Wynne, L., & Carter, G. (2025). Exploring the stigma experienced by people affected by Parkinson's disease: A systematic review. *BMC Public Health*, *25*(25), 1–13. DOI: 10.1186/s12889-024-21236-8 PMID: 39754132

Custodio, N., Allegri, R., Lopera, F., & Carmelli, P. (2024). Need to adapt Alzheimer's disease criteria in Latin America. *Alzheimer's & Dementia, 20*(11), 8206–8208. DOI: 10.1002/alz.14289 PMID: 39369281

Damsgaard, L., Pedersen, J., Laursen, T. M., Nabe-Nielsen, K., Hendriks, S., Jensen-Dahm, C., Waldemar, G., & Janbek, J. (2024). Patterns of sick leave and unemployment before a diagnosis of young-onset Alzheimer's disease. *Alzheimer's & Dementia, 21*(e14607), 1–10. DOI: 10.1002/alz.14607 PMID: 39988964

Davis, G., Baboolal, N., McRae, A., & Stewart, R. (2018). Dementia prevalence in a population at high vascular risk: The Trinidad national survey of aging and cognition. *BMJ Open, 8*(e018288). Advance online publication. DOI: 10.1136/bmjopen-2017-018288 PMID: 29472257

Davis, G., Baboolal, N., Tripathi, V., & Stewart, R. (2019). Health status and risk factors for dementia in the middle-aged in Trinidad: Findings from a nationally representative survey. *Caribbean Medical Journal*, •••, 29–32.

Davis, G., Baboolal, N., Tripathi, V., & Stewart, R. (2021). Health status risk factors and quality of life in 75–84-year-old individuals accessed for dementia using the short 10/66 dementia diagnostic schedule. *PeerJ, 9*(e12040), 1–21. DOI: 10.7717/peerj.12040 PMID: 34466297

. Davis, G. K., Baboolal, N., McRae, A., & Stewart, R. (2016). P1-379: Dementia Prevalence in The Oldest Old in Trinidad and Medical Comorbidities. *Alzheimer's & Dementia*, 12(7S): P5770P577. https://doi.org/DOI: 10.1016/j.jalz.2016.06.1131

Davis, G. K., Baboolal, N. S., Seales, D., Ramchandani, J., McKell, S., & McRae, A. (2007). Potential biomarkers for dementia in Trinidad and Tobago. *Neuroscience Letters, 424*, 27–30. DOI: 10.1016/j.neulet.2007.07.011 PMID: 17703882

Dosso, J. A., Bandari, E., Malhotra, A., Guerra, G. K., Hoey, J., Michaud, F., Prescott, T. J., & Robillard, J. M. (2022). User perspectives on emotionally aligned social robots for older adults and persons living with dementia. *Journal of Rehabilitation and Assistive Technologies Engineering, 9*, 1–15. DOI: 10.1177/20556683221108364 PMID: 35782883

Du Plessis, M. R., Smeekens, S., Cillessen, A. H. N., Whittle, S., & Gurolgu, B. (2019). Bullying the Brain? Longitudinal Links Between Childhood Peer Victimization, Cortisol, and Adolescent Brain Structure. *Frontiers in Psychology, 10*(9), 1–9. DOI: 10.3390/jcm7080200 PMID: 30692951

Edahiro, A., Okamura, T., Arai, T., Ikeuchi, T., Ikeda, M., Utsumi, K., Ota, H., Kakuma, T., Kawakatsu, S., Konagaya, Y., Suzuki, K., Tanimukai, S., Miyanaga, K., & Awata, S. (2023). Initial symptoms of early-onset dementia in Japan: Nationwide survey. *Psychogeriatrics*, *23*(3), 422–433. DOI: 10.1111/psyg.12949 PMID: 36814116

Egdell, V., Stavert, J., & McGregor, R. (2018). The legal implications of dementia in the workplace: Establishing a cross-disciplinary research agenda. *Ageing and Society*, *38*, 2181–2196. DOI: 10.1017/S0144686X17000642

Equal Opportunity Commission of Trinidad and Tobago (EOCTT). (2018). *Guidelines for Employers on Disability in Trinidad and Tobago*. EOC.

Espinoza Jeraldo, R. I., Zahibi, S., Miranda-Castillo, C., Marshall, C. R., & Cooper, C. (2024). Lifestyle interventions for dementia prevention in Low-and Middle-income countries: A systematic review. *MedRxiv*, 1-36. https://doi.org/DOI: 10.1101/2024.03.01.24303612

EU-Joint Programme-Neurodegenerative Disease Research (JPND). (2023). Alignment of National Strategies. https://neurodegenerationresearch.eu/initiatives/jpnd-alignment-actions/alignment-of-national-strategies/

Evans, D. (2016). An exploration of the impact of younger-onset dementia on employment. *Dementia (London)*, *0*(0), 1–20. DOI: 10.1177/1471301216668661 PMID: 27609937

Feng, Y., Yu, S., Van de Mortel, D., Barakova, E., Hu, J., & Rauterberg, M. (2019). *LiveNature: Ambient display and social robot-facilitated multi-sensory engagement for people with dementia*. In, *Proceedings of the 2019 on Designing Interactive Systems Conference*, 1321-1333. https://doi.org/DOI: 10.1145/3322276.3322331

Garcia, A., & Johann, F. (2025). *ND: Can digital speech testing help reduce global health inequity?* Cologny, Switzerland: World Economic Forum (WEF).

Gerlowska, J., Furtak-Niczyporuk, M., & Rejdak, K. (2020). Robotic assistance for people with dementia: A viable option for the future. *Expert Review of Medical Devices*, •••, 1–12. DOI: 10.1080/17434440.2020.1770592 PMID: 32511027

Ghadimi-Moghadam, A., Mortazvi, S. M. J., Hosseini-Moghadam, A., Haghani, M., Taeb, S., Hosseini, M. A., Rastegariyan, N., Arian, F., Sanipour, L., Aghajari, S., Mortazavi, S. A. R., Soofi, A., & Dizavandi, M. R. (2018). Does Exposure to Static Magnetic Fields Generated by Magnetic Resonance Imaging Scanners Raise Safety Problems for Personnel? *Journal of Biomedical Physics & Engineering*, *8*(3), 333–336. PMID: 30320037

Gibson, A., Coccia, K., & Hasten, L. (2024). The Experiences of Disclosing Alzheimer's Diseases in the Workplace. *Innovation in Aging*, *8*(S1), 1007. DOI: 10.1093/geroni/igae098.3241

. Hansson Mild, K., Mattsson, M-O., Jeschke, P., Israel, M., Ivanova, M., & Shalamanova, T. (2023). Occupational Exposure to Electromagnetic Fields—Different from General Public Exposure and Laboratory Studies. *International Journal of Environmental Research and Public Health*, 20(16): 6552. J. https://doi.org/DOI: 10.3390/ijerph20166552

Harrison, D., Rorot, W., & Laukaityte, U. (2022). Mind the matter: Active matter, soft robotics, and the making of bio-inspired artificial intelligence. *Frontiers in Neurorobotics*, *16*(880724), 1–19. DOI: 10.3389/fnbot.2022.88072 PMID: 36620483

Harvard Biodesign Lab. (n.d.). *Soft Exosuits*. https://biodesign.seas.harvard.edu/soft-exosuits

Hung, L., Liu, C., Woldum, E., Au-Yeung, A., Berndt, A., Wallsworth, C., Horne, N., Gregorio, M., Mann, J., & Chaudhury, H. (2019). The Benefits of and Barriers to Using a Social Robot PARO in Care Settings: A Scoping Review. *BMC Geriatrics*, *19*(232), 1–10. DOI: 10.1186/s12877-019-1244-6 PMID: 31443636

Hurtado, J. V., Londono, L., & Valada, A. (2021). From Learning to Relearning: A Framework for Diminishing Bias in Social Robot Navigation. *Hypothesis and Theory*, *8*(650325), 1–19. DOI: 10.3389/frobt.2021.650325 PMID: 33842558

Jerry, M., Arcona, S., McMorrow, D., Schwartz, H., Princic, N., & Sasance, R. (2023). Work Loss and Direct and Indirect Costs Associated with Parkinson's Disease. *ClinicoEconomics and Outcomes Research*, •••, 309–319. DOI: 10.2147/CEOR.S398509 PMID: 37138588

Joint Programme-Neurodegenerative Disease Research (JPND). (2023). *About JPND*. Paris, France: JPND. https://neurodegenerationresearch.eu/about/

Keci, A., Tani, K., & Xhema, J. (2019). Role of Rehabilitation in Neural Plasticity. *Open Access Macedonian Journal of Medical Sciences*, *7*(9), 1540–1547. DOI: 10.3889/oamjms.2019.295 PMID: 31198470

Klepac, N., Habek, M., Adamec, I., Barusic, A. K., Bach, I., Margetic, E., & Lusic, I. (2013). An update on the management of young-onset Parkinson's disease. *Degenerative Neurological and Neuromuscular Disease*, *5*, 53–62. DOI: 10.2147/DNND.S34251 PMID: 30890879

Koh, W. Q., Felding, S. A., Budak, K. B., Toomey, E., & Casey, D. (2021). Barriers and facilitators to the implementation of social robots for older adults and people with dementia: A scoping review. *BMC Geriatrics*, *21*(351), 1–17. DOI: 10.1186/s12877-021-02277-9 PMID: 34107876

Koh, W. Q., Vandemeulebroucke, T., Gastmans, C., Miranda, R., & Van den Block, L. (2023). The eithics of pet robots in dementia care settings: Care professionals' and organizational leaders' ethical intuitions. *Frontiers in Psychiatry*, *14*(1052889), 1–12. DOI: 10.3389/fpsyt.2023.1052889 PMID: 36756218

Koutentakis, D., Pilozzi, A., & Huang, X. (2020). Designing Socially Assistive Robots for Alzheimer's Disease and Related Dementia Patients and Their Caregivers: Where We Are and Where We Are Headed. *Health Care*, *8*(73), 1–15. DOI: 10.3390/healthcare8020073 PMID: 32225117

Kumai, M., Kumai, K., Nakamura, K., Kinomura, S., & Meguro, K. (2021). Difficulties with Workplace Support for a Patient with Frontotemporal Dementia: A Case report. *International Medical Case Reports Journal*, *14*, 621–625. DOI: 10.2147/IMCRJ.S321244 PMID: 34522144

La Foucade, A., Davis, G., Baboolal, N., Bethelmie, D., Laptiste, C., Ali-Sisbane, H., & Theodore, K. (2020). The socio-economic determinants of multimorbidity among the elderly population in Trinidad and Tobago. *PLoS One*, *15*(9), e0237307. DOI: 10.1371/journal.pone.0237307 PMID: 32915825

Lamptey, R. N., Chaulagain, B., Trivedi, R., Gothwal, A., Layek, B., & Singh, J. (2022). A Review of the Common ND: Current Therapeutic Approaches and the Potential Role of Nanotherapeutics. *International Journal of Molecular Sciences*, *6*(23), 1–18. DOI: 10.3390/ijms23031851 PMID: 35163773

Lee-Cheong, S., Amanullah, S., & Jardine, M. (2022). New assistive technologies in dementia and mild cognitive impairment care: A PubMed review. *Asian Journal of Psychiatry*, *73*, 103135. DOI: 10.1016/j.ajp.2022.103135 PMID: 35569363

Lekoubou, A., Echouffo-Tcheugui, J., & Kengne, A. P. (2014). Epidemiology of ND in Sub-Saharan Africa: A systematic review. *BMC Public Health*, *14*(654), 1–32. DOI: 10.1186/1471-2458-14-653 PMID: 24969686

Li, Y., Liang, N., Effati, M., & Nejat, G. (2022). Dances with Social Robots: A Pilot Study at Long-Term Care. *Robotics (Basel, Switzerland)*, *11*(96), 1–20. DOI: 10.3390/robotics11050096

Lopera, F., Custodio, N., Rico-Restrepo, M., Allegri, R. F., Barrientos, J. D., Batres, E. G., Calandri, I. L., Moscoso, C. C., Caramelli, P., Quiroz, J. C. D., Jansen, A. M., Alvarado, A. J. M., Nitrini, R., Parodi, J. F., Ramos, C., Slachevsky, A., & Brucki, S. M. D. (2023). A task force for diagnosis and treatment of people with Alzheimer's disease in Latin America. *Frontiers in Neurology*, *14*(1198869), 1–17. DOI: 10.3389/fneur.2023.1198869 PMID: 37497015

Mantri, S., Purks, J. L., Kinel, D., Hosamath, A., Allen, A., Amara, A., Anderson, K., Chahine, L. M., Eberly, S., Mathur, S., Standaert, D., Oakes, D., Weintraub, D., Shoulson, I., & Marras, C. (2024). In Their Own Words: Fears Expressed by People with Parkinson's Disease in an Online Symptom Database. *Journal of Parkinson's Disease*, *14*, 865–872. DOI: 10.3233/JPD-230305 PMID: 38669556

Mazumdar, H., Khondakar, K. R., Das, S., & Kaushik, A. (2025). Soft Robotics for Parkinson's Disease Supported by Functional Materials and Artificial Intelligence. *BMEF Frontiers*, *6*(0143), 1–22. DOI: 10.34133/bmef.0143 PMID: 40606522

McRae, A., Baboolal, N., Davis, G., & Ramchandani, J. (2006). P3-082: Risk factors and biomarkers in Alzheimer's disease in Trinidad and Tobago. *Alzheimer's & Dementia*, *2*(3S Part 13), S397–S397. DOI: 10.1016/j.jalz.2006.05.1349

Micco, F. D., Grassi, S., Tomassini, L., Palma, G. D., Ricchezze, G., & Scendoni, R. (2024). Robotics and AI into healthcare from the perspective of European regulation: Who is responsible for medical malpractice? *Frontiers in Medicine*, *11*(1428504), 1–10. DOI: 10.3389/fmed.2024.1428504 PMID: 39309674

Ministry of Justice (MOJ). (2025). National Strategy for Alzheimer's Disease and Other Dementias Act. https://laws-lois.justice.gc.ca/PDF/N-16.75.pdf

Ministry of the Attorney General and Legal Affairs (MAGLA). (2001). *Equal Opportunity Act, chapter 22: 03, Act 69 of 2000*. Port-of-Spain, Trinidad and Tobago: MAGLA.

Murphy, R., Tubridy, N., Kevelighan, H., & O'Riordan, S. (2013). Parkinson's disease: How is employment affected? *Irish Journal of Medical Science*, *182*, 415–419. DOI: 10.1007/s11845-013-0902-5 PMID: 23325501

Nakamura, A. E., Opaleye, D., Tani, G., & Ferri, C. P. (2015). Dementia underdiagnosis in Brazil. *Lancet*, *385*(9966), 418–419. DOI: 10.1016/S0140-6736(15)60153-2 PMID: 25706975

O'Reilly, A. (2007). The right to decent work of persons with disabilities. Geneva, Switzerland: International Labor Organization (ILO). https://www.ilo.org/media/323516/download

Pan American Health Organization (PAHO). (2023). *Dementia in Latin America and the Caribbean: Prevalence, Incidence, Impact and Trends over Time*. PAHO., DOI: 10.37774/9789275126653

Pauly, M. V., Nicholson, S., Xu, J., Polsky, D., Danzon, P. M., Murray, J. F., & Berger, M. L. (2002). A general model of the impact of absenteeism on employers and employees. *Health Economics*, *11*, 221–231. DOI: 10.1002/hec.648 PMID: 11921319

. Proietti, T., Gerez, L., Cole, T., Mendelowitz, S., Nuckols, K., Hohimer, C., Lin, D., Paganoni, S., & Walsh, C. (2023). Restoring arm function with a soft robotic wearable for individuals with amyotrophic lateral sclerosis. *Science Translational Medicine*, 15(681): eadd1504. https://doi.org/DOI: 10.1126/scitranslmed.add1504

Public Health Agency of Canada (PHAC). (2019). A Dementia Strategy for Canada. https://www.canada.ca/content/dam/phac-aspc/images/services/publications/diseases-conditions/dementia-strategy/National%20Dementia%20Strategy_ENG.pdf

Qiu, C., Fratiglioni, L., Karp, A., Winblad, B., & Bellander, T. (2004). Occupational exposure to electromagnetic fields and risk of Alzheimer's disease. *Epidemiology (Cambridge, Mass.)*, *15*(6), 687–694. DOI: 10.1097/01.ede.0000142147.49297.9d PMID: 15475717

Ribero, F., Teixeira-Santos, A. C., Caramelli, P., & Leist, A. K. (2022). Prevalence of dementia in Latin America and Caribbean countries: Systematic review and meta-analyses exploring age, sex, rurality, and education as possible determinants. *Ageing Research Reviews*, *81*(101703), 1–24. DOI: 10.1016/j.arr.2022.101703 PMID: 35931410

Riso, S., & Adăscăli ei, D. (2024). Human–robot interaction: What changes in the workplace? Dublin, Ireland: Eurofound. https://www.eurofound.europa.eu/sites/default/files/2024-07/ef23010en.pdf

Ritchie, L., Egdell, V., Danson, M., Cook, M., Stavert, J., & Tolson, D. (2022). Dementia, Work and Employability: Using the Capability Approach to Understand the Employability Potential for People Living with Dementia. *Work, Employment and Society*, *36*(4), 591–609. DOI: 10.1177/0950017020961929 PMID: 35935451

Ritchie, L., Tolson, D., & Danson, M. (2017). Dementia in the workplace case study research: Understanding the experiences of individuals, colleagues and managers. *Ageing and Society*, *38*(10), 2146–2175. DOI: 10.1017/S0144686X17000563

Robertson, J., Evans, D., & Hornell, T. (2013). Side by Side: A workplace engagement program for people with younger-onset dementia. *Dementia (London)*, *12*(5), 666–674. DOI: 10.1177/1471301212473881 PMID: 24337338

Rodriguez-Salgado, A. M., Acosta, I., Kim, D., Zitser, J., Sosa, A. L., Acosta, D., Jiménez-Velázquez, I. Z., Guerra, M., Salas, A., Valvuerdi, A., Llibre-Guerra, J. C., Jeyachandran, C., López Contreras, R., Hesse, H., Tanner, C., Llibre-Rodríguez, J., Prina, M., & Llibre-Guerra, J. J. (2023). Prevalence and impact of neuropsychiatric symptoms in normal ageing and neurodegenerative syndromes: A population-based study from Latin America. *Alzheimer's & Dementia*, *19*(12), 5730–5741. DOI: 10.1002/alz.13384 PMID: 37427840

Rosano, C., Karikari, T. K., Cvejkus, R., Bellaver, B., Ferreira, P. C. L., Zmuda, J., Wheeler, V., Pascoal, T. A., & Miljkovic, I. (2024). Sex differences in Alzheimer's disease blood biomarkers in a Caribbean population of African ancestry: The Tobago Health Study. *Alzheimer's & Dementia: Translational Research & Clinical Interventions*, *10*(2), 1–9. DOI: 10.1002/trc2.12460 PMID: 38617114

Roto Cataldo, A. V., Collimore, A. N., Spangler, J., Ribeirinha-Braga, L., Hutchinson, K., Wang, Q. M., Thompson, L., & Awad, L. N. (2023). Enhancing Neuroplasticity in the Chronic Phase After Stroke: Effects of a Soft Robotic Exosuit on Training Intensity and Brain-Derived Neurotrophic Factor. *IEEE Open Journal of Engineering in Medicine and Biology*, *4*, 284–291. DOI: 10.1109/OJEMB.2023.3313396 PMID: 38196979

Rouaix, N., Retru-Chavastel, L., Rigaud, A.-S., Monnet, C., Lenoir, H., & Pino, M. (2017). Affective and Engagement Issues in the Conception and Assessment of a Robot-Assisted Psychomotor Therapy for Persons with Dementia. *Frontiers in Psychology*, *8*(950), 1–15. DOI: 10.3389/fpsyg.2017.00950 PMID: 28713296

Saha, R. (2024). The Exciting Frontier of Neuroplasticity: Innovations in Brain Health and Recovery. *Journal of Behavioral and Brain Science*, *15*(3), 47–80. DOI: 10.4236/jbbs.2025.153003

Salin, D. (2003). Ways of Explaining Workplace Bullying: A Review of Enabling, Motivating and Precipitating Structures and Processes in the Work Environment. *Human Relations*, *56*(10), 1213–1232. DOI: 10.1177/00187267035610003

Simao, H., & Guerreiro, T. (2019). Maty: Designing an assistive robot for people with Alzheimer's. In *Extended Abstracts of the 2019 CHI Conference on Human Factors in Computing Systems*. New York, United States: ACM, pp. 1–6. https://doi.org/DOI: 10.1145/3290607.3313016

Smyth, K. A., Fritsch, T., Cook, T. B., McClendon, M. J., Santillan, C. E., & Friedland, R. P. (2004). Worker functions and traits associated with occupations and the development of AD. *Neurology*, *63*(3), 498–503. DOI: 10.1212/01.WNL.0000133007.87028.09 PMID: 15304581

Soler, V. M., Olazaran Rodriguez, J., Mendoza Rebolledo, C., Perez Munoz, A., Rodriguez Perez, I., Osa Ruiz, E., Barrios Sanchez, A., Herrero Cano, V., Carrasco Chillon, L., Felipe Ruiz, S., Lopez Alvarez, J., Leon Salas, B., Canas Plaza, J. M., Martin Rico, F., & Martinez Martin, P. (2015). Social robots in advanced dementia. *Frontiers in Aging Neuroscience*, *7*(133), 1–12. DOI: 10.3389/fnagi.2015.00133 PMID: 26388764

Spitale, M., Axelsson, M., & Gunes, H. (2023). Robotic Mental Well-being Coaches for the Workplace: An In-the-Wild Study on Form. In, *Proceedings of the 2023 ACM/IEEE International Conference on Human-Robot Interaction (HRI '23)*, pp. 301-310. New York, USA, Association for Computing Machinery. https://doi.org/ DOI: 10.1145/3568162.3577003

Steinmetz, J. D., Seeher, K. M., Schiess, N., Nichols, E., Cao, B., Servili, C., & Atalell, K. A. (2024). Global, regional, and national burden of diseases affecting the nervous system, 1990–2021: A systematic analysis for the Global Burden of Disease Study 2021. *Lancet Neurology*, *23*(4), 344–381. DOI: 10.1016/S1474-4422(24)00038-3 PMID: 38493795

Stowell, R. D., Sipe, G. O., Dawes, R. P., Batchelor, H. N., Lordy, K. A., Whitelaw, B. S., Stoessel, M. B., Bidlack, J. M., Brown, E., Sur, M., & Majewska, A. K. (2019). Noradrenergic signaling in the wakeful state inhibits microglial surveillance and synaptic plasticity in the mouse visual cortex. *Nature Neuroscience*, *22*(11), 1782–1792. DOI: 10.1038/s41593-019-0514-0 PMID: 31636451

Subramaniam, S. (2019). Selective Neuronal Death in ND: The Ongoing Mystery. *The Yale Journal of Biology and Medicine*, *92*(4), 695. https://pmc.ncbi.nlm.nih.gov/articles/PMC6913821/ PMID: 31866784

Sylvester, A., Temesgen, E., Etori, N., & Gini, M. (2023). Ethical robot design considerations for individuals suffering from a neurodegenerative disease. Presented in Workshop on, *Geriatronics: AI and Robotics for Health & Well-Being in Older Age*, pp.1-3. Detroit, Michigan, United States.

Thomson, L., Stanyon, M., Dening, T., Heron, R., & Griffiths, A. (2018). Managing employees with dementia: A systematic review. *Occupational Medicine*, *69*(2), 89–98. DOI: 10.1093/occmed/kqy161 PMID: 30481339

United Nations-Department of Economic and Social Affairs Disability (UN-DESA). (n.d.) Article 27-Work and Employment. https://www.un.org/development/desa/disabilities/convention-on-the-rights-of-persons-with-disabilities/article-27-work-and-employment.html

United Nations Human Rights Office of the High Commissioner (UN-OHCHR). (2006). Convention on the Rights of Persons with Disabilities. https://www.ohchr.org/en/instruments-mechanisms/instruments/convention-rights-persons-disabilities

Van Assche, M., Petrovic, M., Cambier, D., Calders, P., Van Gelder, P., Werner, F., & Van de Velde, D. (2024). Socially Assistive Robots in Aged Care: Expectations of Older Adults with MCI in Assisted Living Facilities and Their Caregivers. *International Journal of Social Robotics*, *16*, 687–698. DOI: 10.1007/s12369-024-01115-3

Varrasi, S., Di Nuovo, S., Conti, D., & Di Nuovo, A. (2019). Social Robots as Psychometric Tools for Cognitive Assessment: A Pilot Test. In Ficuciello, F., Ruggiero, F., & Finzi, A. (Eds.), *Human Friendly Robotics. Springer Proceedings in Advanced Robotics* (Vol. 7, pp. 99–112). Springer., DOI: 10.1007/978-3-319-89327-3_8

Vila-Castelar, C., Fox-Fuller, J., Guzman-Velez, E., Schoemaker, D., & Quiroz, Y. T. (2022). A cultural approach to dementia-insights from US Latino and other minoritized groups. *Nature Reviews. Neurology*, *18*(5), 307–314. DOI: 10.1038/s41582-022-00630-z PMID: 35260817

Vishwanath, A., Singh, A., Chua, Y. H. V., Dauwels, J., & Magnenat-Thalmann, N. (2019). Humanoid co-workers: How is it like to work with a robot? In, *Proceedings of the 2019 IEEE International Conference on Robot and human Interactive Communication (RO-MAN)*, pp. 1-6. New Delhi, India. https://doi.org/DOI: 10.1109/RO-MAN46459.2019.8956421

Walsh, J. L., Persky, L. R., & Pinnock, K. (2019). The Effect of High Performing Bullying Behavior on Organizational Performance: A Bullying Management Dilemma. *Global Journal of Business Research*, *13*(1), 71–81.

Wang, S., Jiang, Y., Yang, A., Meng, F., & Zhang, J. (2024). The Expanding Burden of ND: An Unmet Medical and Social Need. *Aging and Disease*, •••, 1–16. DOI: 10.14336/AD.2024.1071 PMID: 39571158

Whitesides, G. M. (2018). Soft Robotics. *Angewandte Chemie International Edition*, *57*(16), 4258–4273. DOI: 10.1002/anie.201800907 PMID: 29517838

World Health Organization (WHO). (2012). *Dementia A Public Health Priority*. WHO.

World Health Organization (WHO). (2017a). *Global action plan on the public health response to dementia 2017-2025*. WHO.

World Health Organization (WHO). (2017b). *WHO Global Dementia Observatory Provisional Country Profile 2017 Trinidad and Tobago*. WHO.

World Health Organization (WHO). (2025a). *Over 1 in 3 people affected by Neurological conditions, the leading cause of illness and disability worldwide*. WHO.

World Health Organization (WHO). (2025b). Aging and Health. https://www.who.int/news-room/fact-sheets/detail/ageing-and-health

Xiang, Y., Vilmenay, K., Poon, A. N., Ayanian, S., Aitken, C. F., & Chan, K. Y. (2021). Systematic Review Estimating the Burden of Dementia in the Latin America and Caribbean Region: A Bayesian Approach. *Frontiers in Neurology, 12*(628520), 1–9. DOI: 10.3389/fneur.2021.628520 PMID: 34393965

Zhou, Z., Shi, M., Liu, M., Gu, J., Tarimo, C. S., Wu, J., & Miao, Y. (2021). Multimorbidity in Hospitalized Patients Admitted to General Practice Departments and Its Implications for the General Practice Healthcare System: A Four-Year Longitudinal Study in China. *Frontiers in Public Health, 9*(760792), 1–10. DOI: 10.3389/fpubh.2021.760792 PMID: 34988048

Chapter 13
Transitioning Into Industry 5.0:
Integrating Neuroplasticity Principles Into Human Capital Development Strategies in Trinidad and Tobago

Roshnie Anita Doon
https://orcid.org/0000-0002-3285-0355
Global Labor Organization (GLO), Essen, Germany

ABSTRACT

Unlike Industry 4.0, which focuses on automation and digitalisation, Industry 5.0 places greater emphasis on the collaborative efforts between humans and machines, which fosters higher levels of productivity and efficiency through innovation, and competitiveness. Industry 5.0 technologies, when combined with neuroplasticity and human capital strategies, have helped developed countries enhance decision-making, improve performance, and streamline operations. There is limited information on neuroplasticity in developing countries like Trinidad and Tobago. This study aims to explore its role in human capital development and strategies to enhance workers' neuroplasticity, ultimately improving performance for both workers and organizations.

INTRODUCTION

Human capital is a concept rooted in the classical history of economic thought and has been recognised in various contexts over time. One of the earliest references can be found in Smith's (1977) work on "The Wealth of Nations," where he explains

DOI: 10.4018/979-8-3373-2292-6.ch013

the composition of fixed capital, where he notes that one of its main components is the skills and abilities that workers acquire throughout their lifetime. Since then, the concept of human capital has been explored by several renowned economists, but the most well-known and accepted views in the education economics literature are those put forward by Mincer (1958). Who explained that human capital may not only reflect the relationship between a worker's earnings and working experience, but also a form of investment through education and training to improve their skills and abilities that may bring about higher earnings in the future.

It is in this context that neuroplasticity, i.e., the ability of the human brain to evolve and adapt to new experiences and environments through changes in how neurons and neural networks communicate, can alter how workers learn and acquire these new skills (Dan, 2019). Such changes in human capital accumulation occur because plasticity directly impacts workers' capacity to rewire and reorganise neural connections within their brain in response to new learning and working environments (Chapman et al., 2019). These new learning activities can include professional education and training programs and repetitive skill-building exercises. As a result, through such training, not only do the cognitive performance (cognitive capital) and strategic thinking of workers improve, but their emotional cognition (emotional capital) and overall life functionality also improve (Vance et al., 2010). This, in turn, can enhance the productivity of workers, as well as encourage both a better work-life balance and mental health outcomes in a rapidly changing working environment.

Within the literature on human capital and neuroplasticity, several scholars like Staneiu (2023) has explored the connection that brain health has to human capital employment, in which plasticity has been a tool to optimise and expand human cognitive performance and mental agility of workers and even students, by encouraging a growth mindset and lifelong learning. As a result, when plasticity is combined with industry 4.0 technologies such as neural networks, particularly deep learning networks, it has been observed to improve the learning outcomes of workers by identifying skill gaps, building better teams, and better managing the workforce, but also predicting future skills demand.

In this context, countries, such as Finland, Singapore, and Japan, have begun to successfully leverage such techniques to enhance their human capital asset base through innovative education, health initiatives, and workforce development programs, while integrating cutting-edge technologies like virtual reality (VR), Augmented Reality (AR), and Gamification (Kumar et al., 2023). Even though organizations within these highly developed countries have exploited neuroplasticity for economic gain, the same cannot be said for developing countries like Trinidad and Tobago, where there is a dearth of knowledge on its application and importance. Such a gap in the literature is likely to exist because of systemic biases in neuroscience research that underrepresent data from socio-economically diverse countries like

Trinidad and Tobago. As a result of this, there is a limited understanding of how these strategies in the workplace are likely to operate under the socio-economic conditions prevalent in developing countries like Trinidad and Tobago. Therefore, using a secondary research methodology, this study aims to conceptually discuss the role of neuroplasticity in human capital development, as well as the strategies that can be employed to enhance the neuroplasticity of workers in Trinidad and Tobago.

A study of this nature is of critical importance for the human capital development of developing countries like Trinidad and Tobago due to its potential to enhance learning, adaptability, and productivity, as well as encourage economic growth and reduce socio-economic inequalities. From a developing country perspective, the implementation of neuroplasticity-based strategies in the workplace using continuous learning strategies, Multi-Modal Learning (MML), Multi-Sensory Training (MST), Augmented Reality (AR), Virtual Reality (VR), and Gamification presents organizations in Trinidad and Tobago with a transformative pathway to enhance their performance and competitiveness. As these advancements may not only improve employee engagement and skill acquisition but also enable organizations to respond effectively to market demands to drive their operational efficiency. Ultimately, this study emphasizes the importance of adopting holistic learning frameworks that leverage neuroplasticity within the context of technological evolution, paving the way for a more resilient workforce.

HUMAN CAPITAL AND ITS IMPORTANCE TO HUMAN DEVELOPMENT

Central to the theories of productivity and growth, the concept of human capital has become a key driver behind some of the most crucial aspects of human development. Having been influenced by several seminal pieces of economics literature from the neoclassical view of human capital that perceives human capital in itself to be a form of a factor of production, to its role in endogenous growth models developed by macroeconomists like Romer (1987), the concept has conventionally become a cornerstone behind technical advancement, labor market productivity, and its ultimate part in economic growth and sustainability of many economies around the world.

Throughout the literature, human capital has been defined in a multitude of ways. As noted by Goldin (2016), it reflects the diverse range of skills that the labor force possesses, which contributes to its overall productivity and innovation. While the World Bank Group (2025a), in their Human Capital Project, seeks to encourage and accelerate the advancement of human capital investments in regions in the Global South, such as Sub-Saharan Africa (SSA), the Middle East and Northern Africa, by helping workers better realise their full potential. This is done through the provision and access of education, training, healthcare and employment, so that the value of human capital base, i.e., the economic value of workers' experience, skills, and other characteristics that potential employers value, can be improved.

The strengthening of the human capital asset base of countries, especially those that are considered to be economically less-developed and which are home to many pockets of vulnerable communities, women, children and the differently-abled, has become essential to not only helping these countries manage against global economic shocks but also preparing and creating a highly skilled and educated workforce. Who can then easily adapt to the technological change and innovations of both industry 4.0 and industry 5.0 (WBG, 2025b). Amid the education and learning crisis that many countries are yet to come to terms with because of the negative fallout of the COVID-19 pandemic, which impacted billions of children worldwide and led to plummeting student achievement. This implies that such losses may have had a profound impact on both the human capital aspects of economic development, as well as those of global sustainability, undermining any hopes of economic recovery.

It is within this context that the seminal work of Romer (1989), on human capital accumulation and lifelong learning by doing becomes a crucial factor in sustaining the long-term viability of economic growth, as the value of intangible assets like education and training relies on the worker himself demonstrated through their knowledge, skills development, abilities, and experience, rather than a physical asset. Thus, it is with this investment in these areas that countries can enhance their economic output and productivity across sectors, as the accumulation of human capital returns on the investments made does not appear to weaken over time but becomes reinforced. Such strengthening in the human capital returns base can be attributed to numerous factors, from the previously mentioned accumulation of skills and experience that lead to compounding productivity benefits when combined with physical assets such as technology and infrastructure, to knowledge creation through technological progress and growth makes it even more valuable over time (Ben-Porath, 1970).

At the national level, the continued development of human capital accumulation has been established as a key driver of economic growth. However, while human capital may reflect a combination of the transmission of intergenerational abilities, i.e., can either be cognitive or non-cognitive skills, through parent to child, and that of acquired or learned abilities, at a national level human capital tends to focus more on the latter of acquired abilities which can be influenced by the workers' access to a high-quality form of education, and socio-economic mobility. While, at the level of the individual worker, Mincer (1996), explains that the concept of human capital is a much more complex phenomenon, which is not only the result of the demand for skilled labor, but a reflection of a much wider range of factors such as household income/stability, parent background/education, as well family demographics such as fertility, marriage, and family size/structure.

In this framework, the mincerian earnings function developed by Mincer (1958), and used widely throughout the field of empirical labor economics, demonstrates that factors such as education/schooling and experience tend to have a significant and multiplying impact on human capital, which allows workers to maximize their future income. As with most theoretical models in economics, while the mincerian earnings function has formed a crucial aspect of the discussion on human capital, the theory isn't without its limitations, as the earnings profiles or life cycle earnings of workers tend to differ across different levels of education, fields of expertise, amount of working experience, and even generations. Furthermore, it can also be influenced by selection bias, and unobservable external factors like social networks, cognitive ability, healthcare, and even peer effects that may not be captured in standard labor market datasets across countries.

Regardless of these criticisms, human capital has become an integral aspect of not only supporting the earnings capacity and job prospects of potential and current workers but also enhancing their resilience and compliance with changes in the labor market, while nurturing a more inclusive form of economic growth that promotes social inclusion and protection (WBG, 2025a). Thus, studying human capital has noteworthy policy implications that can influence the socio-economic development and public sector investment programs of countries worldwide. Human capital policies that focus on the development of workers and the overall labor market, as discussed by Rossi and Weber (2024), should capitalize on strategic investments in human capital made across the life cycle at different stages of life, so human capital accumulation can begin during early childhood and be sustained throughout late adulthood and older age. In doing so, the returns associated with human capital investments during early childhood are more likely first to make education and training more effective during adolescence, and second to encourage the adaptability and upskilling of the workforce. Thus, these changes during the working age are not only less costly and better aligned to the technological transitions of sectors brought upon by Industry 4.0 and 5.0 but also prevent the prevalence of educational mismatch within the labor market.

Further to this, human capital policies need to address the developmental challenges faced by workers to minimize socio-economic inequalities and disparities. By ensuring that their intended use, whether it be for the adoption of technology, accumulation of specialized skills, or economic transition, have indeed contributed to economic growth and development. When achieved, as explained by Roder-Dewan et al. (2019), such implications of human capital policies are likely to manifest far beyond the initial investments in education and training to bring about positive changes in the physical and mental healthcare outcomes that also impact their overall productivity and earnings in the future. This, in turn, can minimize the negative implications of inequality, discrimination, and biasness experienced by vulnerable segments of society, allowing them equal access to opportunities that support their movement to a higher social standing, or intergenerational upward mobility, and strengthening their human capital over time (Rossi & Weber, 2024).

BRAIN FUNCTION AND ADAPTABILITY THROUGH NEUROPLASTICITY

Neuroplasticity

Neuroplasticity, also referred to as either neural or brain plasticity, was first introduced in 1980 by William James to describe the changes in the nervous pathways in response to a person's behavior. The term, however, was later redeveloped by Polish neurophysiologist Jerzy Konorski in 1948 and popularized by Donald Hebb in 1949 to symbolize the ability of the brain through structural and functional changes to modify its structure, function and connection to either internal or external factors (Berlucchi & Butchel, 2009). The presence of these factors, such as neurodegenerative conditions like Dementia, Alzheimer's, and Parkinson's disease, which may lead to degeneration in the neurological pathways, can lead to a wide range of neuroplasticity mechanisms which support the ability of the brain to adapt to become especially impaired. This then diminishes not only the brain's overall ability to function, but also the neural pathways associated with learning and memory, which are crucial to key cognitive processes such as attention, perception, language, motor skills, emotional regulation, and other executive functions. Notwithstanding these negative implications, in cases where there are no neurodegenerative conditions present, there are several other circumstances. These include traumatic brain injuries, developmental disorders, sensory impairments, and mental health conditions where positive outcomes, i.e., either through the recovery/regeneration of function or the creation of new neural pathways, have been observed (Dabrowski et al., 2019).

Mechanisms and Types of Neuroplasticity

Within the neuroscience literature, several mechanisms influence the development of neuronal connections and the plasticity of the brain. For purposes of this study, emphasis will be placed on the two main types of mechanisms, i.e., neuronal regeneration and functional reorganization, and the different types of neuroplasticity associated with each that are used to classify the changes in the brain. According to Mateos-Aparicio and Rodriguez-Moreno (2019), one of the main concepts in neuronal regeneration is synaptic plasticity, which refers to the ability of neurons to change the strength and efficiency of synaptic transmission using different activity-dependent mechanisms.

For this reason, they are crucial for brain development over time, as platicity often involve several different processes such as (1) synaptic plasticity (strengthening of synaptic connections), and (2) structural plasticity that includes (a) synaptic pruning (elimination of inactive synapses), (b) collateral sprouting, synaptogenesis and neurogenesis (birth of new synapses and neurons to compensate for lost connections or enhance existing networks), as well as (c) other biological processes that include the branching and growing of dendrites and axons, transformations in the spine and postsynaptic density in response to stimuli, that all allow for the development and refinement of neural circuits overtime (Krauss, 2024).

Accordingly, these processes are thought to play an integral role in not only shaping the brain during different developmental stages but are the fundamental mechanism that forms the foundation for the formation of memories, together with specific neural circuits during the postpartum period after the first 6-8 weeks after childbirth (Imai, 2025). For this reason, any alterations to these processes can lead to the development of neurodevelopmental disorders such as autism spectrum disorder (ASD), attention-deficit/hyperactivity disorder (ADHD) and intellectual disability formed during the early childhood period can persist into adulthood, leading to the emergence of developmental delays (Antolini & Colizzi, 2023). When such processes are altered in adults, either through aging or the presence of neurodegenerative diseases, it can harm their synaptic plasticity, as the decline in brainstem neuromodulators can contribute to cognitive decline, sleep, and emotional dysregulation.

Another key mechanism is that of functional reorganization, which concerns the ability of the brain to alter its structure and function in response to a wide range of internal/external stimuli. Such changes resulting from new experiences/situations, brain injuries like stroke, physical changes in the size/density of the brain, the learning of new skills and tasks, and neurological changes associated with physical injuries such as amputations, spinal cord and peripheral nerve injuries (Kikkert et al., 2016). Considered to be a form of adaptive change, where in functional reorganization, there are several processes present such as equipotentiality, which occurs when one area of the brain has sustained an immediate injury, the corresponding opposite side can compensate for the lost function. Diaschisis which occurs when there is a loss in brain function in distance or interconnected regions of the brain that is connected to the immediate area of injury, and cross-modal reassignment which occurs when brain functions are controlled by different sensory modalities so that they can now process information from multiple senses, rather than just one sense (Nudo, 2003). These functional changes are primarily observed in the motor cortex and are thought to be linked to functional neuroplasticity because they influence the brain's ability to adapt and reorganize itself in response to injuries, in addition to restoring balance by altering local synaptic strengths (Pascoa dos Santos & Verschure, 2022). Functional plasticity in a broader sense enables the brain to modify biological neural networks, not only through the shifting of functions across different areas of the brain but also by streamlining and strengthening synaptic connections that support the overall adaptability and relearning to enhance the receiving and processing of information after injury (Alexiou, 2024).

Stages of Neuroplasticity

Based on the structural and functional changes occurring within the brain, neuroplasticity goes through 3 different phases to support brain adaptation and recovery after injury. In the immediate phase following an injury, there is cellular death in the affected areas of the brain, leading to the loss of cortical pathways, i.e., pathways that neurons take to transmit information between different cortical regions, due to the death of neurons (Su et al., 2016). As a result, the brain may attempt to maintain its function by utilizing secondary neuronal networks and changing synaptic strength to compensate for lost functions.

Following this, during the subacute phase, within weeks of injury, in an attempt at enabling synaptic plasticity, these cortical pathways begin shifting their neural activity from one where there is a decrease in inhibitory neurons (neurons that act as a break on neuronal activity) to one where there is an increase in excitatory neurons (neurons that promote action and communication between neurons), while neuronal proliferation (generating new neurons), and synaptogenesis (creating new synaptic connections between neurons) occurs (Zotey et al., 2023). In the chronic phase, which occurs several months after an injury, neuronal regeneration increases through structural plasticity. During this time, functional reorganization and plasticity also take place, allowing the brain to adapt and restore lost abilities by transferring functions from injured areas to unaffected areas of the brain (Kreber, 2025).

Factors Influencing Neuroplasticity

Although neuroplasticity can bring about both positive and negative changes to the brain, its natural tendency to do so is often influenced by a multitude of factors such as (1) genetics such as gene expression, and genetic variability, (2) lifestyle features such as physical activity, diet and nutrition, access to stimulating environments, sleep quality, as well as (3) harmful stimuli such as chronic stress, substance abuse, and exposure to environmental hazards (Pickersgill et al., 2022).

Beginning with genetic factors, i.e., gene expression and genetic variability, are all thought to have a significant impact on the ability of the brain to adjust to changes.

As explained by McClung and Nestler (2008), in the case of gene expression, transcription factors such as the cAMP response element-binding protein (CREB), when stimulated at a cellular level, can pave the way for protein stability, which improves through the activation of secondary pathways and DNA binding. It also supports molecular change through the generation of new growth factors such as synaptic connections, neurons, blood vessels, non-neuronal glial cells, structural proteins, and molecules, while improving cerebral blood flow and brain tissue. Such changes are crucial to either the sensitization, or desensitization of interconnected neurons in the brain, as they can eventually influence the processing and transformation of information over time, as well as regulatory and cognitive functions that underlie much positive neuroplasticity (Tau & Peterson, 2010).

In addition, modifications to gene variability have also been observed to impact long-term synaptic changes at the cellular level through long-term potentiation and depression (LTP and LTD) (McClung & Nestler, 2008). Epigenetic changes, i.e., changes in gene activity, and specific genetic variations in proteins such as Brain-derived Neurotrophic Factor (BDNF), and Apolipoprotein E, which are important to the growth and continued existence of neurons, the transport and metabolism of lipids/cholesterol, and the signalling and communication of dopamine receptors and transports, all play an important role in creating positive neuroplasticity (Stewart & Cramer, 2017). This is because such changes influence how well a person responds to neuroplasticity-based intervention used to treat stroke victims, through their learning and recovery, acquisition and refinement of motor skills, and overall cognitive and behavioral changes.

Moreover, interacting with a person's genetic predispositions are non-genetic factors such as lifestyle features and harmful stimuli. Features of a person's lifestyle (physical activity, diet and nutrition, access to stimulating environments, sleep quality) and consistent exposure to harmful stimuli (chronic stress, substance abuse, and environmental hazards). As alluded to by Pickersgill et al., (2022), they all have the potential to change both the structure and function of the brain, and its inclination towards either positive or negative neuroplasticity. This is because having a poor diet (high in calories, saturated fats and processed foods, as well as insufficient intake of fruits, vegetables and whole grains), together with a lack of physical activity (inadequate aerobic and resistance exercise, with extended periods of lying or sitting down) and insufficient sleep (irregular sleep patterns), can significantly diminish not only the expression of plasticity biomarkers such as BDNF but also neurophysiological adaptations, growth factors such as insulin (IGF-1) and vascular endothelial (VEGF) and ones' overall motor and cognitive function (Nanda et al., 2013).

On the account of these adverse lifestyle choices, the presence of chronic stress and the lack of healthy coping mechanisms for cognitive and emotional regulation, substance abuse (drug abuse, smoking and the consumption of alcohol), and exposure to environmental hazards are likely to worsen the resilience of the brain to reorganize and modify its structure and function in response changes in the internal and external environment (Phillips, 2017). This therefore creates the perfect conditions for negative neuroplasticity to thrive, as it reduces the cognitive reserves, i.e., the ability of the brain to withstand age-related decline in cognitive function, as it may no longer be able to form, strengthen and reorganize dendrodendritic synapses which are essential to receiving, integrating, processing and transmitting signals from neurons for various brain functions (Vance et al., 2010).

THE ROLE OF NEUROPLASTICITY FOR ADVANCING HUMAN CAPITAL DEVELOPMENT IN TRINIDAD AND TOBAGO

Bearing in mind the crucial role that neuroplasticity plays in the brain's ability to adapt to changes throughout a worker's lifespan as Industry 5.0 emerges, it is integral to examine the role that it plays in human capital development in Trinidad and Tobago. To briefly reiterate, neuroplasticity enables the nervous system of individuals to adapt to changes in the brain by developing and modifying new skills and abilities through the reinforcement of neural pathways in response to intrinsic stimuli arising from within an individual, or extrinsic stimuli that are external to the individual (Cramer et al., 2011). Such adaptability is critical for fostering sustainability and resilience in the workplace, especially as the labor market in Trinidad and Tobago evolves due to the presence of digital technologies such as Artificial Intelligence (AI), Edge Computing, and Human-machine interactions.

Industry 5.0 in the Workplace

Given that industry 5.0 technologies have sought to focus on strengthening sustainability and resiliency using a more human-machine-centric method, several key technologies have been developed, including collaborative robots (cobots), human-machine interface (HMI) technologies, green technologies, sixth-generation mobile technology (6G), and digital twins (Ghobakhloo et al., 2024). The main technological doctrines guiding the development of Industry 5.0 machinery include the identification, collaboration, processing of data analytics, coordination, automation and communication between humans and machines. These doctrines are fundamental not only to the organization and management of the workplace at different levels from entry to executive positions but also play a key role in forming/maintaining communities based on shared interests, driving innovation between stakeholders, and evaluating the performance/productivity of workers.

Using the existing Industry 4.0 technologies such as Artificial Intelligence (AI), Machine Learning (ML), Internet of Things (IoT), Block Chain, Cloud Computing, and Robotics and Automation, employers have begun using Industry 5.0 technologies to merge the human element of intuition into digitized processes to create a stimulating working environment that attracts skilled workers, supports the growth and development of advanced technical skills/abilities and jobs that necessitate them, and to enhance creativity/personalization of products based on customer demand, while encouraging critical thinking, improved job satisfaction and employment retention (Alves et al., 2023; Ghobakhloo et al., 2024).

Most importantly, by taking a human-machine-centric approach, Industry 5.0 seeks to create a more personalized form of support and experience for workers, who can then tailor the technology to meet their specific needs/abilities and preferences (Alves, 2023). In doing so, workers will be better able to maximize not only their productive efficiency and performance, safety, precision and training, but the integrated use of industry 5.0 technologies will be able to create a workplace that is more inclusive, sustainable and resilient. This then forms the basis for Industry 5.0 as the emphasis is placed on creating a sustainable society that acknowledges the merits of value creation and the well-being of workers. Thus, to enhance sustainability and resilience, Industry 5.0 can encourage workplaces to adopt sustainable production processes that have a minimal impact on the environment, use ethical, transparent industrial practices, and create resilient systems that can adapt quickly to changes in the production process.

For example, the use of blockchain-enabled smart contracts and decentralized applications in the manufacturing industry that are not only tamper-proof but support the transparency, autonomy, traceability and verifiability in supply chains, which encourages sustainability and confidence in the technology used. Another example is the use of cobots, which support sustainability and resilience through a blended approach of human-machine precision using both human creativity and cognition with machinery/robots handling repetitive, often labour-intensive tasks, to allow for greater customization, personalization of individual products using human touch in the manufacturing process (Hibbert, 2020). Even though these technologies are equipped with sensors, AI and edge computing that allow them to better learn and collaborate with workers to improve collaboration and efficiency, De Giovanni (2023) explains that it is important to note that with such digital transformation, there are several negative implications. These include the high consumption of energy, the loss of jobs, disruptions to the supply chain, and data security and privacy issues, which need to be addressed simultaneously with this digital transition.

Despite these negative implications, Industry 5.0 still plays a significant role in advancing the integration of neuroplasticity within the workplace. Within the workplace industry 5.0 based technologies fosters an environment conducive to the continuous learning and development for employees. It frees workers from repetitive tasks to focus on problem-solving and creativity, it is personalized to each worker based on their needs, and it encourages meaningful work that focuses more on creativity and sustainability. By integrating advanced technologies and human-centred approaches, Industry 5.0 not only enhances productivity but also enriches the cognitive and emotional capabilities of workers.

Industry 5.0, Neuroplasticity and Human Capital Development in Trinidad and Tobago

Notwithstanding the importance of Industry 5.0 to sustainability and resilience, it also plays an important part in neuroplasticity, in that in the workplace, it can be used to support the learning, resilience and adaptability of workers in response to changing environments and situations within the industry. In that, the technologies embedded within Industry 5.0 can be used to restructure how industries particularly those in the areas of manufacturing, healthcare, education and agriculture in Trinidad and Tobago approach (1) human resource management and development in areas such as talent management, recruitment and onboarding, personalized compensation and benefits based on performance and understanding worker performance and efficiency, (2) education and training of current and potential workers to boost their learning, motivation and emotional well-being, (3) the establishment of smart hospitals, remote monitoring of patients in real-time, and the development of personalized treatment plans, and finally (4) the integration of circular manufacturing practices that focuses on optimizing resources, reducing waste, and solidifying human-machine collaboration within the manufacturing process (Erro-Garces & Aramendia-Muneta, 2023; Roy & Ateeq, 2024).

For these reasons, human capital in Trinidad and Tobago can be enhanced by addressing learning poverty through compulsory education at the secondary school level, promoting high university enrolment rates, and investing in healthcare and talent management. These efforts are crucial for supporting economic growth and the development of the labor market. Thus, with this in mind, the main areas in which it is likely to play a significant role in creating booming human-centric industries in Trinidad and Tobago that prioritize the well-being of workers and incorporate human values while integrating technology, and will be examined briefly in this study are (1) Continuous Learning and Development, (2) Reskilling and Upskilling & Individual Growth, (4) Human-Machine Connection, and (4) Mental and Emotional Health.

Continuous Learning and Development

The ability of workers to learn and develop new skills and abilities is a crucial feature needed to meet the technological demands of Industry 5.0. The concept of neuroplasticity has become equally as important because besides being the foundation of learning, it also facilitates the adaptation and strengthening of neural pathways that may allow workers in Trinidad and Tobago to acquire, improve, and expand new skills and competencies to adapt to the new technologies being integrated into the labor market (Gamberini & Pluchion, 2024). Such an expansion through continuous learning and adaptability implies that it may become one of the most vital cornerstones of lifelong learning, which, when applied may help workers in Trinidad and Tobago to comply and work effectively with Industry 5.0 changes. Where such changes include the emergence of new sectors, changing job requirements, the shifting to soft skills, and the focusing on ethical labor market practices and sustainability.

Moreover, by employing the benefits of Industry 5.0 together with that of neuroplasticity, industries in Trinidad and Tobago can now foster greater levels of creativity, and problem-solving amongst their workers using continuous lifelong learning, as they can design more effective continuous learning and adaptive training programs to create a better learning experience for their workers (Gamberini & Pluchion 2024). In doing so, workers in Trinidad and Tobago may be better able to maintain their competitive edge as continuous learning and skill development as lifelong learning together with a growth mindset can help workers maintain updated competencies in the workplace, sustain key cognitive function and improve brain health, which may reduce the prevalence of neurodegenerative conditions later in life (Staneiu, 2023). This is essential as it enhances cognitive skills such as attention (sustained, selective and divided), memory (short and long-term memory), executive functions (short-term memory, cognitive flexibility), auditory and visual processing, perception, as well as logic and reasoning. These skills are all needed to maintain a high level of performance and productivity in the workplace for effective problem-solving, decision-making, leadership, and critical thinking, which may help workers become more adaptive and resilient to rapid changes in the working environment.

Reskilling/Upskilling and Individual Growth

As Industry 4.0 and 5.0 technologies continue to evolve in response to the rapid growth of digital and data-driven interconnected systems used for predictive maintenance, there is an increasing focus on developing more resource-efficient solutions that are energy efficient, reduce waste and support the move towards a more circular economy in Trinidad and Tobago (Rame et al., 2024). Thus, with the production of these solutions organizations in Trinidad and Tobago are more likely to not only monitor their carbon footprint, creating a manufacturing process that is both efficient and flexible, but also produce high-quality innovative products that align with consumer preferences for personalized items.

Hence, such a strategy may also enhance the competitive advantage of domestic organizations, and as a result, may require workers in Trinidad and Tobago to participate in reskilling and upskilling training opportunities. This is where the concept of neuroplasticity becomes imperative to workers in Trinidad and Tobago, as it may help them to adapt to new technologies to meet the changing demands of their roles on the job, as well as make the transition between new roles, tasks and industries (Nardo et al., 2020). This, in turn, allows the worker to embrace a continuous learning mindset that encourages not only knowledge sharing, but the development of a diverse range of capabilities, abilities, and skills that may all be essential for career longevity while promoting feelings of strength, confidence, and control to manoeuvre the trials and openings presented by Industry 5.0 (Alves et al., 2023). Thus, the advent of Industry 5.0 and the integration of enriching positive neuroplasticity further allows workers to experience heightened levels of personal growth, development and fulfilment. This can then strengthen their ability to adapt to new and challenging situations throughout their lifetime and make more meaningful contributions to their communities.

Human-Machine Connection

Given that Industry 5.0 is founded upon the collaborative efforts between human-machine/robots, an understanding of neuroplasticity can strengthen these interactions by bridging the minds and the machines used in the industries across Trinidad and Tobago. To do this it can be used to support the learning experience of workers by (1) optimizing cognitive skills making their attention and decision-making skills more efficient, (2) using targeted plasticity training that is personalized to help workers to easily adapt to change in the labor market, making it easier to work with human-machine interface (HMI) technologies and allowing them to make more meaningful decisions, and finally (3) creating user interfaces that are more intuitive and has a smaller cognitive load that requires minimal mental energy, which then allows for effective learning and emotional regulation (Rueda-Castro et al., 2024; Ghobakhloo et al., 2023; Gamberini et al., 2024). The use of these techniques in a hybrid human-machine environment may then advance the creativity and innovativeness of workers in Trinidad and Tobago using AI and automation, as well as their overall productivity and well-being. For industries to transition into a more human-machine-centric system, there is a need for organizations to go beyond the human-machine collaboration to integrate empathy and coevolution.

Mental and Emotional Health

Finally, in consideration of the human-centric approach that Industry 5.0 is built upon, organizations are also likely to benefit as it may assist employers in better understanding how the mechanisms, and most importantly, the factors influence the performance of the worker in Trinidad and Tobago. In instances, where workers exhibit high signs of psychological distress, poor coping mechanisms, destructive tendencies, and a reduced sense of belonging in response to challenging situations or stressful events within the workplace, an understanding of plasticity can be used to inform strategies to improve mental health outcomes in the workplace, which can reduce employee turnover and burnout, while improving retention in the long-term (Foster et al., 2024).

This is of particular significance because neuroplasticity often influences the development of socio-emotional or soft skills such as communication and leadership skills, emotional and interpersonal skills, management and organizational skills, as well as teamwork and problem-solving skills, which are becoming increasingly important in Industry 5.0 as these are the skills that set workers apart from machine/robots and which cannot be replaced by the latter (Castrillo, 2024). With the coming of Industry 5.0, as workplaces in Trinidad and Tobago become more automated and technologically advanced to boost their productivity and efficiency the demand for soft skills is likely to grow because it (1) supports the collaboration between humans and machines, allowing for greater productivity and efficiency, (2) empowers workers to adapt and become more resilient to change, and (3) enables organizations to design and implement programs to target stress, improve performance and well-being (Polakova et al., 2023).

NEUROPLASTIC STRATEGIES AS A CATALYST FOR TRANSFORMATIVE HUMAN CAPITAL DEVELOPMENT IN TRINIDAD AND TOBAGO

Given that neuroplasticity plays such a fundamental role in the development of human capital across industries and sectors, by supporting the learning and development of workers to adapt to the digital and interconnected industrial landscape that is now Industry 5.0, organizations in Trinidad and Tobago must begin to leverage the benefits of neuroplasticity-based interventions and principles in the workplace. These principles according to Kleim & Jones (2008), include the concepts of (1) interference, where the learning of new activities can disrupt already established patterns of knowledge, (2) transference and time, where the skills and techniques from learning of new activities can be transferred and positively influence the learning and performance in other existing activities if implemented early, (3) age, repetition, salience, where neuroplasticity based interventions are designed based on not only the workers' age and ability to adapt and engage in repetitive practices to strengthen synaptic connections but also its importance, and (4) specificity, where the approach to learning new activities, must be specific and targeted to not only improve the functioning of specific neural circuits but reduce their weakening over time, so to bring about the desired outcome, are important in creating and sustaining a dynamic and vibrant workforce.

In essence, whether it is through informal or formal training, neuroplasticity-based interventions can significantly benefit organizations in Trinidad and Tobago, as well as their workers. These interventions can enhance workers' cognitive performance, decision-making, critical thinking and leadership skills, and nurture their emotional intelligence, creativity, and innovation, which are all needed to develop effective communication skills when navigating diversity in Trinidad and Tobago's workforce (Khaneja & Arora, 2024) In doing so, organizations in Trinidad and Tobago can gain a competitive edge when it comes to differentiating themselves from the competition, making them more attractive based on their unique approach to achieving long-term success. Thus, for this study focus will be placed on specific neuroplasticity-based strategies such as (1) Continuous Learning Techniques, (2) Multi-Modal Learning (MML) and Multi-Sensory Training, (3) Virtual and Augmented Reality, and finally (4) Gamification, their benefits and how they can be used to expand the human capital development of Trinidad and Tobago's workforce. However, before exploring these areas, it is integral that organizations have ethical guidelines and frameworks in place to guide the implementation of these neuroplasticity-based strategies. The need for such a protocol stem from the concerns about privacy, data security, and informed consent when such strategies are used, which often collect sensitive information from the workers. This can become a problem if such sensitive information is used by employers to manipulate workers, if strategies are used to manipulate workers into making decisions that are not in their best interest.

Continuous Learning Techniques

As discussed in the previous section, continuous learning and development are essential in cultivating human capital in developing countries like Trinidad and Tobago. This process involves not only acquiring new skills but also fostering a culture of adaptability and innovation within the organization, however, Guglielman (2012) explains that even with such importance, as worker ages within the workplace, their participation in mature learning activities often declines regardless of the incentives and the initiatives put in place. Such aversion to participation in adult education and training is often a result of the brain's weakening plasticity, which is reflected by workers' weakened mental capacity to acquire and learn new skills, as well as their inability to concentrate on complex tasks for prolonged periods (Lazarov et al., 2024). Notwithstanding this, cutting-edge research in neuroplasticity and modern neuroscience shows that the deterioration in brain plasticity observed alongside ageing can be reversed and become more plastic throughout a worker's lifetime. According to Mishra et al. (2015), such advances include changes in the temporal lobe that processes auditory information needed for understanding complex sounds, language processing and speech recognition, as well as the posterior and inferior parietal cortex, which controls spatial reasoning, social cognition, movement and attention through an increase in grey matter.

Additionally, while there are several interrelated factors such as age, gender, physical health, diet, lifestyle choices, socioeconomic background, the presence of multimorbidity, and neurodegenerative conditions, that impact the ability of a worker's capacity for plasticity. One of the most deciding factors is the stimuli from one's external environment particularly those exposed to during the implementation of continuous learning techniques within the working environment (Voss et al., 2017). Organizations in Trinidad and Tobago can implement several educational and workplace strategies to promote continuous learning among their workers, by encouraging a growth mindset through lifelong learning, providing access to ongoing training opportunities and certifications to help workers develop their technical skills, and offering coaching and mentorship programs to support employees in building their confidence and autonomy in the workplace, as the labor market continues to evolve.

Multi-Modal Learning (MML) & Multi-Sensory Training (MST)

Multi-Modal learning (MML) involves the use of multiple forms of sensory modalities such as visual (images, videos, charts, info-graphs, slides), auditory (discussions, lectures, seminars, workshops, podcasts, audio and verbal instruction), reading (written material such as textbooks, articles, manuals, transcripts and written assignments), and kinesthetics (real-world scenarios and simulations, role-playing), to accommodate the different learning and communication styles of workers (Bouchey et al., 2021). This then gives rise to a key component of MML, i.e., Multi-Sensory Training (MST) where workers can interact with different sensory modalities simultaneously during learning activities to build stronger neural networks.

Besides the more traditional types of MML techniques, with the integration of Industry 4.0 and 5.0-based technologies within the labor market, there are several innovative learning educational tools that organizations in Trinidad and Tobago can begin to adopt to create a richer learning and working environment. These include the use of multimedia and Information Communications Technology (ICT) solutions such as digital storytelling, online and blended learning (e-learning), flipped classrooms, game-based techniques (gamification), and Virtual and Augmented reality (VR & AR) that allow for the creation of interactive simulations, diagrams, quizzes and diagrams (Ryoo & Winkelmann, 2021). The combination of traditional and digital MML techniques, together with MST, enables workers in a culturally diverse workplace to have a flexible means of learning and communicating with each other, regardless of their age, gender, cultural background, level of education, and expertise. The use of these techniques in workplaces across Trinidad and Tobago may foster inclusivity and help workers stay engaged with the material being taught. By appealing to multiple senses, these techniques enhance knowledge retention, understanding and recollection, as the information can be applied to real-world situations. Furthermore, MML changes perceptions of learning, making it easier for workers to absorb information as reflected by improved attention, performance and productivity rates (Gilakjani et al., 2011).

In the context of an aging workforce, the combined use of different MML and MST techniques and programs has been found to significantly improve cognitive function more than the use of a singular modality because these techniques can use a more targeted approach to designing a program that focuses on the different principles of neuroplasticity (Cespon et al., 2018). As a result of this, these programs can be used to design effective interventions that can strengthen synaptic connections and improve cognitive function, based on the needs of workers in Trinidad and Tobago. These techniques have been found to induce both structural changes, such as an increased volume of grey matter in different brain regions, as well as functional changes, such as enhanced brain connectivity between regions, which can lead to improved cognitive and functional performance (Banaszkiewicz et al., 2020).

To illustrate, such changes have been demonstrated by MML musical training programs, which, according to Lappe et al. (2008), has had a positive impact on multimodal brain plasticity. This is achieved by engaging multiple sensory modalities, such as sensorimotor and auditory systems, which bring about plasticity changes across different areas of the brain. Due to the interconnectedness of multiple sensory modalities, when combined with physical training exercises such as aerobic exercise, to bring about changes in motor systems, workers can then experience improvements across several cognitive domains such as engaging in higher-level cognitive processes like decision-making, acquiring, recollecting and applying new information, processing sensory information, and improved communication through a better understanding of multiple language constructs and social cues (Ward et al., 2017). Such plasticity enhances the workers' ability to develop multimodal and cross-modal responses to multiple forms of modalities, which improves not only their long-term memory and recall, but also perception.

Augmented and Virtual Reality (AR & VR)

Focusing on the digital MML techniques, emerging technologies such as VR and AR have been found, according to Kumar et al. (2023), to be operative in enhancing cognitive abilities through improvements in memory and attention, as well as promoting neuroplastic changes such as increased thickness in the brain's outer layer. The use of VR and AR exercises through the creation of immersive virtual environments, and mental training exercises that are not only interesting but stimulating, can assist workers in Trinidad and Tobago to improve both their motor function and develop their cognitive abilities. The integrated use of AR and VR, together with Machine Learning (ML) can become a transformative mechanism for relaxing interventions and treating a wide range of conditions such as mental health conditions such as anxiety disorders, obsessive-compulsive disorders (OCD), Attention-Deficit Hyperactivity Disorders (ADHD), Post-Traumatic Stress Disorders (PTSD) and depression, as well as neurodegenerative disorders such as dementia, within the workplace. In addition to workplace-related challenges such as difficult relationships, performance anxiety, and work-life balance, these digital solutions can provide workers with personalized immersive/non-immersive treatment plans using data and predictive analysis (Halkiopoulos & Gkintoni, 2024).

Further to this, human resource departments, and in-house counselling departments, can use AR and VR (together with the relevant gear such as Xbox and Wii consoles/remotes, controllers, Kinect cameras, 2D display screens, human mobile devices, exercise equipment such as treadmills, dance platforms, and driving simulators), as a part of their Mental Health Awareness Initiatives, Employee Assistance Programs (EAPs), and Virtual Reality Rehabilitation Program, to strengthen traditional modes of treatment methods such as Cognitive Behavioral Therapy (CBT), Dialectical Behavioral Therapy (DBT), and exposure therapy (Carroll et al., 2021). By exposing distressed workers to trauma-related stimuli in a controlled environment using either 2D VR, non-immersive/immersive VR, or AR, they can safely re-process realistic scenarios to develop resilient coping mechanisms and improve their mental health in the context of the working environment (Tay et al., 2023). In doing so, workers better reframe either negative or unhealthy thought patterns, become aware of excessive critical self-thinking, and foster effective communication and problem-solving skills. In addition to this, workers may be better able to set realistic but achievable goals, to manage stressful situations within the workplace in Trinidad and Tobago, while enhancing their well-being, resilience, work performance, and engagement.

Although mature workers with underlying mental health disorders such as schizophrenia can become prone to experiencing cybersickness, i.e., a form of visually induced motion sickness because of exposure to immersive AR and VR environments, resulting in headaches, nausea, eyestrain, and disorientation, the benefits far outweigh the costs (Chang et al., 2009). Mature adults and workers who engaged in VR-based cognitive training programs were more likely to exhibit gains in mental function, reduced levels of stress, anxiety, irritation, burnout, and despondency, as indicated by reduced heart rates and relaxed breathing patterns (Lawson & Stanney, 2021). With such positive uses of digital technologies, AR and VR-based neuroplastic interventions can be anticipated to reduce the stigma surrounding mental health in workplaces in Trinidad and Tobago, allowing for a change in perception through open dialogue, education and empowerment.

In addition to the societal and workplace benefits, integrating VR and AR cognitive training programs in the workplace significantly impacts neuroplasticity. Engaging multiple sensory inputs can trigger changes in synaptic processes such as the strengthening of synaptic connections, structural plasticity, neurogenesis, and dendritic branching in response to stimuli particularly to the parts of the brain such as the hippocampus, frontal, and sensorimotor, and parietal cortices, as well as occipital areas that are responsible for memory, visual processing, and higher-order cognitive functions such as decision-making, and motor planning (Drigas & Sideraki, 2024). This can lead to an improvement in the workers' executive functions, cognitive perceptions and motor skills, as they are more aware of their performance within the workplaces in Trinidad and Tobago. The responsiveness of workers, combined with the efforts of human resources and counselling departments implementing VR and AR interventions, allows for the implementation of adaptive learning systems as modifications of tasks based on workers' responses to stimuli and their needs can be made in real-time (Gangemi et al., 2023). This approach better supports targeted neuroplasticity, particularly in cases where workers show signs of mental health issues or neurodegenerative disorders, as workers can track their progress and identify areas for development.

Notwithstanding the many benefits that VR and AR have, several ethical considerations need to be made when using these technologies in the workplace. To encourage diversity and representation, virtual training environments need to accurately reflect a diverse workforce, must be accessible by all workers regardless of their technical know-how and abilities and must be tailored to the learning journey of the individual worker. In doing so, it reduces inequality in the workplace, as it promotes inclusion. Thus, creating a safe space for highlighting and addressing sensitive topics, without encountering the consequences of the real world.

Gamification

The incorporation of game design elements into contexts outside of the traditional gaming arena in areas such as education, health, and business is known as Gamification (Groh, 2012). This technique has been used in a multitude of applications, from improving the marketability of products to augmenting the engagement and motivation of students in the classroom. To create value, the technique tends to focus on the experiential nature of games i.e., the users' experience. By integrating game-design elements such as game mechanics (puzzle-solving, interactive challenges), progression and rewards (new abilities, points, leaderboards, and digital badges and trophies following the successful completion of a level), feedback mechanisms/ audio and visual design (visual cues, sound effects and SMS messages), narration and immersion (captivating storyline with the use of AR and VR effects), as well as User Interface and Experience (UI & UE) to maintain the ease at which workers can navigate and interact with the gaming application (Kim, 2015).

For instance, examples of gamification in workplaces that organizations in Trinidad and Tobago can adopt include the design of motivational cards grounded in motivational psychology, the integration of rewards and recognition aspect of gamification into Microsoft's Making Agents Great, the initiative to reduce absenteeism, building better teams, and improve knowledge and performance of agents (Schmidt et al., 2015; Microsoft, 2019). Apart from the United States (US), other companies such as L'Oreal which originated in France, have used Gamified tools such as Reveal the Game, platforms such as Reveal, business games such as L'Oreal Brandstorm, and gamified advertising to build brand engagement, and to drive innovation and competitiveness amongst workers (Rocha et al., 2017). They have also integrated gamification in their recruitment and training process to appear as an attractive employer to attract and retain workers (Rocha et al., 2017).

For example, within the workplace in countries like Italy, Finland and Spain, gamification has become embedded into several organizational processes such as (1) human resource management such as the onboarding of new workers, talent management and selection, corporate communications training, learning management systems, competitive, cooperative and performance-driven work, and knowledge management, (2) positive psychological outcomes within the workplace, resulting from workers improved awareness of company values, increased level of participation and engagement, heightened levels of motivation and morale leading to less stress and distress, and (3) organizational outcomes such as a greater interest in taking advantage of upskilling opportunities by workers, enhanced performance and productive efficiency, as well as creativity and innovativeness by both workers and teams (Behl et al., 2023).

As a result of using the progression and rewards systems ingrained within Gamification, it is then possible that workers and teams, as well as organizations across industries in Trinidad and Tobago, can benefit from increased knowledge retention and application within the workplace, as well as encourage cross-functional collaboration. In the case of the former, the use of the progression and rewards system may help workers to preserve new skills and improve their level of job satisfaction as they can be recognized for their accomplishments within the game, leading to a deeper sense of satisfaction, accomplishment and self-worth (Capatina et al., 2024). In a well-designed game, workers may also use the rewards system as a means of setting personal goals, tracking personal performance and establishing themselves as an expert in a particular field or skill. In addition to this, workers can redeem the rewards earned for tangible incentives such as company memberships, frequent flyer programs, health and fitness plans, and gift cards (Grunewald et al., 2019). In the case of the latter, increased social interaction and communication mediated by gamification can encourage workers with different skill sets and from different departments to work together. As a result, there can be greater collaboration and formation of teams within organizations in Trinidad and Tobago, by workers who, through their collective expertise, can develop new approaches to problem-solving and decision-making, which can effectively change both the organizations' culture and their environment.

Organizational Perspective of Gamification

From the perspective of organizations in Trinidad and Tobago, the integrated use of gamification mechanisms into their digital platforms and business processes, apart from being used as a form of product branding to attract and retain workers, as in the case of L'Oréal, can be used to bring about several positive business outcomes (Panda, 2024). In that gamification with the onset of industry, 5.0 technologies can be mobilized to streamline and enrich complex business processes such as sales and performance management, to incentivize workers to meet sales targets. It can also be used to build empowering marketing strategies that will sustain customers' loyalty to the brand in the future and boost customer engagement with e-commerce and online platforms (Sharma, 2024).

Most importantly, gamification can help organizations in Trinidad and Tobago to conduct market research through the collection of high-quality data in a more cost-effective manner by transforming survey questionnaires into enjoyable games that customers are more likely to engage with and complete, providing meaningful responses (Bailey, 2015). The richness of the data collected is crucial to understanding customer behavior and gaining a competitive edge within the industry because it provides organizations in Trinidad and Tobago with more detailed and valid insights into customer needs and industry/market trends. Furthermore, with the use of smartphones and wearable smart devices, organizations in Trinidad and Tobago can make more informed marketing decisions by monitoring the fluctuations in consumer demand by making the experience more personalized and immersive based on their behavior, and preferences (Luccasen & Jansen, 2015).

Notwithstanding the important role that gamification plays in maintaining customer loyalty and employee performance, it is expected that the technique of gamification itself is likely to have an impact on the neuroplasticity of workers. The process of gamification is said to have a positive impact on improving the overall brain plasticity of workers within the workplace in Trinidad and Tobago because the process of engaging with challenging tasks that are both repetitive and interactive is likely to stimulate not only the learning capacity of workers but also their physical activity using games that are engaging and pleasurable (Serice, 2023). These, as game design elements, often support continuous learning, cognitive flexibility, as well as better decision-making and problem-solving skills, which in turn strengthen the development of neural synapses and connectivity.

Further to this, the regions within the brain which are often activated by gamification that brings about the previously mentioned changes include the pre-frontal cortex (influences the workers' ability to plan, solve problems, and make goals in response to challenges set forth by the game in a relaxed mental state), amygdala (influences the workers' emotional responses to challenges encountered within the game, as well as to feelings of collaboration and competition created with co-workers who may also be playing), striatum (influences the reward centre of the worker's brain, which is activated in response to completing tasks, and receiving rewards), and finally the hippocampus and anterior cingulate cortex are both activated based on the workers gaming experience, which then influences their memory, adaptive learning, responsiveness and changes in behavior (Li & Tai, 2023). These changes were further supported by Ong (2013), who explains that the growth in prefrontal oxygenation of workers was observed to have grown when they were more engaged in completing tasks to earn points or rewards. Since workers' intrinsic motivation was noted to be linked to the narrative mechanics of the game, this may be indicative of favourable changes in the workers' cognition and behavior, particularly inhibitory control (response and attentional) needed for goal setting, decision making and self-regulation and Visuospatial skills (Wang, 2025).

Similar positive outcomes in worker engagement and performance were observed in the short term, as the ability of the work to utilise gamification in setting goals and objectives based on feedback from game mechanics directly influenced their intrinsic motivation in the long run (Passalacqua, 2020). When considering the age of workers, gamified cognitive-based changes were observed by Wang (2024) to have produced significant activity in the left inferior frontal junction (IFL), left inferior parietal lobule (IPL) and superior parietal lobule (SPL), but were more pronounced in older than in young workers. This implies that the integration of gamification in the workplace has the potential to reduce the rate of cognitive decline and the prevalence of cognitive impairments in older workers (Ganter-Argast, 2024).

THE OBSTACLES FACED IN APPLYING NEUROPLASTICITY-BASED APPROACHES IN TRINIDAD AND TOBAGO'S LABOR MARKET

The study of neuroplasticity, as demonstrated throughout this study, has undoubtedly proven to be one of the most influential means of improving the quality of the workforce in developing countries like Trinidad and Tobago, as well as the overall performance of its labor market. When implemented effectively, the strategies highlighted in the previous section have the potential to unlock a wide range of benefits for high-performing workers such as the (1) learning of new skills and honing of technical abilities, (2) optimizing cognitive abilities through heightened development of problem-solving skills, memory retention and recall, and (3) creation of a comfortable working environment where there is a decline in emotional, physical and mental exhaustion, and burnout.

Whereas, organizations and the labor market alike can benefit from having, (1) workers who are resilient and can easily adapt to technological shifts, (2) workers who are more compliant with engaging in programs that support continuous training and development as they age, and (3) elevated gains in productivity, innovativeness, and competitiveness, that are directly related to their workers' growth mindset (Al Halyan et al., 2019). Regardless of the positive outcomes, these strategies are likely to have on the human capital development of organizations across Trinidad and Tobago. However, the implementation of these techniques may be hindered by two major issues, i.e., limited resources and infrastructure and workers' perceptions.

Limited Resources and Infrastructure

Within business organizations, several resource-related barriers may influence the adoption of strategies such as continuous learning techniques, AR, VR, MML, MST, and Gamification. From the perspective of business organizations in Trinidad and Tobago, the inability to successfully implement these strategies on either a continuous or, phased basis may be due to the unavailability of sufficient funding to support the provision of the learning implements (access to online courses and trained instructors, professional networking communities, books/modules, physical manuals, AR and VR hardware, multimedia materials, learning management systems, software development, and access to digital platforms and tools), as well as infrastructure (adequate IT infrastructure, network capacity, device management and technical support) (Ceker & Ozdaml, 2017; Yazdi, 2024).

Despite large corporate business entities in Trinidad and Tobago may be able to support the often-excessive initial costs of implementing such programs involving MML and MST, AR and VR in training, as well as their integration with existing systems, small and medium-sized businesses may not be able to do the same. Smaller businesses may have less access to capital to invest in providing adequate equipment, content creation needed for simulations, and personnel needed to design and conduct an effective training course tailored to either the individual needs of the workers or teams (Gasteiger et al., 2022). As a result, to smaller resource-constrained business organizations in countries such as South Africa, even though the cost may decline over time, the initial costs associated with the setting up of these strategies may be considered an additional cost and become a barrier, rather than a form of a long-term investment in the education and training of their workforce (Akinradewo et al., 2025). This, therefore, restricts the provision and access to comprehensive training efforts.

Even though investing in these strategies has been associated with reduced training time, improved productivity and engagement amongst workers that may drive long-term growth, performance, innovation, and competitiveness, the inability of business organizations in Trinidad and Tobago to make such an investment in the short-term to create diverse training scenarios and opportunities, may limit their ability to adapt to changes in the labor market brought upon by Industry 5.0. In addition to this, it may imply that the level of technology readiness and acceptance may be low amongst workers and business organizations. Business organizations in Trinidad and Tobago may not only lack adequate infrastructure and financial resources to support these strategies, but also the strategic alignment needed for the adoption of these strategies, to meet the overall goals and objectives of the organization (Hamid, 2022).

Notwithstanding these constraints, there are several strategies that can be implemented to facilitate the integration of neuroplasticity-based strategies and principles into workplaces in Trinidad and Tobago. The first strategy will be to encourage public-private sector partnerships between the government of Trinidad and Tobago and private sector companies across industries to finance and develop both the physical and technological infrastructure needed to implement neuroplasticity-based strategies in workplaces. This has many benefits, as increased access to finances, improved efficiency through access to specialized skills and technical know-how, as well as risk sharing. The second strategy includes the integration of neuroplasticity principles into Trinidad and Tobago's existing national initiatives dealing with health, training and development, as well as organizational frameworks that address organizational change, talent management, and well-being.

Workers' Perception

Furthermore, instances where these neuroplasticity-based strategies are available in Trinidad and Tobago, but workers are reluctant to dedicate sufficient time to become fully immersed and engaged with the information and material provided, may reflect not only the workers' overwhelming workload but much deeper management and organization problems. These issues may include the workers' poor time management, deteriorating work-life balance, and reduced job satisfaction and motivation, which may lead to increased employee turnover and absenteeism. In Canada, it was found that these problems may limit not only the worker's participation but also the transfer of knowledge and skills, i.e., learning transfer, to the situations in the workplace, if the facilitator does not have sufficient time to prepare materials and activities to implement AR, VR, and Gamification strategies effectively (Brown & McCraken, 2009). Thus, reducing the overall impact of neuroplasticity-based strategies on the performance, productivity, and efficiency of organizations in Trinidad and Tobago.

Moreover, workers' resistance to becoming engaged with these strategies may also be in response to their perceptions such as preferring traditional workplace systems and procedures that encourage departments to work separately with limited information sharing within organizations in Trinidad and Tobago. Although the division of departments may be a cultural feature of most organizations due to different workflows, tasks, locations, communication styles, and technologies used, they can lead to inefficiencies within departments resulting from structural communication barriers, wasted resources, conflicting priorities, reduced innovation and collaboration, and complications in achieving departmental targets (Bento et al., 2020). This silo mentality amongst workers can become a barrier to implementing neuroplasticity-based strategies within the workplace. This is because such an attitude hinders both the adoption of these strategies and cross-functional collaboration and learning across departments, which is a major benefit of implementing strategies such as Gamification, MST, and MML (De Waal et al., 2019). This is particularly important as the latter can assist workers from different departments and styles of organizational structures to form better teams by encouraging better communication, collaboration, and coordination of activities and learning across different knowledge domains. This then helps workers and teams to stay in line with long-term objectives, rather than focusing on immediate results and short-term targets.

Fear of Failure and Learning (FoF and FoL)

In addition, workers' perceptions of their Fear of Failure and Learning (FoF and FoL), and concerns about being reprimanded for not completing specific training programs, which stem from educational gaps, and a lack of familiarity with Industry 4.0 and 5.0 technologies and equipment may influence the uptake and effective use of these techniques within organizations in Trinidad and Tobago. While in some instances, workers' FoF and FoL can be a driving factor behind their motivation to engage and complete courses, it can also create significant barriers to learning. This is because workers may hesitate to either take risks, make mistakes or ask for clarification during training programs using neuroplasticity-based strategies, so as not to appear incompetent to their colleagues (Cannon & Edmondson, 2005). Such perceived fears can result in workers like Brazilian Civil Servants becoming disinterested and disengaged during training programs, leading to less information and knowledge on the application of new skills in the workplace being absorbed. This then reduces the impact of the techniques being taught to workers as they may perform poorly during training sessions, which in turn hampers overall performance and internal dynamics within the organization (De Souza & Tomei, 2016).

Even more, workers' educational gaps and lack of familiarity with Industry 4.0 and 5.0 technologies can also lead to feelings of frustration and demotivation amongst those who because of their limited knowledge, may find it difficult to not only progress easily through such training program, but also feel disconnected from the content being taught as the training material may be too advanced than their initial knowledge (Yang & Rao, 2023). Such feelings, when combined with the workers' perceptions of FoF and FoL can create several problems such as (1) resistance to using neuroplasticity strategies to enhance their knowledge and development, (2) aversion against learning new skills even though it may be beneficial due to fears of job redundancies, and (3) increased workplace anxiety and stress for workers who may not be technologically proficient or competent to engage easily with AR, VR, and Gamification based systems and equipment (Bonekamp & Sure, 2015). As a result, persons with disabilities and those who lack advanced levels of technological training are more likely to experience heightened levels of discrimination, bias, and exclusion associated with elevated levels of inequalities in the workplace.

CONCLUSION

The study of neuroplasticity has begun to increasingly capture the attention of researchers in several fields within the social sciences, such as economics and human resource management. Evidence within the literature suggests that such an attraction stems from the potential benefits that organizations stand to gain from integrating neuroplasticity principles into human capital development strategies. Neuroplasticity simply refers to the ability of the human brain to adapt and reorganize itself through the formation of neural pathways in response to new internal and external stimuli. It is during this process of neural reworking that organizations can use neuroplasticity-based strategies to improve productivity and adaptability by inducing structural and functional changes in the human brain.

This study, which undertook a secondary research methodology, discusses not only the mechanics and types of neuroplasticity but, first and foremost, the role that it plays in advancing human capital development in Trinidad and Tobago. Most importantly, it explains conceptually that in considering the technological advancements of industry 4.0 and 5.0, neuroplasticity was found to influence the ability of workers in Trinidad and Tobago to engage in continuous learning and development, through reskilling and upskilling, as well as their mental and emotional health. To support the learning and development of workers and to reap the benefits of enhanced cognitive performance, organizations in Trinidad and Tobago can implement several neuroplasticity-based strategies such as Continuous Learning Techniques, Multi-Modal Learning (MML), Multi-Sensory Training (MST), Augmented and Virtual Reality (AR & VR), and Gamification.

While this study highlights the promising outcomes of these neuroplasticity-based strategies, there are several barriers that limit their implementation, such as limited resources and infrastructure, company culture, limited technical know-how, as well as workers' perception. As a result of this, such strategies are likely to be adopted more slowly by organizations in Trinidad and Tobago. In addition, this study was limited due to the lack of case studies investigating the implementation of neuroplasticity-based strategies in the working environment of Trinidad and Tobago. This study is the initial step in exploring thoroughly the relationship that neuroplasticity has with the changing workplace in Trinidad and Tobago. It lays the groundwork for future research and a deeper understanding of neurodegenerative diseases and their implications for workplaces in Trinidad and Tobago. To explore the technological neurorehabilitation of workers with neurodegenerative diseases using Socially Assistive Robotics and Soft Robotics.

Competing Interests

The author of this publication declares there are no competing interests.

Funding

This research received no specific grant from any funding agency in the public, commercial, or not-for-profit sectors. Funding for this research was covered by the author of the article.

REFERENCES

Akinradewo, O., Hafez, M., Aliu, J., Oke, A., Aigavboa, C., & Adekunle, S. (2025). Barriers to the Adoption of Augmented Reality Technologies for Education and Training in the Built Environment: A Developing Country Context. *Technologies*, *13*(62), 1–15. DOI: 10.3390/technologies13020062

Al Halyan, A. Al badi, A., & Al Hosani, B. (2019). *Applying Brain Science to Achieve Organizational Success*. In Conference Proceedings, Abu Dhabi International Petroleum Exhibition & Conference, Abu Dhabi, UAE. DOI: 10.2118/197532-MS

Alexiou, E. (2024). The Role of Neuroplasticity in Recovery from Brain Injuries. Neuroscience and *Psychiatry*. *Open Access*, *7*(6), 301–203.

Alves, J., Lima, T. M., & Gaspar, P. D. (2023). Is Industry 5.0 a Human-Centered Approach? *Systematic Reviews*, *11*(193), 1–15. DOI: 10.3390/pr11010193

Antolini, G., & Colizzi, M. (2023). Where Do Neurodevelopmental Disorders Go? Casting the Sye Away from Childhood towards Adulthood. *Health Care*, *11*(1015), 1–22. DOI: 10.3390/healthcare11071015 PMID: 37046942

Bailey, P., Pritchard, G., & Kernohan, H. (2015). Gamification in Market Research. *International Journal of Market Research*, *57*(1), 17–28. DOI: 10.2501/IJMR-2015-003

Banaszkiewicz, A., Matuszewski, J., Bola, L., Szczepanik, M., Kossowski, B., Rutkowski, P., Szwed, M., Emmorey, K., Jednorog, K., & Marchewka, A. (2020). Multimodal imaging of brain reorganization in hearing late learners of sign language. *Human Brain Mapping*, *42*(2), 382–397. DOI: 10.1002/hbm.25229 PMID: 33098616

Behl, A., Pereira, V., Jayawardena, N., Nigam, A., & Mangla, S. (2023). Gamification as an innovation: A tool to improve organizational marketing performance and sustainability of international firms. *International Marketing Review*, *41*(1), 107–137. DOI: 10.1108/IMR-05-2022-0113

Ben-Porath, Y. (1970). The Production of Human Capital Over Time. Cambridge, MA: National Bureau of Economic Research (NBER).

Bento, F., Tagliabue, M., & Lorenzo, F. (2020). Organizational Silos: A Scoping Review Informed by a Behavioral Perspective on Systems and Networks. *Societies (Basel, Switzerland)*, *10*(3), 1–27. DOI: 10.3390/soc10030056

Berlucchi, G., & Butchel, H. A. (2009). Neuronal plasticity: Historical roots and evolution of meaning. *Experimental Brain Research, 192*, 307–319. DOI: 10.1007/s00221-008-1611-6 PMID: 19002678

Bonekamp, L., & Sure, M. (2015). Consequences of Industry 4.0 on Human Labour and Work Organization. *Journal of Business and Media Psychology*, *6*(1), 33–40.

Bouchey, B., Castek, J., & Thygeson, J. (2021). Multimodal Learning. In Ryoo, J., & Winkelmann, K. (Eds.), *Innovative Learning Environments in STEM Higher Education*. Springer Briefs in Statistics., DOI: 10.1007/978-3-030-58948-6_3#DOI

Brown, T. C., & McCracken, M. (2009). Building a bridge of understanding: How Barriers to Training Participation become Barriers to Training Transfer. *Journal of European Industrial Training*, *33*(6), 492–512. DOI: 10.1108/03090590910974392

Cannon, M. D., & Edmondson, A. C. (2005). Failing to Learn and Learning to Fail (Intelligently): How Great Organizations Put Failure to Work to Innovate and Improve. *Long Range Planning*, *38*(3), 299–319. DOI: 10.1016/j.lrp.2005.04.005

Capatina, A., Juarez-Varon, D., Micu, A., & Micu, A. E. (2024). Leveling up in corporate training: Unveiling the power of gamification to enhance knowledge retention, knowledge sharing, and job performance. *Journal of Innovation & Knowledge*, *9*(3), 1–15. DOI: 10.1016/j.jik.2024.100530

Carroll, J., Hopper, L., Farrelly, A. M., Lombard-Vance, R., Bamidis, P. D., & Konstantinidis, E. I. (2021). A Scoping Review of Augmented/Virtual Reality Health and Wellbeing Interventions for Older Adults: Redefining Immersive Virtual Reality. *Frontiers in Virtual Reality*, *2*(655338), 1–26. DOI: 10.3389/frvir.2021.655338

Castrillo, V. I. (2024). *Job Skills: What are they and which ones are most in demand?* IDB.

Ceker, E., & Ozgaml, F. (2017). What "Gamification" is and what it's not. *European Journal of Contemporary Education*, *6*(2), 221–228. DOI: 10.13187/ejced.2017.2.221

Cespon, J., Miniussi, C., & Pellicciari, M. C. (2018). Interventional programmes to improve cognition during healthy and pathological aging: Cortical modulations and evidence for brain plasticity. *Ageing Research Reviews*, *43*, 81–98. DOI: 10.1016/j.arr.2018.03.001 PMID: 29522820

Chang, C. L. F., Ngai, E. K. Y., Leung, P. K. H., & Wong, S. (2009). Effect of the adapted virtual reality training program among Chinese older adults with chronic schizophrenia: A pilot study. *International Journal of Geriatric Psychiatry*, *25*(6), 643–649. DOI: 10.1002/gps.2403 PMID: 19806599

Chapman, S. B., Cook, L. G., Vas, A. K., & Robertsson, I. H. (2019). Enhancing Human Cognitive Capital by Harnessing the Brain's Inherent Neuroplasticity. In *Human Performance Optimization: The Science and Ethics of Enhancing Human Capabilities*, eds Michael D. Matthews and David. M Schnyer. Oxford University Press. New York: United States.

Cramer, S. C., Sur, M., Dobkin, B. H., O'Brian, C., Sanger, T. D., Trojanowski, J. Q., Rumsey, J. M., Hicks, R., Cameron, J., Chen, D., Chen, W. G., Cohen, L. G., DeCharms, C., Duffy, C. J., Eden, G. F., Fetz, E. E., Filart, R., Freund, M., Grant, S. J., Haber, S., Kalivas, P. W., Kolb, B., Kramer, A. F., Lynch, M., Mayberg, H. S., McQuillen, P. S., Pascual-Leone, A., Reuter-Lorenz, P., Schiff, N., Sharma, A., & Shkim, L., stryker, M., Sullivan, E. V., & Vinogradov, S. (2011). Harnessing neuroplasticity for clinical applications. *Brain*, *134*(6), 1591–1609. PMID: 21482550

Dan, B. (2019). Neuroscience underlying rehabilitation: What is neuroplasticity? *Developmental Medicine and Child Neurology*, *61*(11), 1237–1348. DOI: 10.1111/dmcn.14341 PMID: 31576561

De Giovanni, P. (2023). Sustainability of the Metaverse: A Transition to Industry 5.0. *Sustainability*, *15*(6079), 1–29. DOI: 10.3390/su15076079

De Souza, D. O. S., & Tomei, P. A. (2016). Fear of Failure in the workplace among Brazilian employees. *Academia (Caracas)*, *29*(4), 407–418. DOI: 10.1108/ARLA-11-2015-0299

De Waal, A., Weaver, M., & Day, T. (2019). Silo-Busting: Overcoming the Greatest Threat to Organizational Performance. *Sustainability*, *11*(23), 1–21. DOI: 10.3390/su11236860

Drigas, A., & Sideraki, A. (2024). Brain Neuroplasticity Leveraging Virtual Reality and Brain-Computer Interface Technologies. *Sensors (Basel)*, *24*(5725), 1–36. DOI: 10.3390/s24175725 PMID: 39275636

Erro-Garces, A., & Aramendia-Muneta, M. E. (2023). The role of human resource management practices on the results of digitalization. From Industry 4.0 to Industry 5.0. *Journal of Organizational Change Management*, *36*(4), 585–608. DOI: 10.1108/JOCM-11-2021-0354

Foster, K., Shakespeare-Finch, J., Schochet, I., Mayberry, D., Bui, M. V., Steele, M., & Roache, M. (2024). Psychological distress, well-being, resilience, posttraumatic growth, and turnover intention of mental health nurses during COVID-19: A cross-sectional study. *International Journal of Mental Nursing Health*, *33*, 1542–1552. DOI: 10.1111/inm.13354 PMID: 38747675

Gamberini, L., & Pluchino, P. (2024). Industry 5.0: A Comprehensive insight into the future of work, social sustainability, sustainable development and career. *Australian Journal of Career Development*, *33*(1), 5–14. DOI: 10.1177/10384162241231118

Gangemi, A., De Luca, R., Fabio, R. A., Lauria, P., Rifici, C., Pollicino, P., Olivo, A., Quartarone, A., & Calabro, R. S. (2023). Effects of Virtual Reality Cognitive Training on Neuroplasticity: A Quasi Randomized Clinical Trial in Patients with Stroke. *Biomedicines*, *12*(3225), 1–14. DOI: 10.3390/biomedicines11123225 PMID: 38137446

Ganter-Argast, C., Schipper, M., Shamsrizi, M., Stein, C., & Khalil, R. (2024). The light side of gaming: Creativity and brain plasticity. *Frontiers in Human Neuroscience*, *17*(1280989), 1–12. DOI: 10.3389/fnhum.2023.1280989 PMID: 38249576

Gasteiger, N., Van der Veer, S. N., Wilson, P., & Dowding, D. (2022). How, for Whom, and in Which Contexts or Conditions Augmented and Virtual Reality Training Works in Upskilling Health care Workers: Realist Synthesis. *JMIR Serious Games*, *10*(1), e31644. DOI: 10.2196/31644 PMID: 35156931

Ghobakhloo, M., Iranmanesh, M., Fathi, M., Rejeb, A., Foroughi, B., & Nikbin, D. (2024). Beyond Industry 4.0: A systematic review of Industry 5.0 technologies and implications for social, environmental and economic sustainability. *Asia-Pacific Journal of Business Administration*, •••, 1757–4323. DOI: 10.1108/APJBA-08-2023-0384

Gilakjani, A. P., Ismail, H. N., & Ahmadi, S. M. (2011). The Effect of Multimodal Learning Models on Language Teaching and Learning. *Theory and Practice in Language Studies*, *1*(10), 1321–1327. DOI: 10.4304/tpls.1.10.1321-1327

Goldin, C. (2016). Human Capital. In Diebolt, C., & Haupert, M. (Eds.), *Handbook of Cliometrics*. Springer., DOI: 10.1007/978-3-642-40406-1_23

Groh, F. (2012). Gamification: State of the Art Definition and Utilization. In *Proceedings of the 4th Seminar on Research Trends in Media Informatics*, Institute of Media Informatics, Ulm University, Germany.

Grunewald, H., Kneip, P., & Kozica, A. (2019). The Use of Gamification in Workplace Learning to Encourage Motivation and Engagement. In *The Wiley Handbook of Global Workplace Learning*, eds by Vanessa Hammler Kenon, and Sunay Vasant Palsole, pp. 557-575.

Guglielman, E. (2012). The Aging Brain: Neuroplasticity and Lifelong Learning. *eLearning Papers*, 29: 1-7. ISSN: 1887-1542.

Halkiopoulos, C., & Gkintoni, E. (2024). Leveraging AI in e-learning: Personalized learning and adaptive assessment through cognitive neuropsychology-A systematic analysis. *Electronics (Basel)*, *13*(18), 3762. DOI: 10.3390/electronics13183762

Hamid, R. A. (2022). The Role of Employees' Technology Readiness, Job Meaningfulness and Proactive Personality in Adaptive Performance. *Sustainability*, *14*(15696), 1–16. DOI: 10.3390/su142315696

Hibbert, L. (2020). *Cobots and HMI: A Look at Industry 5.0.* https://www.allaboutcircuits.com/industry-articles/a-look-at-industry-5.0-cobots-and-hmi/

Imai, T. (2025). Activity-dependent synaptic competition and dendrite pruning in developing mitral cells. *Frontiers in Neural Circuits*, *19*(1541926), 1–11. DOI: 10.3389/fncir.2025.1541926 PMID: 40034992

Khaneja, S., & Arora, T. (2024). The potential of neuroscience in transforming business: A meta-analysis. *Future Business Journal*, *10*(1), 1–14. DOI: 10.1186/s43093-024-00369-7

Kikkert, S., Kolasinski, J., Jbabdi, S., Tracey, I., Beckmann, C. F., Johansen-Berg, H., & Makin, T. R. (2016). Revealing the neural fingerprints of a missing hand. *eLife*, *5*(e15292), 1–19. DOI: 10.7554/eLife.15292 PMID: 27552053

Kim, B. (2015). Gamification. *Library Technology Reports*, *51*(2), 10–18. DOI: 10.5860/ltr.51n2 PMID: 38600768

Kleim, J. A., & Jones, T. A. (2008). Principles of experience-dependent neural plasticity: implications for rehabilitation after brain damage. *Journal of Speech, Language, and Hearing Research*, 51(1): S225-S239. (2008/018DOI: 10.1044/1092-4388

Krauss, P. (2024). Building Blocks of the Nervous System. In: *Artificial Intelligence and Brain Research.* Springer, pp, 19-26. Berlin, Heidelberg: Springer. DOI: 10.1007/978-3-662-68980-6_3

Kreber, L. (2025). *Neuroplasticity.* Centre for Neuro Skills.

Kumar, J., Patel, T., Sugandh, F., Dev, J., Kumar, U., Adeeb, M., Kachhadia, M. P., Puri, P., Prachi, F., Zaman, M. U., Kumar, S., Varrassi, G., & Syed, A. R. S. (2023). Innovative Approaches and Therapies to Enhance Neuroplasticity and Promote Recovery in Patients with Neurological Disorders: A Narrative Review. *Cureus*, *15*(7), e41914. DOI: 10.7759/cureus.41914 PMID: 37588309

Lappe, C., Herholz, S. C., Trainor, L. J., & Pantev, C. (2008). Cortical Plasticity Induced by Short-Term Unimodal Musical Training. *The Journal of Neuroscience: The Official Journal of the Society for Neuroscience*, *28*(39), 9632–9639. DOI: 10.1523/JNEUROSCI.2254-08.2008 PMID: 18815249

Lawson, B. D., & Stanney, K. M. (2021). Editorial: Cybersickness in Virtual Reality and Augmented Reality. *Frontiers in Virtual Reality*, *2*(759682), 4–7. DOI: 10.3389/frvir.2021.759682

Lazarov, O., Gupta, M., Kumar, P., Morrissey, Z., & Pham, T. (2024). Memory circuits in dementia: The engram, hippocampal neurogenesis and Alzheimer's disease. *Progress in Neurobiology*, *236*(102601), 1–14. DOI: 10.1016/j.pneurobio.2024.102601 PMID: 38570083

Li, L., & Tai, C. (2023). Neuroscience supports gamification improving memory. In *Proceeding of the 2023 7th International Seminar on Education, Management and Social Sciences (ISEMSS 2023). Advances in Social Science, Education and Humanities Research*, *779*, 1291–1303. DOI: 10.2991/978-2-38476-126-5_14

Mateos-Aparicio, P., & Rodriguez-Moreno, A. (2019). The Impact of Studying Brain Plasticity. *Frontiers in Cellular Neuroscience*, *13*(66), 1–5. DOI: 10.3389/fncel.2019.00066 PMID: 30873009

McClung, C. A., & Nestler, E. J. (2008). Neuroplasticity Mediated by Altered Gene Expression. *Neuropsychopharmacology: Official Publication of the American College of Neuropsychopharmacology*, *33*, 3–17. DOI: 10.1038/sj.npp.1301544 PMID: 17728700

Microsoft. (2019). *Using Gamified Performance and Learning to Drive Call Center Agents.* https://news.microsoft.com/en-nz/2019/05/16/using-gamified-performance-learning-to-drive-call-center-agents/

Mincer, J. (1958). Investment in Human Capital and Personal Income Distribution. *Journal of Political Economy*, *66*, 281–302.

Mincer, J. (1996). Economic development, growth of human capital, and the dynamics of the wage structure. *Journal of Economic Growth*, *1*, 29–48. DOI: 10.1007/BF00163341

Mishra, J., De Villers-Sidani, E., Merzenich, M., & Gazzaley, A. (2015). Adaptive Training Diminishes Distractibility in Aging across Species. *Neuron*, *84*(5), 1091–1103. DOI: 10.1016/j.neuron.2014.10.034 PMID: 25467987

Nanda, B., Balde, J., & Manjunatha, S. (2013). The acute effects of a single bout of moderate-intensity aerobic exercise on cognitive functions in healthy adult males. *Journal of Clinical and Diagnostic Research: JCDR, 7*(9), 1883–1885. DOI: 10.7860/jcdr/2013/5855.3341 PMID: 24179888

Nardo, M., Forino, D., & Murino, T. (2020). The evolution of man-machine interaction: The role of human industry 4.0 paradigm. *Production & Manufacturing Research, 8*(1), 20–34. DOI: 10.1080/21693277.2020.1737592

Nudo, R. J. (2003). Functional and structural plasticity in motor cortex: Implications for stroke recovery. *Physical Medicine and Rehabilitation Clinics of North America, 14*(1), S57–S76. DOI: 10.1016/S1047-9651(02)00054-2 PMID: 12625638

Ong, M. J. K. (2013). Gamification and its effect on employee engagement and performance in a perceptual diagnosis task. [Unpublished Master's Thesis]. University of Canterbury.

Panda, D., Raut, S. K., Rana, S., & Shamsudin, M. N. (2024). Leveraging gamification to enhance persuasive behaviour and streamline online product returns: An approach grounded in multiple case studies. *Journal of Business and Industrial Marketing, 39*(12), 2684–2698. DOI: 10.1108/JBIM-01-2024-0007

Pascoa dos Santos, F., & Verschure, P. F. M. J. (2022). Excitatory-Inhibitory Homeostasis and Diaschisis: Tying the Local and Global Scales in the Post-stroke Cortex. *Frontiers in Systems Neuroscience, 15*(806544), 1–19. DOI: 10.3389/fnsys.2021.806544 PMID: 35082606

Passalacqua, M., Leger, P.-M., Nacke, L. E., Fredette, M., Labonte-Lemoyne, E., Lin, X., Caprioli, T., & Senecal, S. (2020). Playing in the backstore: Interface gamification increases warehousing workforce engagement. *Industrial Management & Data Systems, 120*(7), 1309–1330. DOI: 10.1108/IMDS-08-2019-0458

Phillips, C. (2017). Lifestyle Modulators of Neuroplasticity: How Physical Activity, Mental Engagement, and Diet Promote Cognitive Health during Aging. *Hindawi Neural Plasticity, 2017*(3589271), 1–22. DOI: 10.1155/2017/3589271 PMID: 28695017

Polakova, M., Suleimanova, J. H., Madzik, P., Copus, L., Molnarova, I., & Polednova, J. (2023). Soft skills and their importance in the labor market under the conditions of Industry 5.0. *Heliyon, 9*(e18670), 1–20. DOI: 10.1016/j.heliyon.2023.e18670 PMID: 37593611

Rame, R., Puranto, P., & Sudarno, S. (2024). Industry 5.0 and sustainability: An overview of emerging trends and challenges for a green future. *Innovation and Green Development, 3*(100173), 1–20. DOI: 10.1016/j.igd.2024.100173

Rocha, A., Correia, A.-M., Adeli, H., Reis, L. P., & Costanzo, S. (2017). Gamification in the Workplace: A Systematic Literature Review. In *Recent Advances in Information Systems and Technologies, Advances in Intelligent Systems and Computing*, eds (A. Rocha et al.,), 571. DOI: 10.1007/978-3-319-56541-5_29

Roder-Dewan, S., Akala, F. A., & Veillard, J. (2019). *Human Capital and Health*. Washington, DC: World Bank (WB). https://blogs.worldbank.org/en/health/human-capital-and-health

Romer, P. M. (1987). Growth Based on Increasing Returns to Specialisation. *The American Economic Review*, *77*(2), 56–62.

Romer, P. M. (1989). Human Capital and Growth: Theory and Evidence. [Cambridge, MA: National Bureau of Economic Research] [NBER]. *Working Paper Series*, *3173*, 1–51.

Rossi, F., & Weber, M. (2024). The Accumulation and Utilization of Human Capital over the Development Spectrum. [Washington, DC: World Bank Group] [WBG]. *Working Paper*, *10891*, 1–40.

Roy, R., & Ateeq, A. (2024). Moving Towards Industry 5.0: Opportunities and Challenges in Bahrain Higher Education. In Awwad, B. (Ed.), *The AI Revolution: Driving Business Innovation and Research. Studies in Systems, Decision and Control* (Vol. 524, pp. 425–435). Springer., DOI: 10.1007/978-3-031-54379-1_37

Rueda-Castro, V., Azofeifa, J. D., Chacon, J., & Caratozzola, P. (2024). Bridging minds and machines in Industry 5.0: Neurobiological approach. *Frontiers in Human Neuroscience*, *18*(1427512), 1–13. DOI: 10.3389/fnhum.2024.1427512 PMID: 39257699

Ryoo, J., & Winkelmann, K. (2021). *Innovative Learning Environments in STEM Higher Education Opportunities, Challenges and Looking Forward*. Springer Briefs in Statistics., DOI: 10.1007/978-3-030-58948-6

Schmidt, R., Schmidt, B., Lattenkamp, K., Scheja, S., & Masuch, M. (2015). *Motivational design cards: a practical approach for game-based motivational design at the workplace*. In *Proceedings of the 19th International Academic Mindtrek Conference*, pp. 10-17. New York, United States (US): Association for Computing Machinery. DOI: 10.1145/2818187.2818299

Serice, L. (2023). Prisms of neuroscience: Frameworks for thinking about educational gamification. *Intech Open Journals, AI. Computer Science and Robotics Technology*, *13*(2), 1–30. DOI: 10.5772/acrt.13

Sharma, W., Lim, W. M., Kumar, S., Verma, A., & Kumra, R. (2024). Game on! A state-of-the-art overview of doing business with gamification. *Technological Forecasting and Social Change, 198*(122988), 1–21. DOI: 10.1016/j.techfore.2023.122988

Smith, A. (1776). *An Inquiry into the Nature and Causes of the Wealth of Nations*. The Clarendon Press.

Staneiu, R.-M. (2023). Nuturing Neuroplasticity as an Enabler for Growth Mindset through Lifelong Learning and Knowledge Dynamics. In *Proceedings of the 17th International Conference on Business Excellence*, Budapest, Romania, 2023, 1265-1274. DOI: 10.2478/picbe-2023-0113

Stewart, J. C., & Cramer, S. (2017). Genetic Variation and Neuroplasticity: Role in Rehabilitation after Stroke. *Journal of Neurologic Physical Therapy; JNPT, 41*(3), S17–S23. DOI: 10.1097/npt.0000000000000180 PMID: 28628592

Su, Y. R. S., Veeravgu, A., & Grant, G. (2016). Neuroplasticity after Traumatic Brain Injury. In Laskowitz, D., & Grant, G.Taylor and Francis Group (Eds.), *Translational Research in Traumatic Brain Injury*.

Tau, G. Z., & Peterson, B. (2010). Normal Development of Brain Circuits. *Neuropsychopharmacology Reports, 35*, 147–168. DOI: 10.1038/npp.2009.115 PMID: 19794405

Tay, J. L., Xie, H., & Sim, K. (2023). Effectiveness of Augmented and Virtual Reality-Based Interventions in Improving Knowledge, Attitudes, Empathy and Stigma Regarding People with Mental Illnesses-A Scoping Review. *Journal of Personalized Medicine, 13*(112), 1–16. DOI: 10.3390/jpm13010112 PMID: 36675773

Vance, D. E., Roberson, A. J., McGuinness, T. M., & Fazeli, P. L. (2010). How Neuroplasticity and Cognitive Reserve: Protect Cognitive Functioning. *Journal of Psychosocial Nursing and Mental Health Services, 48*(4), 23–30. DOI: 10.3928/02793695-20100302-01 PMID: 20349891

Voss, P., Thomas, M. E., Cisneros-Franco, J. M., & De Villers-Sidani, E. (2017). Dynamic Brains and the Changing Rules of Neuroplasticity: Implications for Learning and Recovery. *Frontiers in Psychology, 8*(1657), 1–11. DOI: 10.3389/fpsyg.2017.01657 PMID: 29085312

Wang, P., Guo, S.-J., & Li, H.-J. (2024). Brain imaging of a gamified cognitive flexibility task in young and older adults. *Brain Imaging and Behavior, 18*, 902–912. DOI: 10.1007/s11682-024-00883-w PMID: 38627304

Wang, P., Zhou, Z.-X., Gao, P., Hou, H.-Y., Zhang, J.-X., Li, H.-J., & Zuo, X.-N. (2025). Gamified working memory intervention enhances prefrontal neurocognitive plasticity during aging. *Psychology and Aging*. Advance online publication. DOI: 10.1037/pag0000895 PMID: 40323840

Ward, N., Paul, E., Watson, P., Cooke, G. E., Hillman, C. H., Cohen, N. J., Kramer, A. F., & Barbey, A. K. (2017). Enhanced Learning through Multimodal Training: Evidence from a Comprehensive Cognitive, Physical Fitness, and Neuroscience Intervention. *Scientific Reports*, *7*(5808), 1–8. DOI: 10.1038/s41598-017-06237-5 PMID: 28724914

World Bank Group (WBG). (2025a). Human Capital Project. Washington, DC: World Bank Group (WBG).

World Bank Group (WBG). (2025b). World Development Report 2019. The Changing Nature of Work. Washington, DC: World Bank Group (WBG).

Yang, Y., & Rao, N. (2023). The status, pathways and discourses of professionalism for early childhood education teachers in Chinese policies. *International Journal of Educational Development*, *99*, 102760. DOI: 10.1016/j.ijedudev.2023.102760

Yazdi, M. (2024). Augment Reality (AR) and Virtual Reality (VR) in Maintenance Training. In: *Advances in Computational Mathematics for Industrial System Reliability and Maintainability*, pp. 169-183. Springer Series in Reliability Engineering. Springer, Cham. DOI: 10.1007/978-3-031-53514-7_10

Zotey, V., Andhale, A., Shegekar, T., & Juganavar, A. (2023). Adaptive Neuroplasticity in Brain Injury Recovery: Strategies and Insights. *Cureus*, *15*(9), 1–11. DOI: 10.7759/cureus.45873 PMID: 37885532

About the Contributors

Bryan Christiansen is a highly indexed author of 50 Reference books on topics ranging from cultural studies to organizational sociology. He is currently an Adjunct Professor of business at Southern New Hampshire University and the Chief Executive Officer of the management consultancy, Soluvex, LLC in New Mexico, USA. Professor Christiansen is fluent in Chinese, Japanese, and Spanish, and is currently learning Russian. He has presented his work at several universities in numerous countries including Oxford University in the United Kingdom in 2016. He earned an MBA from Capella University in 2003 and a BS in Marketing from the University of the State of New York in 1996.

Anatoly Zhuplev is a professor of international business and entrepreneurship at Loyola Marymount University (Los Angeles, California) and former editor- in- chief at the Journal of East West Business (2011- 2013). He taught for ten years at the Moscow Management Institute, and subsequently at the Advanced Training Institute of the State Committee for Printing and Publishing in Moscow; in Bonn, Germany in 1994, 1998, 2009; in Warsaw, Poland (as a Fulbright scholar) in 2005; in Paris, France in 2004- 2007, and at Northeastern University in Boston, Massachusetts in 1989- 1990. His books, book chapters, and articles on International Management, International Entrepreneurship, International Business, European Energy Security and Corporate Governance (around 100 overall) have been published in the U.S., Canada, Western Europe, Russia, and the former USSR. He received his PhD from the Moscow Management Institute, Russia, in 1981, and his BS from the Moscow Engineer-Economics Institute in 1974.

***

Muralidhar L B is a professionally qualified dental surgeon. He holds Ph.D. from Canara Bank School of Management Studies, Bangalore University after his

MBA from Christ University. He has 18 years of total teaching experience spanning across dental education, healthcare management and general management and 4 years of corporate experience. He has worked as research associate in Institute for Social and Economic Change, Bangalore in a project entitled as "Management By and For Common Man: An Exploration of the Concept and Direction for Future"post Ph.D. He had secured 28th rank in S.S.L.C examination for Karnataka State in 1993. His 26 patents have been published and 2 design patents have been granted. He has 41 research publications to his credit out of which 17 are Scopus-indexed and 5 are ABDC indexed. He has co-authored 11 books. He is currently working as Assistant Professor in Department of Business Management Studies, School of Commerce, Jain Deemed to be University, Bangalore.

Rathi Priya Bollu Yashwant is a research scholar currently pursuing a Ph.D. in Medical Science at Taylor's University. With a strong academic foundation that includes a Master's in Neuroscience and another Master's in Counselling Psychology, Rathi Priya bridges the gap between brain and mind. The main research focuses on phytochemicals and neurodegenerative disease, e.g., exploring the potential role of phytochemicals underlying neuroprotective mechanisms in neurodegenerative disease like Alzheimer's and Parkinson's disease. Rathi Priya is experienced in scientific writing with more than 10 years of experience in crafting articles related to therapeutic areas like stroke, depression, and neurodegeneration. Rathi Priya combines expertise in neuroscience and psychology to uncover insights that contribute to theoretical knowledge and practical applications in mental health and wellness.

Benny Godwin J. Davidson is a seasoned academic and experienced Management Consultant with over 12 years of expertise in teaching, research, and industry. He currently serves as an Assistant Professor at the University of the Fraser Valley and is a Sessional Faculty member at University Canada West and Acsenda School of Management. Through his ongoing research, Dr. Davidson continues to contribute to knowledge advancement, bringing fresh perspectives to both his consulting and teaching practices. His research interests include consumer behavior, marketing strategy, and real estate purchase decisions. Additionally, Dr. Davidson serves as a member of the Research Ethics Board, ensuring the integrity of research practices at his institutions.

Roshnie A. Doon holds a PhD in Economic Development Policy, MSc. Economics, and BSc. Economics from the University of the West Indies (UWI), St. Augustine, Trinidad. She has completed a Ph.D. Master Class in International Business from the Henley School of Business at the University of Reading, U.K.

She currently serves as the Secretary of the Caribbean Academy of Sciences (CAS) Regional Executive and is a research affiliate of the Global Labor Organization (GLO), Essen Germany. Her independent academic interests are primarily focused on Applied and Empirical Economics, in the areas of Labor, Gender, and Education Economics. She has undertaken research and published on a wide range of areas, which includes Labor Economics, Gender Education, and Education Economics, in the areas of Educational Mismatch, STEM Education, Social Policy, Migration, COVID-19, Climate Change, Apart from being an academic editor, she has published her research in a wide array of journal and book publications, while sharing her research work through guest lectures, and conference presentations domestically, regionally, and internationally.

Jossy P. George is a distinguished academic and leader, currently serving as the Director of CHRIST (Deemed to be University) at the BGR Campus in Bangalore. With a unique combination of qualifications in both Computer Science and Human Resources, he holds dual master's degrees in these fields. His educational journey includes completing the Faculty Development Programme (FDPM) from the prestigious Indian Institute of Management (IIM), Ahmedabad, and obtaining a Doctorate in Computer Science from Christ University. His career is marked by extensive research, particularly in the area of "Technology in Higher Education," where he has led numerous significant projects. His contributions to the academic world extend beyond research, as he has authored several research papers that have been published in both national and international journals.

Evangelina Rekha Joseph is a research scholar specializing in Human Resource Management (HRM) at CHRIST (Deemed to be University), Bengaluru. Holding an MBA in Human Resource Management, her academic foundation is complemented by a deep and current engagement with the field. Her primary research interests focus on the interdisciplinary aspects of Human Resource Development (HRD), Organizational Behaviour (OB), and Organizational Psychology. Her work aims to explore the empirical links between these areas to develop actionable strategies for enhancing workforce performance and organizational effectiveness. Furthermore, she actively contributes to academic discourse as a reviewer for peer-reviewed journals and book chapters, demonstrating a commitment to scholarly rigor and the advancement of knowledge within the domains of Management and Industrial-Organizational (I-O) Psychology.

V Lava Kumar is an accomplished academician with a diverse educational background and over 13 years of enriching experience in the field of education. Holding an impressive array of qualifications including an MBA, M. Com, PhD in

Management, PGDTTM, along with certifications from APSET in Management and APSET in Commerce. His area of specialization is Finance and Accounting. Currently serving as an Assistant Professor at GITAM School of Business, GITAM Deemed to be University, Bengaluru, Dr. Lava Kumar is dedicated to nurturing the minds of future business leaders. His teaching methodology is grounded in both theoretical knowledge and practical insights, ensuring that students receive a holistic understanding of their subjects. Dr. Lava Kumar's contributions extend beyond the classroom, evident through his prolific research endeavours. He has authored numerous research papers, with 3 publications in prestigious Scopus-indexed and ABDC journals, and an impressive tally of 10 research articles in international and peer-reviewed journals. He co-authored a Book which is published in Surana publications. His research interests primarily revolve around contemporary issues in management and finance, where he strives to contribute to the academic discourse with valuable insights and innovative perspectives.

Srishti Muralidharan is a UGC NET qualified postgraduate in Clinical Psychology and an Assistant Professor of Psychology in PES University, RR Nagar Bangalore. Her areas of interest include Clinical Psychology, Indian Psychology, Philosophy, Gender and Sexuality, Social Psychology and Neuropsychology. She has published a paper titled 'Mental Health Literacy for Depression and Schizophrenia among Undergraduate Students' in the International Journal of Indian Psychology. She has also authored and co-authored several book chapters. She is a practicing psychologist and believes in making mental health care accessible to all. She has also obtained Level I Certifications in Pranic Healing and Reiki and strongly believes in integrating alternative healing practices into modern day psychotherapy, customising it according to clients and delivering quality treatment.

Peter Varghese Muttungal is Professor of Social Sciences and the Campus Administrator & Director of Student Affairs at CHRIST (Deemed to be University) Delhi NCR. He has a Ph.D in Educational Psychology from Bharathiar University, Coimbatore. He also completed an EdLEAP Executive Leadership Program from Indian Institute of Management (IIM), Kolkata and is a UB Fellow for 2022-23 from Singapore Management University. His research interest is in the areas of Educational Leadership in Higher Education.

Usha Prabhu, is a senior academician having 23+ years of teaching experience – at UG & PG levels, aware of new changes in the education sector like the implementation of the National Education Policy, a member of the Syllabus Revision Committee -PG Level MBA at Bangalore University. She holds three Master degrees - MA, M. Phil, MBA, UGC- NET qualified and holds Ph.D from Bangalore

University, presently working at Jain(Deemed-to-be University.· She has authored 2 books - Organization Behaviour for MBA and Entrepreneurship Skill for III semester B. Com published by Sapna Book House. She has published 12 research papers in International Journals, presented 9 research papers at International Conferences and 4 research Book Chapters(Scopus indexed).

Nishi Priya is a research scholar at Christ University, Delhi NCR. She holds a Master's in Counseling Psychology, where she graduated with the first rank in her class. Her academic and research interests lie at the intersection of applied psychology, organizational behavior, and artificial intelligence.

Varanasi Rahul is an Assistant Professor at JAIN University (Bangalore), a semi-qualified Cost Accountant, holding three postgraduate degrees: M.Com (Accounting & Taxation), M.Com (Management), and an MA (History). He has cleared the UGC NET in Commerce twice and has also cleared the AP SET in Commerce, History, and Management, as well as the TS SET in Commerce. With 3 years of experience as an Assistant Professor and 6 months as a Corporate Tax Associate for Australian Taxation at PwC (Big 4), he brings both academic and professional expertise to his field. Additionally, he is a freelance archaeologist, practicing epigraphist in Brahmi script, and a numismatist. His scholarly and creative contributions include over 30 newspaper articles published in Deccan Chronicle, Times of India, and other local newspapers. He has presented and participated in 25 national and international seminars and has published 2 research articles in ABDC Journals, 3 in SCOPUS-indexed journals, 2 in Web of Science and 4 in peer-reviewed journals. As a story writer, 5 of his stories have been published in top magazines of Andhra Pradesh. Furthermore, he has authored 3 books. He has also contributed 13 book chapters to various academic publications. In addition to his extensive achievements, he has applied for 16 patents, of which 2 has been granted in India, 2 in UK and 11 have been published.

Rismawati is a distinguished academic and consultant with a robust focus on sustainability reporting. She currently holds the position of lecturer and Head of the Quality Assurance Institute at Universitas Muhammadiyah Palopo, where she has been pivotal in ensuring the quality of university processes and outputs since February 2020. Additionally, she serves as an Assurer for Sustainability Reporting at the National Center for Sustainability Reporting in Jakarta, specializing in the evaluation of sustainability reports for mining companies across Indonesia. Her academic qualifications are extensive, comprising a Bachelor's degree in Economics and Accounting, a Master's degree in Accounting Science, and a Doctorate in Accounting from Universitas Brawijaya, Malang. Rismawati is also a certified specialist in sustainability reporting, risk management, and tax planning, holding several prestigious certifications including Certified Sustainability Reporting Specialist (CSRS), Certified Risk Management Practitioner (CRMP), Certified Management Accountant (CMA) and Certified Enterprise Risk Analysis (CERA). In her professional capacity, she has provided consultancy services in financial management and tax compliance for companies such as PT. Reski Barito Timur and PT. Helay Raya. Furthermore, she has authored multiple publications on Social Accounting, Sustainability Accounting, Management Accounting and CSR, contributing significantly to the academic community. Her interests in reading, writing, and travel reflect her commitment to lifelong learning and professional growth in her field.

N Sathyanarayana He has over 15 years of teaching experience at both undergraduate and postgraduate levels. He holds a Master of Business Administration degree from Visvesvaraya Technological University (VTU) and a Master of Commerce degree from Periyar University. He has completed his Ph.D. in Banking and Finance from Dayananda Sagar University. Currently, he serves as an Assistant Professor in the School of Commerce at Jain (Deemed-to-be University), Bangalore. He has published 24 research papers in international journals, including three indexed in Scopus, and seven papers in national journals. Additionally, he has presented 14 research papers at international conferences and 11 at national conferences. He has co-authored two books at the national level and contributed nine Scopus-indexed book chapters.

Nidhi Suresh is a highly motivated undergraduate psychology student at PES University, India, with a perfect CGPA of 10. Passionate about clinical psychology and neuroscience, she has demonstrated academic excellence, earning awards and a prestigious MRD scholarship. She has actively contributed to mental health initiatives, including organizing various events in her department. Her leadership experience spans event planning, compering, and managing school and university activities. She is certified in Psychological First Aid and Psychometric Analysis. With strong interpersonal and organizational skills, Nidhi continues to explore opportunities that bridge academic learning and practical experience in psychology.

Joanne Vázquez de Príncipe, DM, is recognized as a scholar, practitioner, and industry leader. Joanne's career spans over three decades of multidisciplinary success in business and academia. An Associate Faculty of Doctoral Organizational Behavior, Faculty Assessment Liaison, and industry SME at the College of Doctoral Studies for the University of Phoenix. She's a Fellow for the Centers of Leadership & Entrepreneurial Research and Org'l Wellness, Engagement & Belonging. Príncipe holds a Doctorate in Management, specializing in Organizational Leadership. Dissertation concepts, models, and principles of her research were applied to the following book: Vázquez de Príncipe, J. (Ed.). (2024). Resilience of Multicultural and Multigenerational Leadership and Workplace Experience. IGI Global Scientific Publishing. ISBN13: 9798369318027. Dr. Príncipe is the Founder and CEO of Purposeful Advising & Coaching, a boutique firm offering faith-based business operations advisory services, executive leadership coaching, and OD custom holistic culture designs.

Wong Eng Hwa is a leading academic in medical sciences, currently serving as Professor at the School of Medicine, Faculty of Health and Medical Sciences. She has been the Programme Director for the School's Postgraduate Programme since 2018 and has held key academic leadership roles, including Associate Dean for Postgraduate, Research and Innovation. With a PhD in Medical Sciences from the University of Malaya and a BSc in Microbiology from Universiti Putra Malaysia, her work focuses on advancing postgraduate education, translational research, and interdisciplinary collaboration in health sciences.

Index

A

B

C

D

E

F

H

I

J

L

M

N

O

P

R

S

T

W

www.ingramcontent.com/pod-product-compliance
Lightning Source LLC
LaVergne TN
LVHW062336300926
849161LV00007B/1714